TO Michael:

Best regards,

Patrick Wong
July, 1998

RECENT ADVANCES IN PROSTAGLANDIN, THROMBOXANE, AND LEUKOTRIENE RESEARCH

ADVANCES IN EXPERIMENTAL MEDICINE AND BIOLOGY

Recent Volumes in this Series

Volume 424
THE FATE OF THE MALE GERM CELL
Edited by Richard Ivell and Adolf-Friedrich Holstein

Volume 425
CHEMISTRY AND BIOLOGY OF SERPINS
Edited by Frank C. Church, Dennis D. Cunningham, David Ginsburg, Maureane Hoffman, Stuart R. Stone, and Douglas M. Tollefsen

Volume 426
PHYSIOLOGY AND PATHOPHYSIOLOGY OF THE ISLETS OF LANGERHANS
Edited by Bernat Soria

Volume 427
DIETARY FIBER IN HEALTH AND DISEASE
Edited by David Kritchevsky and Charles Bonfield

Volume 428
OXYGEN TRANSPORT TO TISSUE XIX
Edited by David K. Harrison and David T. Delpy

Volume 429
BRAIN PLASTICITY: Development and Aging
Edited by Guido Filogamo, Antonia Vernadakis, Fulvia Gremo, Alain M. Privat, and Paola S. Timiras

Volume 430
ANALYTICAL AND QUANTITATIVE CARDIOLOGY
Edited by Samuel Sideman and Rafael Beyar

Volume 431
PURINE AND PYRIMIDINE METABOLISM IN MAN IX
Edited by Andrea Griesmacher, Peter Chiba, and Mathias M. Müller

Volume 432
HYPERTENSION AND THE HEART
Edited by Alberto Zanchetti, Richard B. Devereux, Lennart Hansson, and Sergio Gorini

Volume 433
RECENT ADVANCES IN PROSTAGLANDIN, THROMBOXANE, AND LEUKOTRIENE RESEARCH
Edited by Helmut Sinzinger, Bengt Samuelsson, John R. Vane, Rodolfo Paoletti, Peter Ramwell, and Patrick Y-K Wong

---

A Continuation Order Plan is available for this series. A continuation order will bring delivery of each new volume immediately upon publication. Volumes are billed only upon actual shipment. For further information please contact the publisher.

RECENT ADVANCES IN PROSTAGLANDIN, THROMBOXANE, AND LEUKOTRIENE RESEARCH

Edited by

Helmut Sinzinger
University of Vienna and
Wilhelm Auerswald Atherosclerosis Research Group Vienna
Vienna, Austria

Bengt Samuelsson
Karolinska Institutet
Stockholm, Sweden

John R. Vane
The William Harvey Research Institute
London, England

Rodolfo Paoletti
University of Milan
Milan, Italy

Peter Ramwell
Georgetown University Medical Center
Washington, D.C.

and

Patrick Y-K Wong
University of Medicine and Dentistry of New Jersey
Stratford, New Jersey

PLENUM PRESS • NEW YORK AND LONDON

Library of Congress Cataloging in Publication Data

Recent advances in prostaglandin, thromboxane, and leukotriene research / edited by Helmut Sinzinger ... [et al.].
p. cm. -- (Advances in experimental medicine and biology ; v. 433)
"Proceedings of the 10th International Conference on Prostaglandins and Related Compounds, held September 22-27, 1996, in Vienna, Austria"--T.p. verso.
Includes bibliographical references and index.
ISBN 0-306-45776-8
1. Prostaglandins--Physiological effect--Congresses. 2. Thromboxanes--Physiological effect--Congresses. 3. Leukotrienes--Physiological effect--Congresses. I. Sinzinger, H. (Helmut) II. International Conference on Prostaglandins and Related Compounds (10th : 1996 : Vienna, Austria) III. Series.
[DNLM: 1. Prostaglandins--congresses. 2. Thromboxanes--congresses. 3. Leukotrienes--congresses. W1 AD559 v.433 1998]
QP801.P68R43 1998
612'.01575--dc21
DNLM/DLC
for Library of Congress 97-45692
CIP

Proceedings of the 10th International Conference on Prostaglandins and Related Compounds, held September 22 – 27, 1996, in Vienna, Austria

ISBN 0-306-45776-8

A Division of Plenum Publishing Corporation
233 Spring Street, New York, N.Y. 10013

http://www.plenum.com

10 9 8 7 6 5 4 3 2 1

Printed in the United States of America

PREFACE

The International Symposium on Prostaglandins and Related Compounds, first held in Vienna 1972, revisited the city after 24 years for the 10TH Symposium. For the many researchers working in this multi-disciplinary field it was an opportunity to exchange their experiences and share new data with colleagues from all around the world. This scientific exchange was largely encouraged by the unseasonably cold and rainy weather. For the first time, there was quite a large attendance from the former Communist countries.

Eugene Garfield prepared a key note address delivered during the meeting (The Scientist 1996, 12) reviewing the contribution of the Nobel Laureates U.S. von Euler, J.R. Vane, S.K. Bergström, and B.I. Samuelsson, discussing the relevance of the more than 40,000 papers in this area published since 1991. Overall, there is still a rapidly growing interest, and in particular a great variety of clinical applications of this family of compounds which were discussed in detail during the meeting.

Beside the lectures there were 19 workshops covering nearly all the topics of key interest. All the speakers were invited to prepare a manuscript which has resulted in the volume now in your hands.

Special thanks to Dr. Patrick Wong and the new publisher of this series who helped to publish the proceedings in the usual quality and reasonable period of time.

Looking forward to seeing all of you again in Florence in 1999, hopefully with much more sun.

The Editors

CONTENTS

SOME RECENT ADVANCES IN LEUKOTRIENE RESEARCH

Bengt Samuelsson

Department of Biochemistry and Biophysics
Division of Physiological Chemistry II
Karolinska Institutet
S-171 77 Stockholm
Sweden

INTRODUCTION

The leukotrienes are formed by transformation of arachidonic acid into an unstable epoxide intermediate, leukotriene A_4 (LTA_4), which can be converted enzymatically by hydration to LTB_4, and by addition of glutathione to LTC_4. This last compound is metabolized to LTD_4 and LTE_4 by successive elimination of a γ-glutanyl residue and glycine [1, 2].

LTB_4 seems to be an important mediator of inflammation. Thus, it is a potent chemotaxin for neutrophils and increases leukocyte adhesion to the vessel wall [3]. The biological effects of LTB_4 are mediated via a specific receptor (BLT) [4]. Recently, LTB_4 has also been reported to bind to the peroxisome proliferator-activated receptor (PPARα) [5]. It has been suggested that the binding of LTB_4 to this receptor, which is a transcription factor regulating gene expression in fatty acid degradation, could affect the duration of an inflammatory response induced by LTB_4.

LTB_4 has also been suggested to have a number of roles in immunoregulation [6-8]. In monocytes, LTB_4 stimulates production of various cytokines [9, 10] and activates c-fos and c-jun [11]. B-lymhocyte functions may also be regulated by LTB_4 [12].

The cysteinyl-leukotrienes (LTC_4, D_4 and E_4) are potent constrictors of smooth muscle in the airways and in the microvasculature [2]. They are considered to be important mediators in asthma [13]. Recent developments of 5-lipoxygenase inhibitors and Cys LT_1 antagonists as anti-asthmatic drugs support this view [14-17].

5-LIPOXYGENASE

Lipoxygenases introduce molecular oxygen into polyunsaturated fatty acids with one or more *cis*, *cis*-1,4-penta-diene moiety in the structure to form hydroperoxy acids. 5-lipoxygenase catyzes the dioxygenation of arachidonic acid into 5(S)-hydroperoxy-6-*trans*-

Recent Advances in Prostaglandin, Thromboxane, and Leukotriene Research
Edited by Sinzinger *et al.*, Plenum Press, New York, 1998

8,11,14-*cis*-eicosatetraenoic acid, (5-HPETE). The same enzyme also catalyzes the transformation of 5-HPETE into the allylic epoxide, LTA_4.

In analogy to soybean lipoxygenase, active 5-lipoxygenase contains 1 mol iron/mol protein. Crystal structures of soybean lipoxygenase have been published by two groups. In one case three histidine residues and the carboxyl group of the C-terminal isoleucine were found to be ligands to the iron [18]. In the other case Asn 694 also served as ligand [19]. Recent mutagenesis studies with human 5-lipoxygenase indicate that four amino acids function as ligands to iron [20]. These are His 372, His 550, and the C-terminal Ile 673 as permanent ligands to iron, and His 367 as a possible replaceable ligand. For the permanent ligands, mutagenesis led to complete loss of both acitivity and iron. For the suggested replaceable ligand, mutations led to complete loss of the lipoxygenase activity, while iron was partially retained [20].

The genomic structure of human 5-lipoxygenase has been determined [21]. The gene comprises more than 82 kb DNA consisting of 14 exons divided by 13 introns. It is located on chromosome 10. The promoter region of the human gene has been sequenced and found to contain multiple GC boxes, potential sp1 and Egr-1 binding sites, but no TATA or CCAAT boxes. This is characteristic for "house keeping" genes. It has recently been demonstrated that asthmatic subjects exhibited polymorphism in the 5-lipoxygenase gene promoter consisting of deletions or additions of GC boxes [22].

Recently, a functional retinoid Z receptor (RZR) response element has been identified in the promoter of the human 5-lipoxygenase gene. 5-lipoxygenase expression in RZRα–expressing B–lymphocytes was repressed by stimulation with melatonin. In other cells, lacking RZRα, 5-lipoxygenase m RNA levels were not affected [23].

Transforming growth factor β (TGF-β) has been found to upregulate cellular 5-lipoxygenase activity in dimethyl sulfoxide differentiated HL-60 cells [24]. In the human monocytic cell line Mono Mac 6, 5-lipoxygenase activity was upregulated by combined treatment with TGF-β and 1,25-dihydroxy-vitamin D_3 (VD_3). The induction of 5-lipoxygenase protein was obvious after 1 day, but the increase in activity was delayed and did not appear until the second day. These and additional experiments indicated the presence of additional factors that affect the activity of 5-lipoxygenase [25].

The expression of 5-lipoxygenase in keratinocytes has been controversial [26]. Recently, it was found that in undifferentiated keratinocytes 5-lipoxygenase gene expression was low or undetectable. However, after shift to culture conditions that promote conversion of prokeratinocytes into a more differentiated phenotype, 5-lipoxygenase gene expression was markedly induced in HaCaT cells and, to a lesser degree in normal keratinocytes [27].

Studies on the subcellular distribution of 5-lipoxygenase have shown that in some resting cells as peripheral blood polymorphonuclear leukocytes (PMNs) and peritoneal macrophages [28, 29] it is predominantly cytosolic. However, in alveolar macrophages and mast cells, the lipoxygenase is found in both the nucleus and cytosol [28, 30, 31]. Upon stimulation, both cytosolic and nuclear 5-lipoxygenase can translocate to the nuclear envelope [30, 32, 33]. It has also recently been shown that recruitment of PMNs into sites of inflammation is associated with nuclear import of 5-lipoxygenase [34]. The role of these translocations of 5-lipoxygenase in the regulation of leukotriene formation is not fully understood.

It is of interest in this context that 5-lipoxygenase contains a proline-rich sequence resembling a Src homology region SH3-binding site [35]. 5-lipoxygenase bound to fusion products of Grb2 and a truncated version of Grb2 containing its SH3 domain. It was also reported that tyrosine kinase inhibitors block the translocation of 5-lipoxygenase from cytosol to membranes during cellular activation. Moreover, phosphorylated 5-lipoxygenase was present in the nuclear fraction but not in the membrane or cytosolic fractions [36].

Leukotriene A_4 hydrolase

Leukotriene A_4 hydrolase is a bifunctional zinc metalloenzyme which catalyzes the hydrolysis of the unstable intermediate LTA_4 into the proinflammatory substance LTB_4 and also exhibits peptidase activity [37, 38]. Several important elements of the active site corresponding to the two enzymatic activities have been identified. Thus, the intrinsic zinc atom was shown to be catalytic and complexed to His-295, His.299 and Glu-318. Anions selectively stimulate the peptidase activity and site-directed mutagenesis studies have identified Glu-296 as a catalytic residue which can act as a general base in the peptidase reaction [39].

During catalysis, LTA_4 hydrolase is covalently modified and inactivated by its endogenous substrate LTA_4 [40, 41]. This suicide reaction may be important for the regulation of the biosynthesis of LTB_4, c f [42]. We have recently shown that suicide inactivation of LTA_4 hydrolase by its substrate, LTA_4, occurs via an irreversible covalent binding of LTA_4 to the protein. Differential lysine-specific peptide mapping of unmodified and suicide-inactivated LTA_4 hydrolase was used to identify a heneicosa peptide (residues 365-385) of human LTA_4 hydrolase which is involved in the binding of LTA_4 to the native enzyme. A modified form of the peptide using LTA_4 ethyl ester for the inactivation was isolated and subjected to Edman degradation. The sequence analysis showed that the binding of the LTA_4 derivative occurred via Tyr-378 in LTA_4 hydrolase. Enzyme inactivation and peptide modification could be prevented by preincubation of LTA_4 hydrolase with the competive inhibitor bestatin [43].

In order to detail the importance of Tyr-378 for catalyses and mechanism based inactivation we have employed site directed mutagnesis. Thus, Tyr-378 was exchanged for Phe or Gln in two separate mutants. The mutated enzymes were expressed as (His)6-tagged fusion proteins in E. coli, purified and characterized. The results showed that mutants in position 378 were no longer inactivated or co-valently modified by LTA_4. Moreover in (Y378F) LTA_4 hydrolase, the value of kcat for epoxide hydrolysis was increased 2.5-fold over that of the wild type enzyme. By mutation of Tyr-378 in LTA_4 hydrolase, catalysis and covalent modification / inactivation could hence be dissociated, yielding an enzyme with increased turnover and resistance to inactivation [44].

Additional work has also provided new insight into the role of Tyr-378 in the catalytic mechanism [45]. The mutant forms, (Y378F) and (Y378Q), of LTA hydrolase convert LTA_4 not only into LTB_4 but also into a second, previously unknown, product of LTA_4. The structure of this derivative was 5S, 12R-dihydroxy - 6,10-*trans*-8,14-*cis*-eicosatetraenoic acid, i.e. Δ^6-*trans*-Δ^8-*cis*-LTB_4. The relative formation of Δ^6-trans-Δ^8-*cis*-LTB_4 versus LTB_4 by (Y378F)- and (Y378Q) LTA_4 hydrolase, was 18 % and 32 %, respectively. These findings indicate that Tyr-378 is located at the active site where it assists in the formation of the correct double-bond geometry in the product LTB_4 [45].

The role of another tyrosine residue in LTA_4 hydrolase, viz. Tyr 383, has also been studied by site directed mutagenesis [46]. The mutant enzymes, (Y383F), (Y383H) and (Y383Q) LTA_4 hydrolase were assayed for epoxide hydrolase and peptidase activities. Tyr-378 was found to be essential for peptidase activity but was not critical for epoxide hydrolase activity. However, the mutants in position 383 were found to convert LTA_4 into 5S, 6S-dihyroxy-7,9-*trans*-11,14-*cis*-eicosatetraenoic acid (5S, 6S-DHETE). From the stereochemical configuration of the vicinal diol it could be inferred that 5S, 6S-DHETE is formed via an S_N1 mechanism involving a carbocation intermediate. Since soluble epoxide hydrolase also accepts LTA_4 as substrate and converts it into an epimeric 5,6-DHETE, the results suggest a functional link between LTA_4 hydrolase and xenobiotic epoxide hydrolase[47].

Acknowledgments

Work from the author's laboratory was financially supported by the Swedish Medical Research Council (03X-00217) and The European Union (BMH4-CT960229).

REFERENCES

1. B. Samuelsson, Leukotrienes: Mediators of immediate hypersensitivity reactions and inflammation, *Science*, 220:568-575, (1983).
2. B. Samuelsson, S.-E. Dahlén, J.-Å. Lindgren, C.A. Rouzer, and C.N. Serhan, Leukotrienes and lipoxins: Structures, biosynthesis, and biological effects, *Science*, 237:1171-1176, (1987).
3. S.-E. Dahlén, G. Hansson, P. Hedqvist, T. Björck, E. Granström, and B. Dahlén, Allergen challenge of lung tissue from asthmatics elicits bronchial contraction that correlates with the release of leukotrienes C_4, D_4 and E_4, *Proc. Natl. Acad. Sci. USA*, 80:1712-1716, (1983).
4. R.A. Coleman, R.M. Eglen, R.L. Jones, S. Narumiya, T. Shimizu, W.L. Smith, S.E. Dahlén, J.M. Drazen, P.J. Gardiner, W.T. Jackson, T.R. Jones, R.D. Krell, and S. Nicosia, Prostanoid and leukotriene receptors: a progress report from the IUPHAR working parties on classification and nomenclature, in: *Adv. Prostaglandin Thromboxane Leukotriene Res*, 23:283-285, (1995).
5. P.R. Devchand, H. Keller, J.M. Peters, M. Vazquez, F.J. Gonzalez, and W. Wahli, The ppar-alpha-leukotriene b-4 pathway to inflammation control, *Nature*, 384:39-43, 1996.
6. A.W. Ford-Hutchinson, Leukotriene B_4 in inflammation, *Immunology*, 10:1-12, (1990).
7. M. Rola-Pleszczynski, L. Gagnon, and P. Sirois, Leukotriene B_4 augments human natural killer cytotoxic cell activity, *Biochem. Biophys. Res. Commun.*, 113: 531-537, (1983).
8. M. Rola-Pleszczynski, Differential effects of leukotriene B_4 on $T4^+$ and $T8^+$ lymphocyte phenotype and immunoregulatory functions, *J. Immunol.*, 135:1357-1360, (1985).
9. M. Rola-Pleszczynski and I. Lemaire, Leukotrienes augment interleukin-1 production by human monocytes, *J. Immunol.*, 135:3958-3961, (1985).
10. M. Rola-Pleszczynski, P.-A. Chavaillaz, and I. Lemaire, Stimulation of interleukin-2 and interferon gamma production by leukotriene B_4 in human lymphocyte cultures, *Prostagl. Leukotr. Med*, 23:207-210, (1986).
11. J. Stankova and M. Rola-Pleszcynski, Leukotriene B_4 stimulates c-*fos* and c-*jun* gene transcription and AP-1 binding activity in human monocytes, *Biochem. J.*, 282:625-629, (1992).
12. K.A. Yamaoka, H.-E. Claesson, and A. Rosén, Leukotriene B_4 enhances activation, proliferation, and differentiation of human B lymphocytes, *J. Immunol.*, 143: 1996-2000, (1989).

13. M. Kumlin, B. Dahlén, T. Björck, O. Zetterström, E. Granström, and S.-E. Dahlén, Urinary excretion of leukotriene E_4 and 11-dehydro-thromboxane B_2 in response to bronchial provocations with allergen, aspirin, leukotriene D_4, and histamine in asthmatics, *Am. Rev. Respir. Dis*, 146:96-103, (1992).
14. W.R. Henderson, New modalities for the pharmacotherapy of asthma - leukotriene inhibitors and antagonists, *Immunology & Allergy Clinics of North America*, 16:797 ff., (1996).
15. J.M. Drazen, Pharmacology of leukotriene receptor antagonists and 5-lipoxygenase inhibitors in the management of asthma, *Pharmacotherapy*, 17:S22-S30, (1997).
16. J.S. Larsen and S.K. Jackson, Antileukotriene therapy asthma, *American Journal of Health-System Pharmacy*, 53:2821-2830, (1996).
17. P.M. Obyrne, Exercise-induced bronchoconstriction - elucidating the roles of leukotrienes and prostaglandins, *Pharmacotherapy*, 17:S 31-S 38, (1997).
18. J.C. Boyington, B.J. Gaffney, and L.M. Amzel, The three-dimensional structure of an arachidonic acid 15-lipoxygenase, *Science*, 260:1482-1486, (1993).
19. W. Minor, J. Steczko, J.T. Bolin, Z. Otwinowski, and B. Axelrod, Crystallographic determination of the active site iron and its ligands in soybean lipoxygenase L1, *Biochemistry*, 32:6320-6323, (1993).
20. T. Hammarberg, Y.-Y. Zhang, B. Lind, O. Rådmark, and B. Samuelsson, Mutations at the C-terminal isoleucine and other potential iron ligands of 5-lipoxygenase, *European J Biochem*, 230:401-407, (1995).
21. C.D. Funk, S. Hoshiko, T. Matsumoto, O. Rådmark, and B. Samuelsson, Characterization of the human 5-lipoxygenase gene, *Proc. Natl. Acad. Sci. USA*, 86: 2587-2591, (1989).
22. K.H. In, K. Asano, D. Beier, J. Grobholz, P.W. Finn, E.K. Silverman, E.S. Silverman, T. Collins, A.R. Fischer, T.P. Keith, K. Serino, S.W. Kim, G.T. Desanctis, C. Yandava, A. Pillari, P. Rubin, J. Kemp, E. Israel, W. Busse, D. Ledford, J.J. Murray, A. Segal, D. Tinkleman, and J.M. Drazen, Naturally occurring mutations in the human 5-lipoxygenase gene promoter that modify transcription factor binding and reporter gene transcription, *Journal of Clinical Investigation*, 99:1130-1137, (1997).
23. D. Steinhilber, M. Brungs, O. Werz, I. Wiesenberg, C. Danielsson, J.-P. Kahlen, S. Nayeri, M. Schräder, and C. Carlberg, The Nuclear Receptor for Melatonin Represses 5-Lipoxygenase Gene Expression in Human B Lymphocytes, *J. Biol. Chem*, 270:7037-7040, (1995).
24. D. Steinhilber, O. Rådmark, and B. Samuelsson, Transforming growth factor-β upregulates 5-lipoxygenase activity during myeloid cell maturation, *Proc. Natl. Acad. Sci. USA*:in press, (1993).
25. M. Brungs, O. Rådmark, B. Samuelsson, and D. Steinhilber, Sequential induction of 5-lipoxygenase gene expression and activity in Mono Mac 6 cells, by transforming growth factor-β and 1,25-dihydroxyvitamin D_3., *Proc. Natl. Acad. Sci.*, 92:107-111, (1995).
26. A.W. Ford-Hutchinson, 5-Lipoxygenase activation in psoriasis: a dead issue?, *Skin Pharmacol*, 6:292-297, (1993).
27. U. Janssen-Timmen, P.J. Vickers, U. Wittig, W.D. Lehmann, H.-J. Stark, N.E. Fusenig, T. Rosenbach, O. Rådmark, B. Samuelsson, and A.J.R. Habenicht, Expression of 5-lipoxygenase in differentiating human skin keratinocytes., *Proc. Natl. Acad. Sci.*, 92:6966-6970, (1995).

28. T.G. Brock, R. Paine, III, and M. Peters-Golden, Localization of 5-lipoxygenase to the nucleus of unstimulated rat bosophilic leukemia cells, *J. Biol. Chem*, 269: 22059-22066, (1994).
29. M. Peters-Golden and R.W. McNish, Redistribution of 5-lipoxygenase and cytosolic phospholipase A_2 to the nuclear fraction upon macrophage activation., *Biochem. Biophys. Res. Commun.*, 196:147-153, (1993).
30. J.W. Woods, M. Coffey, T. Brock, I. Singer, and M. Peters-Golden, 5-Lipoxygenase is located in the euchromatin of the nucleus in resting human alveolar macrophages and translocates to the nuclear envelope upon cell activation. *J. Clin. Invest.*, 95:2035-2040, (1995).
31. X.-S. Chen, T.A. Naumann, U. Kurre, N.A. Jenkins, and N.G. Copeland, C.D, *J. Biol. Chem*, 270:17993-17999, (1995).
32. J.W. Woods, J.F. Evans, D. Ethier, S. Scott, P.J. Vickers, L. Hearn, J. A. Heibein, S. Charleson, and I.I. Singer, 5-lipoxygenase and 5-lipoxygenase activating protein are localized in the nuclear envelope of activated human leukocytes., *J. Exp. Med.*, 178:1935-1946, (1993).
33. T.G. Brock, R.W. McNish, and M. Peters-Golden, *J. Biol. Chem.*, 270:21652-21658, (1995).
34. T.G. Brock, R.W. McNish, M.B. Bailie, and M. Peters-Golden, Rapid Import of Cytosolic 5-Lipoxygenase into the Nucleus of Neutrophils after *in Vivo* Recruit-ment and *in Vitro* Adherence, *J. Biol. Chem*, 272:8276-8280, (1997).
35. R.A. Lepley and F.A. Fitzpatrick, 5-Lipoxygenase Contains a Functional Src Homology 3-binding Motif that Interacts with the Src Homology 3 Domain of GrB2 and Cytoskeletal Proteins, *J. Biol. Chem*, 269:24163-24168, (1994).
36. R.A. Lepley, T. Muskardin, and F.A. Fitzpatrick, "Tyrosine Kinase Activity Modulates Catalysis and Translocation of Cellular 5-Lipoxygenase, *J. Biol. Chem*, 271:6179-6184, (1996).
37. J.Z. Haeggström, A. Wetterholm, B.L. Vallee, and B. Samuelsson, Leukotriene A_4 hydrolase: An epoxide hydrolase with peptidase activity, *Biochem. Biophys. Res. Commun.*, 173:431-437, (1990).
38. M. Minami, N. Ohishi, H. Mutoh, T. Izumi, H. Bito, H. Wada, Y. Seyama, H. Toh, and T. Shimizu, Leukotriene A_4 hydrolase is a zinc-containing aminopeptidase, *Biochem. Biophys. Res. Commun.*, 173:620-626, (1990).
39. J.Z. Haeggström, A. Wetterholm, J.F. Medina, and B. Samuelsson, Leukotriene A_4 hydrolase: structural and functional properties of the active center, *J. Lipid. Med.*, 6:1-13, (1993).
40. J. McGee and F. Fitzpatrick, Enzymatic hydration of leukotriene A_4: Purification and characterization of a novel epoxide hydrolase from human erythrocytes, *J. Biol. Chem.*, 260:12832-12837, (1985).
41. J.F. Evans, D.J. Nathaniel, R.J. Zamboni, and A.W. Ford-Hutchinson, Leukotriene A_3: A poor substrate but a potent inhibitor of rat and human neutrophil leukotriene A_4 hydrolase, *J. Biol. Chem.*, 260:10966-10970, (1985).
42. F.F. Sun and J.C. McGuire, Metabolism of arachidonic acid by human neutrophils. Characterization of the enzymatic reactions that lead to the synthesis of leukotriene B_4, *Biochim. Biophys. Acta*, 794:56-64, (1984).
43. M.J. Mueller, A. Wetterholm, M. Blomster, H. Jörnvall, B. Samuelsson, and J.Z. Haeggström, Leukotriene A_4 hydrolase: Mapping of a henicosapeptide involved in mechanism-based inactivation, *Proc. Natl. Acad. Sci. USA*, 92: 8383-8387, (1995).

44. M.J. Mueller, M. Blomster, U. Oppermann, H. Jörnvall, B. Samuelsson, and J.Z. Haeggström, Leukotriene A_4 hydrolase: Protection from mechanism-based inactivation by mutation of tyrosine-378, *Proc. Natl. Acad. Sci. USA*, 93:5931-5935, (1996).
45. M.J. Mueller, M. Blomster Anderberg, B. Samuelsson, and J.Z. Haeggström, Leukotriene A_4 hydrolase: mutation of tyrosine-378 allows conversion of leukotriene A_4 into an isomer of leukotriene B_4, *J. Biol. Chem*, 271:24345-24348, (1996).
46. M. Blomster, A. Wetterholm, M.J. Mueller, and J.Z. Haeggström, Evidence for a catalytic role of tyrosine 383 in the peptidase reaction of leukotriene A_4 hydrolase, *Eur. J. Biochem*, 231:528-534, (1995).
47. M. Blomster-Andberg, Hamberg, M and Haeggström, J.Z., Mutation of tyrosine 383 in leukotriene A_4 hydrolase allows conversion of leukotriene A_4 into 5*S*,6*S*-DHETE, implications for the epoxide hydrolase mechanism, *J. Biol. Chem.*, subm. for publ. (1997).

PRACTICAL ASPECTS OF PROSTAGLANDIN E_1 BEFORE AND AFTER SOLID ORGAN TRANSPLANTATION

H. Müller, MD; A. Wasler, MD; B. Petutschnigg, MD; T. Allmayer, MD; B. Grasser, MD; G. Prenner, MD; S. Schaffellner, MD; KH. Tscheliessnigg, MD

Karl-Franzens-University Graz
Department of Transplantation Surgery
Auenbruggerplatz 29, A-8036 Graz, Austria

INTRODUCTION

Prostaglandins and analogues for therapeutic use gained importance in the early eighties, when a substantial number of clinical studies about their therapeutical effects was published[6, 12, 17, 22]. Out of the huge family of prostanoids, prostaglandins E_1 (PG E_1) and I_2 and their synthetical analogues are the most important substances in clinical practice in the field of solid organ transplantation. Organ procurement and organ preservation[4, 5, 9, 15, 19] as well as treatment of primary graft failure[8, 10] have been described as possible indications for PG E_1. Vasodilating effects of prostaglandins are mediated by increasing cAMP levels in vascular smooth muscle cells. This mediation by cAMP is shared with beta-agonists and phosphodiesterase-inhibitors, but not with nitrates and nitric oxide[3, 14, 21]. Treatment of elevated pulmonary vascular resistance (PVR), which is frequently needed after heart transplantation (HTX), and in the pretransplant evaluation as well as bridging to transplantation[1, 7, 20] has been the indication for PG E_1 in our patient cohort.

ASPECTS OF PRACTICAL USE OF PROSTAGLANDIN E_1

Despite the numerous possible indications for PG E_1, the questions of optimal dosage and possible contraindications have not yet been solved. There is evidence that an inappropriate application of PG E_1 can cause severe problems, especially in critically ill patients. An important change of the relationship between dosage and effects and between desired effects and side effects has to be expected under certain conditions. The following deliberations are based on our experience with 60 patients before and after orthotopic HTX, who received PG E_1 mainly to improve their hemodynamic situation.

Metabolization and Elimination of PG E_1

PG E_1 has a very short biological half-life of about 30 seconds[2, 8, 17, 18]. After

Recent Advances in Prostaglandin, Thromboxane, and Leukotriene Research
Edited by Sinzinger *et al.*, Plenum Press, New York, 1998

intravenous administration, 80-90% of the given PG E_1 is usually metabolized during the first passage of the pulmonary circuit. In case of pulmonary hypertension, this first metabolism has been described to be distinctly reduced, which can be explained by functional disorder of the endothelium and a loss of capillaries and endothelial cells at a later stage[8]. One of the metabolites of PG E_1 (PG E_0) contributes to the effects of PG E_1 with a similar range of actions[2, 17]. This metabolite has a much longer half-life than the primary substance and is partly responsible for prolonged effects after PG E_1-administration.

11 alpha-hydroxy-9,15-dioxo-2,3,4,5,20-pentanor-19-carboxyprostanoic acid (PG E_M), which is almost inactive biologically, is eliminated by the kidneys[16]. An accumulation of PG E_M has been shown in patients with chronic renal failure.

Renal failure and PG E_1-therapy

Despite the fact that accumulation of the biologically active metabolite PG E_0 in case of renal failure has not yet been proved, we have observed that patients who develop renal failure tend to show a lower tolerance of continuously given PG E_1 and an earlier occurrence of arterial hypotension. A possible explanation might be a change in the balance of metabolites caused by the cumulation of PG E_M. Especially in case of infection and renal failure this potentially higher sensibility to PG E_1 has lead to extremely low values of systemic vascular resistance and consecutive arterial hypotension at dosages which had initially been tolerated.

Fixed Pulmonary Hypertension / Elevated PVR

Chronic secondary pulmonary hypertension (PH) usually becomes increasingly fixed due to arteriosclerosis, arteriolosclerosis and necrotizing arteriolitis. With patients in the later stages of illness the PVR therefore has to be expected to be to a high degree irreversibly elevated[8]. When given in such a situation, PG E_1 can cause critical arterial hypotension and acute decompensation even at low dosages because of the relatively slow metabolization and the missing effect on the pulmonary circuit.

DISCUSSION

Secondary PH and consecutively elevated PVR is a frequent complication of chronic heart failure and can be observed in about 40 % of HTX candidates in our center. An elevated PVR in heart recipients increases the risk of right heart failure. An irreversibly elevated PVR of more than 4-6 Wood Units is therefore regarded to be a contraindication for orthotopic HTX[7, 8, 11]. In HTX-candidates PG E_1 has been used for therapy of elevated PVR and in the transplantation evaluation. The reported dosages of PG E_1 differ to a high degree[11, 13]. For moderation of secondarily elevated PVR in the transplantation evaluation of potential heart recipients dosages up to 300 ng/kg/min have been given[11].

For therapeutic application of PG E_1 in patients before HTX different concepts of dosage finding have been described[7, 8, 13]. Generally the occurrence of intolerable side effects has been defined as the endpoint of a stepwise increase and the maintenance dosage has depended on the patients′ individual tolerance of PG E_1.

Since it has never been proved that these concepts provide optimum therapeutic effects, we have tried to find a hemodynamic parameter for a dosage-definition. For this method right heart catheterization is necessary. When PG E_1 is given alone or with dopamine in a low dosage (3 microg/kg/min)[13], the effects of PG E_1 on both the pulmonary and the systemic vascular resistance seem to be responsible for hemodynamic improvement. With a decrease of mean pulmonary pressure (mPAP) that exceeds the loss of arterial pressure (AP) and an increase of cardiac index (CI), a hemodynamic improvement by PG E_1 can be assumed. A calculation of mPAP divided by (CI multiplied by mean AP)

$$\frac{m\,PAP}{CI\,.\,mAP}$$

can help to estimate therapeutical effects. With increasing dosages of PG E_1 in patients with secondarily elevated PVR, the parameter usually decreases until a minimum is reached. In some patients the above mentioned parameter (mPAP/CI.mAP) shows an inverse development with PG E_1 application even at a low dosage. This occurrence of hypotension without a comparable effect on the pulmonary artery pressure and without an increase of cardiac index strongly suggests a predominantly fixed elevated PVR. Thus, when mPAP/CI.mAP at a dosage of 10 ng/kg/min PG E_1 is higher than the calculated value before the application, a fixed elevation of PVR is presumed. In those patients PG E_1-application for hemodynamic reasons is regarded contraindicated. This hemodynamic definition may facilitate the differentiation between therapeutic responders and non-responders even at low dosages of PG E_1.

Intraoperatively and in the early postoperative period, PG E_1 has often been used to prevent right heart failure in patients at risk. For this application, dosages up to 120 ng/kg/min have been reported[8]. Since prostaglandins in high dosages cause edema[21] and a decrease of systemic vascular resistance which is not always desired, alternative medicaments may provide a more suitable solution in some cases. The possibility of highly selective pulmonary vasodilation with nitric oxide, the whole range of actions of all potentially indicated therapeutic agents including their side effects and the biological half-lives have to be taken into consideration for an optimum choice.

CONCLUSION

PG E_1 with its complex range of actions can be helpful in problems with solid graft function after transplantation and can help to improve the hemodynamics of patients on the waiting list and in the perioperative period. Therapy of elevated PVR is regarded to be the most important indication for PG E_1 in the field of transplantation and intensive care at the moment. Postoperatively or intraoperatively given PG E_1 should be tapered down whenever the need of catecholamines steadily increases, especially in case of renal failure.

In our experience, an improvement of graft function after solid organ transplantation and of kidney function by continuous PG E_1 can be expected only in case of hemodynamic stability. Despite the beneficial range of actions, situations such as lung edema, high need of predominantly vasoconstricting catecholamines and renal failure due to low arterial pressure should be regarded as contraindications for PG E_1 and as indications for an alternative regimen. In our opinion an individually taylored concept of dosage and the performance of right heart catheterization in unclear hemodynamic situations are necessary for an optimal therapy regimen with PG E_1 in critically ill patients. A calculation of mPAP/CI.mAP after right heart catheterization can be helpful in locating patients with a predominantly fixed elevation of PVR and in avoiding an inappropriately high dosage of PG E_1, as well as finding the optimum combination of therapeutic agents especially in the early postoperative period.

REFERENCES

1. Armitage G., Hardesty R. et al., Prostaglandin E_1: An effective treatment of right heart failure after cardiac transplantation. *J Heart Transplant* 6:348-51 (1987)

2. Cawello W., Leonhardt A., Schweer H. et al., Dose proportional pharmacokinetics of alprostadil (prostaglandin E_1) in healthy volunteers following intravenous infusion. *Br J Pharmacol.* 40(3):273-6 (1995)
3. Fullerton D., Hahn A. et al., Pulmonary vascular smooth muscle cell relaxation by cGMP- versus cAMP-mediated mechanisms. *J Surg Res.* 57(2):259-63 (1994)
4. Harjula A., Baldwin J. et al., Clinical heart lung preservation with Prostaglandin E_1. *Transpl Proc.* 19:4101 (1987)
5. Hide EJ., Ney P., Piper J. et al., Reduction by prostaglandin E_1 or prostaglandin E_0 of myocardial infarct size in the rabbit by activation of ATP-sensitive potassium channels. *Br J Pharmacol.* 116(5):2435-40 (1995)
6. Higenbottam T, Wheeldon D et al., Long-term treatment of primary pulmonary hypertension with continuous intravenous epoprostenol (Prostacyclin). *Lancet* 12:1046-7 (1984)
7. Iberer F., Wasler A. et al., Prostaglandin E_1 induced moderation of elevated pulmonary vascular resistance. Survival on waiting list and results of orthotopic heart transplantation. *J Heart Lung Transplant.* 12(2):173-8 (1993)
8. Iberer F., Müller H. et al., *Prostaglandin E1 in der Therapie der Abstoßung und des pulmonalen Hypertonus im Rahmen der Herztransplantation. Immundiagnostik und Immuntherapie in der modernen Intensivmedizin,* Vogt S., ed., Pabst-Verlag, Lengerich (1993)
9. Kim Y., Kai T. et al., Hepatoprotection by a PG I_2 analogue in complete warm ischemia of the pig liver. Prostanoid release from the reperfused liver. *Transplantation* 58(8):875-9 (1994)
10. Moran M., Mozes M. et al., Prevention of acute graft rejection by the prostaglandin E_1 analogue misoprostol in renal-transplant recipients treated with cyclosporine and prednisone. *N Engl J Med.* 322:1183-88 (1990)
11. Murali S., Uretsky B., Armitage J. et al., Utility of Prostaglandin E_1 in the Pretransplantation Evaluation of Heart Failure Patients with Significant Pulmonary Hypertension. *J Heart Lung Transplant.* 11:716-23 (1991)
12. Naeije R., Melot C et al., Reduction in pulmonary hypertension by prostaglandin E_1 in decompensated chronic obstructive pulmonary disease. *Am Rev Despir Dis.* 125:1-5 (1982)
13. Pacher R., Globits S., Wutte M. et al., Beneficial hemodynamic effects of prostaglandin E_1 infusion in catecholamine-dependent heart failure: results of a prospective, randomized controlled study. *Crit Care Med.* 22(7):1084-90 (1994)
14. Prielipp R., McLean R. et al., Hemodynamic profiles of prostaglandin E_1, isoproterenol, prostacyclin and nifedipin in experimental porcine pulmonary hypertension. *Crit Care Med.* 19(1):60-7 (1991)
15. Quiroga J., Prieto J., Liver cytoprotection by prostaglandins. *Pharmacol Ther.* 58(1):67-91 (1993)
16. Schweer H., Cawello W., Seyberth HW, Gas chromatogryphy negative ion chemical ionization triple quadrupole mass spectometric determination and pharmacokinetics of 11 alpha-hydroxy-9,15-dioxo-2,3,4,5,20-pentanor-19-carboprostanoic acid in plasma. *Prostaglandins Leukot Essent Fatty Acids.* 51(6):401-5 (1994)
17. Sinzinger H., Rogatti W., Prostaglandin E_1 in der Therapie der peripheren arteriellen Durchblutungsstörung. *Wiener Klinische Wochenschrift* 103/18:558-565 (1991)
18. Sorbara C., Pitarello D. et al., Prostaglandine in der Intensivmedizin und bei Herztransplantationen. Translation from *A.P.I.C.E;* Vol. 8:327-337 (1993)
19. Ueno T., Horita K. et al., The effect of Prostaglandin E_1 on heart preservation. *Transplantation* 54(3):418-22 (1992)
20. Vincent J., Carlier E. et al., Prostaglandin E_1 infusion for right ventricular failure after after cardiac transplantation. *J Thorac Cardiovasc Surgery.* 103:33-39 (1992)
21. Warren J., Wilson A. et al., Opposing roles of cyclic AMP in the vascular control of edema formation. *FASEB J.* 7(14):1394-400 (1993)
22. Watkins W., Peterson M. et al., Prostacyclin and Prostaglandin E_1 for severe idiopathic pulmonary hypertension. *Lancet* I. 1083 (1981)

PROSTAGLANDINS IN LIVER TRANSPLANTATION

Robert M. Merion

Department of Surgery
University of Michigan Medical School
Ann Arbor, Michigan
United States of America

SUMMARY

This review summarizes experimental studies and clinical experiences with prostaglandins in liver transplantation emphasizing two randomized, double blinded placebo controlled clinical trials of prostaglandin E_1 (PGE_1) involving nearly 300 liver transplant recipients. Resource utilization and pharmacoeconomic aspects are also discussed. In the randomized trials, PGE_1 did not affect patient and graft survival. Acute cellular rejection and primary allograft non-function were not reduced by PGE_1. Postoperative renal failure was significantly less frequent among PGE_1 patients in both trials and in one study perioperative blood product utilization was significantly lower. PGE_1 treated patients had reduced intensive care unit length of stay, shorter hospitalization, and significantly lower total health care charges. Despite its failure to show improvements in patient and graft survival, rejection, or primary non-function, PGE_1 use in hepatic allograft recipients reduces morbidity and results in notable reduction in the cost of liver transplantation.

INTRODUCTION

Liver transplantation has evolved from its origins as a risky experimental therapy with uncertain results for desperately ill patients to a nearly routinely offered mainstream therapy for a wide variety of acute and chronic hepatic diseases. In the United States, 3923 liver transplants were performed in 1995, yet the number of patients on the waiting list grew to 5691[1]. The disparity between the increasing numbers of patients eligible for liver transplantation and the desperate shortage of organs made available each year has resulted in large numbers of patients dying of their underlying disease before a transplant can be performed. Transplant programs and organ procurement organizations, faced with the pressure exerted by a mismatch in supply and demand. have increasingly turned to donors of less than optimal quality as an alternative to seeing otherwise salvageable recipients die. Thus, issues of organ quality and adequacy of post-transplant liver function have become progressively more important in recent years. The need to decrease the likelihood of graft failure (and consequent need for retransplantation) has led to new approaches such as pharmacological intervention with the eicosanoid prostaglandin PGE_1 for use in organ preservation and for the prophylaxis and treatment of allograft rejection.

The rationale for using PGE_1 derives from a range of experimental and clinical studies[2]. Improvement in survival following extended liver allograft preservation time has been demonstrated in a rat model of PGE_1[3] and has been demonstrated to reduce hepatic preservation injury in isolated perfused rat liver[4] and canine liver allografts[5]. However, despite improved organ preservation and post-transplant function, theoretical effects on the hepatic microcirculation have not translated into measurable improvements in overall hepatic allograft blood flow in animals[5] or in humans[6]. Other cytoprotective effects such as maintenance of endothelial cell integrity, reduction in oxidative stress, and directly or indirectly mediated vasodilatory effects have been proposed[2].

The most extreme form of liver allograft ischemia-reperfusion injury is manifest in the syndrome of primary non-function, where coagulation factors are not produced, naturally occurring anticoagulant proteins are not eliminated, and toxic intermediates remain in circulation. A patient whose liver allograft fails due to primary non-function is in grave danger of early death from hepatic failure, cerebral edema, and massive hemorrhage.

In the non-transplant setting, this may occur as fulminant hepatic failure due to any of a variety of hepatitides, drugs, toxins, and miscellaneous conditions whose collective case fatality rate is 60 to 100%. A series of mostly uncontrolled trials of prostaglandins in fulminant hepatic failure reported between 1989 and 1992 failed to clearly demonstrate any benefit (reviewed in Quiroga and Prieto[2].

In the setting of primary non-function after liver transplantation, Grieg et al.[7] reported on the benefits associated with PGE_1 infusion in 13 patients. Graft survival was 80% in the treated group and 17% in the untreated group. These rather spectacular results generated much interest in the use of PGE_1 in clinical liver transplantation and suggested that prophylactic use might hold promise.

Intraoperative use of PGE_1 has also been proposed as a means to improve the impaired platelet aggregation that is seen following reperfusion in liver transplantation[(8-9)]. In an open-label prospective randomized trial, PGE_1 infusion was associated with significantly higher platelet aggregability. In addition, the drop in platelet count seen after reperfusion was ameliorated in the PGE_1 group. In this small series, blood product transfusion requirements were similar in the treated and untreated groups.

The known immunosuppressive properties of prostaglandins of the E and I series make prevention and treatment of liver allograft rejection attractive potential indications for PGE_1 use. Inhibition of cell-mediated lymphocytotoxicity, diminished splenic T cell hepatocytotoxicity, and inhibition of interleukin-2-driven alloproliferation have been associated with prostaglandin administration. In one clinical trial, the PGE_1 analogue misoprostol was used in cyclosporine and steroid treated renal transplant recipients and was associated with a significantly lower incidence of acute rejection when compared to placebo-treated patients[10]. However, as detailed below, randomized clinical trials of PGE_1

in liver transplantation have not demonstrated a statistically significant decrease in the incidence of liver allograft rejection.

RANDOMIZED TRIALS OF PGE_1 IN LIVER TRANSPLANTATION

The most comprehensive and scientifically rigorous examinations of PGE_1 in clinical liver transplantation have come from two randomized double-blind placebo controlled trials reported in the past two years[(11-12)]. The first randomized trial of PGE_1 in liver transplantation was reported by the University of Michigan in 1995[11]. In this double-blind, randomized, placebo-controlled study, 78 patients were randomized to PGE_1 and 82 to placebo. Patients were randomized prior to the anhepatic phase of the liver transplant so that intravenous infusion could begin intraoperatively, using a stepwise dose escalation to 40 μg/hr (approximately 0.6 μg/kg/hr). The treatment period ran for 21 days or until patient discharge, retransplantation, or death. Primary study end points were patient and graft survival and the incidence of primary non-function, need for retransplantation and rejection. In a recently reported multicenter trial of PGE_1, Klein et al.[12] reported on the results of a similarly designed trial carried out under the auspices of the South Eastern Organ Procurement Foundation (SEOPF). In this double-blind, randomized, placebo-controlled trial, 58 patients were treated with PGE_1 and 60 with placebo. Randomization occurred following allograft revascularization, dose escalation proceeded to 1 μg/kg/hr, and patients were treated for seven days or until transfer out of the intensive care unit. Primary end points were the incidence of primary non-function and the occurrence of post-transplant renal dysfunction. The demographic characteristics of patients in the two arms of the Michigan study as well as the multicenter report were similar. Patients underwent liver transplantation for a variety of specific hepatic pathological conditions in the two groups and, typical of North American patients undergoing this procedure, many of the recipients had alcoholic cirrhosis.

There were no significant differences in patient or graft survival in either study when PGE_1-treated patients were compared to placebo-treated controls. Fewer PGE- treated recipients than placebo-treated patients had primary non-function in the Michigan study (4% vs 7%), but this difference was not statistically significant. Primary non-function occurred in 7% of the experimental and control groups in the SEOPF study. The incidence of rejection was reduced by about one-third in the PGE_1-treated group in the Michigan study (26% vs 38%), although this was not statistically significant. The incidence of rejection was not reported in the SEOPF study.

The most striking effects of PGE_1 were related to preservation of renal function and to reduced utilization of health care resources. The incidence of renal dysfunction in the PGE_1 group was 2% compared to 14% in the control group in the Michigan study and 14% and 27%, respectively, in the SEOPF study. This difference (statistically significant in the Michigan report) was accompanied by a smoother overall post-operative course. Intensive care unit stay was shortened by 40% and 64% in the Michigan and SEOPF studies, respectively, and total transplant hospitalization length of stay was reduced 23% amongst patients receiving PGE_1 in the Michigan study. The precise reasons for improved renal function in liver transplant recipients receiving PGE_1 are unknown, but it has been noted that fewer fungal and opportunistic infections occurred in the PGE_1 treated patients, and this factor may have resulted in a lower incidence of multisystem organ failure, including renal failure. Vasodilatory effects of prostaglandins may have also played a role in preserving renal function, but these parameters have not been carefully studied.

Blood product utilization was less extensive in patients receiving PGE_1 in the Michigan study. Although operative times were similar for the PGE_1 and control groups, red blood cell transfusion requirements were significantly lower in the PGE_1 group starting with the anhepatic phase of surgery, and overall cryoprecipitate and platelet transfusion requirements were significantly lower in the PGE_1 group (Table 1). These effects are unlikely to be due to hepatic synthesis of coagulation factors, but may result from effects on platelet aggregation (9), persistence of increased circulating endothelin-l[13], or prostaglandin E_1 mediation of endothelin release as in intestinal ischemia reperfusion injury[14].

Table 1. Blood component utilization during various phases of liver transplantation

Time subset	Group	Red blood cells	Cryo-precipitate†	Plasma	Platelets††
Hepatectomy	Prostaglandin E1	7.7 ± 0.9	6.6 ± 0.7	0.2 ± 0.1	0.2 ± 0.1
	Placebo	8.5 ± 1.0	8.4 ± 1.0	0.2 ± 0.1	0.5 ± 0.1
Anhepatic	Prostaglandin E1	5.2 ± 0.4*	5.1 ± 0.4	0.2 ± 0.1	0.4 ± 0.1
	Placebo	6.9 ± 0.7	5.8 ± 0.6	0.4 ± 0.1	0.4 ± 0.1
Reperfusion†††	Prostaglandin E1	6.1 ± 0.7 ***	11.9 ± 1.5	2.4 ± 0.2	2.9 ± 0.2**
	Placebo	10.3 ± 1.6	15.0 ± 1.8	3.0 ± 3.3	3.9 ± 0.4
Total	Prostaglandin E1	19.1 ± 1.5*	22.7 ± 1.8	2.8 ± 0.3*	3.5 ± 0.3**
	Placebo	25.1 ± 3.0	28.0 ± 3.0	3.6 ± 0.4	4.8 ± 0.6

† 10-packs; †† 6-packs; ††† Reperfusion to 48hr * $P<0.05$; ** $P<0.025$; ***$P< 0.01$

PHARMACOECONOMIC ANALYSIS

Modern medical practice invariably compares the potential risks of a treatment or intervention against the potential benefit. In recent years, however, comparable treatments have been put to an additional economic validity test in order to ensure that precious health care resources are not squandered, In the case of PGE_1, a detailed cost analysis of alprostadil in liver transplantation was carried out by Smith et al. using data generated from the Michigan randomized trial[15]. They examined data from all single-transplant surviving patients (42 PGE_1, 49 controls) for whom complete billing records were available through 150 post-transplant days. Patients given PGE, had total hospital and professional charges that were US$ 50,375 lower than the charges for patients who received placebo ($p = 0.043$).

When these data were further examined, it was apparent that hospital charges, rather than professional fees, were selectively affected. Every major hospital charge category was significantly reduced in PGE_1 patients, including room charges, blood bank, pathology lab, respiratory therapy, general equipment, and radiology. These pharmacoeconomic data suggest that the reduction in morbidity associated PGE_1 use in liver transplantation is accompanied not only by significant improvements in patient outcome as evidenced by reduced hospital stay, but also by marked reduction in the cost of providing liver transplantation services.

CONCLUSIONS

Despite many years of experimental study, prostaglandins continue to occupy a position of uncertainty in clinical liver transplantation practice. Two well-designed, prospective, double blinded, randomized, clinical trials have failed to demonstrate reductions in allograft dysfunction and graft rejection that were suggested by experimental work and early anecdotal experience[16]. The incidence of primary non-function, perhaps the most feared immediate complication of liver transplantation, has not been reduced by PGE_1. Neither has there been a significant reduction in the incidence of allograft rejection. Despite these disappointments, morbidity after liver transplantation has been reduced, and accompanying reductions in the incidence of renal failure have led to shorter intensive care unit requirements and lower overall lengths of transplant hospitalization. These, in turn, have been associated with important reductions in the use and cost of ancillary hospital resources.

REFERENCES

1. United Network for Organ Sharing World Wide Web site. http://www.unos.org

2. J. Quiroga and J. Prieto, Liver cytoprotection by prostaglandins. *Pharmac Ther*. 58:67 (1993)

3. H.S. Xu, L.K. Rosenlof, T.L. Pruett, and R.S. Jones, Prostaglandin E_1 increases survival with extended anhepatic phase during liver transplantation. *Ann Surg*. 220:53 (1994).

4. K.M. Olthoff, E. Wasef, and P. Seu, et al,. PGE_1 reduces injury in hepatic allografts following preservation. *J Surg Res*. 50:595 (1991).

5. K. Takahaski, N. Yamamoto, and H. Egawa, et al., Effect of prostaglandin E_1 on preservation of canine liver grafts preserved with UW solution. *Transpl Int*. 6:245 (1993).

6. K.D. Chavin, L. Uber, P. Baliga, P. Rajagopalan, and J. Cofer, The effects of prostaglandin E_1 on hepatic allograft vascular inflow: a prospective randomized double-blind study. *Am Surg*: 62:184 (1996).

7. P. D. Greig, G. M. Woolf, and S. B. Sinclair, et al., Treatment of primary liver graft nonfunction with prostaglandin E_1. *Transplantation*. 18:447 (1989).

8. G. Himmelreich, K. Hundt, P. Neuhaus, W. O. Bechstein, P. Roissant, and H. Reiss, Influence of prostaglandin E_1 infusion on hemostasis in orthotopic liver transplantation. *Semin Thromb Hemos*. 19:273 (1993).

9. G. Himmelreich, K. Hundt, P. Neuhaus, W. O. Bechstein, R. Roissant, and H. Reiss, Evidence that intraoperative prostaglandin E_1 infusion reduces impaired platelet aggregation after reperfusion in orthotopic liver transplantation. *Transplantation*. 55: 819 (1993).

10. M. Moran, M. F. Mozes, and M. S. Maddux, et al., Prevention of acute graft rejection by the prostaglandin E_1 analog misoprostol in renal-transplant recipients treated with cyclosporine and prednisone. *New Engl J Med*. 322:1183 (1990).

11. K. S. Henley, M. Lucey, and D. P. Normolle, et al., A double-blind randomized placebo-controlled trial of prostaglandin E_1 in liver transplantation. *Hepatology*. 21:366 (1993).

12. A. S. Klein, J. B. Cofer, and T. L. Pruett, et al., Prostaglandin E_1 administration following orthotopic liver transplantation A randomized prospective multicenter trial *Gastroenterology*. 111:710 (1996).

13. M. Asbert, A. Gines, and P. Gines, et al., Circulating levels of endothelin in cirrhosis *Gastroenterology*. 104:1485 (1993).

14. A. O. Aktan, O. Buyukgebiz, and C. Yegen, et al., Does PGE_2 act as a mediator for endothelin release? *Prostaglandins Leukot Essent Fatty Acids*. 50:37 (1994).

15. D. G. Smith, K. S. Henley, C. S. Rernmert, S. L. Haas, D. A. Campbell, and I. D. McLaren, A cost analysis of alprostadil in liver transplantation. *Pharmacoeconomics.* 9:517 (1996).

16. D. F. Schafer, and M. F. Sorrel, Prostaglandins in liver transplantation: a promise unfulfilled (editorial). *Gastroenterology*. 111:819 (1996).

PROSTAGLANDINS IN HEART TRANSPLANTATION

F. Iberer, A.Wasler, B.Petutschnigg, H.Müller, T.Allmayer, B.Grasser, G.Prenner, S.Schaffellner, G.Hipmayr and KH Tscheliessnigg

Karl-Franzens-University of Graz, Austria, Department of Surgery, Division of Transplantation Chirurgische Klinik, Fach 37, A-8036 Graz
Phone ++43 316 385 4444, Fax ++43 316 385 4446

PATIENT SELECTION FOR ORTHOTOPIC HEART TRANSPLANTATION AND ESTIMATION OF PULMONARY VASCULAR RESISTANCE

Heart transplant recipients with secondary pulmonary hypertension (PH) are prone to acute right ventricular (RV) graft failure after orthotopic heart transplantation (oHTX). A suitable donor heart of a healthy individual is not adapted to elevated RV afterload caused by PH. In contrast, the majority of potential heart transplant recipients suffer from chronic left ventricular (LV) failure. LV insufficiency requires elevated filling pressures to maintain cardiac output (CO), (LV backwards failure) and causes systemic hypotension (LV forward failure) which induces systemic and pulmonary vasoconstriction. This increases systemic (SVR) and pulmonary vascular resistance (PVR), induces RV hypertrophy and secondary PH. After oHTX, the unadapted transplanted RV is exposed to the recipients PVR, RV afterload mismatch results in acute RV failure when the patient is weaned from cardiopulmonary bypass. RV volume overload, dilatation and structural damage are followed by RV failure and death. The preoperative estimation of PVR and of the reactivity of the pulmonary vascular bed to pulmonary vasodilators permits the selection of patients with reversible PH that are still suitable for oHTX. Many attempts failed to define a clear borderline beyond which oHTX is not feasible. In fact, RV failure after oHTX is caused by both, the elevated RV afterload of the recipient and by insufficient RV performance. Reduction of RV afterload by pulmonary vasodilator therapy reduces the risk of RV failure resulting from RV forward failure. The risk of inadequate myocardial function remains.

Prostaglandin E_1 (PGE_1) has been shown as the most potent pulmonary vasodilator when compared to nitro-glycerine and sodium nitroprusside (1,2). It is the only drug that significantly lowered transpulmonary pressure gradient (3). Preoperative estimation of the reversibility of PVR (1,3) or moderation of PVR by combined PGE_1 and catecholamine therapy (4) improves overall survival by excluding patients at high risk for postoperative RV failure.

Recent Advances in Prostaglandin, Thromboxane, and Leukotriene Research
Edited by Sinzinger *et al.*, Plenum Press, New York, 1998

TREATMENT OF RIGHT VENTRICULAR FAILURE AFTER ORTHOTOPIC HEART TRANSPLANTATION

Strategies to overcome RV failure are appropriate matching of the donor size, the use of oversized donor hearts, rapid decompression of both ventricles after aortic cross clamping during organ harvesting, short ischemic time, avoidance of tricuspid insufficiency caused by surgery, prevention of RV volume overload and dilatation by catecholamine support, reduction of RV preload, ventilation without positive end-expiratory pressure, increase of heart frequency and pulmonary vasodilatation, which supports the adaptation of the transplanted RV to the recipients pulmonary afterload. During postoperative ventilation, the selective pulmonary vasodilator properties of Nitric Oxide are used to maintain adequate LV preload (5), after extubation, PGE_1 remains the therapy of choice (6,7).

The time course of resolution of PH and RV remodelling after HTX has been investigated by echocardiography studies. With appropriate pulmonary vasodilator therapy, the adaptation of the transplanted RV to the recipients PVR needs up to 2 weeks (8). Diastolic dysfunction may take as long as 3 months to resolve completely (9). As a result of RV structural damage, highly elevated Troponin -T levels (up to 3µg/l) have been found by our group (10). A correlation was found between the preoperative systolic pulmonary artery pressure and the cumulative Troponin T release after oHTX. Troponin may be helpful for the management of RV function and for the guidance of dosage and duration of pulmonary vasodilator therapy.

The long-term outcome of RV function after oHTX in patients with elevated PVR has been investigated by our group (11). One year after oHTX, right heart recatheterisation was performed. We compared 63 patients with preoperatively elevated PVR who were treated with PGE_1 to 27 recipients with normal pulmonary pressure. (PVR 3.9 ± 1.2 versus 1.4 ± 0.7 Wood units, mean PAP syst/ dia/mean 63/29/41 versus 27/10/18 mm Hg). One year after oHTX no difference was found between the groups (PVR 0.9 ± 0.3 versus 1.3 ± 0.5, PAP 26/11/18 versus 25/8/16, respectively). These data suggest, that PH and PVR before oHTX mainly depends on LV function. After oHTX, appropriate LV function is provided and PH resolves.

INFLUENCE OF PGE_1 ON RENAL FUNCTION, CYCLOSPORIN A AND ON REJECTION AFTER HEART TRANSPLANTATION

PGE_1 acts as free oxygen scavenger, protect organs from reperfusion injury and influence the immune system. A reduction of the frequency and the severity of rejection episodes during the application of PGE_1 after oHTX has been reported (7). Effects on organ preservation might contribute to this result.

Beneficial effects of PGE_1on renal function have been reported. We compared patients after oHTX with and without postoperative PGE_1 therapy (12). During CyA treatment patients showed lower Cyclosporine A (CyA) levels (124 + 54 versus 198 +83 ng/ml) and higher serum creatinine levels were found. After discontinuation of PGE_1 therapy no differences could be found. RV failure after oHTX increases central venous pressure, impairs perfusion of small bowel and kidneys and promotes malabsorbtion of CyA and elevated serum creatinine. Daily adjustment of CyA levels is mandatory for appropriate CyA therapy during PGE_1 application after oHTX.

TREATMENT OF SECONDARY PULMONARY HYPERTENSION PHARMACOLOGICAL BRIDGING TO HEART TRANSPLANTATION

In patients with dilated cardiomyopathy, PGE_1 reduces SVR and PVR and the LV responds by a redistribution of diastolic filling to later in diastole and in an increase of stroke volume (13). Pulmonary vascular pressure flow plots are displaced to equal or lower pressures and to higher flows by PGE_1 before as well as after oHTX (14). The effect of maximum catecholamine support further can be improved by additional PGE_1 infusion (15). In patients waiting for oHTX, one week of combined therapy with PGE_1 (up to 40 ng/kg/min) and Catecholamines resulted in a lasting decrease of PVR and an improvement of survival time on waiting list (Graz-Protocol), (4). PGE_1 induced peripheral and pulmonary vasodilatation. Consecutive improvement of LV filling pattern and stroke volume and a reduction of sympathetic influence on SVR might have led to a persisting improvement of LV function.

Currently, Pacher investigates a new method for bridging to transplant in catecholamine-dependent patients. PGE_1 is administered permanently via a Hickman catheter and portable pumps. During right heart catheterisation, PGE_1 dosage is increased, 50% of the maximum tolerated dosage are used for maintenance therapy (Vienna-Protocol), (15). This results in higher CO, an improvement of kidney and liver function and an increased survival on waiting list to oHTX.. In a prospective, randomized parallel trial, PGE_1 was compared to Prostacyclin and Dobutamine. More patients of the group reached oHTX (16). Moreover, reduction of PVR by permanent PGE_1 application seems to reduce the risk of RV graft failure after oHTX.

PGE_1 AS A BRIDGE TO LEFT VENTRICULAR ASSIST DEVICE

The lack of donor organs urged the development of bridging methods to oHTX. One the one hand, the goal is to prolong survival on waiting list, on the other hand is to improve transplant survival by providing more stable patients for surgery. No reports have been presented so far on the use of PGE_1 as a bridge to left ventricular assist device (LVAD). A 60 year old patient received PGE_1 therapy (Graz -Protocol), (4). The cardiac index (CI) and PVR before therapy was 1.8 l/min and 4.9 Wood units, after 2.6 l/min and 1.5 Wood units, the patient was put on waiting list for oHTX. Three months later, cardiac decompensation (CI of 1.8 l/min, PVR 3.5 Wood) indicated permanent PGE_1 therapy (Vienna-Protocol), (16). One month later, the CI was 2.5 l/min and PVR was 2.2 Wood units. Catecholamine refractory heart failure indicated LVAD implantation. Despite appropriate LVAD-function, PGE_1 could not be weaned without the instantaneous appearance of acute right heart failure. Serum Troponin T levels were comparable to those obtained after oHTX. Catecholamine support and stepwise reduction of PGE_1 infusion over a period of two weeks enabled the readaption of the RV.

LVAD is an effective bridge for patients with predominant LV failure. In contrast to oHTX, during LVAD implantation the RV is not replaced, the gain of CO is limited by the function of the remaining RV. Bridging to LVAD with permanent PGE_1 infusion reduces PVR and lowers the adaptation of the RV to overcome elevated RV afterload. Long term decrease of PVR is unadvantageaous for bridging to LVAD implantation. After bridging to LVAD implantation with PGE_1, RV readaptation to the elevated RV afterload appears to be very similar to the adaptation of the transplanted RV after oHTX.

REFERENCES

(1) Murali S, Uretzky BF, Armitrage JM et al. Utility of Prostaglandin E_1 in the pretransplantation evaluation of heart failure patients with significant pulmonary hypertension J.Heart Lung Transplant; 11: 716-23 (1992)

(2) Coddens J, Deloof T, Housmans P. Haemodynamic effects of sodium nitroprusside versus ProstaglandinE_1 in heart transplant candidates. British Journal of Anaestesia 76; Suppl 1: Abstract A26 (1996)

(3) Murali S, Uretzky BF, Reddy PS et al. Reversibility of pulmonary hypertension in congestive heart failure in patients evaluated for cardiac transplantation: Comparative effects of various pharmacologic agents. Am Heart J; 122(5):1375-81 (1991)

(4) Iberer F, Wasler A, Tscheliessnigg KH et al. ProstaglandinE_1 - induced moderation of elevated pulmonary vascular resistance. Survival on waiting list and results of oHTX. J Heart Lung Transplant; 12: 173-8 (1993)

(5) Kieler-Jensen N, Lundin S, Ricksten SE. Vasodilator therapy after heart transplantation: effects of inhaled nitric oxide and intravenous prostacyclin, prostaglandin E_1, and sodium nitroprusside. J. Heart Lung Transplant; 14: 436-43 (1995)

(6) Vincent JL, Carlier E, Pinsky MR et al. Prostaglandin E_1 infusion for right ventricular failure after cardiac transplantation. J.Thorac Cardiovasc Surg ; 103: 33-9 (1992)

(7) Iberer F, Wasler A, Tscheliessnigg KH et al. Prostaglandin E_1 reduces the frequency of rejection after heart transplantation. J Heart Lung Transplant; 11: 727-32 (1992)

(8) Bathia S, Kirshenbaum J, Shemin R et al. Time course of resolution of pulmonary hypertension and right ventricular remodeling after orthotopic cardiac TX. Circulation; 4: 819-26(1987)

(9) Hausdorf G, Banner NR, Mitchell A, Khaghani A, Marin M, Yacoub M. Diastolic function after cardiac and heart-lung transplantation. Br Heart J; 62:123-32 (1989)

(10)Halwachs G, Iberer F, Pieber T et al.:Troponin T as a marker for posttransplantation adaptional problems of the donor heart. J Heart Lung Transplant;15:451-455 (1996)

(11)Iberer F, Müller H, Wasler A et al. Prostaglandin E1 and heart transplantation. In Körner MM, Posival H, Körfer R (eds): Thoracic organ transplantation. Routine as a challenge. Elsevier Science B.V. (1994)

(12)Iberer F, Vujicic R, Rödl S et.al. Effects of Prostaglandin E_1 therapy on Cyclosporine A and creatinine levels after orthotopic heart transplantation. Transplantation proceedings 26;6:3246-3248 (1994)

(13)Barbieri E, Perini P, Marino P, Zardini P. Combined hemodynamic echocardiographic Doppler evaluation of Prostaglandin E_1 effects in patients with severe dilated cardiomyopathy undergoing evaluation for HTX. J Heart Lung Transplant; 14 (3): 572-8 (1995)

(14)Naeije R, Lipski A, Abramowicz M et al. Nature of pulmonary hypertension in congestive heart failure. Effects of cardiac transplantation. Am J Respir Crit Care Med; 194: 881-7 (1994)

(15)Pacher R, Globits S, Wutte M et al. Beneficial hemodynamic effects of Prostaglandin E_1 infusion in catecholamine-dependent heart failure: Results of a prospective, randomized, controlled study. Crit Care Med; 22: 1084-90 (1994)

(16)Pacher R, Stanek B, Hülsman M et.al. Prostaglandin E_1 bridge to cardiac transplantation: technique, dosage, results. British Heart J; in press

EICOSANOIDS IN PERIODONTAL DISEASES: POTENTIAL FOR SYSTEMIC INVOLVEMENT

Suzanne M. Damaré,[1] Steve Wells,[2] and Steven Offenbacher[3]

[1]Research Associate
[3]Professor of Periodontology
School of Dentistry, CB #7455
[2]School of Medicine
Department of Obstetrics and Gynecology
University of North Carolina at Chapel Hill
Chapel Hill, North Carolina 27599

ABSTRACT

The potential application of gingival crevicular fluid (GCF) levels of prostaglandin E_2 (PGE_2) for predicting periodontal disease severity and reflecting disease activity has been described by several investigators. The potential application of GCF levels of PGE_2 for predicting amniotic fluid levels of PGE_2, and thereby estimating risk of preterm low birth weight, has not been explored until recently. The fact that intraamniotic levels of many inflammatory mediators, such as prostaglandin E_2, increase during pregnancy and peak at delivery suggests that these mechanisms play an important role in normal physiologic parturition. There is increasing evidence to suggest that the presence of a subclinical endogenous Gram-negative infection of periodontal disease may present a systemic challenge sufficient to initiate the onset of premature labor, either as a source of lipopolysaccharide (LPS, endotoxin) and/or through stimulation of secondary inflammatory mediators such as PGE_2 and interleukin 1 beta (IL-1β). This pilot investigation determines associations among the levels of PGE_2 in gingival crevicular fluid, within serum, and within amniotic fluid of 18 women who were undergoing routine amniocentesis during early midtrimester, in an attempt to reveal new baseline data regarding the relationship between periodontal inflammatory mediators and intraamniotic mediators during normal parturition. Full-mouth periodontal examinations were performed and samples of GCF, serum and amniotic fluid were collected from each subject for the quantitative assessment of PGE_2 by radioimmunoassay (RIA). Results of a pairwise regression analysis model reveal that PGE_2 levels within the GCF are positively associated with intraamniotic PGE_2 levels at the $P = 0.018$ level of significance. Thus, gingival crevicular levels of PGE_2 may be used to provide an indirect estimate of the amniotic fluid levels of PGE_2.

Recent Advances in Prostaglandin, Thromboxane, and Leukotriene Research
Edited by Sinzinger *et al.*, Plenum Press, New York, 1998

INTRODUCTION

Preterm birth is the leading perinatal problem in the United States. The incidence of preterm birth (less than 37 weeks of gestation), which accounts for a large proportion of infant morbidity and mortality, increased to 10.8 percent in 1991, compared with 10.6 percent in 1990. Low birth weight (LBW; less than 2500 grams, or 5 pounds 8 ounces), and in particular, very low birth weight (VLBW; less than 1500 grams, or 3 pounds 5 ounces), - the result of preterm birth, intrauterine growth restriction (IUGR), or both - are strongly associated with neonatal mortality and morbidity and may be related to childhood developmental delays (McCormick et al., 1991). Between 1990 and 1991 the infant mortality rate decreased for 6 of the 10 leading causes of infant death, increased for 3 causes, and remained unchanged for 1 cause. The 3 increases were for 1) disorders relating to short gestation and unspecified LBW; 2) accidents and adverse effects; and 3) infections specific to the perinatal period. More than 60% of the mortality that occurs among infants without anatomic or chromosomal congenital defects is attributable to LBW (Shapiro et al., 1988). According to the National Center for Health Statistics (1993), "The increasing rate for Disorders relating to short gestation and unspecified low birth weight was the principal cause preventing the infant mortality rate from decreasing more than it did."

In addition to the psychological and social impact, preterm birth creates economic problems. In 1981, for example, $1.5 billion was spent on neonatal intensive care in the United States, most of it on LBW babies (Phibbs et al., 1981). Low birth weight often results in long-term illness or infant death; even small variations have important consequences for the populations involved. The underlying cause of preterm birth is not evident in most cases, and currently, physicians are left with marginally effective attempts to delay delivery by arresting contraction. The emphasis of medical care then, has been placed on costly tertiary care rather than prevention. Breakthroughs in the prevention of preterm birth will require an improved understanding of its causes followed by appropriate strategies to prevent these causes. Evidence from several sources suggests a link between subclinical infection and prematurity. The pathogenesis of preterm labor and premature rupture of membranes (PROM) appears to be mediated by maternally secreted inflammatory agents, especially prostaglandin E_2 (PGE_2), in response to endotoxin challenge.

The focus of this investigation is to contribute to an understanding of the factors that activate premature rupture of membranes, premature labor, and low birth weight by determining whether the gingival crevicular fluid (GCF) levels of PGE_2 can be correlated to the intraamniotic fluid level of this eicosanoid. It remains to be determined, however, whether a direct effect of periodontitis - induced LPS exposure and maternal cytokines release exists or whether periodontal disease is simply a risk marker for some underlying trait, placing the patient at risk for both periodontal disease and preterm low birth weight (PLBW). It is important to note that 50-74% of women of child-bearing age (i.e. 18-34 years) have at least some form of periodontal disease (U.S. Department of Human Services, PHS, 1987), but severe disease is less common affecting 7-14% of the total population.

PROSTAGLANDIN E_2 AND PERIODONTAL DISEASES

Evidence is well-documented that the primary etiologic agents responsible for periodontal diseases are the bacteria and their products which are found in dental plaque (Genco et al., 1969, Ellison, 1970, Socransky, 1970). Inflammatory mediators, including prostaglandin E_2, produced by the gram-negative bacteria associated with progressive disease, act in concert to destroy periodontal supporting connective tissue and bone; gingival crevicular fluid levels of these mediators are elevated in periodontal diseases. GCF is a serum transudate that physiologically appears around the neck of the tooth at the gingival margin. In the presence of gingival inflammation the GCF flow increases and the fluid contains inflammatory mediators which have diffused from the adjacent inflamed gingival tissues. Gingival crevicular fluid levels of PGE_2, for example, are elevated in patients who are at high risk for certain periodontal diseases (Goodson et al., 1974, Loning et al., 1980, Offenbacher et al., 1981); in health, they are at low levels or non-detectable.

Heasman, Collins, et al., (1993) observed an elevation in GCF-PGE_2 during the induction of experimental gingivitis in humans, increasing from 20.5 ng/mL to 53.5 ng/mL at 4 weeks. Offenbacher and colleagues (1986) demonstrated that periodontitis patients who have received traditional conservative therapy (i.e. scaling and root planing) can be placed into either a high-risk group or a low-risk group, based upon the GCF-PGE_2 level. The odds of a patient having an attachment loss episode within 6 months of detection are elevated 47-fold if the GCF-PGE_2 level is above 66.2 ng/mL. It is estimated that a level of 300 ng PGE_2/mL GCF reflects an underlying disease activity of 0.5 to 1.0 mm of periodontal attachment loss per year (Offenbacher, Collins, Heasman, 1993). Furthermore, Nakashima and colleagues (1994) have recently shown that within the GCF, there is significant correlation between PGE_2 levels and 2 markers of bone resorption, alkaline phosphatase and osteocalcin. This indicates that the GCF-PGE_2 level not only reflects ongoing subclinical disease activity, but can also be used to predict future clinical changes to identify patients at high risk.

Recent data suggest that the GCF-PGE_2 levels are higher in patients with risks such as refractory periodontitis (RP), early onset periodontitis (EOP) or insulin-dependent diabetes mellitus (IDDM) (Offenbacher, Collins, et al., 1994; Yalda, Collins, et al., 1994), (Fig. 1).

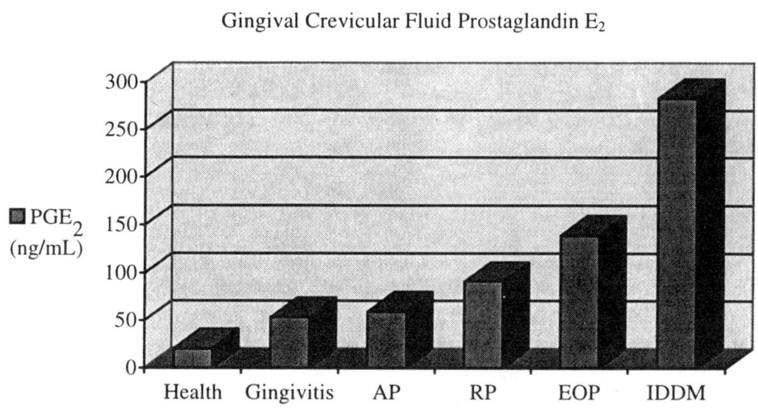

Figure 1. GCF-PGE_2 Levels (ng/mL) in health, gingivitis, and in various periodontal diseases [Adult Periodontitis (AP), Refractory Periodontitis (RP), Early Onset Periodontitis (EOP), Insulin-Dependent Diabetes Mellitus-associated Periodontitis (IDDM)].

Especially noteworthy is that the GCF-PGE_2 levels at clinically healthy sites (i.e., 0 to 3 mm, no inflammation) within these special patient groups are higher than those levels associated with clinically similar sites in patients with the more prevalent form of adult periodontitis, or in health. Even low levels of endotoxin challenge within the periodontal pocket seem to induce high levels of PGE_2 secretion at these sites (Offenbacher, Collins, Yalda, and Haradon, 1994). This suggests that the GCF-PGE_2 level reflects the collective response of the patient's periodontium, not as a collection of sites which function independently, but rather as an organ which is reflecting the patient's systemic response to a local infection. Previous data have demonstrated that GCF-PGE_2 levels increase in synchrony at many periodontal sites simultaneously during periods of active attachment loss, even at sites which are stable and relatively healthy (Salvi, Yalda, et al, 1997; Offenbacher, Collins, Yalda, and Haradon, 1994). This finding, that the inflammatory response is synchronized at multiple sites including healthy sites, suggests that it is an underlying systemic response, rather than a local site response, which is driving the inflammatory process.

PROSTAGLANDIN E_2 AND PRETERM LOW BIRTH WEIGHT

Inflammatory mediators such as prostaglandin E_2 are present not only in the periodontal inflammatory process but also regulate the normal physiologic process of parturition, as well as pathologic prematurity. Amniotic fluid levels of PGE_2 rise steadily throughout pregnancy until a critical threshold level is reached to induce labor, cervical dilation and delivery. Romero, Wu and co-workers (1988), observed elevated levels of PGE_2 as a consistent and reproducible fact of preterm low birth weight even in the absence of clinical or subclinical genitourinary tract infection and they postulated that the majority of PLBW are "probably caused by an infection of unknown origin."

Several risk factors have been associated with preterm delivery and LBW (Simpson JL and Elias S, eds., 1993). Evidence from several sources suggests a link between subclinical infection and prematurity (Gibbs et al., 1992). This evidence suggests that the effects of infection on the fetus are not mediated directly by the bacteria per se, but rather, by the maternal cytokines. The bacteria from a urinary tract infection (UTI), for example, rarely invade the fetoplacental unit; nevertheless, such infections are known risk factors for PLBW (Gibbs et al., 1992), preterm labor (PTL) and premature rupture of membranes (PROM) (Niswander and Gordon, 1972, Minkoff et al., 1984, Martius et al., 1996). The current model of the pathogenesis of urinary tract infection, or of bacterial vaginosis in PLBW, is that the endotoxins from this maternal gram negative infection cause the release of inflammatory mediators, especially PGE_2, which in turn induce preterm labor through premature rupture of membranes and the delivery of a PLBW infant.

Bacterial vaginosis (BV), a condition in which the normal, lactobacillus-predominant vaginal flora is replaced with anaerobic bacteria, Gardnerella vaginalis, and Mycoplasm hominis (Hillier, Krohn, Rabe, et al., 1993; Martius, Krohn, et al., 1988), is a common infectious cause of preterm deliveries, however, the maternal inflammatory response is the effector mechanism that leads to prematurity (Gibbs, Romero, et al., 1992; Hillier, Nugent, et al., 1995; Hillier, Krohn, et al., 1995; Romero, Baumann, Gomez, et al., 1993; Romero, Baumann, Gonzalez, et al., 1994; Romero, Hobbins, and Mitchell, 1988; Romero, Munoz, Gomez, et al., 1996; Romero, Wu, Mazor, et al., 1988; Romero, Brody, Oyarzum, et al., 1989; Romero, Durum, et al., 1989; Romero, Manogue, et al., 1989; Romero, Mazor, et al., 1989). The similarities in the mixed anaerobic infection of vaginosis and periodontitis and the striking parallels in inflammatory mediator responses, suggest that there is a common underlying pathophysiologic pathway or mechanism that warrants further investigation of the linkage between these infections and PLBW. It is important to note, however, that while enteric endotoxins are capable of inducing placental necrosis, spontaneous abortions, fetal organ damage, fetal death and malformations in animal models (Haesaert and Ornoy, 1986; Lanning et al., 1983), the experimental data from enteric Gram-negative organisms, including those associated with bacterial vaginosis, pelvic inflammatory disease and other sexually transmitted organisms might not be readily extrapolated to expected outcomes from Gram-negative oral infections (Refer to discussion in Offenbacher, Jared, et al., In press).

Animal studies examining the effects of various localized, non-disseminating challenges of *Porphyromonas gingivalis* (*P.g.*), an oral pathogen associated with periodontitis, on inflammatory mediator production and pregnancy outcome in the golden hamster have been completed by Collins, Smith, et al., (1994). In these studies, a subcutaneous tissue chamber model (described by Genco et al., 1991; Genco and Arko, 1994) served as the site for bacterial inoculation (10 CFU/mL) and the infection confined within this chamber was demonstrated to cause fetal growth retardation and embryolethality. Chronic exposure to the oral pathogens did not afford protection, but actually enhanced the feto-placental toxicity of exposure during pregnancy (Collins, Windley, et al., 1994). Of critical importance to these early experiments was the demonstration that these low-grade infections with low numbers of pathogens were not of sufficient magnitude to induce maternal malaise or fever. There was, however, a measurable local increase in PGE_2, as well as a 15-18% decrease in fetal weight (Collins, Smith, et al., 1994). Furthermore, the magnitude of the PGE_2 response was inversely related to the weight of the fetuses, mimicking the intraamniotic changes observed in humans with preterm labor or PROM (Hillier, Witkin, et al., 1993; Romero, Baumann, Gomez, et al., 1993).

PERIODONTAL DISEASES AND PRETERM LOW BIRTH WEIGHT

The experiments described above were extended to include experimental models of periodontitis in pregnant hamsters (Collins, Kirtland, et al., 1995). Two groups of animals were fed either control chow or plaque-promoting chow for 8 weeks to induce experimental periodontitis prior to mating. Each of these groups were also challenged with exogenous *P. gingivalis,* (strain A7476, 10^8 CFU/mL every other day by 1 mL gastric gavage), using methods described by Lasfargues and Saffar (1983) to form 4 treatment groups. There were no significant effects on fetal weight of *P. gingivalis* challenge under either diet conditions in which the organism can not colonize to induce periodontal infection (normal control diet chow). In contrast, both groups of animals with the plaque promoting chow (allowing colonization) developed periodontitis and had a resulting 19% decrease in fetal weight that was significant at $P = 0.003$ by t-test. The group of 11 dams on both plaque-promoting chow and *P. gingivalis* gavage had a 22.5% reduction in fetal weight, as compared to controls at $P = 0.037$ by t-test. Exogenous *P. gingivalis* challenge by gastric gavage did not promote either more severe periodontal disease or more severe fetal growth restriction. Nonetheless, in both periodontitis groups of animals there was a significant difference in mean fetal weight. Furthermore, in these pooled periodontitis groups there is a statistically significant elevation of intraamniotic fluid levels of PGE_2, several fold higher, from 3.31 ± 1.07 mg/mL to 13.5 ± 4.08 mg/mL at $P < 0.03$. This provides direct evidence that periodontal infection can result in a change in the fetal environment. It is possible that PGE_2 is produced by the periodontium, appears in the systemic circulation, eventually crosses the chorioamniotic barrier and finally appears in the amniotic fluid. However, it seems more likely that blood borne bacterial products, especially LPS, target the chorioamniotic plexus to trigger local PGE_2 synthesis. Irrespective of targeting mechanism, it is clear that experimental periodontitis in an animal model can provide sufficient challenge to elicit LBW.

Based upon these animal experiments, Offenbacher, Katz, and colleagues, (1996), conducted a matched case control study on 124 pregnant or post-partum women to examine periodontal disease status and birth outcomes. PLBW cases were defined as a mother with a birth weight of less than 2500 grams and one or more of the following: gestational age < 37 weeks, preterm labor (PTL), or preterm premature rupture of membranes (PPROM). Controls were all normal birth weight infants (NBW). Assessments included a broad range of known obstetric risk factors, such as tobacco usage, drug use, and alcohol consumption, level of prenatal care, parity, genitourinary infections, and weight gain during pregnancy, as well as a full mouth periodontal examination to determine clinical attachment levels. It was found that PLBW cases and primiparous (first births) PLBW cases had significantly worse periodontal disease, than the respective NBW controls. Multivariate logistic regression models, controlling for other risk factors and covariates, demonstrate that periodontitis is a statistically significant risk factor for PLBW with adjusted odds ratios of 7.9 and 7.5 for all PLBW cases and primiparous PLBW cases, respectively. The association between periodontal disease and PLBW may be a reflection of the patient's altered inflammatory trait that places the patient at risk for both conditions.

GCF-PGE_2 LEVELS AND BIRTH OUTCOME

New experiments from 48 case-control subjects have measured gingival crevicular fluid levels of PGE_2 to determine whether the levels were related to current pregnancy outcome (Offenbacher, Jared, et al., In press). Since these are also sensitive markers of current periodontal disease status, we secondarily sought to determine whether these would provide additional information with regard to the current pregnancy outcome, in addition to the periodontal status, which reflects a historical record of all previous periodontal disease experience. Results indicate that GCF-PGE_2 levels are significantly higher in PLBW mothers when compared with NBW controls (Figure 2).

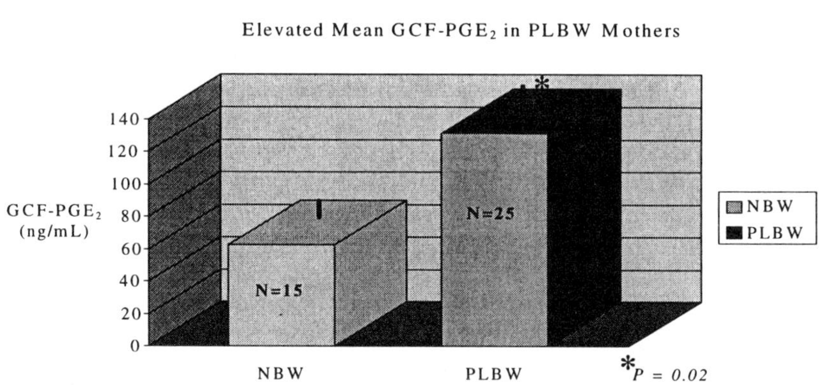

Figure 2. Dose-response relationship for increasing GCF-PGE_2, as a marker of current periodontal disease activity and decreasing birth weight. The pooled mean GCF-PGE_2 level for case mothers (131.4 ± 21.8) is over twice the level for control mothers (62.6 ± 10.3); this is significant at P = 0.02, by non-paired T-test.

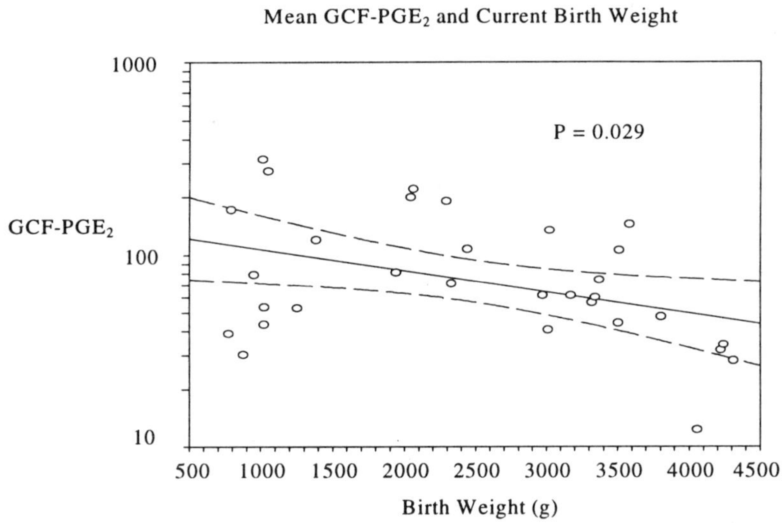

Figure 3. Relationship between the log GCF-PGE_2 and birth weight among primiparous mothers. The GCF-PGE_2 levels are inversely related to current birth weight (P=0.029). Smaller and more premature babies were born to primiparous mothers with higher GCF-PGE_2.

Furthermore, within primiparous PLBW mothers there was a significant inverse association between birth weight (as well as gestational age) and GCF-PGE_2 levels at P = 0.029 (Figure 3). These date suggest a dose-response relationship for increasing GCF-PGE_2, as a marker of current periodontal disease activity, and decreasing birth weight.

GCF-PGE_2 IS CORRELATED WITH INTRAAMNIOTIC FLUID- PGE_2 LEVELS

We also recently completed a pilot investigation of eighteen pregnant women undergoing routine amniocentesis in the Obstetrics-Gynecology clinics at University of North Carolina Hospitals to determine whether the amniotic fluid levels of PGE_2 rise in synchrony with the GCF levels of this eicosanoid in order to further elucidate the association between periodontal status and birth weight.

Methods and Materials

Patient Selection and Clinical Assessment. The sample population consisted of eighteen women who were undergoing amniocentesis for reasons related to age, familial history, or results of blood tests, except for one who elected to have amniocentesis for personal reasons. Since about 1 in 200 amniotic fluid sampling procedures results in pregnancy loss, none of these patients were having amniocentesis performed solely for the purpose of this investigation. No volunteers had to be excluded due to medical risk factors such as bacterial endocarditis, which would require prophylactic antibiotics for periodontal examination. In addition, these women were not necessarily high risk for either preterm labor or periodontal disease.

Upon completion of a medical history and informed consent for planned obstetric procedures as well as consent for this study, the subject underwent amniocentesis and small portions (about 3.0 mL each) of the drawn amniotic fluid and blood were supplied to the examiner of this study. Thereafter, gingival crevicular fluid samples were collected and a full-mouth periodontal examination was performed by a single examiner. Four gingival crevicular fluid samples for each mediator were collected per patient using periopaper strips (Pro Flow ™ Incorporated, Amityville, NY) and volumes were determined immediately by a calibrated Model 1 Periotron™ 6000 (Pro Flow™ Incorporated, Amityville, NY). After the 16-second measuring cycle, a reading falling between 30 and 180 was recorded. Readings outside this range have been found to be unreliable with regard to volume conversions and sampling was repeated in such cases. Acceptable samples were secured, frozen in liquid nitrogen and transferred the same day to the Dental Research Center at the University of North Carolina for further analysis. Blood samples were also transferred and separated on the same day.

Periodontal measurements were recorded following crevicular fluid collection by a single examiner. The periodontal examination included dichotomous measures of erythema, plaque score and bleeding upon probing (BOP). Probing measurements were performed at 6 sites per tooth using a UNC-15 manual probe to obtain attachment levels, (the distance between the periodontal fiber attachment and the cementoenamel junction). Periodontal status was classified into mean attachment level and the extent of attachment level, using a 4 mm threshold as defined by Carlos et al., (1986). BOP was expressed as a percentage of sites that bled upon probing.

Measurement of Mediator Levels in Biofluids. Filter paper strips with crevicular fluid samples were taken from liquid nitrogen storage immediately prior to assay and PGE_2 was measured by radioimmunoassay (RIA) as described previously by Offenbacher (1986). Each gingival crevicular fluid sample was assayed independently and data was pooled to form patient mean ± standard error in ng/mL GCF. Prior to analyzing the lipid metabolite PGE_2, amniotic fluid was extracted by reverse phase chromatography according to manufacturers directions (Sep-Pak™, Millipore Corporation/Waters Chromatography Division, Milford, MA).

Analyses of Data. Previous experience has shown that the distribution of this eicosanoid tends to be skewed (Collins, Windley, et al., 1994) so the data was log transformed to achieve a normal distribution. Pairwise comparisons among gingival crevicular fluid, serum, and amniotic fluid PGE_2 levels were performed to examine for an association between eicosanoid levels testing for slope not equal to 0 at $P < 0.05$. A linear regression model was also developed to use GCF-PGE_2 levels to predict amniotic fluid levels of this mediator. An alpha level of less than 0.05 was considered statistically significant.

RESULTS

Of the eighteen subjects, fifteen women were greater than or equal to 33 years old, three were less than or equal to 20 years old. African-Americans and Caucasians were almost equally represented with 7 and 8 respectively; 3 belonged to other races. The reasons that amniocentesis were performed were increased maternal age for 12 subjects, increased alpha-fetal-protein for 5 subjects, and one subject requested it for her own reasons (previous spontaneous abortion and uterine fibroids). Of the nine subjects who were in their second pregnancy (Gravida II), three had previous abortions (spontaneous or elective); of the four subjects who were in Gravida ≥ IV, each had previous abortions with 1, 1, 5, and 9 per subject; one of two subjects in her third pregnancy had a previous abortion; and the three subjects in their first pregnancy never had a previous abortion. Two of the subjects smoked throughout the pregnancy, one had quit smoking during pregnancy, one was Type 1 Diabetic, three were taking Synthroid throughout the pregnancy, two had received systemic antibiotic therapy before amniocentesis was performed. Due to the low number of patients in this study, there were no sequential associations between any of these demographic variables and the mediator level data.

Table 1. Means of PGE_2 levels

	ng/mL	± SE
GCF	8.13	± 1.27
SERUM	0.50	± 0.07
AMNIO	1.22	± 0.86

Table 2. PGE_2 correlation matrix

	GCF	AMNIO	SERUM
GCF	1		
AMNIO	0.55*	1	
SERUM	0.17	0.24	1

* Significantly correlated at $P = 0.018$

Table 1 shows the mean levels of PGE_2 in GCF, serum, and amniotic fluid in all subjects. Generally, prostaglandin levels in gingival crevicular fluid and amniotic fluid were greater than serum levels.

Table 2 shows the correlation matrix examining the relationship among PGE_2 levels in the three compartments: GCF, serum and amniotic fluid. The Spearman rank-order coefficient comparing GCF and amniotic fluid levels of PGE_2 is 0.55, reflecting a fair to good linear relationship that is significant at a 0.018 level.

A regression analysis between these two parameters is shown in Figure 4. The slope of this regression is positive and greater than zero at $P = 0.001$. Thus, there is a significant positive association between the PGE_2 levels in the GCF and in the amniotic fluid; in addition, it appears that GCF-PGE_2 levels may be used to estimate the intra-amniotic levels of PGE_2.

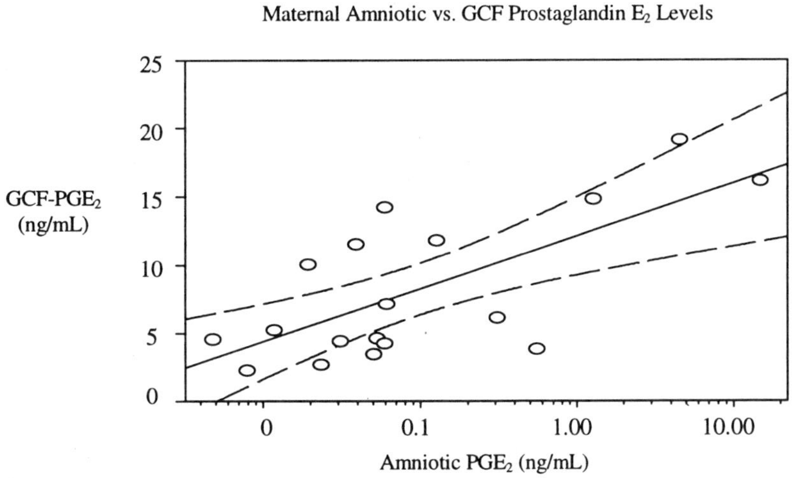

Figure 4. A scattergram in which each datum point represents a single mother, plotting amniotic fluid PGE_2 level (ng/mL) vs. mean GCF-PGE_2 level (ng/mL) collected from four sites. The slope of the line is significantly greater than zero at a P value of 0.001. Note that this is plotted as a logarithmic scale with 95% confidence intervals.

DISCUSSION

Our data indicate that the levels of PGE_2 within the gingival crevice are within the normal limits of health or gingivitis. These data are consistent with our overall findings regarding the periodontal status of these subjects which show generalized mild gingival inflammation or early periodontitis and relatively little loss of attachment. Mean levels of PGE_2 in gingival crevicular fluid at 8.13 ± 1.27 ng/mL are near the levels consistent with health (20.5 ± 7.6 ng/mL) reported by Offenbacher et al., (1986).

Mean mediator levels in the amniotic fluid of the 18 subjects also reflect normal, "healthy" values. In our study, mean prostaglandin E_2 levels in amniotic fluid in midtrimester at 1.22 ± 0.86 ng/mL are consistent with findings noted by Romero et al., (1993) and McDonald et al., (1991) but slightly greater than the mean concentration of < 10 pg/mL in the early stage of normal pregnancies (14 to 24 weeks) reported by Dray and Frydman (1976).

Although levels of prostaglandin E_2 in amniotic fluid varies from subject to subject, within each subject there is a high degree of association of the levels of PGE_2 in GCF and those levels present in amniotic fluid. Since levels of PGE_2 are low in amniotic fluid, it does not appear that the GCF-PGE_2 is passing through the systemic circulation to the fetal placental unit. It cannot be ruled out that there may be an oral endotoxin present in the gingiva which via the circulation is responsible for the amniotic fluid levels of these mediators. However, due to the mild disease shown in these subjects, systemic exposure of LPS would not be expected to play a dominant role. More likely, it appears that the patient-specific systemic monocytic response determines both the PGE_2 level within the GCF and that within the amniotic fluid. The data support the concept that the GCF levels of PGE_2 may serve as indirect indicators of patient-specific, systemic monocytic inflammatory response trait. The synchrony of PGE_2 expression is consistent with a mechanism involving a patient-based, or systemic-based monocytic characteristic that has been demonstrated in periodontal diseases (Offenbacher et al., 1993) and diabetes (Salvi et al., 1997; Yalda et al., 1994). Thus, the levels of PGE_2 within the amniotic fluid and GCF may be a reflection of the state of systemic responsiveness. In fact, it appears that the GCF levels may eventually be used as a diagnostic test to provide an indirect estimate of the amniotic fluid levels of PGE_2, if appropriate studies were conducted to establish sensitivity and specificity. This may provide information for the possible screening of expectant mothers by measuring GCF levels of PGE_2 and IL-1β, known amniotic factors associated with PLBW. This is a far less invasive procedure than amniocentesis and has a negligible risk of affecting the developing fetus. In contrast to many other biochemical markers, which may be indicators of diseases, this product has the potential to serve as a risk factor, since it possesses innate pathobiological activity.

Since amniotic fluid PGE_2 levels can be predictive of preterm delivery, further studies of GCF analyses with regard to pregnancy outcome appear justified. It should be remembered, however, that these relationships were determined using a relatively small group of subjects (n=18). In general, these subjects had very little periodontal involvement, were living in a limited geographic area, were of similar socioeconomic status background and were not at high obstetric risk. It remains to be seen whether these conclusions remain valid in subjects with a broader spectrum of periodontal disease and obstetric complications.

CONCLUSION

This study demonstrates that the GCF levels of PGE_2 can be used to reflect intra-amniotic levels of this mediator. This eicosanoid and other cytokines such as IL-1β are now recognized as major modulators of pregnancy events. The macrophage, an earlier neglected cell type in reproduction, has emerged as a potentially pivotal cell type because of its ability to synthesize and respond to cytokines and other powerful regulatory molecules. PGE_2, a product of both macrophages and other types of uteroplacental cells as well as a product of the metabolism of arachidonic acid, often acts as a critical intermediary effector molecule in the cytokine network. Moreover, other uterine cells and trophoblast cells of the placenta contribute to the complex cellular and molecular interplay required for successful pregnancy.

The potential health benefits of additional eicosanoid/cytokine research are numerous. The non-invasive nature of GCF sampling would be a preferable alternative to an invasive procedure such as amniocentesis, which presents additional risks to mother and fetus. The risk of pregnancy loss with amniotic fluid sampling is 0.5%, including the assistance of sonography; no risk to unborn child or mother has been demonstrated by GCF sampling techniques, which involve the placement of a small strip of paper gently into the space between the gingiva and tooth. In addition to clarifying normal events, further investigation may uncover effective therapeutic modalities for altering cytokine expression that would be useful when inflammatory mediators threaten pregnancy and active phases of periodontal disease.

REFERENCES

Carlos, J., Wolfe, M., and Kingman, A., 1986, The extent and severity index: a simple method for use in epidemiologic studies of periodontal disease, *J. Clin. Periodontol.* 13:500-505.

Collins, J.G., Kirtland, B.C., Arnold, R.R., and Offenbacher, S., 1995, Experimental periodontitis retards fetal hamster growth, *J. Dent. Res.* 74(Spec. issue):158(Abstr. 1171).

Collins, J.G., Smith, M.A., Arnold, R.R., and Offenbacher, S., 1994, Effects of *Escherichia coli* and *Porphyromonas gingivalis* lipopolysaccharide on pregnancy outcome in the golden hamster, *Infect. Immun.* 62:10;4652-4655.

Collins, J.G., Windley, H.W. III, Arnold, R.R., and Offenbacher, S., 1994, Effects of a *Porphyromonas gingivalis* infection on inflammatory mediator response and pregnancy outcome in the hamster, *Infect. Immun.* 62:10;4356-4361.

Dray, F., and Frydman, R., 1976, Primary prostaglandins in amniotic fluid in pregnancy and spontaneous labor, *Am. J. Obstet. Gynecol.* 126:13-19.

Ellison, S.A., 1970, Oral bacteria and periodontal disease, *J. Dent. Res.* 49(Suppl.2):198-202.

Genco, C., and Arko, R.J., 1994, Animal chamber model for study of host-parasite interactions, *Methods Enzymol.* 235:120-140.

Genco, C.A., Cutler, C.W., Kapczynski, D., Maloney, K., and Arnold, R.R., 1991. A novel mouse model to study the virulence of and host response to *Porphyromonas (Bacteroides) gingivalis, Infect. Immun.* 59:1255-1263.

Genco, R.J., Evans, R.T., and Ellison, S.A., 1969, Dental research in microbiology with emphasis on periodontal disease, *J. Am. Dent. Assoc.* 78:1016-1036.

Gibbs, R.S., Romero, R., Hillier, S.L., Eschenbach, D.A., and Sweet, R.L., 1992, A review of premature birth and subclinical infection, *Am. J. Obstet. Gynecol.* 166:1515-28.

Goodson, J., Dewhirst, F.E., and Brunetti, A., 1974, Prostaglandin-E levels and human periodontal disease, *Prostaglandins* 6:81-85.

Haesaert, B., and Ornoy, A., 1986, Transplacental effects of endotoxemia on fetal mouse brain, bone and placental tissue, *Pedriatr. Pathol.* 5:167-181.

Heasman, P.A., Collins, J.G., and Offenbacher, S., 1993, Changes in crevicular fluid levels of interleukin-1 beta, leukotriene B4, prostaglandin E_2, thromboxane B2 and tumour necrosis factor alpha in experimental gingivitis in humans, *J. Periodontal. Res.* 28(4):241-247.

Hillier, S.L., Krohn, M.A., Cassen, E., Easterling, T.R., Rabe, L.K., and Eschenbach, D.A., 1995, The role of bacterial vaginosis and vaginal bacteria in amniotic fluid infection in women in preterm labor with intact fetal membranes, *Clin. Infect. Dis.* 20(Suppl. 2):S276-278.

Hillier, S.L., Krohn, M.A., Rabe, L.K., Klebanoff, S.J., and Eschenbach, D.A., 1993, The normal vaginal flora, H_2O_2-producing lactobacilli, and bacterial vaginosis in pregnant women, *Clin. Infect. Dis.* 16:Suppl 4:S273-S281.

Hillier, S.L., Nugent, R.P., Eschenbach, D.A., Krohn, M.A., Gibbs, R.S., Martin, D.H., Cotch, M.F., Edelman, R., Pastorek, J.G., II, Rao, A.V. McNellis, D., Regan, J.A., Carey, J.C., and Klebanoff, M.A., 1995, Association between bacterial vaginosis and preterm delivery of a low-birth-weight infant, The Vaginal Infections and Prematurity Study Group, *New Engl. J. Med.* 333:1737-1742.

Hillier, S.L., Witkin, S.S., Krohn, M.A., Watts, D.H., Kiviat, N.B., and Eschenbach, D.A., 1993, The relationship of amniotic fluid cytokines and preterm delivery, amniotic fluid infection, histologic chorioamnionitis, and chorioamnion infection, *Obstet. Gynecol.* 81(6):941-948.

Lanning, J.C., Hilbelink, D.R., and Chen, L.T., 1983, Teratogenic effects of endotoxin on the golden hamster, *Teratogenesis Carcinog. Mutag.* 3:145-149.

Lasfargues, J.J., and Saffar, J.L., 1983, Effect of indomethacin on bone destruction during experimental periodontal disease in the hamster, *J. Periodont. Res.* 18:110-117.

Loning, T., Albers, H-K., Lisboa, B.P., Burkhardt, A., and Caselitz, J., 1980, Prostaglandin E and the local immune response in chronic periodontal disease. Immunohistochemical and radioimmunological observations, *J. Periodont. Res.* 15:525-535.

Martius, J., and Roos, T., 1996, The role of urogenital tract infections in the etiology of preterm birth: a review, *Arch. Gynecol. Obstet.* 258:1-19.

Martius, J., Krohn, M., Hillier, S. L., Stamm, W. E., Holmes, K.K., and Eschenbach, D.A., 1988, Relationships of vaginal *lactobacillus* species, cervical *Chlamydia trachomatis*, and bacterial vaginosis to preterm birth, *Obstet. Gynecol.* 71:89-95.

McCormick, M.C., Bernbaum, J.C., Eisenberg, J.M., Kustra, S.L., and Finnegan, E., 1991, Costs incurred by parents of very low birth weight infants after the initial neonatal hospitalization, *Pediatrics* 88(3):533-41.

McDonald, H.M., O' Loughlin, J.A., Jolley, P., Vigneswaran, P., and McDonald, P.J., 1991, Vaginal infections and preterm labor, *Brit. J. Obstet. Gynecol.* 98:427-435.

Minkoff, H., Grunebaum, A.N., Schwartz, R.H., Feldman, J., Cummings, M., Crombleholme, W., Clark, L., Pringle, G., and McCormick, W.M., 1984, Risk factors for prematurity and premature rupture of the membranes: a prospective study of the vaginal flora in pregnancy, *Am. J. Obstet. Gynecol.* 150:965-972.

Nakashima, K., Roerich, N., and Cimasoni, G., 1994, Osteocalcin, prostaglandin E_2 and alkaline phosphatase in gingival crevicular fluid: their relations to periodontal status, *J. Clin. Periodontol.* 21:327-333.

National Center for Health Statistics, 1993, Advance report of final natality statistics, 1991. Monthly vital statistics report, in: *Health, United States,* Vol. 42, no 3., Suppl., Public Health Service, Hyattsville, MD.

Niswander, K.R., and Gordon, M., 1972, The women and their pregnancies, in: *The collaborative perinatal study of the National Institute of Neurological Diseases and Strokes*, p.252-256, Saunders, Philadelphia.

Offenbacher, S., Collins, J.G., and Heasman, P.A., 1993, Diagnostic potential of host response mediators, *Adv. Dent. Res.* 7(2):175-181.

Offenbacher, S., Collins, J.G., Yalda, B., and Haradon, G., 1994, Role of prostaglandins in high risk periodontitis patients, in: *Molecular Pathogenesis of Periodontal Disease*, pp. 203-214, R.J.Genco, S. Hamada, T. Lehner, and S. Mergenhagen, eds., ASM Press, Washington, D.C..

Offenbacher, S., Farr, D.H., and Goodson, J.M., 1981, Measurement of prostaglandin E in crevicular fluid, *J. Clin. Periodont. Res.* 8:359-367.

Offenbacher, S., Heasman, P.A., and Collins, J.G., 1993, Modulation of host PGE_2 secretion as a determinant of periodontal disease expression, *J. Periodontol.* 64(5 Suppl.):432-444.

Offenbacher, S., Jared, H.L., O'Reilly, P.G., Wells, S.R., Salvi, G.E., Lawrence, H.P., Socransky, S.S., and Beck, J.D., 1997, Potential pathogenic mechanisms of periodontitis-associated pregnancy complications, *Ann. Periodontol.* (in press).

Offenbacher, S., Katz, V., Fertik, G., Collins, J.G., Boyd, D., Maynor, G., McKaig, R., and Beck, J., 1996, Periodontal Infection as a risk factor for preterm low birth weight, *J. Periodontol.* 67:1103-1113.

Offenbacher, S., Odle, B.M., and Van Dyke, T.E., 1986, The use of crevicular fluid prostaglandin E_2 levels as a predictor of periodontal attachment loss, *J. Periodont. Res.* 21:101-112.

Phibbs, C.S., Williams, R.L., and Phibbs, R.H., 1981, Newborn risk factors and costs of neonatal intensive care, *Pediatrics* 68:313-321.

Romero, R., Baumann, P., Gomez, R., Salafia, C., Rittenhouse, L., Barberio, D., Behnke, E., Cotton, D.B., and Mitchell, M.D., 1993, The relationship between spontaneous rupture of membranes, labor and microbial invasion of the amniotic cavity and amniotic fluid concentration of prostaglandins and thromboxane B_2 in term pregnancy, *Am. J. Obstet. Gynecol.* 168:1654-1664.

Romero, R., Baumann, P., Gonzalez, R., Gomez, R., Rittenhouse, L., Behnke, E., and Mitchell, M.D., 1994, Amniotic fluid prostanoid concentrations increase early during the course of spontaneous labor at term, *Am. J. Obstet. Gynecol.* 171:1613-1620.

Romero, R., Brody, D.T., Oyarzun, E., Mazor, M., Wu, Y.K., Hobbins, J.C., and Durum, S.K., 1989, Infection and labor. III. Interleukin-1: a signal for the onset of parturition, *Am. J. Obstet. Gynecol.* 160(5 Pt.1):1117-1123.

Romero, R., Durum, S., Dinarello, C.A., Oyarzun, E., Hobbins, J.C., and Mitchell, M.D., 1989, Interleukin-1 stimulates prostaglandin biosynthesis by human amnion, *Prostaglandin* 37(1):13-22.

Romero, R., Hobbins, J. C., and Mitchell, M.D., 1988, Endotoxin stimulates PGE_2 production by human amnion, *Obstet. Gynecol.* 71:227-228.

Romero, R., Manogue, K.R., Mitchell, M.D., Wu, Y.K., Oyarzun, E., Hobbins, J.C., and Cerami, A., 1989, Infection and labor IV. Cachectin-tumor necrosis factor in the amniotic fluid of women with intraamniotic infection and preterm labor, *Am. J. Obstet. Gynecol.* 161:336-341.

Romero, R., Mazor, M., Wu, Y.K., Avila, C., Oyarzun, E., and Mitchell, M.D., 1989, Bacterial endotoxin and tumor necrosis factor stimulate prostaglandin production by the human decidua, *Prostaglandins Leukot. Essent. Fatty Acids* 37:183-185.

Romero, R., Munoz, H., Gomez, R., Parra, M., Polanco, M., Valverde, V., Hasbun, J., Garrido, J., Ghezzi, F., Mazor, M., Tolosa, J.E., and Mitchell, M.D., 1996, Increase in prostaglandin bioavailability precedes the onset of human parturition, *Prostaglandins Leukot. Essent. Fatty Acids* 54(3):187-91.

Romero, R., Wu, Y.K., Mazor, M., Hobbing, J.C., and Mitchell, M.D., 1988, Amniotic fluid prostaglandin E_2 in preterm labor, *Prostaglandins Leukot. Essent. Fatty Acids* 34:141-145.

Salvi, G.E., Yalda, B., Collins, J.G., Jones, B.H., Smith, F.W., Arnold, R.R., and Offenbacher, S., 1997, Inflammatory mediator response as a potential risk marker for periodontal diseases in insulin-dependent diabetes mellitus patients, *J. Periodontol.* 68:127-135.

Shapiro, S., McCormick, M.C., Starfield, B.H., Krischer, J.P., and Bross, D., 1988, Relevance of correlates of infant deaths for significant morbidity at 1 year of age, *Am. J. Obstet. Gynecol.* 136:363-73.

Simpson, J.L., and Elias, S., 1993, Risk factors associated with preterm delivery, in: *Essentials of Prenatal Diagnosis*, Table 10.3., Churchill Livingstone Inc., New York.

Socransky, S.S., 1970, The relationship of bacteria to the etiology of periodontal disease, *J. Dent. Res.* 49(suppl.2):203-222

U.S. Department of Human Services, 1987, Oral Health of U.S. Adults, The National Survey of Oral Health in U.S. Employed Adults and Seniors 1985-1986: National Findings, in: *PHS, NIH Publication No. 87-2868*, August.

Yalda, B., Collins, J.G., Arnold, R.R., and Offenbacher, S., 1994, Monocytic and crevicular fluid PGE_2 and IL-1β in diabetic patients, *J. Dent. Res.* (IADR Abstract No.2337) 73:394.

MITOGEN-ACTIVATED PROTEIN KINASES AND ENDOTHELIAL PROSTACYCLIN SECRETION

John H. Grose, Luc Caron, and Marcel Lebel

Department of Medicine, Faculty of Medicine, Laval University,
and Laval University Research Centre,
Hotel-Dieu de Quebec Hospital,
Quebec, P.Q., Canada G1R 2J6

INTRODUCTION

Endothelial cells play an important role in hemostasis and in the modulation of vascular tone. Prostacyclin (PGI_2), a major eicosanoid produced by endothelial cells, is a potent vasorelaxant and inhibitor of platelet aggregation (1). As with other eicosanoids, PGI_2 is biosynthesized from free arachidonic acid (AA) released from membrane phospholipids, primarily through the enzymatic action of cytosolic phospholipase A_2 ($cPLA_2$) (1,2). It has been suggested that phosphorylation of specific sites on the enzyme by the 42 kDa mitogen-activated protein (p42 MAP) kinase may be important events leading to $cPLA_2$ activation (2). We have shown in cultured bovine aortic endothelial cells that agonist-induced PGI_2 secretion occurs in the presence of increased phosphorylation of the low molecular weight 27 kDa heat shock proteins (HSP27) and its isoelectric variants (3,4). It is now known that the phosphorylation of HSP27 is directly controlled by MAP kinase activated protein kinase-2 (MAPKAP K-2), which itself is modulated by a 38 kDa homologue (p38 MAP kinase) of p42MAP kinase (5). Experiments included in this chapter are focused on studying the interaction between p38 and p42MAP kinases during BK modulation of PGI_2 secretion in the same cell model.

MATERIAL AND METHODS

The maintenance of the CPA-47 cells (American Type Culture Collection), the rather standard phosphorylation (^{32}P-orthophosphate uptake) experiments, conditions for one-dimensional gel electrophoresis, and the radioimmunoassay (RIA) estimation of PGI_2 as 6-keto-$PGF_{1\alpha}$ were conducted as documented (3). Immunoprecipitation experiments were carried out using specific antibodies against p38 and p42MAP kinases obtained from rabbits immunized with sequence-specific peptides (6). After gel electrophoresis and autoradiography, the bands were quantitated using a densitometer (Biomed Instruments, Inc.) interfaced to an IBM computer with a software to calculate the individual peak areas.

Recent Advances in Prostaglandin, Thromboxane, and Leukotriene Research
Edited by Sinzinger *et al.*, Plenum Press, New York, 1998

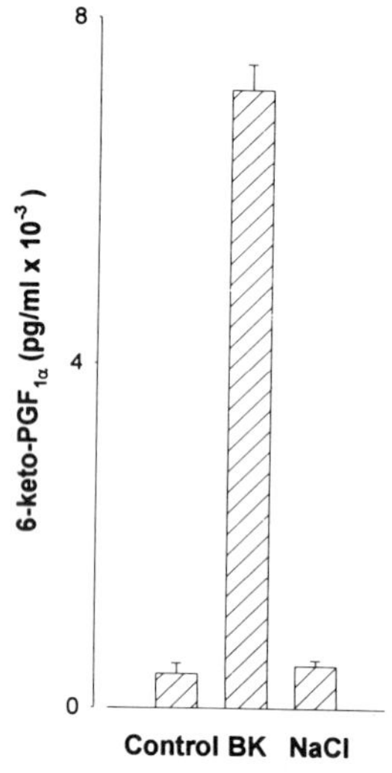

Fig. 1. Effect of BK and NaCl on 6-keto-$PGF_{1\alpha}$ production.

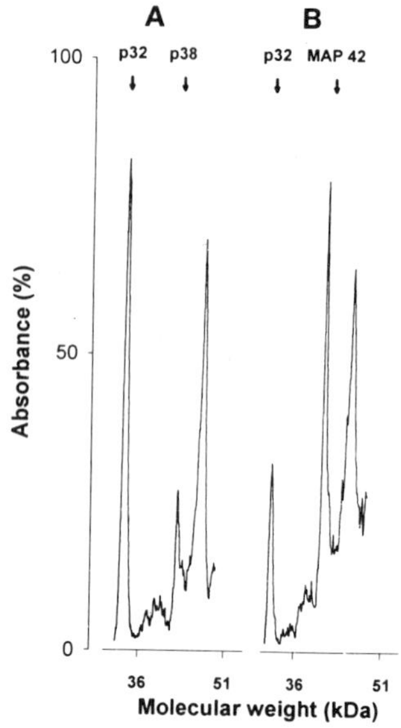

Fig. 2. Densitometric scans showing the presence of p32 in p38 (A) and p42(B) MAP kinase immunoprecipitates after BK (1 ng/ml 15 min) stimulation.

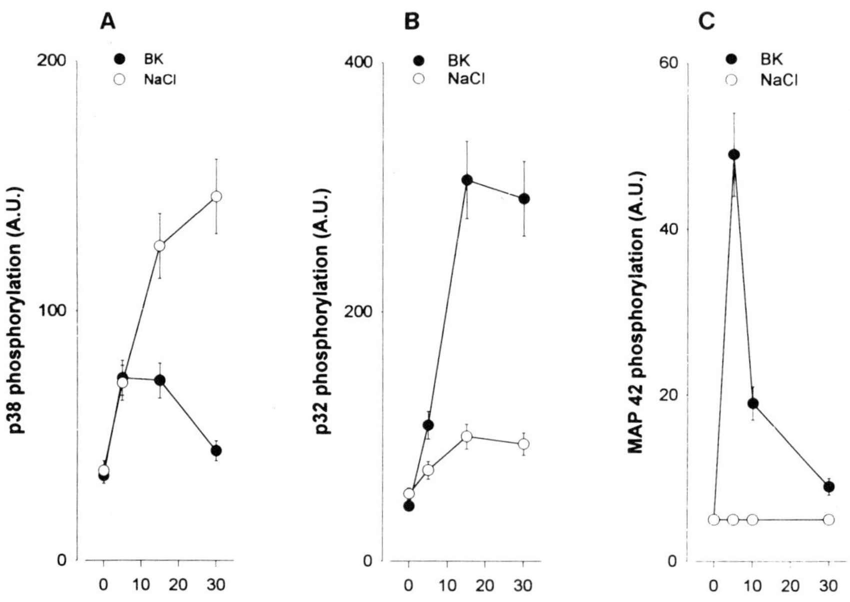

Fig. 3. BK and NaCl effect on quantitative phosphorylation changes in p32 and in p38 (A and B) and p42 (C) MAP kinases.

RESULTS

Low concentration of BK (1 ng/ml, 15 min) was a strong stimulus for PGI_2 secretion (Fig. 1). Mild hyperosmotic shock (0.2M NaCl, 15 min) had no effect on PGI_2 secretion (Fig. 1). Fig. 2A illustrates the presence in BK-stimulated cells (1 ng/ml, 15 min) of a unique phosphorylated protein band in the region of 32 kDa (p32) which coprecipitated with p38MAP kinase, using a specific anti-p38MAP kinase antibody. Under similar experimental conditions, the presence of p32 was also evident in the immunoprecipitate of anti-p42MAP kinase (Fig. 2B). Preincubation of the p38 and p42MAP kinase antisera with the respective peptide antigens completely blocked the appearance of the respective ^{32}P-labeled peaks shown in Fig. 2 (data not shown), indicating the specificity of the two antisera. In immunoprecipitation and Western blot experiments, p38 and p42MAP kinases were clearly visible on the blots but not p32 (data not shown). Taken together, these observations suggest that under the experimental conditions, p32 was physically associated with p38 and p42MAP kinases in the immunocomplexes. BK induced a rapid increased phosphorylation (activation) of p42MAP kinase (Fig. 3C), followed by a subsequent decrease to control levels within 30 min; the changes in PGI_2 secretion also paralleled the variations in p42MAP kinase activation (data not shown). The responsiveness of p38MAP kinase was less dramatic, with moderate phosphorylation increase between 5 and 15 min, followed by a decline to control levels by 30 min (Fig. 3A); a rapid time-dependent rise in the phosphorylation of p32 coprecipitated with p38MAP kinase was evident (Fig. 3B), with sustained peak levels at 30 min. Mild hyperosmotic shock (0.2M NaCl, 15 min) had no effect on p42MAP kinase (Fig. 3C) nor on PGI_2 secretion (Fig. 1). However, the same stimulus induced the expected rapid and sustained activation of p38MAP kinase (Fig. 3A), which is known to be sensitive to stress stimuli (5); compared to the effect of BK, the phosphorylation levels of p32 coprecipitated in the same NaCl treated samples were only moderately enhanced (Fig. 3B).

DISCUSSION

Although numerous mechanisms have been proposed to account for the agonist-mediated biosynthesis of PGI_2 in endothelial cells (1,2), the complex signal transduction processes involved in the $cPLA_2$ activation and AA mobilization step remain unclear. Phosphorylation of specific cellular proteins is an important post-receptor signaling event, and the interplay between phosphorylation-dephosphorylation of these unique proteins are crucial biochemical processes by which extracellular and intracellular information are integrated and amplified, often through direct physical interaction between the activated signal transduction elements. BK is an efficient agonist for endothelial PGI_2 secretion, with a response profile which parallel that of p42MAP kinase activation, implicating a direct effect of BK on the kinase itself. In preliminary studies using the p42MAP kinase kinase inhibitor PD 098059 (7), the BK-mediated stimulation of PGI_2, and also the activation of both p42MAP kinase and $cPLA_2$ were suppressed (data not shown). These observations are in agreement with previous suggestions that the activation of $cPLA_2$, an important step in eicosanoid biosynthesis, is closely coupled to p42MAP kinase activity (2). The NaCl experiments indicate that mild hyperosmotic stress does not affect p42MAP kinase or PGI_2 secretion. Our interest in p38MAP kinase was prompted by our observations that BK-mediated PGI_2 secretion occurred in the presence of increased HSP27 phosphorylation (3,4). Current data indicate that HSP27 phosphorylation is controlled by MAPKAP K-2 isoenzymes, whose activities are modulated by the p38MAP kinases which are highly sensitive to stress stimuli (5). Although p38MAP kinase was activated by BK and mild osmotic stress, there was a higher degree and prolonged responsiveness to the NaCl treatment (Fig. 3A). The response profiles of p38 and p42MAP kinases to BK were also very different; it is possible that the activation of p38MAP kinase can, in part, be mediated through p42MAP kinase. The coprecipitation of the rapidly phosphorylated p32 with p38MAP kinase using the specific p38MAP kinase antibody in the BK experiments is of particular interest. The rapid onset and the sustained high phosphorylation levels of p32 (Fig. 3B), and the attenuation of its phosphorylation levels in cells preincubated with the p42MAP kinase kinase inhibitor PD 098059 under the same experimental conditions (data not shown), indicate that the activation of this unique protein is, at least in part, controlled by p42MAP kinase. Thus, any possible interaction between p38 and p42MAP kinases could well be modulated through the physical association with this phosphoprotein. Exactly how p32 is integrated in the potentially interactive signal transduction pathways remains to be elucidated.

ACKNOWLEDGMENTS

The authors wish to thank Dr. J. Landry for supplying the p38MAP kinase antibody and E. Lemay for secretarial assistance. These studies were supported by a grant from the Medical Research Council of Canada.

REFERENCES

1. W.L. Smith, Prostanoid biosynthesis and mechanism of action, Am. J. Physiol. 263: F181 (1992)
2. J.D. Clark, A.R. Schievella, E.A. Nalefski and L.-L. Lin, Cytosolic phospholipase A_2. J. Lipid Med. Cell Signal. 12: 83 (1995)
3. J.H. Grose, L. Caron, M. Lebel and J. Landry, Prostacyclin secretion and specific intracellular protein phosphorylation, Adv. Prost. Thromb. Leuk. Res. 21: 145 (1990)
4. J.H. Grose, L. Caron, G. Drapeau and M. Lebel, The regulation of prostacyclin secretion in endothelial cells, in: Prostaglandins, Leukotrienes, Lipoxins and PAF, J. M. Bailey, ed., Plenum Press, New York (1991)
5. A.D. Clifton, P.R. Young and P. Coyen, A comparison of the substrate specificity of MAPKAP kinase-2 and MAPKAP kinase-3 and their activation by cytokines and cellular stress, FEBS Lett. 392: 209 (1996)
6. J. Huot, F. Houle, D.R. Spitz and J. Landry, HSP27 phosphorylation-mediated resistance against actin fragmentation and cell death induced by oxidative stress, Cancer Res. 56: 273 (1996)
7. D.R. Alessi, A. Cuenda, P. Cohen, D.T. Dudley and A.R. Saltiel, PD 098059 is a specific inhibitor of the activation of mitogen-activated protein kinase kinase *in vitro* and *in vivo*, J. Biol. Chem. 270: 27489 (1995).

CROSSTALK BETWEEN ELEVATION OF [Ca^{2+}]i, REACTIVE OXYGEN SPECIES GENERATION AND PHOSPHOLIPASE A_2 STIMULATION IN A HUMAN KERATINOCYTE CELL LINE

Rachel Goldman[1], Sandra Moshonov[1], Xiaomei Chen[1], Alexander Berchansky[1], Gerhard Fürstenberger[2] and Uriel Zor[1]

[1]Department of Membrane Research and Biophysics
The Weizmann Institute of Science, Rehovot, Israel 76100
[2]FSP II Tumor Regulation, DKFZ, Heidelberg D6900, Germany

SUMMARY

The aim of the study was to explore the possible interrelationship between reactive oxygen species (ROS) formation and $cPLA_2$ activation and the mediator role that [Ca^{2+}]i may play in these processes in the human keratinocyte cell line, HaCaT. HaCaT cells can be invoked to transiently produce ROS by epidermal growth factor (EGF), thapsigargin (TPG) and the Ca^{2+}-ionophore, A23187. These 3 agonists transiently increase [Ca^{2+}]i with characteristic kinetics and magnitude. TPG and A23187 each activates on its own [^{3}H]AA release from prelabeled cells, whereas EGF on its own has no effect on [^{3}H]AA release. However, EGF augments [^{3}H]AA release invoked by TPG or A23187 several fold. EGF activates MAP kinase cascades in HaCaT cells, leads to ROS formation and induces relatively small (1.6 fold) elevation in [Ca^{2+}]i, whereas A23187 and TPG lead to a substantial elevation in [Ca^{2+}]i (2.5 to 5 fold) and to ROS formation. Both have a minor effect on MAP kinase activation . The synergism in PLA_2 activation by EGF and TPG or A23187, and the sensitivity of [^{3}H]AA release to N-acetylcysteine (NAC) and dithiothreitol (DTT) (potent reducing agents) or to DPI (an inhibitor of FAD-dependent oxidases) lead to the suggestion that ROS formation, elevation of [Ca^{2+}]i and PLA_2 activation are causally related. Since we show that elevation of [Ca^{2+}]i is a prerequisite for both ROS and PLA_2 activation, it is possible that these processes contribute to the toxicity (apoptosis) exerted by chronic elevation of [Ca^{2+}]i .

INTRODUCTION

Reactive oxygen species (ROS) play many roles in normal cell physiology as well as in pathological processes. All cells generate low levels of ROS by a variety of enzyme systems, during normal metabolism or in some cases in response to hormone or growth factor stimulation[1-3]. Recent evidence suggests that ROS may play a second messenger role in the activation of transcription factors[4-6], growth, differentiation and apoptosis[7-9]. We and others have shown that exogenously added and endogenously produced ROS activate the MAP kinase cascades of ERK1/2, JNK and p38[10,11] as well as $cPLA_2$[12-14]. Thus ROS can indirectly serve as second messengers in enzyme activation.

Changes in [Ca^{2+}]i modulate manifold cell functions including secretion, contraction, enzyme activation and cell cycle regulation[15-17]. In human and murine primary keratinocytes a switch from low Ca^{2+} (0.1 mM) in the medium to high Ca^{2+} (> 1 mM) signals terminal differentiation[18-20].

Substantial evidence demonstrates that $cPLA_2$ is activated by phosphorylation. ERK2 is considered the main candidate for a direct physiological activator of $cPLA_2$[21,22], though lately p38, another enzyme of the MAP kinase family, was implicated as the direct $cPLA_2$ activator [23,24]. Thus, activation of protein kinases and inactivation of protein phosphatases (e.g. by ROS) are likely to control the level and duration of $cPLA_2$ activity. An essential role for Ca^{2+} in the regulation of $cPLA_2$ activity was also amply shown. Agents such as the phorbol ester TPA, colony stimulating factor 1 or tumor necrosis factor α (TNFα) led to phosphorylation of $cPLA_2$, and yet little AA was released without a second stimulus to mobilize intracellular Ca^{2+} [25,26]. $cPLA_2$ is a soluble cytosolic protein that translocates to its membrane (plasma & nuclear) phospholipid substrate when free $[Ca^{2+}]i$ levels are raised by cell stimulation [27-29].

MATERIALS AND METHODS

HaCaT cells[30] (obtained from Fusenig, DKFZ) and HaCaT cells transfected with cDNA.$cPLA_2$ ($cPLA_2$.HaCaT; derived in our laboratory) were grown in Dulbecco's modified Eagle's medium containing 10 % fetal calf serum and antibiotics.

For the assessment of ROS formation the cells were trypsinized and suspended (10^6 cells/0.5 ml/assay) in Hank's balanced salt solution containing 0.2 % BSA, luminol (110 μM) and horseradish peroxidase (0.5 to 1 u/ml.). The luminol-dependent chemiluminescence invoked by various agonists was assessed using a Lumac apparatus, model 2080[12]. $[Ca^{2+}]i$ was calculated from measurements of fluorescence (excitation 340 nm; emission 505) of Fura-2/AM loaded cells in suspension in the presence or absence of agonists. Basal and agonist invoked [^{3}H]arachidonic acid ([^{3}H]AA) release from [^{3}H]AA prelabeled cells was assessed on $cPLA_2$.HaCaT cells shown to overexpress $cPLA_2$ in comparison to parent HaCaT cells. Cells grown in 24 well plates were treated with agonists for 45 min and the released label assessed as described[12]. Reducing agents or DPI were added 15 min prior to agonist addition. The nonfluorescent 4-Bromo-Ca^{2+}-ionophore (designated in the following as A23187) was used in order not to interfere with fluorescence measurements.

RESULTS AND DISCUSSION

Ca^{2+}-mobilizing agonists (A23187 and thapsigargin) and EGF invoke a time and dose dependent CL response (formation of H_2O_2) in HaCaT cells (Fig. 1). The peak of the response is attained within 1 to 2 min depending on the agonist and its concentration.

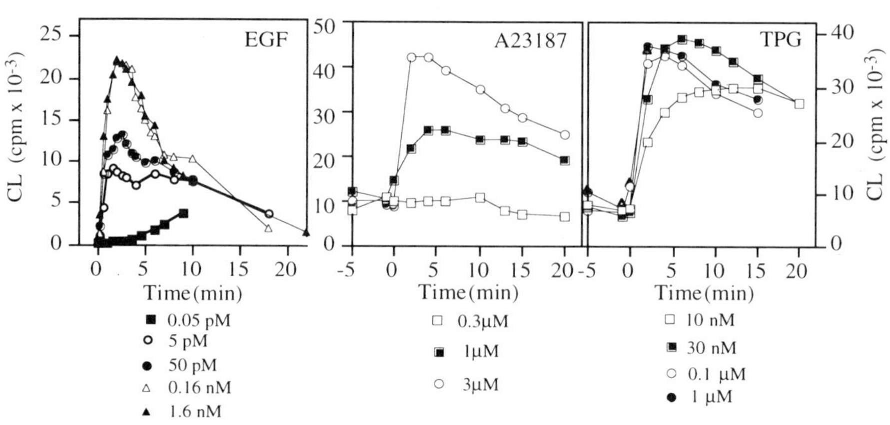

Figure 1. Dose and time dependence of agonist invoked luminol and horseradish peroxidase-dependent chemiluminescence (CL) of HaCaT cells.

Elevation of the intracellular Ca^{2+} concentration ($[Ca^{2+}]i$) by either thapsigargin, which releases Ca^{2+} from internal stores, or A23187, which leads to a rapid influx of Ca^{2+} (Fig. 2), leads to a rapid, sustained and relatively high CL response. The Kd for 50% of the CL response is approximately 80 pM (0.5 ng/ml) for EGF, 1 μM for A23187 and less than 10 nM for thapsigargin. Serum starvation for 24 hours (0.1 % HI-FCS), a procedure that leads to total cessation of growth of HaCaT cells, did not affect the CL response to either EGF or to A23187 (not shown). To test the dependence of the CL response on intra and extracellular Ca^{2+} concentration, we have added Ca^{2+} -chelators. The intracellular Ca^{2+} - chelator BAPTA (20 μM, 20 min before stimulus addition) partially suppressed (32-43 % inhibition) and the extracellular Ca^{2+} - chelator EGTA (3 mM, 5 min before agonist addition) totally suppressed the CL response to A23187, thapsigargin or EGF.

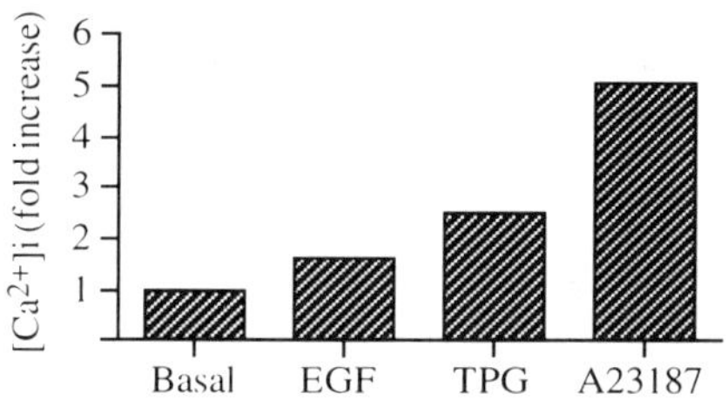

Figure 2. Agonist invoked $[Ca^{2+}]i$ elevation in HaCaT cells. Basal $[Ca^{2+}]i$ amounted to 105±10 nM. The values given represent the peak of the $[Ca^{2+}]i$ transient. EGF, 50 ng/ml; A23187, 3 μM; Thapsigargin (TPG), 0.3 μM.

Preincubation (20 min) with 6 μM DPI led to an almost total inhibition of the CL response suggesting that the H_2O_2 generating enzyme(s) is an FAD dependent enzyme. DPI did not suppress the CL response of the cell free system (H_2O_2+luminol+HRPO) .

We have next explored the possible link between ROS formation and PLA_2 activity. Under our experimental conditions, basal $[^3H]AA$ release was about 4% of the incorporated label. EGF had no effect on $[^3H]AA$ release whereas thapsigargin or A23187 alone increased $[^3H]AA$ release by about 4% over the basal release. Addition of EGF to thapsigargin or A23187 led to a remarkable elevation in $[^3H]AA$ release. Fig. 3 shows that the combination of agonists led to an elevation of 4 to 5 fold in $[^3H]AA$ release over that achieved by A23187 or thapsigargin alone. NAC and DTT had an inhibitory effect on agonist induced $[^3H]AA$ release (Fig. 3). NAC inhibited $[^3H]AA$ release by 30 to 40 % and DTT inhibited by 60 to 70 %. DPI also inhibited by 30 to 40 %.

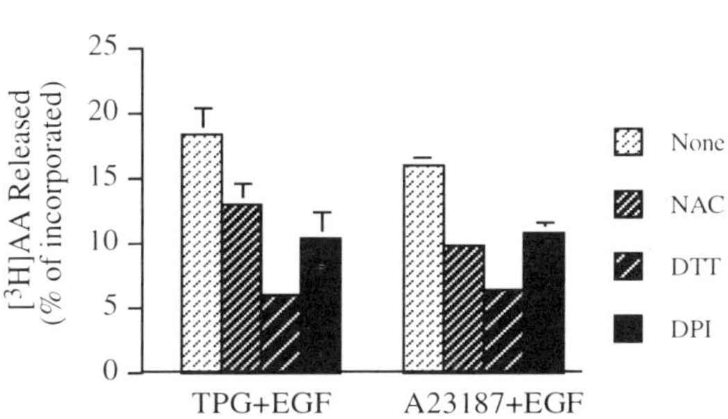

Figure 3. The effect of reducing agents and an inhibitor of FAD-dependent oxidase on agonist invoked $[^3H]AA$ release in $cPLA_2$.HaCaT cells. EGF, 50 ng/ml; A23187, 3 μM; Thapsigargin (TPG), 0.45 μM; DTT, 3 mM; N-acetylcysteine (NAC), 10 mM; DPI, 10 μM.

The synergistic effect of EGF+A23187 or EGF+TPG further support the notion of a need for both elevation of $[Ca^{2+}]i$ and kinase activation (phosphorylation of PLA_2) for optimal

[^{3}H]AA release. The fact that reducing agents such as NAC and DTT as well as an FAD dependent oxidase inhibitor such as DPI, suppress [^{3}H]AA release suggest that ROS play an important role in the cascade leading to activation of cPLA$_2$. The characteristics of the activation of [^{3}H]AA release described above, lead us to assume that the major effect is on cPLA$_2$ and not on other PLA$_2$ enzymes. EGF on its own has no effect (max. 1% release) on [^{3}H]AA release in our system. EGF does however invoke ROS formation and [Ca^{2+}]i elevation. The [Ca^{2+}]i elevation is relatively small (up to 60 %; Fig. 2), peak value is reached within 1 min and decays quickly thereafter. Comparison of the size and duration of the effects of A23187 or thapsigargin on [Ca^{2+}]i elevation to that of EGF support the suggestion that the EGF signal alone does not provide a big enough [Ca^{2+}]i to enable cPLA$_2$ translocation to its target membrane(s).

The findings describe a possible network in which agonist/target interaction result in ROS formation, transient [Ca^{2+}]i elevation and cPLA$_2$ stimulation. These three elements seem to be interconnected; [Ca^{2+}]i elevation is a prerequisite for both ROS formation and cPLA$_2$ activation and ROS formation is a major determinant in cPLA$_2$ activation. H_2O_2 formation may also increase agonist effect on [Ca^{2+}]i elevation (not shown).

The effect of [Ca^{2+}]i elevation on both ROS formation and cPLA$_2$ activation may be relevant to the observed toxicity of chronic [Ca^{2+}]i elevation in cells which was shown to lead to cell apoptosis.

Acknowledgment

This research was supported by a grant from the National Council for Research and Development, Israel and the Deutsches Krebsforschungszentrum (DKFZ).

REFERENCES

1. Y. S. Bae, S. W. Kang, M. S. Seo, I. C. Baines, E. Tekle, P. B. Chock, and S. G. Rhee, Epidermal growth factor (EGF)-induced generation of hydrogen peroxide. Role in EGF receptor-mediated tyrosine phosphorylation. *J Biol Chem.* 272:217 (1997).
2. P. L. Puri, M. L. Avantaggiati, V. L. Burgio, P. Chirillo, D. Collepardo, G. Natoli, C. Balsano, and M. Levrero, Reactive oxygen intermediates mediate angiotensin II-induced c-Jun.c-Fos heterodimer DNA binding activity and proliferative hypertrophic responses in myogenic cells. *J Biol Chem* 270:22129 (1995).
3. M. Ohba, M. Shibanuma, T. Kuroki, and K. Nose, Production of hydrogen peroxide by transforming growth factor-beta 1 and its involvement in induction of egr-1 in mouse osteoblastic cells. *J Cell Biol* 126:1079 (1994).
4. M. Meyer, R. Schreck, J. M. Muller, and P. A. Baeuerle, Redox control of gene expression by eucaryotic transcription factors NF-KB, AP-1 and SRT/TCF. In *Oxidative Stress, Cell Activation and Viral Infection.* Pasquier, ed. Birkhauser Verlag, Basel, p. 217-235 (1994).
5. M. Los, W. Droge, K. Stricker, P. A. Baeuerle, and K. Schulze-Osthoff, Hydrogen peroxide as a potent activator of T lymphocyte functions. *Eur J Immunol* 25:159 (1995).
6. K. N. Schmidt, E. B. Traenckner, B. Meier, and P. A. Baeuerle, Induction of oxidative stress by okadaic acid is required for activation of transcription factor NF-kappa B. *J Biol Chem* 270:27136 (1995).
7. P. A. Cerutti, Prooxidant states and tumor promotion. *Science* 227:375 (1985).
8. D. J. Morre, Hormone- and growth factor-stimulated NADH oxidase. *J Bioenerg Biomembr* 26:421 (1994).
9. S. Shaw, P. Naegeli, J. D. Etter, and P. Weidmann, Role of intracellular signalling pathways in hydrogen peroxide-induced injury to rat glomerular mesangial cells. *Clin Exp Pharmacol Physiol* 22:924 (1995).
10. K. Z. Guyton, Y. Liu, M. Gorospe, Q. Xu, and N. J. Holbrook. Activation of mitogen-activated protein kinase by H2O2. Role in cell survival following oxidant injury. *J Biol Chem* 271:4138 (1996).
11. R. Goldman, U. Zor, R. Meller, S. Moshonov, G. Furstenberger, and R. Seger. Activation of MAP kinases, cPLA2 and reactive oxygen species formation by EGF and calcium mobilizing agonists in a human keratinocyte cell line. In *Eicosanoids and Other Bioactive Lipids in Cancer, Inflammation, and Radiation Injury,* K.V. Honn, S. Nigam, L.J. Marnett, R.L. Jones, and P.Y-K. Wong, eds. (1977).
12. U. Zor, E. Ferber, P. Gergely, K. Szucs, V. Dombradi, and R. Goldman, Reactive oxygen species mediate phorbol ester-regulated tyrosine phosphorylation and phospholipase A2 activation: potentiation by vanadate, *Biochem J* 295:879 (1993)

13. R. Goldman, E. Ferber, and U. Zor, Reactive oxygen species are involved in the activation of cellular phospholipase A2. *FEBS Lett* 309:190 (1992).
14. R. Goldman, E. Ferber, R. Meller, and U. Zor, A role for reactive oxygen species in zymosan and beta-glucan induced protein tyrosine phosphorylation and phospholipase A2 activation in murine macrophages. *Biochim Biophys Acta 1*222:265 (1994).
15. M. J. Berridge, and R. F. Irvine, Inositol phosphates and cell signalling. *Nature* 341:197 (1989)
16. J. Meldolesi, L. Madeddu, and T. Pozzan. Intracellular Ca2+ storage organelles in non-muscle cells: heterogeneity and functional assignment. *Biochim Biophys Acta* 1055:130 (1990).
17. C. W. Taylor and I. C. Marshall, Calcium and inositol 1,4,5-trisphosphate receptors: a complex relationship, *Trends Biochem Sci* 17:403 (1992).
18. L. Li, R. W. Tucker, H. Hennings, and S. H. Yuspa, Chelation of intracellular Ca2+ inhibits murine keratinocyte differentiation in vitro. *J Cell Physiol* 163:105 (1995).
19. S. Pillai, and D. D. Bikle, Role of intracellular-free calcium in the cornified envelope formation of keratinocytes: differences in the mode of action of extracellular calcium and 1,25 dihydroxyvitamin D3. *J Cell Physiol* 146:94 (1991).
20. H. Hennings, K. Holbrook, P. Steinert, and S. Yuspa, Growth and differentiation of mouse epidermal cells in culture: effects of extracellular calcium. *Curr Probl Dermatol* 10:3 (1980).
21. L. L. Lin, M. Wartmann, A. Y. Lin, J. L. Knopf, A. Seth, and R. J. Davis, cPLA2 is phosphorylated and activated by MAP kinase. *Cell* 72:269 (1993).
22. R. A. Nemenoff, S. Winitz, N. X. Qian, V. Van Putten, G. L. Johnson, and L. E. Heasley, Phosphorylation and activation of a high molecular weight form of p hospholipase A2 by p42 microtubule-associated protein 2 kinase and protein k inase C. *J Biol Chem* 268:1960 (1993).
23. W. H. Waterman, T. F. Molski, C. K. Huang, J. L. Adams, and R. I. Sha'afi, Tumour necrosis factor-alpha-induced phosphorylation and activation of cytosolic phospholipase A2 are abrogated by an inhibitor of the p38 mitogen-activated protein kinase cascade in human neutrophils, *Biochem J* 319:17 (1996).
24. R. M. Kramer, E. F. Roberts, S. L. Um, A. G. Borsch Haubold, S. P. Watson, M. J. Fisher, and J. A. Jakubowski, p38 mitogen-activated protein kinase phosphorylates cytosolic phospholipase A2 (cPLA2) in thrombin-stimulated platelets. Evidence that proline-directed phosphorylation is not required for mobilization of arachidonic acid by cPLA2. *J Biol Chem* 271:27723 (1996).
25. N. Nahas, W. H. Waterman, and R. I. Sha'afi, Granulocyte-macrophage colony-stimulating factor (GM-CSF) promotes phosphorylation and an increase in the activity of cytosolic phospholipase A 2 in human neutrophils. *Biochem J* 313:503 (1996).
26. W. H. Waterman, and R. I. Sha'afi, A mitogen-activated protein kinase independent pathway involved in the phosphorylation and activation of cytosolic phospholipase A2 in human neutrophils stimulated with tumor necrosis factor-alpha, *Biochem Biophys Res Commun* 209:271 (1995).
27. S. Glover, T. Bayburt, M. Jonas, E. Chi, and M. H. Gelb, Translocation of the 85-kDa phospholipase A2 from cytosol to the nuclear envelope in rat basophilic leukemia cells stimulated with calcium ionophore or IgE/antigen, *J Biol Chem* 270:15359 (1995).
28. A. R. Schievella, M. K. Regier, W. L. Smith, and L. L. Lin, Calcium-mediated translocation of cytosolic phospholipase A2 to the nuclear envelope and endoplasmic reticulum, *J Biol Chem* 270:30749 (1995).
29. H. Kan, Y. Ruan, and K. U. Malik, Involvement of mitogen-activated protein kinase and translocation of cytosolic phospholipase A2 to the nuclear envelope in acetylcholine-induced prostacyclin synthesis in rabbit coronary endothelial cells. *Mol Pharmacol* 50:1139 (1996).
30. P. Boukamp, R. T. Petrussevska, D. Breitkreutz, J. Hornung, A. Markham, and N. E. Fusenig, Normal keratinization in a spontaneously immortalized aneuploid human keratinocyte cell line, *J Cell Biol* 106:761 (1988).

CLONING, TISSUE-SPECIFIC EXPRESSION AND REGULATION OF THE BOVINE THROMBOXANE A_2 RECEPTOR

Stephanie Muck, and Karsten Schrör

Institut für Pharmakologie, Heinrich-Heine-Universität Düsseldorf, Germany

INTRODUCTION

It is well known that activation of thromboxane A_2 (TP) receptors results in platelet aggregation, vasoconstriction, and mitogenesis of vascular smooth muscle cells[1]. These various effects of thromboxane A_2 (TXA_2) may be exerted through different TP receptor subtypes or a single receptor protein coupled to different intracellular signal transduction pathways. Southern blot analysis of genomic DNA showed that there is only a single copy gene for this receptor. To determine wether the bovine TP receptor differs from receptors identified in human placenta[2], and human endothelium[3], we examined the molecular structure of the bovine receptor. In additional experiments we investigated the tissue distribution of this novel TP receptor in bovine tissues and vascular smooth muscle cells (VSMC) by means of reverse transcriptase-polymerase chain reaction (RT-PCR). Recent studies from our laboratory have shown that TXA_2 potentiates the mitogenic effects of thrombin on VSMC proliferation[4]. As a possible mechanism for these potentiating effects, we studied TP receptor regulation on mRNA-level.

METHODS

Receptor cloning

RT-PCR was performed with heterologous primers to generate a TP receptor-fragment from total RNA of bovine VSMC. About 10^6 plaques of a bovine heart cDNA and about $8x10^5$ plaques of a bovine genomic library (Stratagene, La Jolla, USA) were screened using this homologous TP receptor probe. Two partial clones were isolated, subcloned, and sequenced with gene specific primers.

Cell culture

Bovine smooth muscle cells were isolated enzymatically from coronary arteries of adult female animals. Cells were cultivated in 80% HAM's F12/20% Dulbecco's modified eagle medium supplemented with 15% fetal calf serum, 100 U/ml penicillin and 0.1 mg/ml streptomycin. Subconfluent VSMC were synchronized in serum-free medium for 24 h. Further incubations with thrombin (1 U/ml) were performed in serum-free medium supplemented with 3 μmol/l indomethacin (Luitpold Pharma, München, FRG).

Receptor expression

Total RNA was prepared with trizol reagent (Gibco BRL, Eggenstein, FRG) and first strand cDNA was synthesized from total RNA with Oligo d(T), (Perkin Elmer, Branchburg, USA). Cy5-labeled PCR primers were designed according to the putative fourth and seventh transmembrane domain of the bovine TP receptor. PCR was cycled 30 times at 57°C with 2 mM $MgCl_2$. For semiquantitative analysis, heterologous GAPDH-primers were added. PCR-products were separated by gel electrophoresis and detected by automated laser fluorescence.

RESULTS AND DISCUSSION

Molecular structure of the bovine TP receptor

Sequencing of the two positive subclones revealed that they are overlapping and both together contain the complete receptor sequence. A 1029 bp open reading frame encodes for a protein of 343 amino acids (M_r 37,845 D). The nucleotide sequence has been deposited in the EMBL database under the accession numbers U53484 and U53485. Kyte-Doolittle-hydropathy analysis separated the amino acid sequence in seven transmembrane domains, indicating that this receptor belongs to the family of G-protein coupled receptors (Figure 1).

The amino terminal region showed two potential N-linked glycosylation sites (N-4, N-16). Two potential protein kinase C-dependent phosphorylation sites were present in the second intracellular loop (S-324, T-337) and another one was located in the carboxy terminal tail (S-145), suggesting a possible involvement of protein kinase C in the regulation of the receptor function. Furthermore, a potential cAMP- and cGMP-dependent phosphorylation site was detected in the C-terminus (T-341). The amino acid sequence is homologous with human placenta[2], human endothelium[3], mouse[5], and rat TP receptor[6] at 84.0, 81.4, 71.3, and 70.8%, respectively, indicating that the bovine TP receptor belongs to the splicing-variant of human placenta (Figure 2).

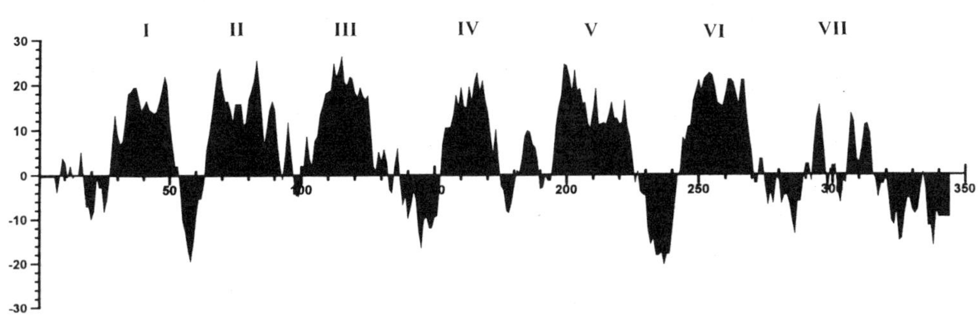

Figure 1. Kyte-Doolittle-hydropathy analysis of the bovine TP receptor. The hydrophobic segments (I-VII) correspond to the seven transmembrane domains of G-protein coupled receptors.

```
bovine        MWPNASSLGPCFRPMNITLEERRLIASPWFAASFCLVGLASNLLALSVLMGARQGSSQSRSSFLTFLCGLVLTDFMGLLVTGAIVVTQHFVLFEWQAVDP  100
human plat.   ..P.GS...P....T....E...L...........VV..A.........A...Q.GSHT.....TF.........L......TI.VS..AA.FE.H.V..  100
human endo.   ..P.GS...P....T....E...L...........VV..A.........A...Q.GSHT.....TF.........L......TI.VS..AA.FE.H.V..  100
mouse         ..P.GT...A....V....Q...A...........AL..G.........A...P.AG-P.....AL.........L......AI.AS..AA.LD.R.T..   99
rat           ..L.ST...A....V....Q...A...........AL..G.........A...P.AG-P.....AL.........L......AV.AS..AA.LD.R.T..   99
                                              I                             II

bovine        GCSLCHFMGVIMVFFGLCPLLLGAAMASERFLGITRPFSRPATASQRRAWTTVGLVWASALALGLLPLLGVGHYTVQYPGSWCFLTLGTDPGDVAFGLLF  200
human plat.   G.R..R...VV.I....S.........S..YL.........AVA.Q....A......AA.LA........V.R.T.............AES...A...L.  200
human endo.   G.R..R...VV.I....S.........S..YL.........AVA.Q....A......AA.LA........V.R.T.............AES...A...L.  200
mouse         S.R..Y...VA.V....C.........S..FV.........TAT.-....A......VA.GA........L.R.S.............TQR...V...I.  198
rat           G.R..H...AA.V....C.........A..FV.........AAT.-....A......VG.GT........L.R.S.............AER...A...M.  198
                  III                                   IV                                           V

bovine        ALLGSISVGMSFLLNTISVATLCHVYHGQATAQQRPRDCEVEMMVQLMGIMVVASICWMPLLVFIAQTVLQSPPAMSPTGQLSRLTERQLLIYLRVATWN  300
human plat.   SM..GL...L.F....V......H...GQEAA.Q....S.....A..L......SV..L......A..V.RNP.A..PA...S.T..KE..........  300
human endo.   SM..GL...L.F....V......H...GQEAA.Q....S.....A..L......SV..L......A..V.RNP.A..PA...S.T..KE..........  300
mouse         AL..SA...L.L....V......R...TREAT.-....C.....V..V......TV..M......M..L.QTP.V..FS...L.A..HQ..........  297
rat           AL..SV...L.L....V......R...AREAT.-....C.....V..V......TV..M......L..L.QTL.V..PS...L.T..RQ..........  297
                                                        VI                                       VII

bovine        QILDPWVYILFRRAVIQRFYPRLSTRSRSLSL------------------------QPQLTRRSTIHX-                               343
human plat.   ............A.LR.LQ..LSTRPRSL..------------------------QPQLTQRSGLQX-                               343
human endo.   ............A.LR.LQ..LSTRPRRV..CGPAWSTVARSRLTATSASRVQAILVPQPPEQLGLQAX                              369
mouse         ............S.LR.LH..FSSQLQAV..RRPP---------------AQAMLSG---------PX                                341
rat           ............S.LR.LH..FTSQLQAV..HSPP---------------TQAMLSG---------PX                                341
```

Figure 2. Amino acid sequence comparison of TP receptors from various species. The deduced amino acid sequences of human placenta (D38081), human endothelium (U11271), mouse (D10849), and rat (D21158) are aligned. Dots indicate identical amino acids, hyphens are inserted to achieve maximum homology.

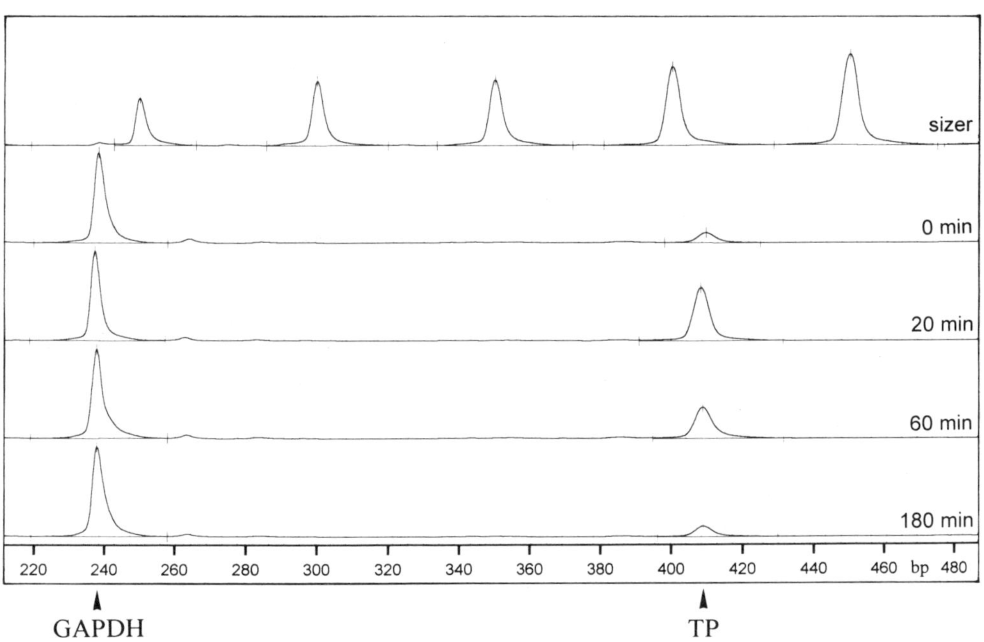

Figure 3. Time-dependent upregulation of TP receptor mRNA by thrombin. SMC were stimulated with thrombin (1 U/ml) for 20, 60, and 180 min. Original registration of Cy5-labeled RT-PCR products separated on a 6% polyacrylamide-gel and detected by automated laser fluorescence. A 410 bp TP receptor-fragment and a 238 bp GAPDH-fragment as an internal standard were amplified. The top lane shows a 50 bp ladder of Cy5-labeled fragments (50-500 bp).

Tissue-specific TP receptor mRNA-expression

Northern-blot analysis failed to detect tissue-specific TP receptor mRNA-expression, probably because of a low-level expression of receptor transcripts. The study of receptor expression in various bovine tissues by RT-PCR revealed a TP receptor transcript in liver, stomach, kidney, heart, ileum, ovary, and VSMC. The presence of TP receptor mRNA in the kidney has been reported in human[3], mouse[5], and rat tissues[6]. TP receptor transcripts in the heart were also detected in mice[5].

Regulation of TP receptor mRNA-expression

Interactions between tissue-specific mitogens and TP receptor expression were studied by semiquantitative RT-PCR. Stimulation of quiescent cultured VSMC with thrombin (1 U/ml) resulted in a marked but transient upregulation of TP receptors (Figure 3). A maximum increase (6-fold) in TP receptor mRNA was seen at 20 min. After continuous stimulation with thrombin for 3 hours, the TP mRNA-expression returned to the level of unstimulated cells.

In summary, we describe a hitherto unkown bovine TP receptor and its mRNA-expression in various tissues. TP receptor upregulation by thrombin might be important for TXA_2-related pathophysiology, i.e. VSMC mitogenesis after vascular injury.

REFERENCES

1. P.V. Halushka, C.J. Allan, and K.L. Davis-Bruno, Thromboxane A_2 receptors, *J Lipid Mediators Cell Signalling 12*:361 (1995).
2. M. Hirata, Y. Hayashi, F. Ushikubi, Y. Yokota, R. Kageyama, S. Nakanishi, and S. Narumiya, Cloning and expression of cDNA for a human thromboxane A_2 receptor, *Nature 349*:617 (1991).
3. M.K. Raychowdhury, M. Yukawa, L.J. Collins, S.H. Mc Grail, K.C. Kent, and J.A. Ware, Alternative splicing produces a divergent cytoplasmic tail in the human endothelial thromboxane A_2 receptor, *J Biol Chem 269*:19256 (1994).
4. T.-P. Zucker, D. Bönisch, and K. Schrör, Thromboxane potentiates thrombin and thrombin receptor activating peptide (TRAP)-induced mitogenesis of coronary artery smooth muscle cells, *Circulation 92 (Suppl)*:J-111 (1995).
5. T. Namba, Y. Sugimoto, M. Hirata, Y. Hayashi, A. Honda, A. Watabe, M. Negishi, A. Ichikawa, and S. Narumiya, Mouse thromboxane A_2 receptor: cDNA cloning, expression and northern blot analysis, *Biochem Biophy Res Commun 184*:1197 (1992).
6. J. Kitanaka, H. Hashimoto, Y. Sugimoto, M. Sawada, M. Negishi, A. Suzumura, T. Marunouchi, A. Ichikawa, and A. Baba, cDNA cloning of a thromboxane A receptor from rat astrocytes, *Biochim Biophys Acta 1265*:220 (1995).
7. D.D. D'Angelo, M.G. Davis, S. Ali, and G.W. II Dorn, Cloning and pharmacologic characterization of a thromboxane A_2 receptor from K562 (human chronic myelogenous leukemia) cells, *J Pharmacol Exp Ther 271*:1034 (1994).

WEAK INHIBITORS OF CYCLOOXYGENASES MAY EXERT THEIR ANTINOCICEPTIVE EFFECT BY MODULATION OF TRANSCRIPTION FACTORS

Nicole Scheuren*, Holger Bang*, Tino Münster, Kay Brune and Andreas Pahl

Institute of Experimental and Clinical Pharmacology and Toxicology, University of Erlangen-Nürnberg, Universitätsstraße 22, 91054 Erlangen, Germany

Nonsteroidal anti-inflammatory drugs (NSAIDs) are widely used in the treatment of pain, fever and inflammation. For these agents inhibition of prostaglandin synthesis via inhibition of cyclooxygenases (COX) is considered to be their principle mode of action.[1] However, inhibition of prostaglandin synthesis alone may not account for all effects of NSAIDs. There is no satisfying correlation between inhibition of prostaglandin synthesis and antinociceptive (analgesic) effects of some of these agents.[2] Weak inhibitors of prostaglandin synthesis, e.g. salicylic acid exert potent antinociceptive effects.[3] Such observations question the conventual concept that NSAIDs act exclusively by inhibiting prostaglandin synthesis. However, neither the cellular target nor cell signaling pathways possibly involved in the antinociceptive effect of weak prostaglandin inhibitors have been identified yet.

Since recent results indicate that sodium-salicylate and aspirin could exert some of their effects via regulation of transcription factor NF-κB,[4] we discuss the possibility that weak inhibitors of cyclooxygenases may exert their antinociceptive effect in part by interfering with the activation of transcription factor NF-κB.

NF-κB was first identified as a regulator of the expression of the kappa light-chain gene in murine B-lymphocytes but has subsequently been found in many different cells and plays a key role in immune and inflammatory responses.[5,6] The activated form of NF-κB is a heterodimer, which usually consists of two proteins, a p65 and a p50 subunit. In unstimulated cells, NF-κB is found in the cytoplasm and is bound to IκB, which prevent it from entering the nucleus.[7] In stimulated cells, specific kinases phosphorylate IκB, and causes its rapid degradation by proteasomes.[8] The release of NF-κB allows translocation of NF-κB into the nucleus, where it binds to specific sequences in the promotor region of target genes.

To distinguish between inhibition of prostaglandin synthesis and additional molecular activities by which weak inhibitors of prostaglandin-synthesis may exert their antinociceptive effect we used the 2-arylpropionic acid ibuprofen, a racemic mixture of R- and S-ibuprofen, with a dissociation between the anti-inflammatory and antinociceptive effect. Only the S (+)-enantiomer of ibuprofen is an effective cyclooxygenase inhibitor, while the R (-)-enantiomer is inactive in this respect.[9] Nevertheless, clinical and experimental data shows that R (-)-ibuprofen has a potent antinociceptive effect.[10]

---

* The first two authors have contributed equally

Recent Advances in Prostaglandin, Thromboxane, and Leukotriene Research
Edited by Sinzinger *et al.*, Plenum Press, New York, 1998

Effect of ibuprofen and its enantiomers on NF-κB activation

Jurkat T-cells were incubated with various concentrations of R- and S-ibuprofen for 2h and subsequently stimulated for 1h with phorbolester (TPA). The effect of R- and S-ibuprofen on NF-κB binding to its DNA response element was investigated with EMSA experiments. Pretreatment of Jurkat cells with R- and S-ibuprofen blocked TPA induced activation of NF-κB binding in a dose dependent fashion (Fig. 1). The effect of ibuprofen was specific because, at the same concentrations, ibuprofen did not affect the heat shock transcription factor and the activation of NF-κB by prostaglandin E_2 (not shown).
Theoretically, inhibition of NF-κB activation by S (+)-ibuprofen could be a side effect resulting from a primary inhibition of prostaglandin synthesis via cyclooxygenase modification. Western blot analysis and RT-PCR however revealed that stimulated T-cells showed no cyclooxygenase expression and did not produce any PGE_2.

Cellular localization studies with confocal immunofluorescence

With immunofluorescence microscopy we further substantiate that ibuprofen inhibits a specific step in the activation of NF-κB. Nuclear translocation of NF-κB is controlled by the association of NF-κB proteins with inhibitor proteins, called IκBs.[11] Phosphorylation of IκB lead to release of NF-κB from inactive complexes and to translocation of active NF-κB into the nucleus.[12] Using p65 antibodies, confocal immunofluorescence microscopy of stimulated T-cells showed the colocalization of NF-κB in the cytoplasm and the nucleus. Pretreatment with R- and S-ibuprofen in concentrations comparable to inhibitory doses for NF-κB in EMSA completely prevented the nuclear translocation of NF-κB after stimulation. NF-κB was exclusively located in the cytoplasm.

DISCUSSION AND CONCLUSION

The 2-arylpropionic acid ibuprofen is the most frequently prescribed class of NSAID to treat different forms of arthritis, migraine, headaches and fever.[10] So far the molecular mechanism by which R (-)-ibuprofen exert the antinociceptive effect is not known. Here, we report that

R (-)-ibuprofen [μM]	-	-	10	1	0.1	-	-	-
S (+)-ibuprofen [μM]	-	-	-	-	-	10	1	0.1
Stimulus (TPA)	-	+	+	+	+	+	+	+

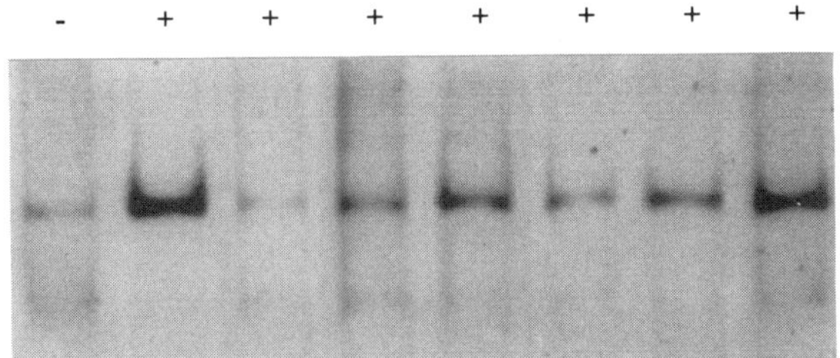

1 2 3 4 5 6 7 8

Figure 1. Effect of antinociceptice concentrations of ibuprofen-enantiomers on the binding of NF-κB to DNA. Nuclear extracts were prepared from Jurkat T-cells and assayed in an EMSA using a κB-binding site as a probe. Jurkat cells were pretreated for 2h with indicated concentrations of ibuprofen-enantiomers (lanes 3-8). Subsequently cells were stimulated with 25 ng/ml TPA for 1h (lanes 2-8). Unstimulated control cells are shown in lane 1.

R (-)-ibuprofen specifically inhibits the activation of transcription factor NF-κB as evidenced by analysis of DNA binding of various transcription factors after different stimulations and characterization of NF-κB transport processes. To our knowledge this is the first time that for R (-)-ibuprofen, a so far regarded useless ballast in the racemic mixture of ibuprofen, a specific interaction with a cellular event was detected. R (-)-ibuprofen inhibition was specific for phorbolester stimulation and effective at concentrations therapeutically achieved in plasma and synovial fluid.[13, 14] It has been demonstrated that in phorbolester stimulated cells protein kinase C (PKC) induces phosphorylation and inactivation of IκB in vitro suggesting PKC as a regulator of NF-κB activation.[15] In contrast to phorbolester mediated NF-κB activation ibuprofen had no effect on PGE_2 mediated NF-κB activtion. Since PGE_2 is an activator of the protein kinase A (PKA) signaling pathway it is interesting to speculate that ibuprofen may specifically interfere with PKC mediated NF-κB activation.[16]

Taken together, analysis of the effects of R (-)-ibuprofen, has led to new insights about the molecular mechanism of this drug. The results have led to the description of a novel mode of action of R (-)-ibuprofen, a so far neglected enantiomer of an important arylpropionic acid, which may contribute to explain the antinociceptive effect.

ACKNOWLEDGMENTS

We thank Margarethe Goppelt-Struebe, Christine Rascher, Gerd Geisslinger, Werner Neupert, and Flora Beiche for providing reagents and helpful discussion during the course of this study. The excellent technical assistance of Ingrid Heubes is gratefully acknowledged. This work was supported by a grant from the DFG-SFB 263, project A7.

REFERENCES

1. J.R. Vane. Towards a better aspirin, Nature 367:215 (1994).
2. K. McCormack and K. Brune, Dissociation between antinociceptive and antiinflammatory effects of the nonsteroidal anti-inflammatory drugs, Drugs 41:533 (1991).
3. K. Brune, W.S. Beck, G. Geisslinger, S. Menzel-Soglowek, B.M. Peskar, and B.A. Peskar, Aspirin like drugs may block pain independently of prostaglandin synthesis inhibition, Experientia 47:257 (1991).
4. E. Kopp, and S. Gosh, Inhibition of NF-κB by sodium salicylate and aspirin, Science 265:956 (1994).
5. R. Sen, and D. Baltimore, Multiple nuclear factors interact with the immunglobulin enhancer sequences, Cell 46:705 (1986).
6. E.B. Kopp, and S. Gosh, NF-κB and rel proteins in innate immunity, Adv. Immunol. 58:1 (1995).
7. A.S. Baldwin Jr. The NF-κB and IκB proteins: new discoveries and insights, Annu. Rev. Immunol. 14:649 (1996).
8. Z.J. Chen, L. Parent, and T. Maniatis, Site specific phosphorylation of IκBα by a novel ubiquitination-dependent protein kinase activity, Cell 84:853 (1996).
9. G. Geisslinger, G.S.H. Ferreira, S. Menzel, D. Schlott, and K. Brune, Antinociceptive actions of R (-)-flurbiprofen - a non-cyclooxygenase inhibiting 2-arylpropionic acid - in rats, Life Sci. 54:PL 173 (1994).
10. P.M. Brooks, and R.O. Day, Nonsteroidal anti-inflammatory drugs - differences and similarities, N. Engl. J. Med. 324:1716 (1991).
11. A.A. Beg, and A.S. Baldwin, The IκB proteins: multifunctional regulators of Rel/NF-κB transcription factors, Genes Dev. 7:2064 (1993).

12. A.A. Beg, T.S. Finco, P.V. Nantermet, and A.S. Baldwin Jr., Tumor necrosis factor and interleukin-1 lead to phosphorylation and loss of IκBα: a mechanism for NF-κB activation, Mol. Cell. Biol. 13:3301 (1993).
13. E.F. Davies, and G.S. Avery, Ibuprofen: a review of its pharmacological properties and therapeutic efficacy in rheumatic disorders, Drugs 2:416 (1971).
14. R.C. Glass, and A.J. Swannell, Concentrations of ibuprofen in serum and synovial fluid from patients with arthritis, Br. J. Pharmacol. 6:453P (1978).
15. I.M. Verma, J.K. Stevenson, E.M. Schwarz, D. van Antwerp, and S. Miyamoto, Rel/NF-κB/IκB family: intimitate tales of association and dissociation, Genes Dev. 9:2723 (1996).
16. J. Feuillard, H. Gouy, G. Bismuth, L.E. Lee, P. Debre, and M. Körner, NF-κB activation by tumor necrosis factor α in the Jurkat T-cell line is independent of protein kinase A, protein kinase C, and Ca^{2+}-regulated kinases, Cytokine 3:257 (1991).

NEW GROUP OF LIPID MEDIATORS CONTAINING ω-HYDROXYARACHADONIC ACID (20-HETE)

Kaiyun Mao, K. Kishta Reddy,[1] J.R. Falck,[1]
John C. McGiff and Michael Balazy

Department of Pharmacology
New York Medical College, Valhalla, NY 10595
[1]Department of Biochemistry and Pharmacology
University of Texas Southwestern Medical Center, Dallas, TX 75235

INTRODUCTION

20-HETE is a biologically active product of arachidonic acid and cytochrome P450 ω-hydroxylase of the 4A gene family and is an abundant microsomal eicosanoid in the rat kidney.[1] However, very little is known about factors controlling the biosynthesis and metabolic fate of 20-HETE in the kidney. Cyclooxygenase metabolizes 20-HETE into ω-OH-PGs, formation of which was proposed to explain the vasoconstrictor activity of 20-HETE in the kidney.[2] Human platelets metabolize 20-HETE into a complex mixture of products of which 11,20-dihydroxyeicosatetraenoic acid (11,20-DiHETE), and 12,20-DiHETE are the major metabolites.[3]. We observed that the stimulation of the rat kidney with hormones or calcium ionophore results in an accumulation of 20-HETE, which is readily detectable by GC/MS.[4,5] The release of 20-HETE from renal slices stimulated by calcium ionophore cannot be inhibited with ETYA, which suggested to us that calcium influx is required for the release of 20-HETE. The isolated perfused kidney, challenged with angiotensin II (ang II) or endothelin-1, rapidly releases 20-HETE and often other CYP450 subterminal HETEs, into the perfusion fluid in amounts 5 to 10-fold higher relative to control levels. Brief stimulation of mTAL cells with ang II leads to a selective release of 20-HETE.[6] These results suggested that 20-HETE could be released from renal tissue phospholipids via a receptor-mediated activation of phospholipase (PL)A_2. In this study we obtained evidence that the release of 20-HETE from renal phospholipids is possible and represents a new component of ang II-dependent effects on renal function.

MATERIALS AND METHODS

Tritium-labeled 20-HETE ([$^3H_2$]-20-HETE) was prepared by a catalytic reduction of 20-hydroxy-5,8,11-eicosatrien-14-ynoic acid with $^3H_2$ (New England Nuclear, Boston MA) and was purified by RP-HPLC prior to *in vitro* experiments. 20-HETE-containing phospholipid standards were prepared by incubating 20-HETE (25 μM) and a *lyso*-phospholipid (*lyso*-PC, *lyso*-PE, *lyso*-PI) in the suspension of rat liver microsomes containing CoASH, GSH, ATP (1 mM each) and an ATP regenerating system. 1-*O*-hexadecyl-2-(ω-hydroxyarachidonyl)-*sn*-glycero-3-phosphatidylcholine (1-*O*-hexadecyl-20-HETE-PC) was prepared from *lyso*-PAF via conjugation with 20-*tert*-butyldiphenyl-

Recent Advances in Prostaglandin, Thromboxane, and Leukotriene Research
Edited by Sinzinger *et al.*, Plenum Press, New York, 1998

silyloxyarachidonic acid[7] followed by hydrolysis of the product with *n*-tetrabutylammonium fluoride.

Total lipids were extracted from rat kidney tissue (slices or whole kidney) and homogenized in methanol/water (2:3) containing BHT (0.005%) as described.[8] Hydrolysis of lipids was accomplished with KOH (0.1 N)/ethanol for 1 hr at 50 °C. 20-HETE was extracted, purified by HPLC, and analyzed by isotopic dilution electron capture GC/MS as described.[9] Total renal lipids were analyzed on normal-phase (NP) HPLC column (Hibar, 250×10 mm, 7 μm, E. Merck) eluted with a gradient of water (0.05% to 0.1%, 0 to 10 min) in hexane/2-propanol (3:4) at a flow rate of 2.5 ml/min. Rat kidney slices or whole kidneys were incubated with [$^3H_2$]-20-HETE (1 μM to 62.5 μM) for 15 to 180 min at 37 °C in HEPES buffer, and the total lipids were extracted and analyzed by NP-HPLC. Rat kidney slices labeled with 20-HETE were stimulated with ang II (5 nM/mg) for 1 to 20 min followed by analysis of total lipids in tissue and supernate by NP-HPLC. Fractions coeluting with 20-HETE-containing diglycerol (20-HETE-DAG) standard were collected and further purified by RP-HPLC. 20-HETE-DAG was derivatized with pentafluorobenzoyl (PFB) chloride and diisopropylethylamine in anhydrous toluene for 30 min at 60 °C and the derivative, 20-HETE-DAG-PFB, was purified on a Microsorb C18 column and analyzed by direct inlet mass spectrometry with negative ion chemical ionization using a Hewlett-Packard 5989A mass spectrometer operating in the electron capture mode with methane flow into the ion source at 1.8 torr. The tip of the probe was heated from 30 °C to 300 °C at 50 °C/min and 20-HETE-DAG-PFB produced maximal ion current at 4 to 5 min.

RESULTS AND DISCUSSION

Table 1. *A.* Distribution of 20-HETE and 19-HETE in rat renal total lipids[1]

Lipid Pool	20-HETE	19-HETE
Neutral Lipids	n.d.	n.d.
Diglycerides	n.d.	n.d.
Fatty acids	11-20	n.d.
Phospholipids	80 -89	100

B. Distribution of 20-HETE and 19-HETE in rat renal phospholipids[1]

Phospholipids	20-HETE	19-HETE
PC	77	25
PE	23	68
PI + PS	n.d.	7
Total	100	100

[1]As percentage of the total found in total lipids or phospholipids following alkaline hydrolysis; n.d. = not detected; average from four kidneys.

Total renal lipids subjected to a mild alkaline hydrolysis yielded 20-HETE and 19-HETE. The concentration of the free acid form of 20-HETE in rat kidney lipids was 0.3 ± 0.1 ng/μmol phosphorus (P) and increased to 2.4 ± 0.4 ng/μmol P upon hydrolysis of renal lipids (n = 9). 19-HETE was not detectable in the lipid extract and increased to 1.8 ± 0.4 ng/μmol P upon hydrolysis. This indicated that 80 to 89% of the endogenous 20-HETE and all of the endogenous 19-HETE were in the esterified form in rat renal tissues (Table 1). Complex renal lipids were further separated by NP-HPLC and collected in fractions, which were analyzed by GC/MS. These experiments revealed that the major forms containing endogenous 20-HETE were PC and PE with a molar ratio of 3.3. The major endogenous phospholipids containing 20-HETE coeluted with biosynthetically prepared 20-HETE-PC and 20-HETE-PE, respectively. The phospholipids containing 20-HETE were more polar and eluted about 3 min after the corresponding phospholipids containing arachidonic acid. The fractions that coeluted with standard 20-HETE-PI did not contain detectable amounts of 20-HETE. 19-HETE was found mostly in PE, and lesser amounts were in PC and PI.

Incubations of radiolabeled 20-HETE with renal slices followed by total lipid extractions and NP-HPLC analyses revealed several metabolites as shown in Fig. 1. Radiolabeled 20-HETE rapidly incorporated into renal slices in a time- and a concentration-dependent manner. The total amount of radioactivity incorporated into the phospholipids was 60 to 70% during 1 hr incubation.

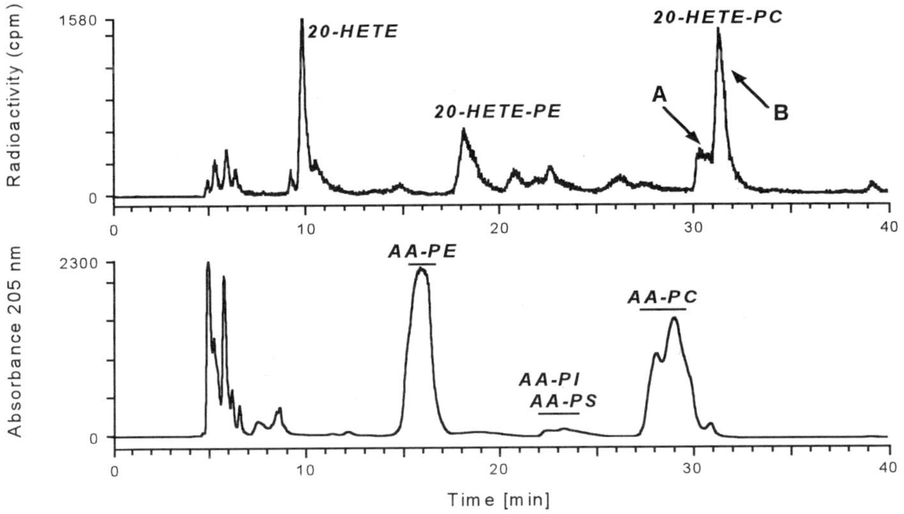

Figure 1. The effect of 30 min (37 °C) incubation of rat kidney slices with tritium-labeled 20-HETE (0.05 μCi/mg, final concentration 10 μM). The total renal lipids were extracted by the method of Bligh and Dyer and analyzed on a Hibar column with detection of radioactivity (top) and UV absorbance (bottom).

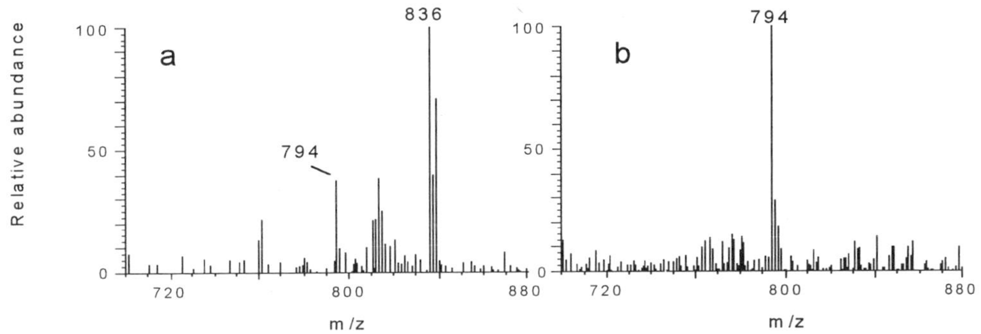

Figure 2. Mass spectra of DAGs (PFB esters) obtained from PLC hydrolysis of renal phospholipids (fractions A+B, Fig. 1)(**a**) and synthetic 1-*O*-hexadecyl-20-HETE-PC (**b**). The abundant peaks correspond to losses of H_2O from the molecular DAG-PFB anions.

The predominant form containing tritium-labeled 20-HETE was PC, and the minor form was PE. The amount of 20-HETE in a fraction coeluting with standard 20-HETE-PI was less that 2% and did not increase with the incubation time. 20-HETE was incorporated into two PC forms A and B (Fig. 1). Treatment of these PCs with PLC generated radiolabeled DAGs which produced mass spectra (Fig. 2) containing major ions at *m/z* 836 and *m/z* 794, corresponding to PFB derivatives of 1-stearoyl-2-(20-HETE)-glycerol and 1-*O*-hexadecyl-2-(20-HETE)-glycerol, respectively. Phospholipid A (Fig. 1) coeluted with synthetic 1-*O*-hexadecyl-20-HETE-PC. Renal incubates also contained minor amounts of 11,20-DiHETE, 12,20-HETE. Ang II stimulation of kidney tissue labeled with 20-HETE released 10 to 20 % of incorporated radioactivity. HPLC analysis of radioactive products released into supernate showed formation of two peaks that were absent in the supernate from unstimulated tissue (Fig. 3).

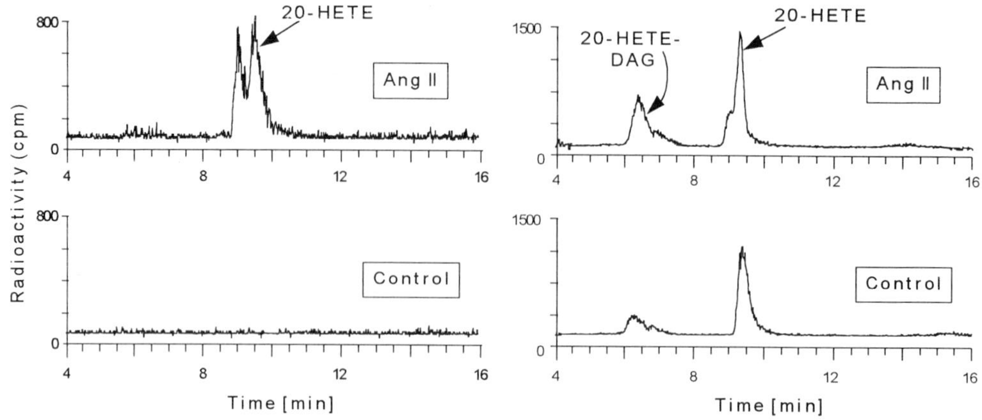

Figure 3. Stimulation of rat renal slices (prelabeled with [$^3H_2$]-20-HETE) with ang II (5 nM/mg tissue for 10 min at 37 °C) resulted in accumulation of 20-HETE (9.5 min) in the supernate (left) and 20-HETE-DAG (6.5 min) in tissue (right). Standard [$^3H_2$]-20-HETE alone eluted as a single peak of radioactivity at 9.5 min.

The peak eluting at 9.5 min was identified by GC/MS as 20-HETE, and the other less polar product eluting before 20-HETE was not characterized. Analysis of radioactive metabolites extracted from tissue (Fig. 3) before and after stimulation with ang II revealed formation of a product at 6.5 min, which was characterized as 20-HETE-DAG based on comparison of chromatographic mobility with standards and mass spectrometry.

Our study demonstrates that endogenous 20-HETE and 19-HETE are esterified in phospholipids. The major pathway of 20-HETE metabolism in the rat renal tissue was via incorporation into phospholipids, predominantly into two PC forms: acyl and alkyl. Very little of 20-HETE was metabolized by cyclooxygenase or lipoxygenase. The incorporation of a HETE molecule into an alkyl form of PC has not been previously described but is not unexpected given the high abundance of alkyl PC in the kidney.[10] 20-HETE showed relative selectivity for incorporation into PC as compared with its uptake into PI and PS. We obtained direct evidence that after 20-HETE accumulation, the stimulation of renal tissue by ang II, which is known to act on the kidney through phospholipase-coupled receptors,[11] released 20-HETE and 20-HETE-DAG. The rapid release of endogenous 20-HETE observed during ang II-stimulation of perfused rat kidney[4,5] may be explained in terms of PLA_2-mediated release from phospholipid stores rather than from released arachidonic acid metabolized by CYP450 ω-hydroxylase. We have shown that increasing extracellular Ca^{2+} from 1.1 to 5 mM elevates the levels of 20-HETE in the thick ascending limb of the rat kidney.[12] PLA_2-dependent hydrolysis of the unique alkyl PC containing 20-HETE raises the possibility that 20-HETE may play a role in the platelet-activating factor cycle within the kidney.[10] Substantial data support an important role of PC in signal transduction mechanisms mediated by ang II, which causes a biphasic increase in DAG accumulation with an initial peak at 15 s and a second sustained peak at 5 to 10 min.[13] The late DAG peak is derived from PC. Ang II also markedly increases phosphatidic acid (PA) production, indicating activation of PLD. Our observations that 20-HETE is found primarily in PC and that the stimulation with ang II results in prolonged accumulation of 20-HETE-DAG suggest that 20-HETE-PC is involved in ang II-mediated signaling. Since PA is the initial product of PC hydrolysis, it will be of particular interest to understand the function of 20-HETE-containing PA. The biological function and activity of oxidized DAG and PA are largely unknown. As various oxidized DAGs may stimulate[14] or inhibit[15] the activity of protein kinase C, further studies are required to elucidate biological function of mediators derived from the phospholipids containing cytochrome P450 metabolites of arachidonic acid.

ACKNOWLEDGMENTS

This work has been supported by grants from the American Heart Association New York State affiliate (950325) and by the National Institutes of Health (HL34300).

REFERENCES

1. M.H. Wang, D.E. Stec, M. Balazy, V. Mastyugin, C.S Yang, R.J. Roman, and M.L. Schwartzman, Cloning, sequencing, and cDNA-directed expression of the rat renal CYP4A2. *Arch. Biochem. Biophys.* 336:240 (1996).
2. M.L. Schwartzman, J.R. Falck, P. Yadagiri, and B. Escalante, Metabolism of 20-hydroxyeicosatetraenoic acid by cyclooxygenase. *J. Biol. Chem.* 264:11658 (1989).
3. E. Hill, F. Fitzpatrick, and R.C. Murphy, Biological activity and metabolism of 20-hydroxyeicosatetraenoic acid in the human platelet. *Br. J. Pharmacol.* 106:267 (1992).
4. B. Oyekan, M. Balazy, and J.C. McGiff, Renal oxygenases: differential contribution to vasoconstriction induced by endothelin-1 and angiotensin II. *Am. J. Physiol.*, in press.
5. M.A. Carroll, M. Balazy, P. Margiotta, D.D. Huang, J.R. Falck and J.C. McGiff, Cytochrome P450-dependent HETEs: profile of biological activity and stimulation by vasoactive peptides. *Am. J. Physiol.* 271:R863 (1996).
6. M. Lu, Y. Zhu, M. Balazy, K.M. Reddy, J.R. Falck, and W.H. Wang, Effect of angiotensin II on the apical K^+ channel in the thick ascending limb of the rat kidney. *J. Gen. Physiol.* 108:537 (1996).
7. S. Manna, J.R. Falck, N. Chacos, and J. Capdevila, Synthesis of arachidonic acid metabolites produced by purified kidney cortex microsomal cytochrome P-450. *Tetrahedron Lett.* 24:33 (1983).
8. Y. Zhu, J.C. McGiff, and M. Balazy, Identification of 20-hydroxyeicosatetraenoic acid in rat kidney phospholipids. *Proc. 43rd Conf. Am. Soc. Mass Spectrometry*, 1148 (1995).
9. Y. Zhu, J.C. McGiff, and M. Balazy, Identification of arachidonate P-450 metabolites in human platelet phospholipids. *Hypertension* 25[part 2]:854 (1995).
10. F. Snyder, Ether-linked lipids and their bioactive species: occurrence, chemistry, metabolism, regulation, and function, in: *Biochemistry of Lipids, Lipoproteins and Membranes*, D.E. Vance and J.E. Vance, eds., Elsevier, Amsterdam (1996).
11. K.K. Griendling, M.U. Fukai, B. Lassegue, and R.W. Alexander, Angiotensin II signaling in vascular smooth muscle - new concepts. *Hypertension* 29[part 2]:366 (1997).
12. W.H. Wang, M. Lu, M. Balazy, and S.C. Hebert, Phospholipase A_2 is involved in mediating the effect of extracellular Ca^{2+} on the apical K^+ channels in the rat TAL. *Am. J. Physiol.* in press.
13. Y. Wen. M.C. Cabot, E. Clauser, S.L. Bursten, and J.L. Nadler, Lipid signal transduction pathways in angiotensin II type 1 receptor-transfected fibroblasts. *Am. J. Physiol.* 269:C435 (1995).
14. S. Takekoshi, Y. Kambayashi, H. Nagata, T. Takagi, Y. Yamamoto, and K. Watanabe, Activation of protein kinase C by oxidized diacylglycerols. *Biochem. Biophys. Res. Commun.* 217: 654 (1995).
15. Y. Cho, and V.A. Ziboh, 13-Hydroxyoctadecadienoic acid reverses epidermal hyperproliferation via selective inhibition of protein kinase Cβ activity. *Biochem. Biophys. Res. Commun.* 201: 257 (1994).

BIOSYNTHESIS OF NOVEL DIVINYL ETHER OXYLIPINS BY ENZYME FROM GARLIC (*Allium sativum* L.) BULBS

Alexander N. Grechkin[1] and Mats Hamberg[2]

[1]Institute of Biology, Russian Academy of Sciences, P.O. Box 30, Kazan, 420503, Russia
[2]Department of Physiological Chemistry II, Karolinska Institutet, Stockholm, S-17177, Sweden

INTRODUCTION

First fatty acid divinyl ethers colneleic and colnelenic acids were detected by Galliard et al.[1,2] in *in vitro* experiments with potato tubers. Formation of these oxylipins, is controlled by 9-lipoxygenase activity and an enzyme, controlling dehydration of 9-hydroperoxide. Such a pathway, being not found in any other plant or animal species, was considered to be the unique property of potato tubers. Our recent *in vitro* studies of lipoxygenase pathway in garlic bulbs led to detection of novel divinyl ethers 12-[1'(*E*)-hexenyloxy]-9(*Z*),11(*E*)-dodecadienoic (etheroleic) and 12-[1'(*E*),3'(*Z*)-hexadienyloxy]-9(*Z*),11(*E*)-dodecadienoic (etherolenic) acids[3], products of 13-hydroperoxides of linoleic and linolenic acids (13-HPOD and 13-HPOT) conversion, respectively.

The present work is concerned with further studies of enzyme properties and mechanism of its action.

MATERIALS AND METHODS

Samples of (9*Z*,11*E*,13*S*)-13-hydroperoxy-9,11-octadecadienoic [13(*S*)-HPOD] and (5*Z*,8*Z*,11*Z*,13*E*,15*S*)-15-hydroperoxy-5,8,11,13-eicosatetraenoic (15-HPETE) acids were prepared by incubations of linoleic and arachidonic acids, respectively, with soybean lipoxygenase. Racemic hydroperoxides, (9*Z*,11*E*,13*R*,*S*)-13-hydroperoxy-9,11-octadecadienoic acid [(9*Z*,11*E*)-13(*R*,*S*)-HPOD] and (9*E*,11*E*,13*R*,*S*)-13-hydroperoxy-9,11-octadecadienoic acid [(9*E*,11*E*)-13(*R*,*S*)-HPOD], were prepared by autoxidation of 100 mg of linoleic acid for 15 hours at 37°C in oxygen atmosphere. The resulting mixture of racemic hydroperoxides, having *Z,E*- and *E,E*-double bond geometry, was separated by reversed phase HPLC (RP-HPLC). Samples of (9*Z*,11*E*)- and (9*E*,11*E*)-13(*R*,*S*)-HPOD were collected and used for incubations with divinyl ether synthase.

Garlic bulbs were homogenized in 0.1 M borate buffer, pH 9.0. Homogenate was filtered through cheesecloth and centrifuged for 16 min at 9300 g. Supernatant was

Recent Advances in Prostaglandin, Thromboxane, and Leukotriene Research
Edited by Sinzinger *et al.*, Plenum Press, New York, 1998

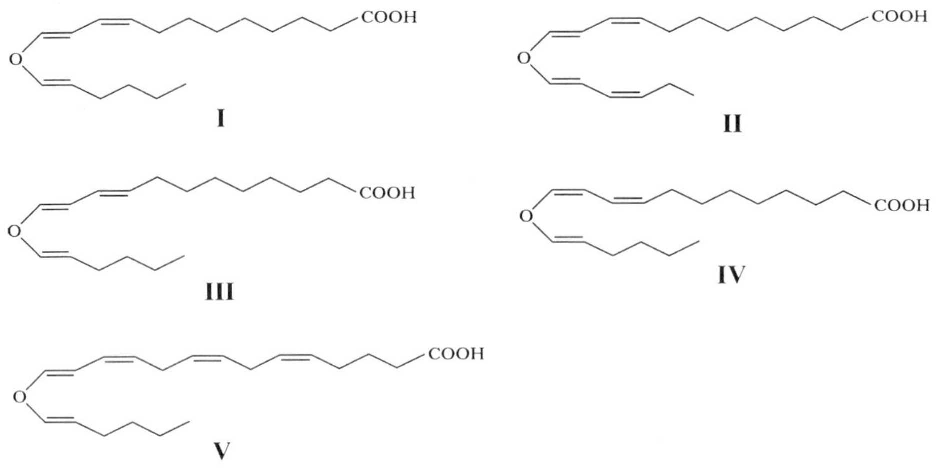

Fig. 1. The structural formulae of divinyl ethers: **I**, etheroleic acid; **II**, etherolenic acid; **III**, all-*trans* isomer of etheroleic acid; **IV**, 9(*Z*),11(*Z*),1'(*E*) isomer of etheroleic acid; **V**, (5*Z*,8*Z*,11*Z*,13*E*,1'*E*)-14-(1'-hexenyloxy)-5,8,11,13-tetradecatetraenoic acid.

decanted and centrifuged for 1 hour at 105,000 g. Microsomal pellet was resuspended in the same buffer and used as enzyme preparation for incubations.

Substrates were incubated with cell-free fractions from garlic (substrate concentrations and incubation times are specified in each case below). After acidification with 2 M HCl to pH 3.5 the incubation mixture was extracted with diethyl ether. Products were separated as free acids either by isocratic RP-HPLC on Macherey-Nagel Nucleosil 5 ODS column (250 x 4.6 mm), solvent mixture acetonitrile - water - acetic acid (60 : 40 : 0.01, by vol.), flow rate 1.5 ml/min, or using gradient elution as described previously[3,4]. Individual divinyl ethers after methylation with diazomethane were separated and purified by normal phase HPLC (SP-HPLC), two columns Daltosil 100 (250 x 4.6 mm; Serva Feinbiochemica, Heidelberg), connected in series, elution by mixture hexane - ethyl acetate (98.5 : 1.5, by vol.), flow rate 0.8 ml/min.

Analyses of hydroperoxide chirality were performed in general as described previously[5]. Chiral and racemic diastereomeric standards were prepared from 13(*S*)-HPOD and 13(*R,S*)-HPOD, respectively. Analyses of diastereomeric derivatives as described[5]. Hydrogenations of divinyl ethers were performed in ethanol solution over Adams catalyst at 23°C.

Ultraviolet topograms were recorded on line during HPLC analyses using RSD 2140 diode array detector (Pharmacia-LKB). Alternatively, UV spectra of purified products were recorded with Hitachi U2000 instrument. Gas chromatography - mass spectrometry (GC-MS) analyses were performed with Hewlett-Packard model 5970B mass selective detector connected to a Hewlett-Packard model 5890 gas chromatograph equipped with a methyl silicone fused capillary column. ^{1}H-NMR spectra (250 MHz, [$^{2}H_3$]acetonitrile), were recorded with Bruker WM-250 spectrometer. The homonuclear double resonance experiments were performed with the same instrument.

RESULTS AND DISCUSSION

Studies of [1-^{14}C]13(*S*)-HPOD metabolism in sub-cellular fractions from garlic revealed than the predominant part (76%) of divinyl ether synthase is localized in microsomes. The remaining part was detected in 105000 g supernatant. The preferential enzyme substrate was 13(*S*)-HPOD. With lesser efficiency enzyme converts (9*Z*,11*E*,13*S*,15*Z*)-13-hydroperoxy-9,11,15-octadecatrienoic acid (13-HPOT) into etherolenic acid (**II**). Unlike divinyl ether synthase from potato tubers, garlic enzyme is ineffective with 9-HPOD as a substrate. At the same time, 15-HPETE is converted to some extent into corresponding divinyl ether, identified as (5*Z*,8*Z*,11*Z*,13*E*,1'*E*)-14-(1'-hexenyloxy)-5,8,11,13-tetradecatetraenoic acid. These results suggest that enzyme uses preferentially (n-6)-hydroperoxides. Position of hydroperoxy group toward carboxyl end is also of critical importance. The observed substrate specificity of garlic divinyl ether synthase correlates with lipoxygenase regiospecificity. As we have recently found [Grechkin et al., to be published]., garlic lipoxygenase possesses high regiospecificity, transforming linoleate and α-linolenate into 13-hydroperoxides.

Racemic hydroperoxide 13(*R,S*)-HPOD was converted into etheroleic acid (**I**) slower than chiral 13(*S*)-HPOD. Remainder of unreacted 13(*R,S*)-HPOD after incubation with enzyme was subjected to stereospecific analysis. This allowed to reveal that remaining hydroperoxide was composed of up to 94% 13(*R*)-HPOD enantiomer. This result shows that 13(*R*)-HPOD is utilized much slower than 13(*S*)-HPOD. At the same time, 13(*R,S*)-HPOD is utilized almost completely when it is incubated with excess amount of enzyme. Thus, enzyme utilizes 13(*S*)-HPOD preferentially, but this requirement to substrate chirality is not absolute. As found previously, "heterolytic" hydroperoxide lyase from tea leaves possesses similar preference to 13(*S*)-HPOD[6]. Unlike these two enzymes, flax seed allene oxide synthase specifically utilizes only 13(*S*)-HPOD, being not able to transform 13(*R*)-HPOD[6].

Reversed phase HPLC analyses of 13(*S*)-HPOD conversion products allowed us to detect the peak of minor products, eluting close to etheroleic acid (**I**) and having slightly longer retention time. This fraction, collected from the reversed phase column, possessed ultraviolet absorption maximum at 251 nm. Components of this fraction were methylated and methyl esters were analysed by GC-MS. This allowed to detect two components **III** and **IV**, having electron impact mass spectra, identical to that of etheroleic acid[3].

Fraction of isomers **III** and **IV**, collected after the reversed phase separation, was subjected to normal phase HPLC analysis. This allowed to achieve a complete separation of two components. Catalytic hydrogenation of both compounds **III** and **IV** over Adams catalyst afforded single product, identified as 12-(1'-hexyloxy)-dodecanoic acid: EIMS (ion attribution, relative intensity, %): 314 (M^+, 0.1), 283 (M^+ - CH_3O, 6), 243 (M^+ - C_5H_{11}, 5), 229 (M^+ - C_6H_{13}, 46), 214 (M^+ - $C_6H_{13}OH$, 34), 197 (229 - CH_3OH, 100). Compounds **III** and **IV** have λ_{max} (MeOH) at 250 and 252 nm, respectively. Their ^{1}H-NMR spectra and spin decoupling data allowed us to reveal double bond geometry. The value of coupling constant $J_{9,10}$ = 15.0 Hz evidently suggests that this double bond in compound **III** has *trans* geometry. Constants of two other double bonds $J_{11,12}$ and $J_{1',2'}$ are close to the same constants of etheroleic acid[3], suggesting *trans* configuration of these double bonds. These results along with the ultraviolet and mass spectral data allow to identify compound **III** as (9*E*,11*E*,1'*E*)-12-(1'-hexenyloxy)-9,11-dodecadienoic acid, i.e., all-*trans* isomer of etheroleic acid. Identification of product **III** was also supported by its specific formation from isomeric hydroperoxide, (9*E*,11*E*)-13(*R,S*)-HPOD.

Spin constants values $J_{11,12}$ = 6.2 Hz and $J_{9,10}$ = 11.0 Hz for compound **IV** clearly suggest that two corresponding double bonds have *cis* geometry. Values of coupling constants $J_{9,10}$ = 11.0 Hz and $J_{1',2'}$ = 12.1 Hz for compound **IV** are nearly equal to that for etheroleic acid (**I**)[3]. These results along with data of ultraviolet spectroscopy and mass spectrometry allow to ascribe the structure of (9*Z*,11*Z*,1'*E*)-12-(1'-hexenyloxy)-9,11-dodecadienoic acid to compound **IV**.

In order to determine the origin of the ether oxygen of etheroleic acid, we performed an incubation of [$^{18}O_2$ -hydroperoxy]13(*S*)-HPOD with preparation of microsomes from garlic bulbs. Analysis of the resulting etheroleic acid by gas chromatography-mass spectrometry and selected ion monitoring (SIM) revealed that one ^{18}O atom was essentially completely incorporated from $[^{18}O_2]$13(*S*)-HPOD into the ether function of etheroleic acid. This result allows to propose a mechanism of etheroleic acid formation, involving epoxyallylic cation as an intermediate. First stage of the proposed mechanism is protonation-dehydration of 13(*S*)-HPOD, which affords an epoxyallylic carbocation. Enzyme assisted deprotonation of epoxyallylic carbocation at C-14, accompanied by C-C bond cleavage of oxirane, affords etheroleic acid. The same mechanism was previously proposed[7,8] for colneleic acid formation.

REFERENCES

1. T. Galliard and D.R. Phillips, Enzymic conversion of linoleic acid into 9-(1,3-nonadienoxy)-8-nonenoic acid, a novel unsaturated ether derivative isolated from homogenates of *Solanum tuberosum* tubers, *Biochem. J.* 129:743 (1972).
2. T. Galliard and D.R. Phillips, Novel divinyl ether fatty acid in the extract of *Solanum tuberosum*, *Chem. Phys. Lipids*. 129:743 (1972).
3. A.N. Grechkin, F.N. Fazliev and L.S. Mukhtarova, The lipoxygenase pathway in garlic (*Allium sativum* L.) bulbs: Detection of the novel divinyl ether oxylipins, *FEBS Lett*. 371:159 (1995).
4. A.N. Grechkin, R.A. Kuramshin, S.K. Latypov, E.Y. Safonova, T.E. Gafarova, and A.V. Ilyasov, Hydroperoxides of α-ketols. Novel products of the plant lipoxygenase pathway, *Eur. J. Biochem*. 199:451 (1991).
5. L.-Y. Zhang and M. Hamberg, A gas-liquid chromatographic method for steric analysis of 2-hydroxy, 3-hydroxy, and 2,3-dihydroxy acids, *Chem. Phys. Lipids* 74: 151 (1994).
6. H.W. Gardner, Recent investigations into the lipoxygenase pathway of plants, *Biochim. Biophys. Acta.* 1084: 221 (1991).
7. P. Fahlstadius and M. Hamberg, Stereospecific removal of the *pro-R* hydrogen at C-8 of (9*S*)-hydroperoxyoctadecadienoic acid in the biosynthesis of colneleic acid, *J. Chem. Soc. Perkin Trans 1*: 2027 (1990).
8. L. Crombie, D.O. Morgan and E.H. Smith, An isotopic study (^{2}H and ^{18}O) of the enzymic conversion of linoleic acid into colneleic acid with carbon chain fracture: the origin of shorter chain aldehydes, *J. Chem. Soc. Perkin Trans. 1*:567 (1991).

CATALYTIC AND SPECTROSCOPIC PROPERTIES OF LINOLEATE DIOL SYNTHASE OF THE FUNGUS *GAÜMANNOMYCES GRAMINIS*

Ernst H. Oliw and Chao Su

Department of Pharmaceutical Biosciences,
Uppsala University Biomedical Center,
S-751 24 Uppsala, Sweden

The fungus *Gaümannomyces graminis* causes "take-all", a devastating root disease of wheat. Our interest in *G. graminis* is due to its biosynthesis of oxylipins, which might be important for sporulation. The fungus can sequentially metabolize linoleic acid (18:2n-6) to 8R-hydroperoxylinoleic acid (8-HPODE) through an 8-dioxygenase and to 7S,8S-dihydroxylinoleic acid (7,8-DiHODE) through a hydroperoxide isomerase[1-3]. A highly purified protein with both enzyme activities was recently found to be a hemoprotein[4]. The protein had an apparent molecular size of 130 kDa on SDA-PAGE and it appeared to be a tetramer on gel filtration. The protein will be referred to as linoleate diol synthase.

Polyunsaturated fatty acids are oxygenated by two major classes of dioxygenases: lipoxygenases with non-heme iron and hemoproteins (PGH synthases, linoleate diol synthase). Dioxygenases introduce oxygen into their substrates in two ways. The ferric form of lipoxygenases and PGH synthases make the substrate active by proton abstraction and the

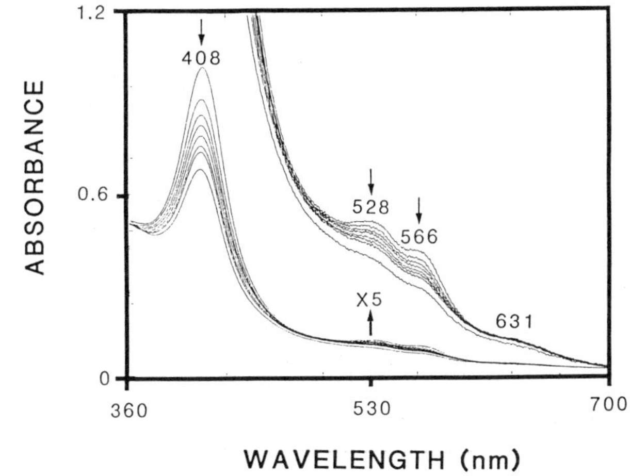

Figure 1. Spectral analysis of linoleate diol synthase. The spectrum was recorded before and 1, 2, 5, 10, 20, 30 min and 9h after addition of 18:2n-6 at room temperature. The concentration of diol synthase was 13 μM (tetramer) and substrate 1.4 mM.

Recent Advances in Prostaglandin, Thromboxane, and Leukotriene Research
Edited by Sinzinger *et al.*, Plenum Press, New York, 1998

carbon centered radical reacts with oxygen[5]. PGH synthases also have peroxidase activity, which initiates the cyclooxygenase activity by generating a Tyr radical of catalytical importance for proton abstraction. Other dioxygenases, for example the ferrous hemoprotein L-tryptophan 2,3-dioxygenase[6], activate oxygen, which then reacts with the substrate.

The main objective of the present study was to determine whether diol synthase contained ferrous or ferric iron and to study the effects of hydroperoxides and other agents on the enzyme activity.

EXPERIMENTAL

Diol synthase was purified from mycelia of *G. graminis* as described[1,2]. Spectroscopy was performed with a dual beam spectrophotometer (Shimadzu UV-2101PC) at room temperature. High and low spin derivatives were prepared by addition of a few grains of NaF, NaN_3, $NaNO_2$ and dithionite, or KCN directly to the sample cuvette[7]. A TLC assay[1,2] or an oxygen electrode was used to measure enzyme activity.

RESULTS

Spectroscopy

The spectrum of purified diol synthase is shown in Figure 1. Absorption maxima were present at about 631 nm, 566 nm (α), 528 nm (β, broad), and 407 nm (γ). The absorption maxima at 631 nm and at 407 nm suggested a ferric hemoprotein in a high spin state[5]. Addition of 18:2n-6 to the purified protein mainly reduced the absorption of the α, β, and γ bands with time (Figure 1).

Ferric iron of hemoproteins can be transformed into low and high spin states by ions and by pH changes[7]. CN^- and N_3^- form low spin derivatives. CN^- markedly changed the spectrum of the protein. A δ band appeared at 364 nm, the γ band decreased in intensity and shifted to 419-421 nm. The α and β bands were replaced with a broad peak with absorption maximum at 550 nm. The absorption maximum at about 631 nm vanished. In the presence of N_3^- the protein showed γ absorption at 411 nm, indicating partial transition to low spin states. The NO derivative showed a decreased extinction and a γ band at 416 nm. High spin states are favoured at low pH and low spin under alkaline conditions. At pH 6.5 (0.1 M $KHPO_4$ buffer) the spectrum was essentially unchanged with γ band at 407.5 nm. This γ band was reduced in intensity and shifted to 412 nm at pH 9.5 (in 0.1 M glycine/OH). F^- can induce a high spin state characterized by a slight blue shift of the γ band, but F^- had little effect on the spectrum of the native protein.

These observations suggested that the native protein was ferric and in a high spin state.

Inhibition and Activation of Linoleate Diol Synthase

Biosynthesis of 8-HPODE by the purified protein occurred with a time lag of about 10s (oxygen consumption monitoring). Addition of 13-HPODE (6 μM) did not reduce the time lag. However, large concentrations of GSH peroxidase and GSH increased the lag time and could totally block the dioxygenase and isomerase[1] activities (TLC assay; Figure 2A), indicating that other hydroperoxides than 13-HPODE might be important for initiating the catalysis. Preliminary experiments suggested that 8-HPODE reduced the lag time to 5s.

Tetranitromethane (TNM), which reacts with Tyr radicals, is a potent time-dependent inhibitor of the cyclooxygenase reaction[5]. We investigated the effect of TNM on linoleate diol synthase. TNM at 0.4 mM caused about 50% inhibition of diol synthase activity (Figure 2B). Diethyl pyrocarbonate (DEPC) is supposed to react with histidine ligands and can inhibit cyclooxygenase[8]. DEPC also reduced the diol synthase activity (Figure 2B). His and Tyr residues might thus be important for the catalytic activity of diol synthase. Some detergents

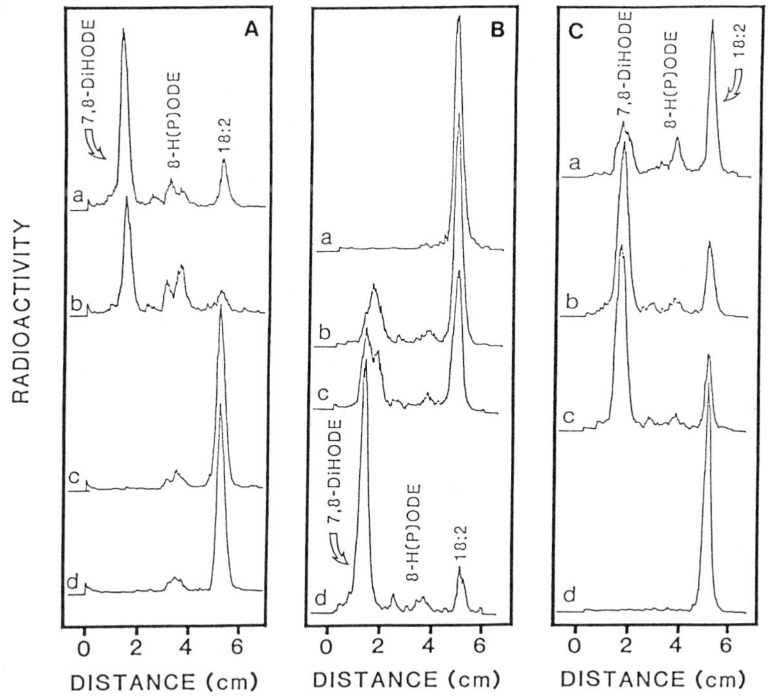

Figure 2. Effect of GSH peroxidase, TNM, DEPC, and detergents on diol synthase activity. The oxidation of [^{14}C]18:2n-6 to [^{14}C]8-H(P)ODE and [^{14}C]7,8-DiHODE was followed by TLC[1,2]. **A**, top trace (a) with enzyme, trace (b) with GSH peroxidase (5 U/ml), trace (c) with GSH peroxidase (5 U/ml) and 1 mM GSH, and trace (d) with 1 mM S-hexaglutathione. **B**, trace (a) control without enzyme, trace (b) after preincubation with 0.4 mM TNM for 30 min, trace (c) after preincubation with 1 mM DEPC for 30 min, and trace (d) enzyme only. **C**, trace (a) enzyme without detergents, trace (b) enzyme with 0.04% Tween-20, trace (c) enzyme with 0.5 mM CHAPS, and trace (d) enzyme blank.

could augment the enzyme activity. The striking effects of Tween-20 (0.04%) and CHAPS (0.5 mM) are shown in Figure 2C. Higher concentrations of detergents were inhibitory. No cofactors appeared to be necessary for the diol synthase activity and addition of heme was without effect on enzyme activity.

In summary, the ternary complex of protein, linoleic acid and molecular oxygen was catalytically competent and detergents augmented the enzyme activity.

DISCUSSION

Spectroscopy of diol synthase suggested that it contained ferric heme in a high spin state. Ferric heme does not bind oxygen. The first step of diol synthase catalysis is therefore likely substrate activation by hydrogen abstraction in analogy with PGH synthases and lipoxygenases. PGH synthases and lipoxygenases are activated by hydroperoxides[5], e.g., by 13-HPODE. We found that 13-HPODE did not reduce the lag time of diol synthase, while 8-HPODE appeared to do so, indicating a link between the 8R-dioxygenase and hydroperoxide isomerase activities.

Linoleate diol synthase shows another similarity with PGH synthases. PGH synthases can oxygenate 5,8,11-eicosatrienoic acid (20:3n-9) to 13R-hydroxyeicosatrienoic acid in an "abortive" cyclooxygenase reaction[9]. This reaction involves hydrogen abstraction and antarafacial oxygen insertion at C-13 of 20:3n-9 in the same way as diol synthase catalyzes hydrogen abstraction and antarafacial oxygen insertion at C-8 of 18:2n-6[1-3]. The hydroperoxide isomerase of diol synthase converts 8R-HPODE into 7S,8S-dihydroxylinoleic acid by elimination of the *pro*-S hydrogen from C-7 and intramolecular suprafacial insertion of oxygen at C-7[1,2]. The hydroxylation step of the isomerase reaction is analogous to hydroxylation by cytochrome P450, which also can use hydroperoxides as oxygen doners in the peroxidase shunt[10].

In summary, the dioxygenase reaction of diol synthase might be mechanistically related to the cyclooxygenase activity of PGH synthases and the isomerase activity to cytochrome P450.

AKNOWLEDGMENTS

Supported by grants from the Swedish Medical Research Council (6523) and Magn. Bergvalls Stiftelse.

REFERENCES

1. ID. Brodowsky, M. Hamberg, and E.H. Oliw, A linoleic acid (8R)-dioxygenase and hydroperoxide isomerase of the fungus *Gaeumannomyces graminis*. *J. Biol. Chem.* 267:14738 (1992).
2. M. Hamberg, L-Y. Zhang, I.D. Brodowsky, and E.H. Oliw, Sequential oxygenation of linoleic acid in the fungus *Gaeumannomyces graminis*: Stereochemistry of dioxygenase and hydroperoxide isomerase reactions. *Arch. Biochem. Biophys.* 309:77 (1994).
3. C.Su, I.D. Brodowsky, and E.H. Oliw, Studies on linoleic acid 8R-dioxygenase of the fungus *Gaeumannomyces graminis*. *Lipids* 30:43 (1995).
4. C. Su, and E.H. Oliw, Purification and characterization of linoleate 8-dioxygenase from the fungus *Gaeumannomyces graminis* as a novel hemoprotein. *J. Biol. Chem.* 271:14112 (1996).
5. W.L. Smith, and L.J. Marnett, Prostaglandin endoperoxide synhase: Structure and catalysis. *Biochim. Biophys. Acta* 1083:1 (1991).
6. J.M Leeds, P.J. Brown, G.M. McGeehan, F.K., Brown, and J.S. Wiseman J. Isotope effects and alternative substrate reactivities for tryptophan 2,3-dioxygenase. *J. Biol. Chem.* 268:17781 (1993).
7. E. Antonini, and M. Brunori, *Hemoglobin and Myoglobin in Their Reactions with Ligands.* North-Holland Publishing Co., Amsterdam (1971).
8. Zhang, X., A-L. Tsai, and R.J. Kulmacz, Chemical modification of prostaglandin H synthase with diethyl pyrocarbonate. *J. Biol. Chem.* 269:5085 (1994).
9. E.H. Oliw, L. Hörnsten, H. Sprecher, and M. Hamberg, Oxygenation of 5,8,11-eicosatrienoic acid by prostaglandin endoperoxide synthase and by cytochrome P450 monooxygenase: Structure and mechanism of formation of major metabolites. *Arch. Biochem. Biophys.* 305:288 (1993).
10. M. Hamberg, Hydroperoxide isomerases. *J. Lipid Mediators Cell Signalling* 12:283 (1995).

AVENOLEIC ACID: A NEW OXYLIPIN FROM OAT SEEDS

Mats Hamberg

Department of Medical Biochemistry and Biophysics,
Division of Physiological Chemistry II,
Karolinska Institutet, S-171 77 Stockholm, Sweden

INTRODUCTION

Several oxylipins, including hydroxyoctadecadienoic acids, epoxyoctadecenoic acids, epoxyhydroxyoctadecenoic acids, and trihydroxyoctadecenoic acids, are formed from linoleic acid upon incubation with preparations of cereal seeds. Lipoxygenases[1] peroxygenase[2,3] as well as cytochrome P-450´s[4] are responsible for these transformations. A pathway in oat seeds for catalytic oxidation of linoleic acid into an anti-fungal trihydroxy acid, 9(*S*),12(*S*),13(*S*)-trihydroxy-10(*E*)-octadecenoic acid, was recently described[5]. During this study, the presence in oat seeds of a new hydroxylated derivative of linoleic acid was discovered[6]. The present paper is a brief account of the structure determination of the new hydroxy acid, for which the trivial name "avenoleic acid" is proposed. In addition, recent data showing that avenoleic acid is mainly localized in the galactolipid fraction of oat seed lipids, and structural studies of the major molecular species of galactolipid containing avenoleic acid, will be described.

EXPERIMENTAL

Seeds of oat (*Avena sativa* L., cv. Freja) were obtained from Svalöf Weibull, Svalöv, Sweden. 3(*R,S*)-Hydroxyhexanoic acid was prepared by sodium borohydride reduction of methyl 3-oxohexanoate followed by saponification. 3(*R*)-Hydroxyhexanoic acid was obtained by enzymatic reduction of 3-oxohexanoic acid using baker´s yeast. Steric analysis[7] showed that the enantiomeric composition of the hydroxy acid was 97.5% "*R*" and 2.5% "*S*". Methyl 3(*R,S*)-(8´-carbomethoxyoctanoyloxy)hexanoate was prepared by acylation of methyl 3(*R,S*)-hydroxyhexanoate (2.3 mmol) with 8-carbomethoxyoctanoyl chloride (2.5 mmol) in 5 mL of toluene-pyridine (9:1, v/v).

For isolation of avenoleic acid (**1**) and its methyl ester (**2**), batches of oat seeds were ground in an electric coffee mill and the powder (100 g) homogenized in acetone (1 l) containing 10 p.p.m. of 2,6-di-*tert*-butyl-4-methylphenol. The product was saponified and subjected to SiO_2 chromatography. Material eluted with diethyl ether-hexane (1:1, v/v) was treated with diazomethane and subjected to a second step of SiO_2 chromatography (elution with diethyl ether-hexane (1:4, v/v)). Preparative reverse-phase HPLC (column, 250 x 10

Recent Advances in Prostaglandin, Thromboxane, and Leukotriene Research
Edited by Sinzinger *et al.*, Plenum Press, New York, 1998

mm of Nucleosil 100-7; solvent system, CH_3CN-H_2O (6:4, v/v)) afforded **2** (179.1-203.8 mL of effluent). Methyl avenoleate (**2**) thus obtained (c:a 50 mg from 100 g of oat seed powder) was homogenous as judged by GLC and TLC. Avenoleic acid (**1**) was obtained following saponification and crystallization from hexane at -80°C.

The major galactolipid molecular species containing avenoleic acid (**3**) was isolated by extracting powder of oat seeds with either $CHCl_3$-MeOH (2:1, v/v) or acetone. The purification steps included SiO_2 chromatography (elution with acetone), reverse-phase HPLC (column, 250 x 10 mm of Nucleosil 100-7; solvent system, MeOH; elution volume, 72.0-77.0 mL), preparative TLC (solvent system, $CHCl_3$-MeOH-H_2O (50:10:1, v/v/v)), and finally a second step of reverse-phase HPLC. Approx. 15 mg of **3** were obtained from 100 g oat seed powder using this procedure.

RESULTS AND DISCUSSION

Structure and Stereochemistry of Avenoleic Acid (1)

The UV spectrum of methyl avenoleate (**2**) was featureless, demonstrating the absence of conjugated double bonds. The FT-IR spectrum showed the presence of a hydroxyl group, a carbomethoxy group and the absence of *E* double bonds. Analysis of the trimethylsilyl (Me_3Si) derivative of **2** showed prominent ions at *m/z* 367 (M^+ - CH_3), 310 (M^+ - OHC-$(CH_2)_2$-CH_3), and 145 (Me_3SiO^+=CH-$(CH_2)_2$-CH_3), suggesting a methyl octadecadienoate substituted with a hydroxyl group at C-15. The ^1H NMR spectrum was in agreement with the presence of two nonconjugated double bonds having the *Z* configuration ($J_{9,10}$ = 10.7 Hz, $J_{12,13}$ = 10.8 Hz) as well as a hydroxyl group at C-15 (δ 3.63). Degradation of the propionyl derivative of **2** by ozonolysis afforded methyl hydrogen azelate as well as 3-propionoxyhexanoic acid. Steric analysis[7] demonstrated that the 3-propionoxyhexanoic acid had the "*R*" configuration (>99.9% *R*). On the basis of these results, avenoleic acid (**1**) was identified as 15(*R*)-hydroxylinoleic acid, a previously unknown hydroxylated derivative of linoleic acid.

Localization of Avenoleic Acid in Oat Seed Lipids

Powder of oat seed (10 g) was extracted with $CHCl_3$-MeOH (2:1, v/v) and the extract fractionated by SiO_2 chromatography into neutral lipids, glycolipids+sulfolipids, and phospholipids by elution with $CHCl_3$, acetone, and MeOH, respectively[8]. To each of the three fractions was added methyl 17-hydroxystearate (124.7 μg) as an internal standard, and the fatty acids were analyzed by GLC following saponification, esterification and trimethylsilylation. The analyses showed that avenoleic acid constituted 0.9, 18.5, and 3.1% of the total fatty acids of the $CHCl_3$, acetone, and MeOH fractions, respectively. Further, by use of the internal standard, it could be calculated that the absolute amounts of avenoleic acid in the three fractions were: $CHCl_3$ fraction, 1.5 mg (29% of total), acetone fraction, 3.2 mg (62% of total), and MeOH fraction, 0.4 mg (9% of total). The finding that avenoleic acid was enriched in the glycolipid/sulfolipid fraction prompted further separation of this material

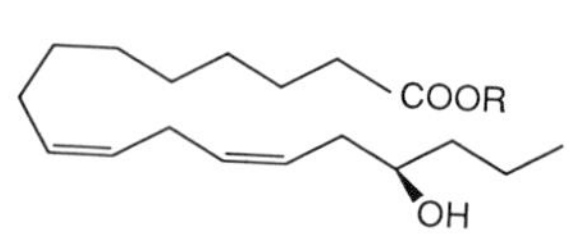

1, R = H

2, R = CH_3

Figure 1. Avenoleic acid (**1**) and its methyl ester (**2**).

by reverse-phase HPLC and preparative TLC as described above. On TLC, the major avenoleic acid-containing material (**3**; $R_f = 0.34$) migrated close to digalactosyldilinoleoylglycerol ($R_f = 0.28$) suggesting that **3** was a digalactosylglyceride.

Structural Studies of the Major Avenoleate-Containing Galactolipid (3) in Oat Seeds

Saponification of **3** liberated linoleic acid and **1** in a ratio of 2:1. Analysis of **3** by negative ion electrospray mass spectrometry showed a molecular ion at *m/z* 1217.836 (M^+ - 1; calculated for digalactosylglycerol acylated with 2 linoleates and 1 avenoleate, *m/z* 1217.837). Identification of the carbohydrate residues was performed by refluxing a sample of **3** with 2% H_2SO_4 in 65% aqueous methanol for 3 h followed by analysis of the trimethylsilylated product by GC-MS. Peaks due to the Me_3Si derivatives of α- and β-galactopyranose, and to the Me_3Si derivatives of methyl α- and β-galactopyranosides, were observed.

In order to determine how the three fatty acid residues were assembled in **3**, an aliquot of the galactolipid was acetylated and subjected to oxidative ozonolysis. GLC analysis of the esterified product demonstrated the presence of methyl 3-(8´-carbomethoxy-octanoyloxy)hexanoate and the absence of methyl 3-acetoxyhexanoate. Accordingly, the C-15 hydroxyl group of the avenoleate residue of **3** did not occur in its free form but was acylated by a linoleoyl residue. In order to localize the linoleate and the linoleoyl-avenoleate residues, an aliquot of **3** was treated with lipase from *Rhizopus arrhizus*. This enzyme specifically hydrolyzes the ester bond at *sn*-1[9] and has been used to characterize molecular species of galactolipids. Treatment of **3** with lipase from *Rhizopus arrhizus* liberated linoleic acid and resulted in the formation of a major galactolipid hydrolysis product upon TLC analysis ($R_f = 0.43$; solvent system, $CHCl_3$-MeOH-H_2O (50:20:2, v/v/v)). This product was identified by chemical degradation and mass spectrometry as a digalactosylglyceride carrying a linoleoyl-avenoleate residue. As judged from the established specificity of *Rhizopus* lipase for the *sn*-1 position, the parent galactolipid **3** was identified as a digalactosylglyceride acylated with a linoleate residue at *sn*-1 and with a linoleoyl-avenoleate residue at *sn*-2 (Fig. 2). This structure has recently been confirmed by NMR spectroscopy[10].

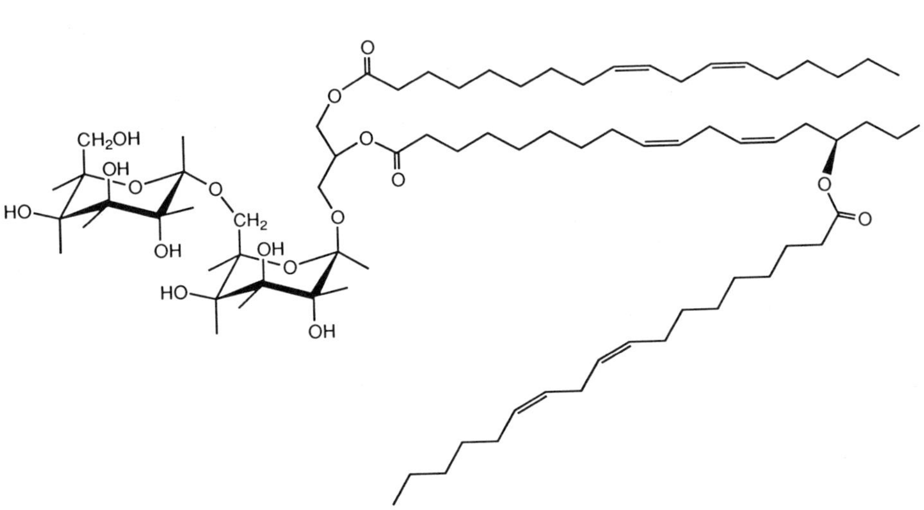

Figure 2. Structure of 3, the major avenoleate-containing galactolipid from oat seeds.

Significance of Avenoleic Acid in Oat

Avenoleic acid is related to ricinoleic acid (12(*R*)-hydroxyoleic acid). Both compounds possess a homoallylic alcohol group, and in both compounds the alcohol group is located at a carbon of the fatty acid chain that in the non-hydroxylated compound may undergo a desaturase reaction. Interestingly, it has recently been reported that an oleoyl hydroxylase from *Ricinus communis*, which catalyzes hydroxylation of oleate into ricinoleate, is homologous to a microsomal ω-6 oleoyl desaturase from *Arabidopsis*, that catalyzes conversion of oleate into linoleate[11]. It can be speculated that the putative hydroxylase involved in the formation of avenoleic acid is related to ω-3 linoleoyl desaturase(s) responsible for the biosynthesis of α-linolenic acid from linoleic acid.

REFERENCES

1. H.W. Gardner, Lipoxygenase pathway in cereals, in: *Advances in Cereal Science and Technology*, Y. Pomeranz, ed., Vol. 9, pp. 161-215. American Association of Cereal Chemists, St. Paul, MN (1988).
2. M. Hamberg and G. Hamberg, Hydroperoxide-dependent epoxidation of unsaturated fatty acids in the broad bean, *Arch. Biochem. Biophys.* 283:409 (1990).
3. E. Blée and F. Schuber, Efficient epoxidation of unsaturated fatty acids by a hydroperoxide-dependent oxygenase, *J. Biol. Chem.* 265:12887 (1990).
4. G.P. Bolwell, K. Bozak, and A. Zimmerlin, Plant cytochrome P450, *Phytochemistry* 37:1491 (1994).
5. M. Hamberg and G. Hamberg, Peroxygenase-catalyzed fatty acid epoxidation in cereal seeds, *Plant Physiol.* 110:807 (1996).
6. M. Hamberg and G. Hamberg, 15(*R*)-Hydroxylinoleic acid, an oxylipin from oat seeds, *Phytochemistry* 42:729 (1996).
7. L.-Y. Zhang and M. Hamberg, A gas-liquid chromatographic method for steric analysis of 2-hydroxy, 3-hydroxy, and 2,3-dihydroxy acids, *Chem. Phys. Lipids* 74:151 (1994).
8. F.D. Gunstone, M. Kates, and J.L. Harwood, Separation and isolation procedures, in: *The Lipid Handbook*, F.D. Gunstone, J.L. Harwood, and F.B. Padley, eds., 2nd ed., pp. 225-248. Chapman & Hall, London (1994).
9. W. Fischer, E. Heinz, and M. Zeus, The suitability of lipase from *Rhizopus arrhizus delemar* for analysis of fatty acid distribution in dihexosyl diglycerides, phospholipids and plant sulfolipids, *Hoppe-Seyler's Z. Physiol. Chem.* 354:1115 (1973).
10. M. Hamberg, G. Otting, E. Liepinsh, and W. Griffiths, manuscript in preparation.
11. F.J. van de Loo, P. Broun, S. Turner, and C. Somerville, An oleate 12-hydroxylase from *Ricinus communis* L. is a fatty acyl desaturase homolog, *Proc. Natl. Acad. Sci. USA* 92:6743 (1995).

DISCOVERY OF NOVEL CANNABINOID RECEPTOR LIGANDS FROM DIVERSE MARINE ORGANISMS

Ken Soderstrom, Thomas F. Murray, Hye-Dong Yoo, Sharon Ketchum, Ken Milligan, William Gerwick, Maria J. Ortega,[1] Javier Salva[2]

[1] College of Pharmacy
Oregon State University
Corvallis, OR 97331 USA
[2] Department of Chemistry
Universidad de Cadiz
Cadiz, Spain

INTRODUCTION

Arachidonic acid ethanolamide (anandamide), a simple nitrogen-containing lipid, has recently been described as an endogenous ligand to the mammalian CNS cannabinoid receptor (CB1)[1]. Since its initial characterization, there has been tremendous interest in exploring a diversity of related structures for activity at this receptor from the viewpoint of defining structure-activity relationships at this receptor, discovering pharmacologically useful agonists and antagonists, and defining the role of anandamide metabolism in the modulation of CB1 activation[2].

As marine life forms are exceptionally rich in structurally-unique lipids, we have initiated a multifaceted program to evaluate diverse marine-derived plant and animal extracts for novel CB1 receptor ligands, or for enzymes capable of converting anandamide into novel substances. In this latter regard, we recently described the production of an novel high affinity ligand to CB1 by conversion of anandamide into "conjugated triene anandamide" (CTA) by action of polyunsaturated fatty acid isomerase, a new enzyme from the red alga *Ptilota filicina*[3]. The screening assay involves displacement[3] of [H]CP-55940 from membranes prepared from CHO cells stably expressing the rat CNS cannabinoid receptor. Pure compounds and extracts evaluated to date in our program have come predominantly from marine algae and sponges.

Another new dimension of our program with anandamide is to explore the effect of oxidation on the activity in the CB1 receptor assay. As anandamide is a minor modification of the arachidonic acid (AA) structure, it is reasonable that it would be a substrate for diverse AA metabolizing enzymes in mammalian systems. Indeed, others have explored the activity of prostaglandin ethanolamides[4], hydroxy-containing arachidonic acid ethanolamides

Recent Advances in Prostaglandin, Thromboxane, and Leukotriene Research
Edited by Sinzinger *et al.*, Plenum Press, New York, 1998

produced by lipoxygenase metabolism[5,6], as well as a leukotriene analog[4] of anandamide for their activity at this receptor. We have taken a slightly different approach, reasoning that with the diversity of lipoxygenase/cytochrome P_{450} systems in mammals, any one of a large number of oxidation products might show affinity at this receptor and be of physiological or pharmacological importance. Thus, we have used non-specific methods to produce a diversity of oxidized anandamide products, separated these by HPLC, and screened them for CB1 affinity.

CB1 BINDING AFFINITY OF UNUSUAL LIPIDS FROM THE SPONGE *Mycale micracanthoxea*

The marine sponge *Mycale micracanthoxea* was recently found to be a rich source of unusual nitrogen-containing lipids. A total of 13 related compounds were isolated and structurally defined. The distinctive feature of these lipids is a pyrrole ring which is attached through C2 to a long chain fatty acid. The pyrrole also possesses a one carbon appendage at C5 as either a primary alcohol or an aldehyde. A variety of carbon chain lengths and double bond unsaturation patterns are present in the fatty acid portion of these structures. The relationship of nitrogen and primary alcohol functionalities in these metabolites is intriguingly homologous to anandamide as one additional carbon atom resides in the sponge metabolites between the carbonyl and nitrogen atom.

Equilibrium binding assays employing all 13 of the sponge metabolites as competitors for [^{3}H]CP-55940 binding showed several to have modest CB1 affinity (Figure 1). Two metabolites in the series showed higher affinity than the others, with IC_{50} values of 12.6 μM and 55.6 μM, respectively. The distinguishing structural features of these two metabolites is the presence of multiple unsaturations and either moderate overall carbon chain length ($<$22 carbons for the fatty acid chain) or the presence of a double bond relatively close to the pyrrole side of the molecule (equivalent to $\Delta 7$ in the most potent sponge compound).

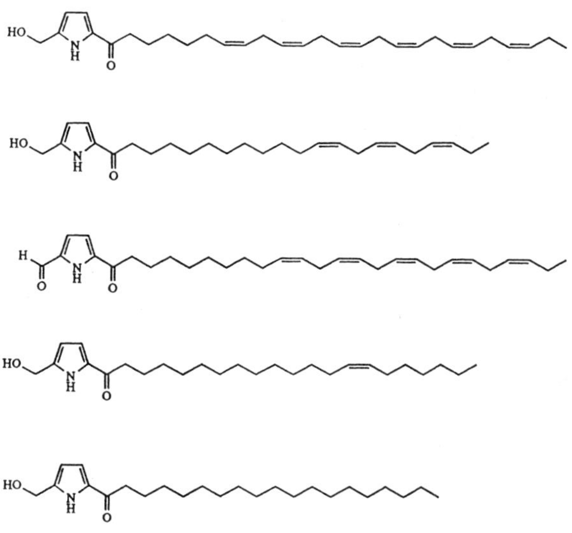

IC50's

12.6 μM

55.6 μM

> 100 μM

> 100 μM

> 100 μM

Figure 1. Structures of several unusual sponge (*Mycale micracanthoxea*) lipids and their activity in the CB1 receptor binding assay. Structures and binding affinity data are shown for five of the thirteen sponge compounds tested.

SCREENING EXTRACTS OF MARINE ALGAE FOR NATURAL PRODUCTS WITH CB1 AFFINITY

Over the past 10 years, our research group in Oregon has amassed approximately 1000 marine algal extracts from all over the world, concentrating on the West Coast of the US and the Caribbean. These are predominately lipid soluble extracts produced with a standard CH_2Cl_2/MeOH extraction protocol. Marine cyanobacterial extracts, which number approximately 130 in our collection, have been additionally extracted for more polar materials (butanol partition of the 70% aqueous MeOH solubles). Due the wealth of known unusual lipids from marine algae, we have undertaken a comprehensive evaluation of our extracts for substances which bind to the CB1 receptor. To date, we have evaluated 86 marine algal extracts with representatives from all of the major taxonomic groups (Figure 2). A total of 10 of these have shown good activity (>40% inhibition of the specific binding component of $[^3H]$CP-55940 to the CB1 receptor preparation at 50 ng of extract/μl of assay volume). The highest incidence of active extracts has been from the green algae (Chlorophyta, 23%), followed by the brown (Phaeophyceae) and blue-green algae (cyanobacteria) with 14% each. The least productive group has been the red algae (Rhodophyta) with only 1 active extract (4%).

We have initiated activity directed isolation efforts for two of these lead extracts; one brown alga *Laminaria angustata* from Northern Japan and one blue-green alga *Lyngbya majuscula* from the Caribbean. In the case of our work with *L. angustata*, we have been led to a very minor constituent which must necessarily bind with very high affinity (2 mg from 1 gm of extract). Based on its chromatographic characteristics, it is substantially less polar than anandamide. Isolation and structure elucidation efforts are continuing and will hopefully define a new ligand for this receptor in the near future.

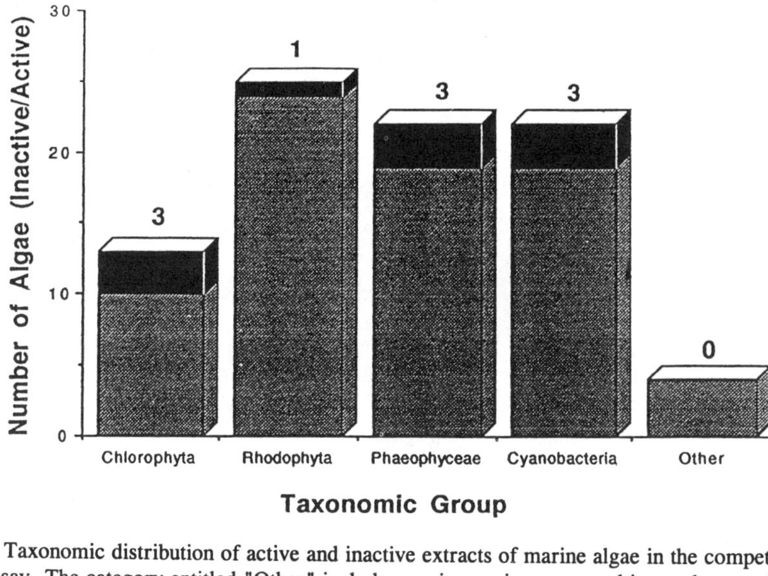

Figure 2. Taxonomic distribution of active and inactive extracts of marine algae in the competitive receptor binding assay. The category entitled "Other" includes marine angiosperms and invertebrate animals.

PRODUCTION, ISOLATION AND CB1-BINDING ACTIVITY OF OXIDIZED ANANDAMIDE DERIVATIVES

Autoxidation conditions were employed to randomly oxidize a small quantity of anandamide to a variety of hydroperoxide derivatives. A portion was reduced with triphenylphosphine and then unreacted anandamide was removed from both the hydroperoxide and hydroxy derivatives by preparative TLC. Both of these preparations showed the presence of high affinity ligands to the CB1 receptor. Consequently, the hydroxy derivatives were produced on a preparative scale and separated without further derivatization by HPLC (Figure 3). A combination of GC-MS, with and without derivatization to TMS ethers, and high field ^{1}H NMR analysis including ^{1}H-^{1}H COSY analysis, have been utilized to identify five of these oxidized products. As previously reported[5,6], the 15-hydroxy derivative was active while the 11-hydroxy derivative was less so. New in this study is the finding that either the 8-hydroxy, the 9-hydroxy (or both) is active (currently tested and analyzed as a mixture). The 5-hydroxy analog, a new compound described in this work, was relatively inactive. It should be noted that in all cases our hydroxyanandamides are racemic and the potential exists that only one of the two enantiomers is active in CB1 receptor binding. This will be the subject of continuing efforts in this area of investigation.

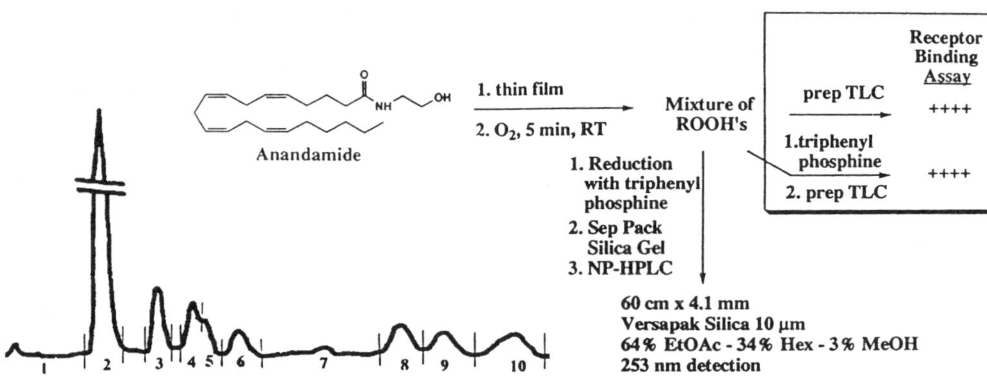

Figure 3. Method for production and HPLC separation of various oxidized anandamide derivatives. Peaks 2 (anandamide, $K_i = 0.51\ \mu M$), 3 (15-hydroxyanandamide, $K_i = 0.74\ \mu M$), 6 (mixture of 8- and 9-hydroxyanandamide, $K_i = 1.83\ \mu M$) and 9 (unknown, $K_i = 2.12\ \mu M$) were active in the CB1 equilibrium competition assay. Fraction 8, a relatively inactive fraction ($K_i = 8.24\ \mu M$) was identified as the 5-hydroxy derivative.

CONCLUSIONS

With its wealth of unusual life forms, the marine environment holds tremendous promise as a rich source of new and pharmacologically active natural products. Our screening for CB1-binding compounds from sponge and algal extracts and natural products supports this contention. Oxidized anandamide products, potential metabolites of anandamide in mammalian systems, has been another source of novel CB1 ligands in this study. From the preliminary efforts to date, the 15-hydroxy compound is most potent followed by either the 8- or 9-hydroxy compound (or both), the 11-hydroxy derivative, and leaving the 5-hydroxy compound as the lowest affinity ligand in the series.

ACKNOWLEDGMENTS

We acknowledge support of the Human Frontiers in Science Program (RG-26/95) for support of this project.

REFERENCES

1. D.A. Devane, L. Hanus, A. Breuer, R.G. Pertwee, L.A. Stevenson., G. Griffin, D. Gibson, A. Mandelbaum, A. Etinger, and R. Mechoulam, Isolation and structure of a brain constituent that binds to the cannabinoid receptor, *Science* 258:1946-1949 (1992).
2. V. Di Marzo, and A. Fontana, Anandamide, an endogenous cannabinomimetic eicosanoid: "Killing two birds with one stone", in: *Prostaglandins, Leukotrienes and Essential Fatty Acids* 53:1-11 (1995).
3. M.L. Wise, K. Soderstrom, T.F. Murray, and W.H. Gerwick, Synthesis and cannabinoid receptor binding activity of conjugated triene anandamide, a novel eicosanoid, *Experientia* 52:88-92 (1995).
4. J.C. Pinto, F. Potie, K.C. Rice, D. Boring, M.R. Johnson, D.M. Evans, G.H. Wilken, C.H. Cantrell, and A.C. Howlett, Cannabinoid receptor binding activity of amides and esters of arachidonic acid, *Mol. Pharmacol.* 46:516-522 (1994).
5. N. Ueda, K. Yamamoto, S. Yamamoto, T. Tokunaga, E. Shirakawa, H. Shinkai, M. Ogawa, T. Sato, I. Kudao, K. Inoue, H. Takizawa, T. Nagano, M. Hirobe, N. Matsuki, H. Saito, Lipoxygenase-catalyzed oxygenation of arachidonylethanolamide, a cannabinoid receptor agonist, *Biochim. Biophys. Acta* 1254:127-134 (1995).
6. A.J. Hampson, W.A.G. Hill, M. San-Phillips, A. Makriyannis, E. Leung, R.M. Eglen, L.M. Bornheim, Anandamide hydroxylation by brain lipoxygenase: Metabolite structures and potencies at the cannabinoid receptor, *Biochim. Biophys. Acta* 1259:173-179 (1995).

ACCUMULATION OF PLATELETS AS A KEY MECHANISM OF HUMAN ERECTION. A SCINTIGRAPHIC STUDY IN PATIENTS WITH ERECTILE DYSFUNCTION RECEIVING INTRACAVERNOUS INJECTION OF PGE_1, PAPAVERINE/PHENTOLAMINE

Margarida Rodrigues,[1] Susanne Granegger,[1] John O'Grady,[2] Helmut Sinzinger,[1] and Walter Stackl,[3]

[1]Department of Nuclear Medicine, University of Vienna
[2]Department of Pharmacology, University of Vienna
[3]Ludwig Boltzmann Institute for Extracorporeal Lithotripsy and Endourology and Department of Urology, Rudolfstiftung Hospital, Vienna

INTRODUCTION

Recent advances in the functional anatomy, neurophysiology and pharmacology of penile erection have fostered a dramatic improvement in the understanding of the physiology of erection and the approaches to correct impotence. However, the complex mechanism of human erection still remains incompletely explained. In vivo animal studies in the Chacma baboon (Dormehl et al., 1984) suggested that an enhanced platelet activity exits in the penis during erection. In this study we evaluate whether a similar accumulation of platelets during erection also occurs in humans.

PATIENTS AND METHODS

Twenty patients, aged from 28 to 67 years, with erectile dysfunction for 0.5 to 20 years, were studied.

Autologous platelets were isolated, radiolabeled in vitro with 3.7 MBq ^{111}In-oxine (Sinzinger et al., 1984), and reinjected intravenously.

Dynamic scintigraphy (5 minutes) of the penile area was performed to determine penile platelet accumulation in the flaccid state. Then, 17 patients received 20 µg prostaglandin E_1 (PGE_1) (The Upjohn Company, Kalamazoo, Michigan, USA) and three patients 15 mg papaverine (Byk Österreich Pharma, Vienna, Austria) plus 0.5 mg phentolamine (Byk Österreich Pharma, Vienna, Austria), all given intracavernously.

To determine the role of plasmatic coagulation in vivo, 37 MBq ^{123}I-fibrinogen (Department of Chemistry, Research Center, Seibersdorf, Austria) were injected intravenously 30 minutes before PGE_1-administration.

For leukocyte imaging, autologous leukocytes were isolated, radiolabeled (Sinzinger and Granegger, 1988) with 3.7 MBq ^{111}In-oxine and reinjected intravenously 30 minutes before PGE_1 administration.

For blood pool imaging, 74 MBq ^{99m}Tc were injected intravenously at full erection.

Static images of the penile area (incline of 30°, 5 minutes each) were performed about 20 minutes after injection of the vasoactive substances, as well as after detumescence (20 minutes to 6 hours). The volume of the penis was estimated by single photon emission computerized tomography (SPECT).

For quantification, standards before and after injection of the radiopharmaceuticals as well blood samples counted for the ^{111}In and ^{99m}Tc peaks were measured under the gamma camera.

The study protocol was approved by an Ethical Committee. Informed consent prior to the studies was obtained from each patient, after full explanation of what was involved in the study.

RESULTS

The accumulation of platelets in the flaccid penis, before PGE_1- or papaverine/phentolamine application, did not differ from background, two to three hours after injection of autologous radiolabeled platelets.

In patients (n = 16) who responded with a full erection, a significant selective and immediate accumulation of platelets occured within one minute after injection, preceding the development of firm erection. During full erection the penile platelet concentration (Table 1) and the penile volume (by between 24 to 407%) increased. PGE_1-induced erection (n = 13) was associated with a comparable platelet accumulation as induced by papaverine/phentolamine (n = 3).

In patients without erectile response (n = 4) penile platelet concentration declined (Table 1).

Table 1. Results of platelet concentration

	PGE_1 n = 17	**Papaverine/phentolamine** n = 3
Full erection (n = 16)	platelet cc increase 16 to 137 % (n = 13)	platelet cc increase 50 to 124 % (n = 3)
Failure in erection (n = 4)	platelet cc decrease -11 to -5 % (n = 4)	

n - number of patients, cc - concentration

No selective accumulation of radiolabeled-leukocytes (n = 6) or radiolabeled-fibrinogen (n = 5) was monitored during full erection.

DISCUSSION

Effective treatment modalities for erectile failure are presently readily available (Godschalk et al., 1996). Intracavernosal agents such as papaverine, phentolamine and PGE_1 have been shown to be useful in eliciting penile erection by increasing arterial inflow and/or relaxing cavernosal smooth muscle (Krane et al., 1989; Lue and Tanagho, 1987). Papaverine increases arterial flow, venous resistance and sinusoidal relaxation, effects similar to those of neurostimulation. Phentolamine increases arterial flow but has no significant effect on venous resistance. When given in combination with papaverine, it potentiates the effect of papaverine (Lue and Tanagho, 1987). PGE_1 has gained outstanding importance within the last 8 years. Because of its advantage of providing the highest response rate with fewer adverse effects as compared to papaverine and phentolamine (Stackl et al, 1988) PGE_1 actually represents the drug of choice for the treatment of erectile failure (Godschalk et al., 1996; Porst, 1996). PGE_1 causes erection by smooth muscle relaxation in a receptor-dependent manner. Specific cell receptors for PGE_1 have been identified in platelets and smooth muscle cells (Schrör et al., 1990).

The demonstration of dynamic hypercoagulability developed in the penile blood, and the accompanying deposition of fibrin onto the endothelium of the deep penile artery and trabecular surface during erection sugested a possible role of platelets in the physiology of erection (Bornman et al., 1983). The Chacma baboon has close anatomical relationship with man (Hampton and Matthews, 1966). In vivo studies performed in the animal model demonstrated an enhanced platelet concentration with respect to red blood cells pooling during erection, and an entrapment of platelets after erection. It was suggested that this enhanced platelet activity during erection could contribute to the hypercoagulability found in the penile blood (Dormehl et al., 1984). The causes and development of hypercoagulability are multifactorial, including changes in platelet function (Bornman et al., 1983; Bornman et al., 1986; Franz and Coetzee, 1981). A possible relationship between hypercoagulability and thrombogenesis could be that the "relative stasis" and prolonged exposure of the endothelium to blood of the corpus cavernosum could trigger the transgression of a hypercoagulable into a thrombotic state. Another possible link suggested is related to the role of prostacyclin (PGI_2) and thromboxane (TXA_2) in haemostasis. TXA_2 is a potent vasoconstrictor and potentiates platelet adhesion and aggregation. An increase in TXA_2 as a result of platelet activation promotes the aggregation of platelets and may contribute to hypercoagulability (Bornman et al., 1986; Kim et al., 1990). Our findings of the lack of selective accumulation of radiolabeled-fibrinogen exclude a relevant influence of plasmatic coagulation. Leukocytes seem not to be involved as well. Increased platelet accumulation is not associated with increased blood flow. Selective penile accumulation of platelets occurs during PGE_1- or papaverine/phentolamine-induced erection. The extent of the platelet accumulation in the penis is comparable to that seen in thrombosis. These data suggest a role of platelets in the corpus cavernosum during erection. Accumulation of platelets in the penis was shown to precede the development of erection, which excludes the possibility that platelet accumulation is only the result of erection achievement. The relative increase of platelets indicates a primary and temporary local platelet activation, inducing reversible accumulation in the penis thus regulating blood flow and erectile response. Probably, a relative local increase of platelets and temporary activation may be sufficient to induce full erection. The local increased TXA_2-production reported (Kim et al., 1990) may also be due to the activation of platelets. Our study suggests that platelets play a major role in the mechanism of human erection, probably by temporary activation, thereby regulating venous outflow. The underlying mechanism causing platelet activation in the penis still remains to be elucidated.

References

Bornman, M.S., du Plessis, D.J., Coetzee, W.J.C., and Franz, R.C., 1986, Penile hypercoagulability during erection:a possible predisposing factor to ageing penile vascular changes, *Br J Urol.* 58: 98.

Bornman, M.S., du Plessis, D.J., Franz, R.C., and Coetzee, W.J.C., 1983, Penile hypercoagulability as a predisposing factor to ageing impotence, *South Afr Med J.* 63: 472.

Dormehl, I.C., Bornman, M.S., du Plessis, D.J., du Plessis, M., and Maree, M., 1984, The use of autologous ^{111}In-labelled platelets and scintigraphy to illustrate enhanced platelet activity during erection in the Chacma baboon, *Nucl Med.* 23: 331.

Franz, R.C., and Coetzee, W.J.C., 1981, Thrombelastographic diagnosis of hemostatic defects, *Surg Ann.* 13: 75.
Godschalk, M., Gheorghiu, D., Chen, J., Gary Katz, P., and Mulligan, T., 1996, Long-term efficacy of a new formulation of prostaglandin E1 as treatment for erectile failure, *J Urol.*155: 915.
Hampton, J.W., and Matthews, C., 1966, Similarities between baboon and human blood clotting, *J Appl Physiol.* 21: 1713.
Kim, S.C., Choi, I.G., Oh, C.H., and Cha, Y.J., 1990, Prostacyclin-to-thromboxane A_2 ratio in arteriogenic impotence, *J Urol.* 144: 1373.
Krane, R.J., Goldstein, I., and Saenz de Tejeda, I., 1989, Impotence, *New Engl J Med.* 321: 1648.
Lue, F.T., and Tanagho, E.A., 1987, Physiology of erection and pharmacological management of impotence, *J Urol.* 137: 829 - 836.
Porst, H., 1996, The rationale for prostaglandin E1 in erectile failure: a survey of worldwide experience, *J Urol.* 155: 802.
Schrör, K., and Smith, E.F., 1990, *Dictionary of Prostaglandins*, Medikon-Verlag, München.
Sinzinger, H., and Granegger, S., 1988, The effect of various antibiotics on the labelling efficiency of human white blood cells with ^{111}In-oxine, *Nucl Med Comm.* 9: 597.
Sinzinger, H., Kolbe, H., Strobl-Jäger, E. and Höfer, R., 1984, A simple and safe technique for sterile autologous platelet labelling using "Monovette" vials, *Eur J Nucl Med.* 9: 320.
Stackl, W., Hasun, D., and Marberger, M., 1988, Intracavernous injection of prostaglandin E_1 in impotent men, *J Urol.* 140: 66 - 68.

CAN SELF-INJECTION THERAPY CURE IMPOTENCE ?

Ira D.Sharlip

Pan Pacific Urology, San Francisco, California, USA

INTRODUCTION

„Cure“ of impotence, sometimes defined as the spontaneous return of natural erections sufficient for satisfactory function in at least 75% of sexual attempts, has been reported to occur in men who use intracavernous injections of papaverine, phentolamine and/or prostaglandin E_1 for erection induction. A review of the urological literature of the past 10 years revealed 14 publications in which this phenomenon was noted. Of the total of 2687 patients in these 14 reports, 2-37% of patients achieved natural erections and no longer required any injections and an additional 1-50% reduced the frequency or dose of injections needed to achieve satisfactory sexual function.

Why does this phenomenon occur? Does self-injection therapy cause only a psychological effect or are there physiologic and anatomical changes which result in improved erectile function? What are the mechnisms by which improvement in natural erectile function might occur during intracavernous self-injection therapy ?

POSSIBLE MECHANISMS OF IMPROVED ERECTILE FUNCTION

There are several possible mechanisms by which improvement in erectile function may occur during intracavernous self-injection therapy.

These mechanisms include :

1. Psychological benefits, e.g. relief of performance anxiety
2. Improved hemodynamic responses resulting from :
   - mechanical dilatation of cavernous arteries
   - mechanical stretching of cavernous and tunical tissues with subsequent improvement of corporovenous occlusive action from improved tissue compliance
3. Angiogenesis and neovascularization of cavernous tissue induced by prostaglandin injection
4. Improved oxygenation of cavernous tissue during induced erection, reducing collagenization and fibrosis and/or producing more efficient biochemical responses to sexual stimulation
5. Hypertrophy of cavernous smooth muscle, resulting in enhanced efficiency of cavernous physiology
6. The waxing and waning natural history of erectile dysfunction, a common observation among impotence experts

Recent Advances in Prostaglandin, Thromboxane, and Leukotriene Research
Edited by Sinzinger *et al.*, Plenum Press, New York, 1998

LABORATORY RESEARCH

Data available from laboratory research form the basis for postulating that smooth muscle hypertrophy or neovascularity induced by prostaglandin E_1 injections contributes to improvement in erectile function from self-injection therapy.

Two studies have shown that smooth muscle hypertrophy may develop in the corpora cavernosa of monkeys given intracavernous injections. Aboseif, Breza, Bosch et al (1989) found that marked smooth muscle hypertrophy occured with papaverine while only minimal hypertrophy occurred with prostaglandin. The hypertrophic response to papaverine was also associated with smooth muscle atrophy and fibrosis in other areas of the corpus cavernosum, while normal muscle architecture was preserved with prostaglandin injections. Hwang, Yang, Ho and Chu (1991) found that papaverine injections produced mitochondrial aggregation and cell atrophy while prostaglandin injection produced smooth muscle hypertrophy but less overall histopathological changes. While these two studies found somewhat different results, both identified that smooth muscle hypertrophy may result from intracavernous injections of vasoactive substances. These papers suggest that smooth muscle hypertrophy may be a factor in producing improved natural erections following chronic self-injection therapy.

In another pertinent research study, Spisni, Manica and Tomasi (1992) investigated the incorporation of radiolabelled sulfate into the capillaries of chick embryo chorioallantoic membranes. They found that prostaglandin E_1 administration significantly increased the number of capillaries in their preparations. This study supports the concept that prostaglandin injection may produce improved natural erections by inducing angiogenesis and neovascularity of the cavernous tissue.

CLINICAL RESEARCH STUDIES

Several clinical research studies are pertinent to the phenomenon of improved natural erectile function with self-injection therapy.

The largest study of the long-term effects of self-injection therapy is that of Virag, Shoukry, Floresco et al (1991). These authors reported on 533 patients followed while on intracavernous injection therapy with papaverine alone, a combination of papaverine with alpha blockers, or a six drug combination. Of the total group, 27.5% were thought to have psychogenic impotence, 25.2% were thought to have organic impotence and 47.3% were thougth to have mixed psychogenic and organic impotence. Of the entire group of patients followed while on therapy, 7.8% stopped using intracavernous injections and were considered „cured" and 7.2% maintained good natural erectile function by monthly „maintenance" intracavernous injection. Another 49% of the patients were partially dependent on self-injections therapy, requiring only occasional injections to maintain adequate sexual function. It is notable that the percentage of psychogenic etiology in the cured group was higher than in the total group (39.3% versus 27.5%), the percentage of organic etiology was lower in the cured group than in the total group (18% versus 25.2%) and the age of the cured group was younger than the total group (38 years versus 48 years). These data suggest weakly that „cure" following self-injection therapy is more likely to occur in younger men with a psychogenic etiology.

Marshall, Breza and Lue (1994) have developed clinical evidence which supports the concept that improved natural erections with self-injection therapy are due to improved penile hemodynamics. They compared 35 patients with arteriogenic and/or venogenic impotence treated with intracavernous injections of prostaglandin alone or a combination of prostaglandin, phentolamine and papaverine to 10 patients with arteriogenic and/or venogenic impotence who were not treated at all. They found that self-injection therapy resulted in a very significant increase in the cavernous artery peak flow velocity in the treatment group, even though there was no difference in cavernous artery diameter between the two groups. Simultaneous with the increased arterial peak flow velocity, 13 of the 35 subjects (37%) reported that they experienced good functional natural erections without the use of intracavernous injections and 19 of the 35 subjects (54%) experienced an improvement in the quality or quantity of their erections achieved without injection. None of the untreated control group experienced functional natural erections. While this study suggests that the development of functional natural erections could be attributed to the improvement in penile hemodynamics, it should be clearly noted that no difference in the arterial peak flow velocity was present in the 13 patients who experienced functional natural erections compared to the 22 patients who did not experience any functional natural erections.

A large retrospective study of patients who claimed „cure" of impotence from self-injection therapy was performed by Basile, Goldstein, Malkevich, et al (1994). They sent questionnaires to 921 patients on chronic self-injection therapy and received responses from 387 patients. The responses occurred after an average of about 13 months of self-injection therapy and an average of 55 injections. These investigators found that 6% no longer needed self-injection therapy, 15% needed injections less often, 17% needed a smaller dose and 22% experienced improved quality of natural or spontaneous erections. They found that the improvement in erectile function correlated to younger age but not to any other clinical parameters, including blood pressure, cholesterol levels, smoking history, or the presence or absence of heart disease and diabetes.

A very well-designed study on the return of spontaneous erections with prostaglandin E_1 self-injection by McMahon (1992) addressed the psychological aspect of this phenomenon. McMahon compared 153 self-injection treated men with arteriogenic, venogenic and/or psychogenic impotence to 53 untreated controls with the same etiologies of impotence. He found that 9% of the treated men with vasculogenic impotence experienced spontaneous natural erections 6 months after initiation of self-injection therapy while 10-12% of the untreated men with vasculogenic impotence also experienced spontaneous erections when evaluated 6 months later. However, there was a major difference due to treatment in the men with psychogenic impotence. Of the 37 men with psychogenic impotence who received self-injection therapy for 6 months, 84% experienced spontaneous natural erections. Of the 18 men with psychogenic impotence who were not treated, 22% experienced spontaneous erections. McMahon concluded that the phenomenon of „cure" of impotence by self-injection therapy occurs predominantly in patients with psychogenic impotence.

CONCLUSION

From this review of the literature, it is not possible to identify definitively the mechanism by which cure of impotence occurs from self-injection therapy. The average incidence of such „cure" seems to be about 10-15%. The most plausible explanations for this phenomenon seem to be the psychological benefits of this therapy and the waxing and waning natural history of erectile dysfunction. Anatomic, physiologic and/or pathologic explanations for „cure" of impotence from self-injection therapy probably occur infrequently.

REFERENCES

1. Aboseif, S.R., Breza, J., Bosch, R.J., et al., 1989, Local and systemic effects of chronic intracavernous injection of papaverine, prostaglandin E_1, and saline in primates, J.Urol. 142:403.
2. Basile, G., Goldstein, J., Malkevich, D., et al., 1994, Patient characteristics of those who claim „cure" from penile injection therapy, Int.J.Impotence Res. 6(Suppl):D118.
3. Hwang, T.I., Yang, C.R., Ho, W.L., and Chu, H.W., 1991, Histopathological change of corpora cavernosa after long-term intracavernous injection, Eur.Urol. 20:301.
4. Marshall, G.A., Breza, J., and Lue, T.F., 1994, Improved hemodynamic response after long-term intracavernous injection for impotence, Urology 43:844.

5. McMahon, C.G., 1992, The return of spontaneous erections after self-injection of prostaglandin E_1, Int.J.Impotence Re. 4:179.
6. Spisni, E., Manica, F., and Tomasi, V., 1992, Involvement of prostanoids in the regulation of angiogenesis by polypeptide growth factors, Prostaglandins Leukot.Essent.Fatty Acids 47:111.
7. Virag, R., Shoukry, K., Floresco, J., et al., 1991, Intracavernous self-injection of vasoactive drugs in the treatment of impotence: 8-year experience with 615 cases, J.Urol. 145:287.

CYTOKINE LEVELS IN FIVE DIFFERENT FLUID COMPARTMENTS DURING AMNIOTIC FLUID INFECTION AND LABOUR

Gundula Hebisch, MD
Department of Obstetrics and Gynaecology
University Hospital
CH-8091 Zurich
Switzerland

INTRODUCTION

There is growing evidence in the recent literature that a large variety of cytokines is involved in the establishment and maintenance of pregnancy (Chen et al,1991; Shorter et al, 1992, Yoon, 1995). In premature delivery, IL-1 and TNF-alpha are considered as stimuli for prostaglandine synthesis and induction of labour via uterine contractions; in chorioamnionitis, these cytokines are also present. IL-6 has been found in amniotic fluid in the third trimester, term labour or amniotic fluid syndrome (AIS), whereas low or nondetectable levels were described in the second trimester or in women not in labour at term. Other authors detected IL-8, TNF-a, G-CSF or GM-CSF for these conditions. In vitro, additional studies were performed using immunohistochemistry on cultures of amniotic membranes or placental and fetal tissues (post abortum or post mortem) showing an expression of IL-1, IL-6, IL-8 and TNF-a in cultured placental tissues (Harty et al,1992; Shorter et al,1992). Little, however, was known about the in vivo kinetics of cytokine production and metabolism, about the interaction of neighbouring compartments as well as the cellular origin of these substances.

MULTICOMPARTMENT MODEL

We therefore developed an in vivo model that allowed us to study several fluid compartments in comparison with placental and placental bed tissues representing deeper and more diagnostic layers of the implantation site. At caesarean section, we nearly simultaneously collected amniotic fluid (AF), umbilical artery (UA) and vein (UV) blood, retroplacental (RB) and maternal (MB) venous blood. In addition, we performed a superficial, but a wide area covering biopsy from the placental bed using a Bumm curette. Fluid samples were centrifuged for 10 mins at 5000 rpm and the supernatants stored at -70°C until analysis. Cytokine levels were determined using conventional enzyme immunoassays (R+D Systems, Minneapolis: Quantikine human G-CSF and from Endogen Inc., Boston: human TNF-α ELISA Kit). Placenta and placental bed biopsy were stained immunohistochemically. In a pilot study, we had tested the study design on 25 patients, and venous blood levels been determined on 100 healthy non-pregnant individuals to serve as non-pregnant controls.

CYTOKINE EXPRESSION PATTERNS IN THE FETO-MATERNAL UNIT

104 women with singleton pregnancies delivered by caesarean section were

included in the study (AIS n=21, labour without infection n=21, preeclampsia n=34, IUGR n=11, healthy controls at term not in labour n=17). Exclusion criteria were multiple pregnancies, other infections such as toxoplasmosis, CMV, HIV or hepatitis, diabetes, maternal tumours as well as fetal malformations which all might interfere with the conditions under study by altered cytokine production.

RESULTS

The highest cytokine levels were found in AF, followed by the RB. The correlation of AF and RB was r = 0.86. Levels within MB and fetal UA and UV and were similar and low, except some during labour or fetal infection/AIS. The highest cytokine expression was that of IL-8, followed by IL-6. As expected, immunohistochemical staining was possible in various tissues of the feto-placento-maternal unit.

TNF-a

No significant differences to non-pregnant values were demonstrated within all compartments of controls nor during labour. During AIS, AF levels increased x 60 and those of the mother x 10 whereas fetal levels remained constant. The differences were highly significant in AF between AIS and controls ($p<0.0001$) or labour and controls ($p\leq 0.001$), and significant between AIS or labour and controls ($p<0.01$). Immunohistochemistry showed positivity at the cytotrophoblast, i.e. the germinal layer of the trophoblastic covering of the chorionic villi in the vicinity of fibrinoid necrosis, and within cells of the so-called islets and septa within the intervillous space. With chorioamnionitis, macrophages linearly arranged within the extraplacental membranes were stained.

IL-1

IL-1 was not detected in MB or UV. AF and RB contained moderate amounts of IL-1 at about the levels of non-pregnant controls. Within MB, IL-1 did not rise, even if AF or RB levels were maximally increased during AIS (x 20; $p<0.0001$/ x 10; $p\leq 0.001$). In labour, changes to the norm were moderate (x 2.5 increase; $p<0.05$). Immunolocalisation was positive for the same cells as for TNF-a (from degenerating chorionic villi), as well as for macrophages of the basal plate of the placenta during AIS.

IL-6

In normal pregnancy at term and in non-pregnant controls, IL-6 was not detectable or in little quantities only within AU, AV and MB. Like with the other cytokines, elevated levels were found within AF and RB. During labour, IL-6 increased x 50 within AF, x 40 within MB and x 8 in RB, but not within the fetus. The increase in RB seemed to be dependent on cervical dilatation, an observation which will be evaluated in further studies. During AIS, the levels increased even more ($p < 0.0001$), and also involved the fetal blood compartments. During AIS, maternal levels were lower than in advanced labour; in RB and within the fetus there was a considerable increase compared to the control patients.

Il-8

Low levels were found in UA, UV and MB in controls, with a significant increase in UA and UV blood during AIS (UA > UV; ($p<0.0001$, and ≤ 0.001, respectively). In RB and AF, IL-8 showed high levels, with an increase x 30/x 50 in AIS though not statistically significant. Within the fetus, the increase was 70-fold, but not within MB. The increase of IL-8 levels in labour was small (n.s.). Immunolocalisation was strongly correlated to granulocytes, macrophages and haematopoietic cells.

G-CSF

G-CSF-levels in the blood of mother and fetus did not differ significantly from each other; yet, they were slightly higher than in non-pregnant controls. RB showed similar levels; only AF levels were x 30 higher than those in blood. In labour, levels were only increased in AF (x 20, $p<0.05$). During AIS, the highest increase rates were found in RB (x 500; $p<0.0001$) which were, however, not reflected in MB, followed by a 100fold increase within UA ($p<0.0001$) and an 65-fold increase in AF ($p<0.0001$). Immunolocalisation of G-CSF was positive for decidual cells (placental macrophages) within the chorionic villi, with an additional diffuse staining of the villous matrix in AIS.

GM-CSF

GM-CSF had not been detected in normal non-pregnant adults; in pregnancy it was not detected in any of the examined compartments with the exception of AF. These levels did not change significantly during labour or under infectious conditions. Immunolocalisation has not yet been attempted.

Labour

During labour, TNF, IL-1 and GM-CSF did not differ significantly to non-pregnant controls. IL-8 was increased in AF and MB, IL-6 significantly increased within AF and MB, but not within the fetus. There are obvious differences between IL-6 levels within retroplacental blood of patients with contractions causing, and those not causing cervical dilatation. G-CSF was significantly increased within AF.

AIS

With the exception of GM-CSF, all of the cytokines examined showed a significant rise during AIS. In AF this was true for TNF, IL-1, IL-6, IL-8 und G-CSF, within fetal and MB IL-6 and G-CSF. IL-8 was increased within fetal, but not maternal blood. For the detection of an AIS IL-6 in MB showed a sensitivity of 100%, specificity 49%, positive predictive value (PPV) 30% and negative predictive value (NPV) 100%; for AF IL-6, however, the values were 92%, 87%, 61% and 98%, respectively, in combination with elevated CRP > 20 mg/l 67%, 97%, 80% and 95%, respectively, whereas the combination of IL-6 and IL-8 showed a values of 90%, 91%, 59% and 99%, respectively.

Preeclampsia and IUGR

Other than expected, cytokine levels did not rise (e.g. TNF-α) during IUGR or preeclampsia, levels were even lower than in controls. Solely IL-8 showed a slight increase in MB during preeclampsia which was not yet statistically significant.

DISCUSSION

The interaction of fetally and maternally derived cells during pregnancy is accompanied by a heavy cytokine release into the different compartments (Cross et al,1994). The highest concentrations of cytokines are found within AF, which is mainly composed of fetal urine but also of secretions from cells of the amniotic epithelium lining the amniotic cavity. After swallowing by thew fetus, substances within AF can pass the intestinum to finally reach fetal blood as well as fetal urine, which again is excreted into AF. Eventually, fetal cytokines might reach maternal peripheral blood. Since IL-1 does not rise within MB, even if AF or RB levels are maximally increased during AIS, there might be a high capacity to degradate this cytokine. The same might be true for IL-6 and G-CSF. Also, IL-8 shows a higher increase in the UA than in the UV, which might indicate production sites within the fetus, whereas maternal levels in AIS remain nearly unchanged. Immunohistochemical staining patterns of various cells and tissues do depict the presence of the cytokines in various conditions, a quantification, however, is so far not possible.

CONCLUSION

Using our multicompartment model, we were able to confirm the findings of other authors detecting high cytokine levels within AF (Hillier et al,1993; Romero,1990; Saito, 1992; Shorter,1992;Yoon,1995). Maternal and fetal blood levels usually do not exceed 1/10 of AF levels. Within venous blood of pregnant controls, cytokines are low or not detectable; moderate to high levels are, however, found in RB. We therefore conclude that the origin of these short-lived cytokines must be close to the placenta or its secondaries and that the maternal organism has a high potential to rapidly degradate them on the first passage. Thus, only peak levels are found within MB. For practical purposes, easily accessible MB does not seem to be useful as a screening tool for the cytokines examined with the possible exception of IL-6 and/or IL-8. Our results also confirm the hypothesis of a complex network function of cytokines even in human pregnancy and permit conclusions concerning production sites. A deeper understanding of the function sites could possibly form the basis for new diagnostic and/ or therapeutic approaches for AIS or premature contractions. The cytokines under study, especially IL-6 and IL-8, alone or in combination with CRP or other known infection parameters, seem to be valuable parameters in the detection of an AIS, the clinical value of which remains to be evaluated.

REFERENCES

Chen, H., Yang, Y., Hu, X., Yelavarthi, K.K., Fishback, J.L., Hunt., J.S., 1991, Tumor necrosis factor alpha mRNA and protein are present in human placental and uterine cells at early and late stages of gestation, Am J Pathol 139:327-35.

Cross, J.C., Werb, Z., Fisher, J.S, 1994, Implantation and the placenta: key pieces of the development puzzle, Science 266:1508-18.

Harty, J.R., 1992, Interleukin 1-beta stimulates colony-stimulating factor-1 production in placental villous core mesenchymal cells, J Clin Endocrinol Metab 75:947-50.

Hillier, S.L., Witkin, S.S., Krohn, M.A., Watts, D.H., Kiviat, N.B., Eschenbach, D.A., 1993, The relationship of amniotic fluid cytokines and preterm delivery, amniotic fluid infection, histologic chorioamnionitis, and choramnion infection, Obstet Gynecol 81:941-8.

Orvieto, R., Ben-Rafael, Z., 1994, The role of cytokines in early detection of preeclampsia. Med Hypotheses 43:315-18.

Romero, R. Avila, C., Santhanam, U., Sehgal, P.B. 1990, Amniotic fluid interleukin 6 in preterm labor. Association with infection. J Clin Invest 85:1392-400.

Saito, S., Kasahara, T., Kato, Y., Ishihara, Y., Ichijo, M.,1992, Elevation of amniotic fluid interleukin 6 (IL-6), IL-8 and granulocyte colony stimulating factor (G-CSF) in term and preterm parturition. Cytokine 5:81-8.

Shorter, S.C., Vince, G.S., Starkey, P.M., 1992, Production of granulocyte colony-stimulating factor at the materno-foetal interface in human pregnancy. Immunology 75:468-74.

Stallmach, T., Karoly, L., 1994, Augmentation of fetal granulopoiesis with chorioamnionitis during the second trimester of gestation. Hum Pathol 25:244-7.

Stallmach, T., Hebisch, G., Joller, H.I., Orban, P., Schwaller, J., Engelmann, M., 1995, Cytokine production and visualised effects in the feto-maternal unit: quantitative and topographic data on cytokines during intrauterine disease. Lab Invest 73:384-92.

Stallmach, T., Hebisch, G., Joller, H., Kolditz, P., Engelmann, M., 1995, Expression Pattern of cytokines in the different compartments of the feto-maternal unit under various conditions. Reprod Fertil Dev 7:1573-80.

Yoon, B., Romero, R., Kim, C.J., Jun, J.K., Gomez, R., Choi, J.H., Syn, H.C., 1995, Amniotic fluid interleukin 6: a sensitive test for antenatal diagnosis of acute inflammatory lesions of preterm placenta and prediction of perinatal morbidity. Am J Obstet Gynecol 172:960-70.

Characterization of A-93178, an Iminoxy-Quinoline Inhibitor of Leukotriene Biosynthesis

R. L. Bell, R. R. Harris, J. B. Bouska, K. I. Hulkower, J. Moore, P. Bhatia, P. E. Malo, T. Kolasa, C. D. W. Brooks and G. W. Carter.

Immunosciences Research Area, Dept. 47 K, 100 N Abbott Park Road, Abbott Laboratories, Abbott Park, Illinois, U.S.A. 60064

Introduction

In recent years a number of leukotriene modulators have shown promise in the treatment of asthma.[1,2] Most of these have been Cys-LT_1 antagonists.[2] Inhibitors of both cysteinyl leukotrienes and LTB_4 however are likely to be more effective given the pro-inflammatory characteristics of LTB_4. In addition, these inhibitors would also inhibit the formation of the mediators formed from double lipoxygenation such as lipoxins.[3] Thus far, most bioavailable inhibitors of 5-lipoxygenase have been related to zileuton or ZD-2138.[4] An alternative to 5-lipoxygenase inhibition is the antagonism of the 5-lipoxygenase activating protein (FLAP).[5] Three compounds with this mechanism have reached the clinic but all three have apparently been withdrawn.[4] We have synthesized a number of leukotriene inhibitors that are not direct inhibitors of 5-lipoxygenase with optimized pharmacokinetic and pharmacological characteristics. The best of these is A-93178.

Results

We have recently observed that the addition of an oxime or iminoxy group in the carboxyl region of an indole series of leukotriene biosynthesis inhibitors gave a modest improvement in *in vitro* potency and significantly improved bioavailability compared with the free acid congener.[6] A similar modification of BAY X 1005 with the additional replacement of the cyclopentyl for a cyclohexyl ring gave A-93176 (figure 1). This compound was racemic and resolution yielded the R and S enantiomers A-93177 and A-93178.[7] *In vitro* these compounds were potent inhibitors of ionophore A23187 induced LTB_4 formation from neutrophils. A-93178 was the most potent giving an IC_{50} of 15 nM, 7 fold lower than A-93177 and 4 fold lower than Bay X1005.[8]

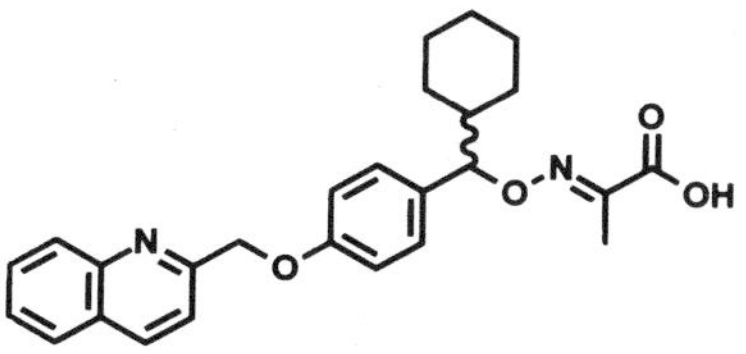

A-93176

Figure 1. Structure of A-93176.

In vitro metabolism studies performed as described[9] indicated that addition of the iminoxy moiety had a dramatic effect on the glucuronidation rate of the carboxyl groups on the compounds. BAY X 1005 gave a glucuronidation rate of 8.1 nmole/min/mg protein in a dog liver microsome preparation while the rate for A-93178 was nearly undetectable (0.05). These differences translated to a significant improvement in the bioavailability of A-93178 versus BAY X 1005 (figure 2). The intravenous half-life in the dog was 0.4 hour for BAY X 1005 and 5 hours for A-93178.

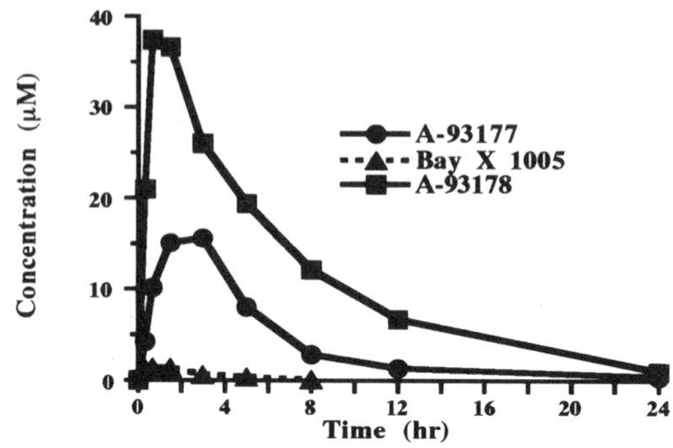

Figure 2. Pharmacokinetic profile of A-93178, A-93177 and BAY X 1005 after oral doing in beagle dogs. Animals were dosed at 5 mg/kg orally, plasma samples taken at times indicated and analyzed for drug concentration.

A-93178 was also found to be a potent inhibitor of leukotriene biosynthesis *in vivo*. In the rat the compound gave an oral ED_{50} of 0.26 mg/kg, 16 fold more potent than BAY X 1005 and 12 fold more potent than zileuton.[10]

A-93178 was also examined in antigen-induced bronchospasm in actively sensitized guinea pigs as described by Malo et al.[11] The compound was a potent inhibitor of the bronchoconstriction induced by aerosolized antigen in the presence of an anti-histamine. The oral ED_{50} was approximately 1 mg/kg, with nearly complete inhibition seen at between 3 and 10 mg/kg. The compound is thus more potent than Bay X 1005, MK-0591 and zileuton in this assay.

The compound was also an extremely potent inhibitor of the leukotriene formation and eosinophil influx induced by intravenous sephadex particles in Brown Norway rats. As

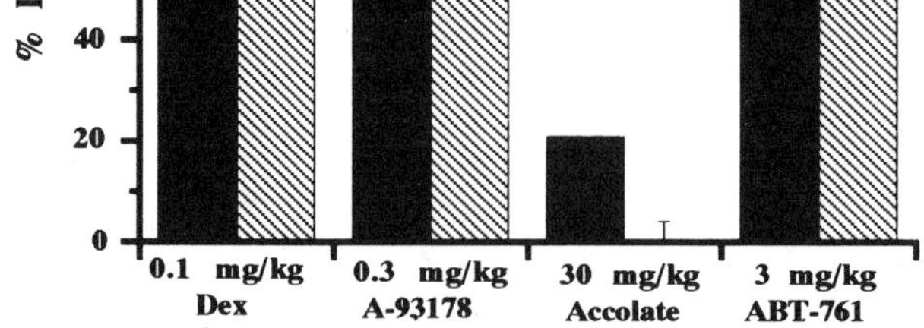

Figure 3. Comparison of the inhibition of leukotrienes and eosinophil influx in Sephadex treated animals dosed with dexamethasone, the sodium salt of A-93178, Accolate (ZD 204,219) and ABT-761. Drugs were given by oral gavage 2 hours prior to Sephadex injection and every 12 hours thereafter through the 4 day time course. Each dose level and the control is the mean ± SEM of 8 animals.

described by Namovic et al.,[12] this model is extremely sensitive to leukotriene inhibitors. The oral ED_{50} against eosinophil influx into the bronchoalveolar lavage (BAL) fluid for the compound was approximately 0.15 mg/kg. Significant inhibition of eosinophil number was seen at all doses above 0.1 mg/kg and leukotriene biosynthesis was inhibited significantly at 0.03 mg/kg. As shown in figure 4, this potency is greater than for ABT-761 and is in the range seen for dexamethasone.

Discussion

The addition of the iminoxy group to the quinoline alkoxy structure described by others was found to give compounds with remarkable improvement in a number of pharmacokinetic and pharmacological characteristics. The addition of the iminoxy group proximal to the carboxyl had a pronounced inhibitory effect on the rate of glucuronidation seen *in vitro*. This decrease in metabolism was also seen *in vivo* in the dog where the plasma levels achieved after oral dosing were significantly greater for the iminoxy compounds. These improvements translated to remarkable increases in *in vivo* potency for A-93178 versus BAY X 1005. In both pulmonary function assays and against pulmonary inflammation, A-93178 was very potent. Interestingly, the compound was much more potent in the bronchospasm models than any other FLAP inhibitor we have examined. The data in the eosinophil influx model also indicated remarkable potency in the range of dexamethasone. In addition, the data with A-93178 confirmed our previous observations that leukotriene biosynthesis inhibitors are effective in this model but Cys-LT_1 antagonists are not.[12]

In conclusion, A-93178 is the most potent of a series of new leukotriene biosynthesis inhibitors. It should be a valuable tool in continuing to explore the role of leukotrienes in inflammation and allergy.

References

1. S. Holgate, P. Bradding and A. Sampson, Leukotriene antagonists and synthesis inhibitors: new directions in asthma therapy, *J. Allergy Clin. Immunol.* 98: 1 (1996).
2. R. Harris, G. Carter, R. Bell, J. Moore and C. Brooks, Clinical activity of leukotriene inhibitors, *Int. J. Immunopharmacology* 17: 147 (1995).
3. C. Serhan, Lipoxin biosynthesis and its impact in inflammatory and vascular events, *Biochim. Biophys. Acta* 1212: 1 (1994).
4. C. D. Brooks and J. B. Summers, Modulators of leukotriene biosynthesis and receptor activation, *J. Med Chem*. 39: 2629 (1996).
5. R. Dixon, R. Diehl, E. Opas, E. Rands, P. Vickers, J. Evans, J. Gillard, and D. Miller, Requirement of a 5-lipoxygenase activating protein for leukotriene biosynthesis, *Nature* 343: 282 (1990).
6. T. Kolasa, P. Bhatia, C. Brooks, H. Hulkower, J. Bouska, R. Harris, and R. Bell, Synthesis of indoylalkoxyiminoalkylcarboxylates as leukotriene biosynthesis inhibitors, *Bioorganic Med. Chem.* 5: 507 (1996).
7. T. Kolasa, D. Gunn, A. Stewart and C. D. Brooks, Synthesis and resolution of 2-(cyclohexyl-4-(2-quinolylmethoxy)phenyl)methoxyiminopropionic acid,leukotriene biosynthesis inhibitors, *Tetrahedron:Asymmetry* 7: 2645 (1996).
8. R. R. Muller-Peddinghaus, H. J. Fruchtmann, B. Ahr, K. Beckermann, B. Buhner, B. Fugmann, M. Junge, C. Matzke, S. Kohlksdorfer, P. Raddatz, P. Theisen-Popp and H. J. Mohrs, BAY X 1005, a new selective inhibitor of leukotriene biosynthesis: pharmacology and pharmacokinetics, *Lipid Mediators* 6:245 (1993).
9. R. Bell, R. Harris, P. Malo, J. Bouska, T. Shaughnessy, K. Hulkower, C. Brooks and G. Carter, ABT-761 attentuates bronchoconstriction and pulmonary inflammation in rodents, *J. Pharmacol. Exp. Therap*. 280: 1366 (1997).
10. G. W. Carter, P. Young, D. Albert, J. Bouska, R. Dyer, R. Bell, J. Summers and D. Brooks, 5-lipoxygenase inhibitory activity of zileuton, *J. Pharmacol. Exp. Ther.* 256:929 (1991)
11. P. Malo, R. Bell, T. Shaughnessy, J. Summers, C. Brooks and G. Carter, The 5-lipoxygenase inhibitory activity of zileuton in *in vitro* and *in vivo* models of antigen-induced airway anaphylaxis, *Pulmon. Pharmacol.* 7: 73 (1994)

12. M. Namovic, R. Walsh, C. Goodfellow, G. Carter, R. Harris, and R. Bell, Pharmacological modulation of eosinophil influx into the lungs of brown norway rats, *Eur. J. Pharmacol*. 315: 81 (1996).

TRANSCELLULAR SYNTHESIS OF CYS-LT: FROM ISOLATED CELLS TO COMPLEX ORGAN SYSTEM

Angelo Sala, Giuseppe Rossoni, Ferruccio Berti
Reiner Müller-Peddinghaus* and Giancarlo Folco

Center for Cardiopulmonary Pharmacology,
Inst. of Pharmacological Sciences,
University of Milan, Milan 20133 Italy

*and Pharma Research Center,
Bayer AG, Wuppertal, Germany,

INTRODUCTION

Cysteinyl-Leukotriene (Cys-LT) formation may proceed either through activation of a single cell type, possessing both 5-lipoxygenase (5-LO) and LTC_4-synthase enzymatic activities (direct synthesis) or it may involve the participation of different cell types whereby donor cells (i.e. polymorphonuclear leukocytes, PMNL), possessing 5-LO activity, synthesize and transfer the unstable intermediate LTA_4 to acceptor cells (i.e. endothelial cells, EC), possessing LTC_4-synthase activity only (transcellular synthesis)(1). This process suggests that the cellular environment (i.e. cell-cell interactions) is an important control in the production of eicosanoids.

Quantitative analysis of 5-lipoxygenase metabolites showed that intact LTA_4 represents the main product released by PMNL upon challenge with A23187 (2, 3). It is therefore clear that the transcellular synthesis of cys-LT may represent a quantitatively relevant phenomenon, resulting in substantial changes of the profile of 5-LO metabolites produced upon activation of a specific cellular specie.

A number of studies addressed the issue of transcellular synthesis of cys-LT using different combinations of isolated cell (4, 5). Fewer groups faced this topic using isolated organ systems perfused with PMNL (6). This latter approach, albeit more complex, represents the best conditions for PMNL-endothelial cell cooperations. Adhesion of activated PMNL to EC may indeed result in optimal transfer of PMNL-derived LTA_4 to endothelial cells.

During the last five years we developed and characterized a model of spontaneously beating, isolated rabbit heart, perfused with purified human

Recent Advances in Prostaglandin, Thromboxane, and Leukotriene Research
Edited by Sinzinger *et al.*, Plenum Press, New York, 1998

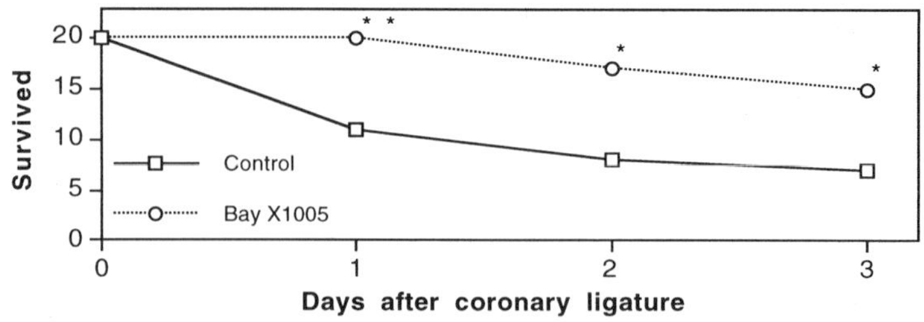

Fig. 1. Cardioprotective effect of BAY-X1005 in the rabbit after permanent coronary artery ligation. Mortality is reported in number of animals survived at a given time. **Chi-square p<0.01, *** p<0.001 vs control (coronary artery ligated) group.

PMNL, showing that PMNL activation caused cys-LT formation through PMNL-EC transcellular synthesis, resulting in increased coronary resistance and cardiac damage (7, 8). Under the working hypothesis that 5-LO inhibition may result in cardioprotection in cardiac ischemia, we studied the effect of a 5-LO inhibitor, (BAY X1005, 2 ter), both in the PMNL-perfused isolated rabbit heart and in an *in vivo* model of cardiac ischemia, induced by permanent ligation of the left descending coronary artery.

METHODS

Rabbit hearts were isolated and perfused retrogradely through the aorta according to Langendorff at constant flow. Coronary perfusion pressure (CPP), left ventricular pressure (LVP) and left ventricular end-diastolic pressure (LVEDP) were monitored continuously. Blood cells were prepared according to Haslett (9). PMNL ($5x10^6$ cells) were supplied with Ca^{++} (2 mM) and Mg^{++} (0.5 mM) and infused into the perfusing medium (50 ml) of recirculating hearts with or without pretreatment of the heart with BAY X1005 (1 μM). At the end of the experiment, the entire heart reservoir (approx. 45 ml) was collected, spiked with 50,000 dpm $^3H-LTD_4$ as well as 25 ng of PGB_2, extracted and analysed by RP-HPLC with diode-array UV-detection.

Male Albino New Zealand rabbits, under anaesthesia and artificial respiration, were opened on the left side of the thorax, the small branch of the left anterior descending coronary artery was separated from the myocardial tissue for a lenght of about 2-3 mm and ligated with a silk 6.0 suture, armed with an atraumatic needle. Mortality was recorded for the following 3 days. BAY X1005 treated animals received 16 mg/kg over a two hours period, starting one hour before ligation, and 8 mg/kg at 8, 24, 32 and 48 hours after ligation.

RESULTS

Challenge with A23187, 0.5 μM, of rabbit hearts perfused under recirculating conditions with human PMNL resulted in a marked increase in coronary perfusion pressure. The increase in CPP was slow in onset and very persistent, resulting in changes of CPP from 46±1.1 mm Hg to 176.2±29.7 mm Hg, 30' after A23187 challenge. These changes were significantly inhibited by pretreatment with BAY-X1005 1 μM (76.7±12.8 mmHg, 30' after A23187 challenge). The total amount of cys-LT formed in control conditions averaged 29.7 ±7.3 pmoles/ml, while the amount of LTB_4 was 3.3±0.5 pmoles/ml. Following pretreatment with compound BAY X1005, the amount of leukotrienes was markedly inhibited (cys-LT 3.2±1.7 pmoles/ml, LTB_4 0.75±0.21 pmoles/ml)

Permanent ligation of the left descending artery resulted in significant mortality over the three days of observation. Pretreatment with BAY-X1005 was able to significantly reduce mortality observed in the control group (Fig. 1).

DISCUSSION

Challenge of PMNL recirculating in a spontaneously beating, perfused rabbit heart, caused a marked and progressive increase in CPP that appeared to be largely dependent from the formation of cys-LT, as shown by the protection afforded by the pretreatment with the 5-LO inhibitor BAY-X1005. These data are in agreement with previously published data (7, 8) and provide further support for a role of cys-LT, locally formed through cooperation between PMNL and EC, in the control of coronary tone.

Pretreatment of rabbits with the specific 5-LO inhibitor BAY X1005 *in vivo*, also afforded significant protection against mortality induced by permanent ligation of the left descending coronary artery, suggesting a potential pathophysiologic role of leukotrienes in the development of cardiac damage following the permanent ischemia resulting from the ligation.

Taken together, our results suggest that 5-LO metabolites, possibly synthesized through the transcellular synthetic pathway involving PMNL-EC cooperation, may play a significant role in alteration of coronary flow and cardiac contractility. In agreement with this hypothesis, cys-LT appear to be released during episodes of myocardial ischemia, as indicated by enhanced urinary LTE_4 excretion (11).

Given the biochemical mechanism proposed for the biosynthesis of cys-LT, the pharmacological control of 5-LO may result of potential interest in those pathological conditions where leukocytes adhesion is enhanced, such as in presence of oxidised LDL.

REFERENCES

1. J. Maclouf, R.C. Murphy, and P. Henson. Transcellular sulfidopeptide leukotriene biosynthetic capacity of vascular cells. *Blood* 74:703 (1989).
2. A. Sala, T. Testa, and G. Folco. Leukotriene A_4, and not Leukotriene B_4, is the main 5-lipoxygenase metabolite released by bovine leukocytes. *FEBS Lett.* 388: 94 (1996).
3. A. Sala, M. Bolla, S. Zarini, R. Müller-Peddinghaus, and G. Folco. Release of Leukotriene A_4 versus Leukotriene B_4 from human polymorphonuclear leukocytes. *J. Biol. Chem.* 271: 17944 (1996).
4. S.J. Feinmark, and P.J. Cannon. Endothelial cell leukotriene C_4 synthesis results from intracellular transfer of leukotriene A_4 synthesized by polymorphonuclear leukocytes. *J. Biol. Chem.* 261:16466 (1986).
5. J.A. Maclouf, and R.C. Murphy. Transcellular metabolism of neutrophil-derived leukotriene A_4 by human platelets. *J. Biol. Chem.* 263:174 (1988).
6. F. Grimminger, B. Kreusler, U. Schneider, G. Becker, and W Seeger. Influence of microvascular adherence on neutrophil leukotriene generation. *J. Immunol.* 144:1866 (1990).
7. A. Sala, G. Rossoni, C. Buccellati, F. Berti, G.C. Folco, and J. Maclouf. Formation of sulfidopeptide-leukotrienes by cell-cell interaction causes

coronary vasoconstriction in isolated, cell-perfused heart of the rabbit. *Br. J. Pharmacol.* 110:1206 (1993).

8. Morphological and functional changes of coronary vasculature caused by transcellular biosynthesis of sulfidopeptide leukotrienes in isolated heart of rabbit. *Blood* 87:1824 (1996).
9. A. Hatzelmann, R. Fruchtmann, K.H. Mohrs, S. Raddatz, and R. Müller-Peddinghaus. Mode of action of the new selective leukotriene synthesis inhibitor BAY X1005 ((R)-2-(4-(quinolin-2-yl-methoxy)phenyl)-2-cyclopentyl acetic acid and structurally related compounds. *Biochem. Pharmacol.*, 45: 101 (1993).
10. C. Haslett, L.A. Guthrie, M.M. Kopaniac, R.B. Johnston, and P.M. Henson. Modulation of multiple neutrophil functions by preparative methods for trace concentrations of bacterial lipolysaccharide. *Am. J. Pathol.* 119:101 (1985).
11. M. Carry, V. Korley, J.T. Willerson, L. Weigelt, A.W. Ford-Hutchinson, and P. Tagari. Incresed urinary leukotriene excretion in patients with cardiac ischemia. In vivo evidence for 5-lipoxygenase activation. *Circulation* 85:230 (1992).

LEUKOTRIENE D_4-INDUCED SIGNALLING EVENTS IN HUMAN EPITHELIAL CELLS : $G\alpha_{i3}$ ACTIVATION AND TRANSLOCATION

John F. Öhd[1], Jonas L. P. Adolfsson[1], and Anita Sjölander[2]

Department of Laboratory Medicine
Division of Experimental Pathology
Lund University, Wallenberg Laboratory, UMAS
S-205 02, Malmö, SWEDEN

INTRODUCTION

Leukotrienes (LTs) are arachidonic acid derivatives that are known to mediate inflammatory reactions[1,2]. LTs have been recognized as potent mediators of vaso- and bronchoconstriction, mucus secretion and smooth muscle contraction[3], and they have been implicated in the pathogenesis of several conditions, e.g. chron´s disease, ulcerative colitis and bronchial asthma[3-5]. LTs exert their effects by binding to receptors on the plasma membranes of their target cells[1,2,6]. Radioligand experiments[6] and functional analyses using specific inhibitors have revealed the presence of LT receptors on a large number of cell types. The functional analyses have indicated the existence of at least two subsets of cysteinyl leukotriene receptors (CysLT receptors), for which a classification was quite recently published[7]. It has been shown that LTD_4 induces a rise in the cytosolic free Ca^{2+} concentration in different types of cells[8,9], and we found that that effect can be blocked by pretreating cells with ICI-198,615 (unpublished data). The G-protein responsible for the influx of Ca^{2+} is sensitive to pertussis toxin (PTX), whereas the one responsible for mobilisation of intracellular Ca^{2+} is *rho* dependent[10].

ACTIVATION OF THE G_{i3}–PROTEIN

G-proteins belong to at least two different families, one heterotrimeric and one monomeric[11,12]. All of the members of both families have one main feature in common: the ability to bind GTP, and thereby become activated and serve as important intracellular signalling elements. A heterotrimeric G-protein comprises three subunits designated α, β and γ. Following receptor activation, the α-subunit binds GTP and dissociates from the β/γ-complex.

---

[1]Both of these authors contributed equally to this work.

[2]To whom correspondence should be addressed; FAX +46-40-33 73 53.

This work was supported by grants from the Swedish Medical Research Council (project 10356), the King Gustav V´s Memorial Foundation, the MAS foundations, and the Crafoord foundation.

For quite some time the activated α-subunit was considered to be the only subunit to take part in intracellular signalling. However, Neer and co-workers showed that the β/γ-subunit also participates in signalling, which suggests that the signal transduction cascade may be bifurcated[13,14].

Our results represent the first evidence that stimulation of epithelial cells with LTD_4 triggers activation of the G_{i3}–protein, with a peak GTP/GDP exchange ratio at around 15 s (Fig. 1A)[15]. G_{i3} was the only one of the known PTX-sensitive G-proteins found in the cells that was activated by LTD_4[15]. This is particularly interesting due to our previous finding that a PTX-sensitive G-protein regulates the LTD_4-induced influx of Ca^{2+} across the plasma membrane of Intestine 407 cells[8]. PTX has been shown to cause an ADP-ribosylation of the α-subunits of the G_i-group, and the proteins are then unable to either interact with a membrane-bound receptor or bind GTP[11]. Therefore, it seems reasonable to conclude that the LTD_4-induced PTX-sensitive influx of Ca^{2+} into INT 407 cells is directly or indirectly attributable to either the activation of the α_{i3}-subunit or the release of its β/γ -complex.

It has been reported that Ca^{2+}-channels can be both activated[16] and inhibited[17] by G-proteins.

More recently, an increasing amount of work has addressed the interaction between β/γ complexes and Ca^{2+}-channels[18,19], leading to the suggestion that β/γ complexes may play a role in Ca^{2+} signalling. However, that possibility is refuted by our observation that the LTD_4-induced influx of Ca^{2+} in Intestine 407 cells is also dependent on the activation of a PTX-sensitive tyrosine kinase, as well as on protein kinase A activity[20,21]. This instead implies an indirect role of the G_{i3}-protein in the regulation of the LTD_4-induced influx of Ca^{2+} into these cells.

TRANSLOCATION OF THE G_{i3}–PROTEIN

Subunits of G-proteins can translocate from one cellular compartment to another. In human neutrophils, α-subunits of G_{i2}-proteins are released from a cytoskeletal fraction upon stimulation with chemotactic peptides[22], whereas in rat hepatocytes stimulation with EGF has been noted to translocate $G\alpha_i$, together with PLCγ, to the cytoskeleton[23]. These translocations probably make it possible for different types of protein-protein interactions to occur and thereby constitute an additional mechanism of signal regulation.

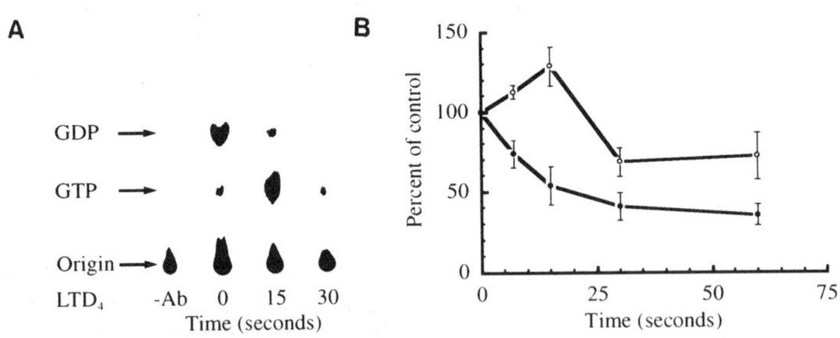

Figure 1. LTD_4-induced activation of G_{i3}-proteins in Intestine 407 cells; representative TLC results are shown in panel **A**, and no unspecific binding to protein A (designated -Ab) can be seen. **B.** The compiled results of several Western blot analyses of the LTD_4-induced changes in the content of $G\alpha_{i3}$-subunits in a membrane fraction (•) and a cytoskeletal pellet fraction (o). Values represent means ± SD of 4 separate experiments. Reprinted from *Biochem. Biophys. Res. Com.* 226:413, 1996, by the kind permission of Academic Press, Inc.

Abbreviations: AC, adenylyl cyclase; ADP, adenosine diphosphate; EGF, epidermal growth factor; GDP, guanosine diphosphate; GTP, guanosine triphosphate; INT407, Intestine 407; IP3, inositol 1,4,5-trisphosphate; LTD_4, leukotriene D_4; LTs, leukotrienes; PIP2, phosphatidylinositol 4,5-bisphosphate; PKA, protein kinase A; PLC, phospholipase C; PTX, pertussis toxin; Tyr K, tyrosine kinase;.

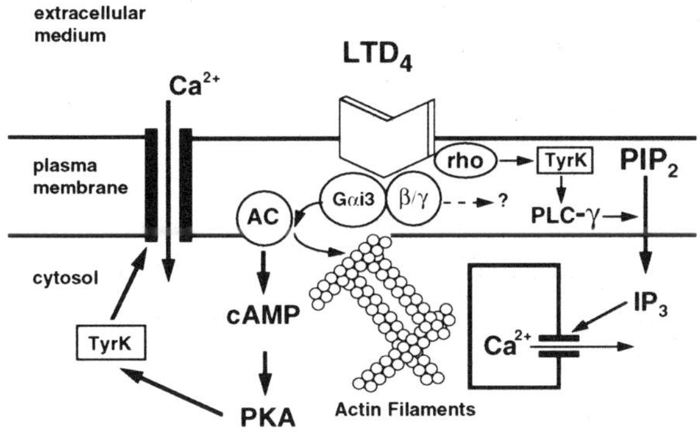

Figure 2. A schematic model of the LTD_4-induced signalling pathways discussed in the present report.

In our study[15], in addition to observing the activation of G_{i3}, we found that, when activated, the α-subunits of this protein translocate from a membrane fraction to a crude cytoskeletal fraction (Fig. 1B). Even if the LTD_4-induced intracellular mobilisation of Ca^{2+} is mediated via a γ-isotype of PLC[21], a coupling between $G\alpha_{i3}$ and PLCγ similar to that seen in rat hepatocytes is unlikely, since the LTD_4-induced mobilisation of Ca^{2+} is insensitive to PTX[8] and instead probably involves the low molecular weight G-protein *rho*[10]. The $G\alpha_{i3}$ translocation might be part of a signalling process that triggers modifications of the cytoskeleton, or it could simply serve to inactivate the $G\alpha_{i3}$-proteins and in doing so may leave an increased amount of signalling β/γ-subunits in the membrane. In the latter case, the translocation of $G\alpha_{i3}$ could enable a more subtle regulation of cell signalling.

SUMMARY

Our model of LTD_4-induced signal transduction in epithelial cells is summarised in Figure 2. Extending what is already known about LTD_4 signalling in epithelial cells, we identified the G_{i3}–protein as the crucial PTX sensitive G-protein and found that it is translocated to what might be a cytoskeletal fraction. This finding suggests a subtle response to LTD_4, mediated via the bifurcation at the α /βγ junction. Although little is known about the role of epithelial cells in inflammation, it has been shown that such cells produce the potent chemoattractant LTB_4 and the proinflammatory 5-HETE in response to intracellular accumulation of Ca^{2+} [24]. The target protein(s) and the effect(s) of the translocation of the activated $G\alpha_{i3}$-proteins, as well as the possible role of the β/γ-subunits of G_{i3}, remain to be elucidated.

REFERENCES

1. B. Samuelsson, S.-E. Dahlén, J.Å. Lindgren, C.A. Rouzer, and C.N. Serhan, Leukotrienes and lipoxins: structures biosynthesis and biological effects, *Science* 237:1171 (1987).
2. C.N. Serhan, J.Z. Haeggström, and C.C. Leslie, Lipid mediator networks in cell signaling: update and impact of cytokines, *FASEB J.* 10:1147 (1996).
3. Z. Marom, J.H. Shelhamer, M.K. Bach, D.R. Morton, and M. Kaliner, Slow–reacting substances, leukotrienes C_4 and D_4, increase the release of mucus from human airways *in vitro*, *Am. Rev. Respir. Dis.* 126:449 (1982).
4. D.M. Hammerbeck, and D.R. Brown, Presence of Immunocytes and sulfidopeptide leukotrienes in the inflamed guinea pig distal colon, *Inflammation* 20:413 (1996).
5. T.D. Wardle, L. Hall, and L.A. Turnberg, Use of coculture of colonic mucosa biopsies to investigate the release of eicisanoids by inflamed and uninflamed mucosa from patients with inflammatory bowel disease, *Gut* 33:1644 (1992).
6. C. Rochette, D.W. Nicholson, and K.M. Metters, Identification and target–size analysis of the leukotriene D_4 recptor in the human THP–cell line, *Biochim. Biophys. Acta* 1177:283 (1993).
7. R.A. Coleman, R.M. Eglen, R.L. Jones, S. Narumiya, T. Shimizu, W.L. Smith, S.–E. Dahlén, J.M. Drazen, P.J. Gardiner, W.T. Jackson, T.R. Jones, R.D. Krell, and S. Nicosia, Prostanoid

and leukotriene receptors: A progress report from the IUPHAR working parties on classification and nomenclature, *Adv. Prostaglandin Thromboxane Leukotriene Res.* 23:283 (1995).

8. A. Sjölander, E. Grönroos, S. Hammarström, and T. Andersson, Leukotriene D_4 and E_4 induce transmembrane signaling in human epithelial cells, *J. Biol. Chem.* 265:20976 (1990).
9. S.T. Crooke, M. Mattern, H.M. Sarau, J.D. Winkler, J. Balcarek, A. Wong, and C.F. Bennett, The signal transduction system of the leukotriene D_4 receptor, *Trends Pharmacol. Sci.* 10:103 (1989).
10. E. Grönroos, T. Andersson, Å. Schippert, L. Zheng, and A. Sjölander, Leukotriene D_4–inducedmobilization of intracellular Ca^{2+} in epithelial cells is critically dependent on activation of the small GTP–binding protein Rho, *Biochem. J.* 316:239 (1996).
11. J.R. Hepler, and A.G. Gilman, G–proteins, *Trends Biochem. Sci.* 17:383 (1992) .
12. A. Hall, Small GTP-binding proteins and the regulation of the actin cytoskeleton, *Annu. Rev. Cell. Biol.* 10: 31 (1994).
13. D.E. Logothetis, Y. Kurachi, J. Galper, E.J. Neer, and D.E. Clapham, The β/γ subunits of GTP-binding proteins activate the muscarinic K^+ channel in heart, *Nature* 325:321 (1987).
14. D.E. Clapham, and E.J. Neer, New roles for G–protein β/γ–dimers in transmembrane signalling, *Nature* 365:403 (1993).
15. J.L.P. Adolfsson, J.F. Öhd, and A. Sjölander, Leukotriene D_4–induced activation and translocation of the G-protein α_{i3}–subunit in human epithelial cells, *Biochem. Biophys. Res. Com.* 226:413 (1996).
16. T.M. Moriarty, E. Padrell, D.J. Carty, G. Omri, E.M. Landau, and R. Iyengar, G_o protein as signal transducer in the pertussis toxin–sensitive phosphatidyl inositol pathway, *Nature* 343:79 (1990).
17. D.A. Ewald, I.-H. Pang, P.C. Sternweis, and R.J. Miller, Differential G–protein mediated coupling of neurotransmitter receptors to Ca^{2+} channels in rat dorsal root ganglion neurons *in vitro*, *Neuron* 2:1185 (1989).
18. S.R. Ikeda, Voltage–dependent modulation of N–type calcium channels by G–protein β/γ subunits, *Nature* 380:255 (1996).
19. S. Herlitze, D.E. Garcia, K. Mackie, B. Hille, T. Scheuer, and W.A. Catterall, Modulation of Ca^{2+} channels by G–protein β/γ subunits, *Nature* 380:258 (1996).
20. E. Grönroos, and A. Sjölander, Leukotriene D_4 induces a rapid increase in cAMP in human epithelial cells: a potential role for this signal in the regulation of calcium influx through the plasma membrane, submitted (1997).
21. E. Grönroos, Å. Schippert, M. Engström, and A. Sjölander, The regulation of leukotriene D_4–induced calcium influx in human epithelial cells involves protein tyrosine phosphorylation, *Cell Calcium* 17:177 (1995).
22. E. Särndahl, G.M. Bokoch, O. Stendahl, and T. Andersson, Stimulus–induced dissociation of α subunits of heterotrimeric GTP–binding proteins from the cytoskeleton of human neutrophils, *Proc. Natl. Acad. Sci.* (USA) 90:6552 (1993).
23. L.J. Yang, S.G. Rhee, and J.R. Williamson, Epidermal growth factor–induced activation and translocation of phospholipase $C\gamma 1$ to the cytoskeleton in rat hepatocytes, *J. Biol. Chem.* 269:7156 (1994).
24. A. Sjölander, Å. Schippert, and S. Hammarström, A human epithelial cell line, Intestine 407, can produce 5–hydroxyeicosatetraenoic acid and leukotriene B_4, *Prostaglandins* 45:85 (1993).

INTERACTIONS OF RENAL CYTOCHROME P450 (CYP), ANGIOTENSIN (AII) AND TUMOR NECROSIS FACTOR-ALPHA (TNF): IMPLICATIONS FOR ION TRANSPORT

John C. McGiff, Nicholas R. Ferreri, Bruno A. Escalante, and Mairead A. Carroll

New York Medical College
Department of Pharmacology
Valhalla, New York 10595

Cytochrome P450-Arachidonic Acid (CYP-AA) in the Medullary Thick Ascending Limb of the Loop of Henle (mTALH)

In cells from the rabbit mTALH, arachidonic acid (AA) is metabolized primarily by the CYP pathway[1,2]. Rabbit mTALH cells, when incubated with ^{14}C-AA for 30 min, form products that segregate into two peaks, designated P_1 and P_2 based on their reverse phase HPLC retention times. Structural analysis by gas chromatography/mass spectrometry (GC/MS) disclosed three AA products generated by the CYP pathway in the mTALH: 20-hydroxyeicosatetraenoic acid (20-HETE), 20-eicosatetraenedioic acid (20-COOH-AA), and 19-HETE[2]. P_2, the more polar peak, was comprised of 20-COOH-AA, an inhibitor of Na^+-K^+-ATPase, that was devoid of vasoactivity. P_1 contained vasoactive material that was also capable of affecting Na^+-K^+-ATPase and yielded two products, 20-HETE and 19-HETE (Fig. 1).

We next examined a potential linkage between changes in ion transport and AA metabolism in isolated cells of the mTALH[3]. AA produced a concentration-dependent inhibition of ^{86}Rb uptake -- an effect that was prevented by selective blockade of CYP. Blockade of the effect of AA on ^{86}Rb uptake by CYP inhibitors could be circumvented by addition of the principal products of CYP-AA metabolism in the mTALH[4].

Changes in ^{86}Rb movement produced by an agent will not discriminate between primary effects of a CYP-AA metabolite on the cotransporter vs. the Na^+ pump as both furosemide (cotransporter inhibitor) and ouabain (Na^+ pump inhibitor) inhibit ^{86}Rb uptake[3,5]. We suggested that CYP-AA metabolites blocked ^{86}Rb uptake by inhibiting Na^+-K^+-ATPase, based on *ex vivo* evidence that 20-HETE and 20-COOH-AA inhibited the activity of purified Na^+-K^+-ATPase. Because furosemide also reduced ^{86}Rb uptake in mTALH cells, inhibition of the Na^+-K^+-$2Cl^-$ cotransporter could also account for reduction of ^{86}Rb uptake.

We addressed possible interactions involving CYP-AA metabolites with the different pathways for K^+ movements in the mTALH by comparing the effects of AA and the

Recent Advances in Prostaglandin, Thromboxane, and Leukotriene Research
Edited by Sinzinger *et al.*, Plenum Press, New York, 1998

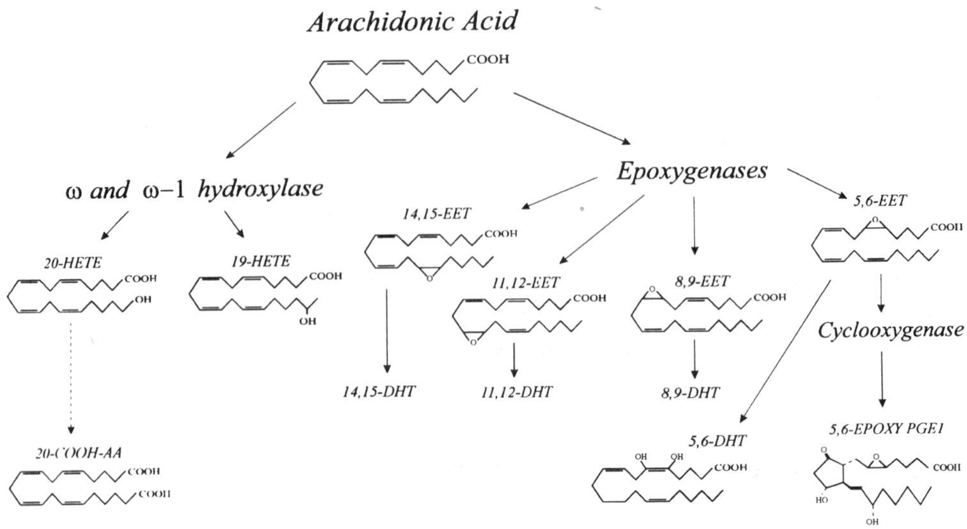

Fig. 1 AA metabolism by CYP-dependent monooxygenases to ω and ω-1 hydroxyeicosatetraenoic acids (HETEs), epoxyeicosatrienoic acids (epoxides/EETs) and dihydroxyeicosatrienoic acids (diols/DHTs). For the DHTs, only the strucuture of 5,6-DHT is given. As 5,6-EET can be a substrate for cyclooxygenase, one of the products -- the vasodilator 5,6-epoxy PGE_1 -- is shown. (With permission, Trends in Cardiovasc. Med., Elsevier Science, Inc.)

principal CYP-AA metabolites, 20-HETE and 20-COOH-AA, on ion content and ^{86}Rb uptake to those same effects produced by ouabain and furosemide[5]. Ouabain increased Na^+ and decreased K^+ content in rabbit mTALH cells, whereas furosemide reduced both $Na^+$and K^+ content. We further tested the possibility of an additional effect of CYP-AA metabolites on Na^+-K^+-ATPase in experiments carried out under conditions in which intracellular Na^+ concentration was not rate limiting for the Na^+ pump. Treatment of mTALH cells with monensin created these conditions, since incorporation of the ionophore into the mTALH cell membrane produced a parallel pathway for Na^+ entry. Under these conditions, furosemide, or a compound acting similarly to furosemide, would not affect ^{86}Rb uptake, whereas ouabain would inhibit monensin-stimulated ^{86}Rb uptake. In contrast to ouabain, AA did not inhibit ^{86}Rb uptake in the presence of monensin, indicating that CYP-derived AA metabolites do not modulate Na^+-K^+-ATPase of mTALH cells[5].

Cytokines and mTALH

An immunomodulator function has also been ascribed to the mTALH segment. Tamm-Horsfall (uromodulin) protein that coats the surface of mTALH cells binds interleukin-1 (IL-1) and TNF with high affinity. A relationship between cytokines and AA metabolism has been demonstrated and includes induction of cyclooxygenase (COX) expression. Synthesis of TNF by mTALH cells has been demonstrated to be low under basal conditions and subject to stimulation by lipopolysaccharide (LPS)[6]. TNF, either endogenous (when stimulated by LPS) or exogenous, can affect transport function of mTALH cells, as evidenced by inhibition of ^{86}Rb uptake. This effect of TNF on ^{86}Rb uptake by mTALH cells required more than 4 h, presumably related to induction of COX synthesis as TNF-mediated inhibition of ^{86}Rb uptake could be abolished by indomethacin. These findings support the following hypothesis: *cytokines can promote COX-dependent AA metabolism increasing the formation of prostanoids that affect transport in the nephron segment of origin (autocrine function) and, possibly, in other segments (paracrine function).*

Interactions of Angiotensin II (AII) and TNF affect blood pressure by a prostaglandin mechanism.

TNF is produced by the renal mTALH in response to either LPS or AII[7,8]. TNF, produced by this nephron segment, acts in an autocrine manner to inhibit ^{86}Rb uptake via a prostanoid-dependent mechanism. We determined whether TNF was produced by the mTALH in AII-induced hypertension which is associated with renal and vascular damage, and infiltration of mononuclear cells. TNF production was approximately 2- to 5-fold higher in tubules obtained from rats made hypertensive by AII infusion (200 ng/min, s.c.) for 10 days, compared to normotensive controls. Production of prostaglandin E_2 (PGE_2) was increased to a similar extent in tubules from hypertensive rats. TNF antisera increased mean arterial pressure (MAP) in AII-hypertensive rats from 150 $\pm$ 3 to 172 $\pm$ 5 mmHg. Pre-immune sera did not affect MAP in normotensive or AII-hypertensive rats; anti-TNF had no effect on MAP of normotensive rats. AII-induced TNF production may not be limited to the mTALH, as TNF release from the aorta also is enhanced by AII infusion. TNF may exert a counter-regulatory influence against the pressor actions of AII via stimulation of PGE_2 formation while decreasing formation.

Hormonal regulation of renal CYP-derived HETEs

We have established a method to profile and quantitate endogenous CYP-HETEs exiting the isolated perfused rabbit kidney in reponse to hormonal stimulation[9]. In response to AII, 16-, 17-, 18-, 19- and 20-HETEs are released from the rabbit kidney in amounts equal to or greater than prostaglandins. The biological activities of the CYP-HETEs are potentially of great importance in view of the preeminent position of 20-HETE in renal regulatory mechanisms, e.g. 20-HETE modulates the mTALH Na^+-K^+-$2Cl^-$ cotransporter and affects rat preglomerular vascular resistance. Our major findings were: 1) AII produced a several-fold increase in the levels of CYP-HETEs above basal levels. An equipressor concentration of arginine vasopressin (AVP) did not elevate CYP-HETEs in the perfusate above basal values; neither did bradykinin (BK), a vasodilator peptide known to stimulate phospholipase activity; 2) authentic standards of 16-, 18-, 19- and 20-HETEs dilated the vasculature of the rabbit isolated kidney, an action dependent on COX activity; 3) the vasoactivity of 16- and 18-HETE, unlike 19-HETE isomers, was stereospecific, in that the (R) configuration was the active isomer; 4) the effects of CYP-HETEs on ion transport were also stereospecific, the (S) enantiomers of 16- and 17-HETEs being potent inhibitors of proximal tubular ATPase , whereas their (R) isomers were virtually inactive.

Separation of venous and ureteral perfusates revealed both a differing profile and levels of CYP-HETEs in each effluent[10]. Except for 19-HETE, urinary efflux of CYP-HETEs exceeded venous efflux by 5- 20-fold. Venous and urinary efflux of 19-HETE were similar. 17-HETE was the principal urinary HETE metabolite (6.74 $\pm$ 1.5 ng/min) and the major venous HETE was 19- (0.46 $\pm$ 0.16 ng/min). AII increased 17-, 19- and 20-HETE efflux from the venous compartment and 20-HETE efflux into the ureter.

Release of renal HETEs by AII is not prevented by inhibiting CYP activity.

Infusion of 2μM 17-octadecynoic acid (17-ODYA), a concentration that was reported to inhibit by greater than 75% AA conversion to CYP metabolites by renal microsomes, virtually abolished basal ureteral efflux of HETEs, whereas venous efflux of HETEs, while reduced, was less affected[10]. We determined whether 17-ODYA (2μM) did, in fact, inhibit CYP activity in the rabbit kidney under these experimental conditions. The conversion of ^{14}C-AA to CYP HETEs by renal homogenates was inhibited by 70 $\pm$ 5% in the kidneys pretreated with 17-ODYA compared to control kidneys. Despite pronounced inhibitory effects of 17-ODYA on basal renal efflux of HETEs, AII continued to stimulate efflux of CYP-HETEs. Clotrimazole (1 μM), a CYP inhibitor of AA metabolism which differs mechanistically from 17-ODYA, also reduced basal efflux of HETEs and, like 17-ODYA, did not prevent AII-induced stimulation of HETE efflux, the exceptions being 16- and 18-HETE venous efflux. The latter , however, constituted a small component of the overall renal HETE response to AII. These results suggest that CYP-HETEs are preformed in the kidney and that their release from storage sites such as phospholipids can be regulated by AII, perhaps by activation of one or more acylhydrolases.

Storage of CYP HETEs in renal lipids.

To obtain definitive evidence that CYP HETEs were preformed in the kidney, the cortical and medullary lipids were extracted, separated by HPLC and CYP-HETEs quantitated following alkaline hydrolysis of individual fractions[10]. Briefly, the cortex and outer medulla were separated and each homogenized and the lipids extracted using the Bligh/Dyer method. The extracts were separated by normal-phase HPLC. This procedure separated neutral lipids (NLs) and the major phospholipid classes; phosphatidylethanolamine (PE), phosphatidylinositol (PI), cardiolipin (CL), phosphatidylserine (PS) and phosphatidylcholine (PC). The molecular species were detected by UV monitoring at 206 nm and identified by comparison of elution times with lipid standards. The fractions were subjected to alkaline hydrolysis to which D_3 19-HETE was added as an internal standard. The incubates were extracted, purified and the HETE fraction derivatized and analyzed by GC/MS.

The CYP-HETEs were incorporated into both NLs and phospholipid fractions in the cortex and medulla (Fig 2).

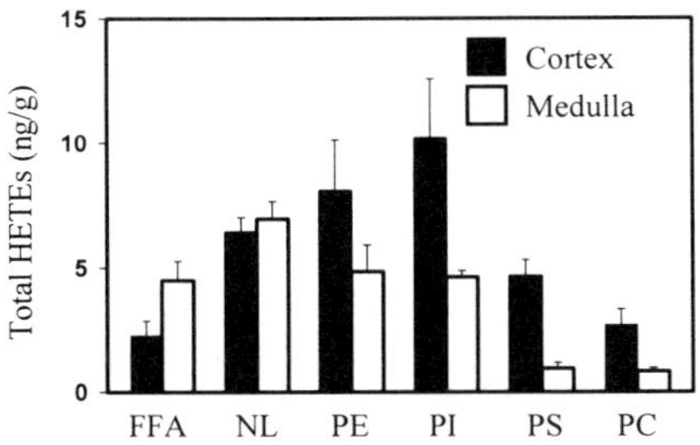

Fig. 2 Distribution of esterified CYP-HETEs. (With permission, Kidney International, Blackwell Scientific Publications, Inc.)

Esterified HETEs were found in larger quantities in the cortex, where their distribution was highest in the PI fraction although they were present in all lipid fractions. In the medulla, the largest quantities of esterified HETEs were found in NLs. The distribution of the individual HETEs was essentially similar in each renal zone. The NL fraction contained all five HETEs, 19- and 20-HETE being in the greatest abundance. Notably, only 16- and 17-HETE were present in PE, whereas 19- and 20-HETE were also present in PI, PS, and PC. PI in both cortex and medulla contained the largest quantities of 19- and 20-HETEs. The incorporation of 18-HETE was different as it was not widely distributed, being localized only to the NL fraction.

Acknowledgments

The authors wish to thank Melody Steinberg for editorial assistance. This work was funded by NIH grant HL 34300.

References

1. M.A. Carroll, E.D. Drugge, C.E. Dunn, and J.C. McGiff: Isolation of rabbit renomedullary cells and arachidonate metabolism. IN: *Methods in Enzymology*. R.C. Murphy and F.A. Fitzpatrick, eds. Academic Press, Inc., San Diego, CA. (1990).
2. M.A. Carroll, A. Sala, C.E. Dunn, J.C. McGiff, and R.C. Murphy. Structural identification of cytochrome P450-dependent arachidonate metabolites formed by rabbit medullary thick ascending limb cells. *J. Biol. Chem.* 266:12306-12312 (1991).
3. B. Escalante, D. Erlij, J.R. Falck, and J.C. McGiff. Effect of cytochrome P450 arachidonate metabolites on ion transport in rabbit kidney loop of Henle. *Science* 251: 799-802 (1991).
4. B. Escalante, D. Erlij, J.R. Falck, and J.C. McGiff. Cytochrome P450-dependent arachidonate metabolites affect renal transport in the rabbit. *J. Cardiovasc. Pharmacol.* 22 (Suppl.2): S106-S108 (1993).
5. B. Escalante, D. Erlij, J.R. Falck, and J.C. McGiff. Cytochrome P-450 arachidonate metabolites affect ion fluxes in rabbit medullary thick ascending limb. *Am. J. Physiol.* 266:C1775-C1782 (1994).
6. B.A. Escalante, N.R. Ferreri, C.E. Dunn, and J.C. McGiff. Cytokines affect ion transport in primary cultured thick ascending limb of Henle's loop cells. *Am. J. Physiol.* 266:C1568-C1576 (1994).
7. C.M. Macica, B.A. Escalante, M.S. Conners, and N.R. Ferreri. TNF production by the medullary thick ascending limb of Henle's loop. *Kid. Inter.* 46:113-121 (1994).
8. N.R. Ferreri, B.A. Escalante, Y. Zhao, S-J. An, and John C. McGiff. Angiotensin II induces TNF production by the thick ascending limb: Functional implications. Submitted (1997).
9. M.A. Carroll, M.Balazy, P Margiotta, D-D. Huang, J.R. Falck, and J.C. McGiff. Cytochrome P-450-dependent HETEs: profile of biological activity and stimulation by vasoactive peptides. *Am. J. Physiol.* 271:R863-R869 (1996).
10. M.A. Carroll, M.Balazy, D-D. Huang, S. Rybalova, J.R. Falck, and J.C. McGiff. Cytochrome P450-derived renal HETEs: storage and release. Submitted (1997).

MODULATION OF LEUKOCYTE-ENDOTHELIAL CELL INTERACTION AND LEUKOTRIENE DEPENDENT VASOCONSTRICTION BY PROSTACYCLIN MIMETICS IN THE ISOLATED RABBIT HEART

Carola Buccellati, Giuseppe Rossoni, Ferruccio Berti and Giancarlo Folco.

Center for Cardiopulmonary Pharmacology, University of Milan

INTRODUCTION

Sulfidopeptide leukotrienes (sLT) are arachidonic acid (AA) metabolites involved in the inflammatory response (Dahlen, 1981). LT production from the unstable metabolic intermediate LTA_4 is cell-type dependent, as polymorphonuclear leukocytes (PMNL) generate predominantly LTB_4 while LTC_4 is the main product synthesized by eosinophils and mast cells; in addition, cooperation between PMNL (donor cells for LTA_4) and endothelial cells (EC, acceptor cells) leads to formation of LTC_4 by "transcellular biosynthesis" (Feinmark, 1986; Maclouf, 1989).

Formation of LTC_4 and LTD_4, due to PMNL cooperation with coronary endothelial cells, occurs in spontaneously beating, perfused rabbit hearts, and correlates with the increase in coronary perfusion pressure (CPP) (Sala, 1993) and cardiac damage (Sala,1996). PMNL adhesion to the wessel wall is, in fact, a promoter for microvascular damage (Lucchesi, 1986).

Prostacyclin (PGI_2) is a potent, biologically active eicosanoid which attenuates cardiac damage (Schror, 1992) and interferes with PMNL adherence to endothelial cells. The PGI_2 analogue Iloprost as well as the PGI_2 releaser Defibrotide are therefore considered as cardioprotective drugs and it was of interest to investigate their capacity to interfere with sLT production, and related changes in coronary function, due to cooperation between PMNL and EC.

Recent Advances in Prostaglandin, Thromboxane, and Leukotriene Research
Edited by Sinzinger *et al.*, Plenum Press, New York, 1998

METHODS

Hearts were isolated from albino rabbits and perfused retrogradely through the aorta according to Langendorff, at a constant flow (20 ml/min) with Krebs-Henseleit buffer at 37°C (Berti, 1988). Hearts were equilibrated, rinsed and then perfused using a recirculating system with a 50 ml volume. Coronary perfusion pressure (CPP) and left ventricular end diastolic pressure (LVEDP) were continuously monitored.

Human PMNL were purified and perfused in the recirculating medium, to obtain a final concentration of $1x10^5$ cells/ml (Sala 1993). Treatment with iloprost (3 nM) or defibrotide (200 µg/ml) was given 15 minutes before PMNL addition. Challenge was performed by 0.5 µM A-23187 given 5 minutes after cells and lasted 30 minutes. Aliquots (1ml) of the recirculating medium were withdrawn at different time intervals and processed for enzyme immunoassay (EIA) of sLT, while the entire reservoir (about 45 ml) was kept for HPLC analysis.

RESULTS

Basal CPP and LVEDP levels were not significantly affected neither by A-23187 nor by cell perfusion alone (CPP: 45±3 mm Hg. LVEDP: 5 mm Hg). On the contrary when A-23187 was given in presence of recirculating PMNL, a progressive increase in CPP reached, within thirty minutes, values approx. six fold higher than control preparation. LVEDP as well, was affected by challenge of circulating cells with A-23187, suggesting onset of an ischemic process (basal: 5.3±3 mm Hg. 30 min after A-23187: 67.9±19 mm Hg, $p<0.01$, $n=6$). These events were accompanied by enhanced sLT formation as measured in the recirculating medium.

Pretreatment of rabbit hearts with 3 nM Iloprost (a concentration that did not cause vasodilation) significantly reduced CPP increase at 30 min (control: 290.5±33 mm Hg, n=6; Iloprost: 75.7±5 mm Hg, n=3) and also ablated sLT in the perfusate (control: 746.2 ±130 pmol; Iloprost: 126.3±77 pmol).

The results obtained with defibrotide fully confirmed those obtained with iloprost (at 30 min: CPP 61.7±5 mmHg, sLT 93.6±22 pmoles), as defibrotide caused endogenous PGI_2 formation (as demonstrated measuring its stable metabolite 6-keto $PGF_{1\alpha}$ - control: 1.03±0.09 ng/ml. Defibrotide: 1.73±0.2 ng/ml, $P<0.05$, $n=4$).

Hearts from rabbits treated with lysine acetylsalicylate (n=3) (60 mg/kg i.v., 2 h before sacrifice) were not protected by defibrotide. Levels of 6-keto $PGF_{1\alpha}$ in the recirculating medium were undetectable (<0.05 ng/ml) while CPP increase (233.3±17 mmHg at 30 min) and sLT levels (445±62 pmoles at 30 min) were close to control values.

DISCUSSION

Our findings support the existence of a cardioprotective mechanism exerted by prostacyclin mimetics, that does not involve a direct inhibition of sLT formation at the level of 5-lipoxygenase enzyme, nor a sLT receptor antagonism. Cardioprotection seems to represent a direct defense mechanism, as shown with iloprost, or an indirect one as a consequence of PGI_2 synthesis, as shown with defibrotide. It is likely that prostacyclin mimetics act by a modulation of PMNL activation and infiltration at the site of injury (Smith, 1984; Simpson, 1987). PGI_2 was in fact demonstrated to reduce PMNL accumulation and to protect the reperfused ischemic myocardium (Schror, 1989), as well as PMNL-induced lung injury (Riva, 1990).

The ability of iloprost and defibrotide to limit coronary vasoconstriction in a model of PMNL-dependent experimental ischemia is suggested to be due to a modulation of early events of neutrophil activation and inhibition of local release of sLT due to transcellular biosynthesis.

REFERENCES

Berti F., Rossoni G., Magni F., Caruso D., Omini C., Puglisi L. and Galli G., 1988, Non-steroidal antiinflammatory drugs aggravate acute myocardial schemia in perfused rabbit heart; a role for prostacyclin, *J Cardiovasc Pharmacol* . 12:438.

Dahlen S.E., Bjork J., Hedqvist P., Arfors K.E., Hammarstrom S., Lingren J.A. and Samuelsson B., 1981, Leukotrienes promote plasma leakage and leukocyte adhesion in postcapillary venules; in vivo effects with relevance to the acute inflammatory response, Proc Natl Acad Sci USA. 78: 3887.

Feinmark S.J. and Cannon P.J., 1986, Endothelial cell leukotriene C_4 synthesis results from intracellular transfer of leukotriene A_4 synthesized by polymorphonuclear leukocytes, J Bio Chem. 261:16466.

Lucchesi B.R. and Mullane K.M., 1986, Leukocytes and ischemia-induced myocardial injury, Annu Rev Pharmacol Toxicol. 26:201.

Maclouf J., Murphy R.C. and Henson P., 1989, Transcellular sulfidopeptide leukotriene biosynthetic capacity of vascular cells Blood, 74 (2):703.

Riva C.M., Morganroth M.J., Ljungman A.G., Schoeneich S.O., Marks R.M., Todd R.F., Ward P.A. and Boxer L.A., 1990, Iloprost inhibits neutrophil-induced lung injury and neutrophil adherence to endothelial monolayers, Am J Respir Cell Mol Biol. 3, 301.

Sala A., Aliev G.M. Rossoni G., Berti F., Buccellati C., Burnstock G., Folco G. and Maclouf J., 1996, Morphological and functional changes of coronary vasculature caused by transcellular biosynthesis of sulfidopeptide leukotrienes in isolated heart of rabbit, Blood, 87 (5):1824.

Sala A., Rossoni G., Buccellati C., Berti F., Folco G. and Maclouf J.,1993, Formation of sulfidopeptide-leukotrienes by cell-cell interaction causes coronary vasoconstriction in isolated, cell-perfused heart of rabbit, Br J Pharmacol. 110:1206

Schrör K., Thiemermann C. and Lobel P., 1989, Stimulation of vascular prostacyclin formation by defibrotide: a new strategy for treatment of acute myocardial ischaemia, in: Limitation of Infarct Size, Schmutzler H, Rutsch W, Dougherty FC(eds). Berlin, Springer-Verlag p83.

Schrör K. Cytoprotective properties of prostacyclin, 1992, in:Prostacyclin: new perspectives for basic research and novel therapeutic indication, Rubanyi GM, Vane J(eds). Amsterdam: Elsevier, p157.

Simpson P.J., Mickelson J., Fantone J.C., Gallagher K.P., Lucchesi B.R., 1987, Iloprost inhibits neutrophil function in vitro and in vivo and limits experimental infarct size in canine heart, Circ Res. 60:666.

Smith E.F.III, Gallenkamper W., Beckman R., et al., 1984, Early and late administration of PGI_2 analogue, ZK 36374:effects on myocardial preservation, collateral blood flow and infarct size, Cardiovasc Res. 18:163-173.

IMPROVEMENT OF RENAL FUNCTION WITH SELECTIVE THROMBOXANE A_2 SYNTHETASE INHIBITOR, DP-1904 IN LUPUS NEPHRITIS

Tadashi Yoshida, Hideto Kameda, Akizuki Masashi, Mitsuo Homma, and Yasuo Ikeda

Department of Internal Medicine
School of Medicine, Keio University
35 Shinanomachi, Shinjuku-ku
Tokyo 160, Japan

INTRODUCTION

An increase in thromboxane A_2 (TXA_2) and a decrease in prostacyclin (PGI_2), which are the metabolites of arachidonic acid, have been reported in patients with lupus nephritis[1, 2]. The renal prostaglandin system locally modulates cortical and medullary function, renal blood flow, glomerular filtration rate, renin release, sodium and water excretion[3]. We therefore undertook the present study to evaluate the abnormalities of prostanoid metabolism, which may adversely affect the renal pathophysiology, in an effort to clarify the pathological significance of these prostanoids in lupus nephritis, using SLE patients without nephropathy and healthy controls.

Moreover, we attempted to correct such abnormalities by administering a TXA_2 synthetase inhibitor. This also enabled us to also study the effects of 4 days' dosing of this selective inhibitor on the prostanoids.

SUBJECTS AND METHODS

1. Subjects

The subjects enrolled in this study were eight lupus nephritis patients (eight females; mean age, Mean±SE: 39.9±12.3 years) who were treated for nephropathy as inpatients between July 1991 and July 1994 at Keio University Hospital. The patients were diagnosed as having SLE as defined by the American College of Rheumatology. Renal biopsy revealed diffuse proliferative nephritis (WHO IVd). None of the patients presented any clinical symptoms or complications other than nephropathy at the commencement of treatment. Informed consent was obtained from each.

Recent Advances in Prostaglandin, Thromboxane, and Leukotriene Research
Edited by Sinzinger *et al.*, Plenum Press, New York, 1998

2. Study Design

DP-1904 [(±) -6-(1-imidazolymethyl)-5,6,7,8-tetrahydronaphthalene-2-carboxylic acid hydrochloride hemihydrate] at a dose of 400 mg/day or indomethacin at dose of 50 mg/day was administered orally for 4 days. The doses and study periods of both drugs were chosen on the basis of comparable inhibition of extrarenal thromboxane A_2 synthetase or cyclooxygenase activities. Each patient was randomly assigned to 4 days of treatment with DP-1904 or Indomethacin at the first study. Three days before the commencement of treatment and 14 days after its completion were considered the predrug and postdrug observation periods, respectively. Two weeks washout period during which no medication was taken was required in order to minimize the potential for carryover effects from previous study period.

We avoided the use of non-steroid anti-inflammatory analgesic antipyretics with a potential effect generated on prostanoid metabolism, such as aspirin, or an anti-platelet drugs for 1 week before the commencement of this study, during the treatment period, and until the end of the study period. Adherence to the treatment was ensured by checking for the appearance of side effects.

3. Biochemical Analysis

(1) Urine samples

Starting with urination at 8:00 a.m., the total urine samples from inpatients were collected in respective polyethylene tanks until 8:00 a.m. the following day. Again, following volume measurement, an aliquot of the total urine sample was stored frozen at -20 C.

(2) Assay methods of various prostanoids (RIA-PEG method)

Under acidic condition, the samples were adsorbed onto an ODS carrier, which was washed successively with hydrochloric acid in ethanol and petroleum ether to remove the protein and lipids. The prostanoids were eluted from the ODS carrier with ethyl acetate. The eluate was fractionated by a mini-column into fractions , each of which contained various prostanoids. The prostanoid fractions were evaporated to dryness, dissolved in buffer and subjected to competitive binding assay with the antibody raised against each prostanoid after the addition of respective ^{125}I-labeled prostanoid. The antibody-antigen precipitate was collected, radioactivity was measured, and the concentration of each prostanoid was calculated.

4. Analysis of patients' clinical features

The clinical features of each patient such as the clinical sign of SLE, data of various immunological examinations, findings in the renal biopsy, data of urinalysis of the 24-hr urine, and medication were investigated under the consent of the patients. A list of these clinical feature comprising four pages was then prepared.

The differences in the prostanoids levels among the groups were examined by the means of a parametric analysis of variance for multiple comparisons and by unpaired T-test. All values are reported as means±SD. P values of less than 0.05 were considered to indicate statistical significance.

RESULTS

Effects of DP-1904 and indomethacin on prostanoid metabolism in lupus nephritis patients

The urinary excretion of TXB_2 in the patients was increased (409.5±123.1 ng/day) during the predrug observation period (Figure 1). It was decreased after 1 to 2 days of treatment with DP-1904 in all patients (61.9±13.6 ng/day, $p<0.001$), and this tendency continued until day 2 to day 3 after the treatment. However, it nearly returned to the

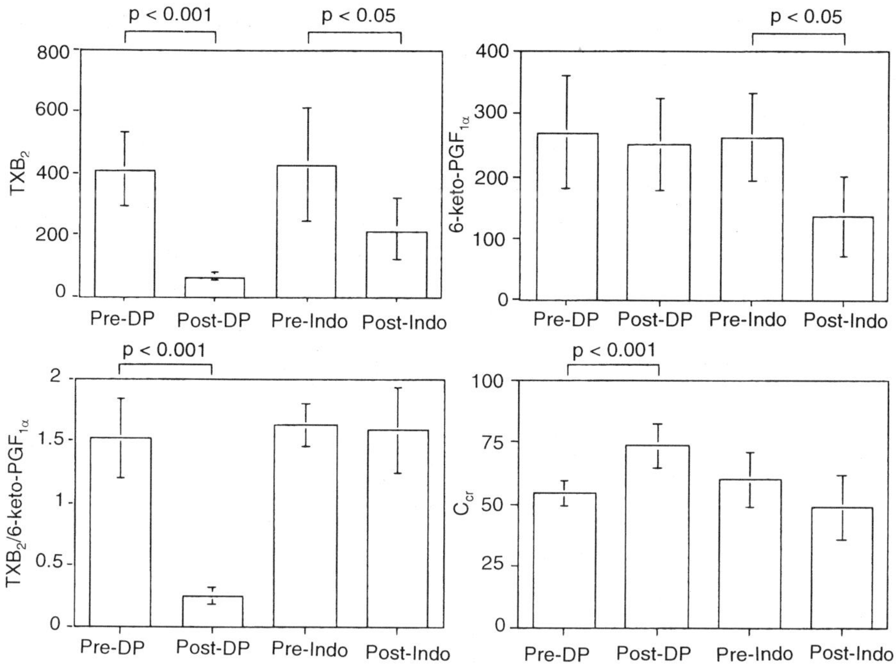

Figure 1. Changes in urinary prostanoid and renal function before and after DP-1904 and indomethacin administration.

Both urinary TXB_2 and $TXB_2/6\text{-keto-PGF}_{1\alpha}$ ratio were decreased after DP administration, showing improvement in creatinine clearance (Ccr), while both TXB_2 and $6\text{-keto-PGF}_{1\alpha}$ were decreased with no changes of urinary $TXB_2/6\text{-keto-PGF}_{1\alpha}$ ratio during the treatment with indomethacin, which revealed no changes in creatinine Ccr.

predrug values two weeks after the completion of the treatment. Changes in the urinary excretion of 11-dehydro-TXB_2 were almost the same as those in the excretion of TXB_2 (1427.2±130.4 to 217.2±95.4 ng/day, $p<0.001$), but urinary $6\text{-keto-PGF}_{1\alpha}$ and urinary 2,3-dinor-$6\text{-keto-PGF}_{1\alpha}$ were not decreased after the treatment. The urinary excretion of PGE_2 showed no prominent tendency throughout the study period in any patients. The urinary $TXB_2/6\text{-keto-PGF}_{1\alpha}$ ratio decreased (1.52 to 0.25, $p<0.001$) after the treatment and it remained decreased throughout the treatment period in all patients. In the patients, creatinine clearance began to improve after the treatment (54.5 to 73.6 ml/min, $p<0.001$) and remained increased until day 4 of post-observation, and a slight increase was noted in sodium excretion (104.3 to 110.2 mEq/day).

On the other hand, the urinary excretion of TXB_2 and 11-dehydro-TxB_2 were decreased after the treatment with indomethacin in all patients (426.6±185.5 to 216.5±101.4 ng/day, $p<0.001$, 1391.2±146.8 to 582.7±171.8 ng/day, $p<0.001$, respectively). Moreover, the urinary excretion of $6\text{-keto-PGF}_{1\alpha}$ and urinary 2,3-dinor-$6\text{-keto-PGF}_{1\alpha}$ decreased throughout the study period in the patients. The urinary excretion of PGE_2 also decreased throughout the study period in all patients (576.4 to 384.0 ng/day, $p<0.05$). The urinary 11-dehydro-TXB_2/2,3-dinor-$6\text{-keto-PGF}_{1\alpha}$ ratio decreased after the treatment (1.25 to 0.78, $p<0.01$), but the urinary $TXB_2/6\text{-keto-PGF}_{1\alpha}$ ratio did not change, which revealed no changes in creatinine clearance.

No particular symptom that was considered to be a side effect was noted during or after the treatment. Furthermore, there were no changes after 4 days of treatment in complete blood count, coagulation time, serum electrolytes, and hepatic, renal chemistry and serological test results.

DISCUSSION

Since abnormal prostanoid metabolism is noted in lupus nephritis patients, vasoactive eicosanoids have been considered to play a role in modulating renal hemodynamics in these patients[1, 2]. The ratio of TXB_2/6-keto-$PGF_{1\alpha}$ in urines of the patients with lupus nephritis was significantly higher than that in the healthy controls. These data demonstrated the possibility that the changes in the renal metabolism of PGI_2 and TXA_2 cause aggravation of lupus nephritis and inflammation.

In addition, increased production of TXB_2 at the renal cortex and medulla of lupus mice has been reported to occur during the proteinuria period[4]. It has also been reported that the amount of protein excreted in the urine is known to correlate with the amount of TXB_2 produced, and that the decrease of TXB_2 production with the injection of PGE_1 results in a decrease of urinary protein. These results also indicate the close correlation between TXA_2 and lupus nephritis[5]. Therefore, the possibility existed that the administration of a thromboxane A_2 synthetase inhibitor would terminate the vicious cycle in lupus nephritis, leading to improved clinical features of this disease.

Recently, the administration of a TXA_2 receptor antagonist, has been reported to result in recovery from the decreased GFR and RPF[6]. Furthermore, it has been stated that the clearance of p-aminohippurate and inulin was improved by 48-hr continuous intravenous drip infusion of a thromboxane antagonist. It was expected that the TXA_2 synthetase inhibitor, DP-1904, would decrease TXA_2 production and improve creatinine clearance when administered orally to lupus nephritis patients. When this drug as a TXA_2 synthetase inhibitor was administered orally to eight lupus nephritis patients to evaluate its effect in the present study, the urinary excretion of TXB_2, the urinary TXB_2/6-keto-$PGF_{1\alpha}$ ratio and 11-dehydro-TXB_2/2,3-dinor-6-keto-$PGF_{1\alpha}$ ratio were found to be markedly decreased and creatinine clearance was increased in the early course of treatment in six of the eight patients, whose renal histologic indices would suggest active renal disease. In other two patients with relatively high chronicity indices, the urinary excretion of TXB_2 and urinary TXB_2/6-keto-$PGF_{1\alpha}$ ratio were decreased and the same tendency was also noted in the clearance.

DP-1904 also had the effects on the expression of adhesion molecules and other mediators[8]. It was possible that the abnormal metabolism of prostaglandins would not only modulate renal hemodynamics, but also cause consequent changes of lupus nephritis.

Finally, impairment of renal function is at least in part mediated hemodynamically and reversible with a thromboxane synthetase inhibitor in lupus nephritis. The duration of treatment was too short for us to evaluate the utility of this drug in lupus nephritis.

REFERENCES

1. C. Patorono, G. Ciabattoni, G. Remuzzi, et al, Functional significance of renal prostacyclin and thromboxane A2 production in patients with systemic lupus erythematosus. J Clin Invest., 76:1011 (1985).
2. T. Yoshida, H. Kameda, M. Akizuki, et al, Abnormal prostanoid metabolism in systemic lupus erythematosus with pulmonary hypertension. J. Rheumatol., 21:2393 (1994).
3. L.A. Scharschmidt, A.E. Lianos, M.J. Dunn, et al, Arachidonate metabolites and the control of glomerular function. Fed Proc., 42:3058 (1983).
4. V.E. Kelley, S. Sneve, S. Musinski, et al, Increased renal thromboxane production in murine lupus nephritis. J Clin Invest., 77:252 (1986).
5. C. Patorono, and M.J. Dunn, The clinical significance of inhibition of renal prostaglandin synthesis. Kidney Int., 32:1 (1987).

6. A. Pierucci, B.M.Simonetti, G. Pecci, et al, Improvement of renal function with selective thromboxane antagonism in lupus nephritis. New Engl J Med., 320:421 (1989).
7. H. Kameda, T. Yoshida, Y. Ichikawa, et al, Thromboxane A2 synthase inhibitor, DP-1904, decreases TNFα secretion from monocytes and inhibits E-selectin and ICAM-1 expression on the endothelial cell surfaces. in : Advances in prostaglandin, thromboxane, and leukotriene research, Vol. 23, Raven Press, New York (1995).

CHARACTERIZATION OF THE PROSTAGLANDIN EP_3-RECEPTOR FROM PORCINE HEART

Jutta Meyer, Thomas Hohlfeld and Karsten Schrör

Institut für Pharmakologie, Heinrich-Heine-Universität Düsseldorf, Germany

INTRODUCTION

Protection of the myocardium from ischemic injury by prostacyclins and PGE_1 has been repeatedly reported. Part of the antiischemic effects may be mediated by cardiac sarcolemmal EP_3-receptors which have been characterized earlier in this laboratory by ligand binding studies. Of particular notice is the presence of multiple EP_3 isoforms, e.g. in mouse, cattle and man which are generated by differential splicing. Here we describe the cloning and subtype characterization of the porcine EP_3-receptor. We obtained two different isoforms of the receptor by screening a porcine heart cDNA library. The predicted receptor proteins consisted of 395 and 387 amino acids, respectively, with differences only in the C-terminal part of the protein. Several overlapping genomic clones were obtained from a porcine genomic library. Sequence comparison of the cDNA with the genomic DNA indicated that an intron was located in the region corresponding to the sixth transmembrane domain of the receptor protein. This position exactly matches the corresponding intron in the human TXA_2-, the human PGD_2- and the mouse PGE-receptor (EP_1-subtype). In further experiments, we expressed the porcine EP_3-receptor in COS-7 cells and performed ligand-binding-studies. In addition the expression of the porcine EP_3-receptor mRNA in different tissues was demonstrated by RT-PCR.

METHODS

Receptor cloning

About 8 x 10^5 plaques of a Clontech oligodt primed λgt11 porcine heart cDNA library were screened with a 1.05 kb *Eco*RI/*Bam*HI-cDNA fragment of the murine EP_3-receptor (1). Hybridization was performed at 37°C in a solution containing 50% formamide. Three positive clones were isolated containing inserts of about 0.9 kb, 1.1 kb and 1.7 kb length, respectively. The inserts were amplified using λgt11 specific PCR primers which comprise sequences flanking the 5′and 3′end of the *Eco*RI cloned cDNA insert (Clontech). The resulting PCR fragments were subcloned into pUC18 (p3909, 0.9 kb insert; p13311, 1.1 kb insert; p6917, 1.7 kb insert) and sequenced in an ALF*express* sequencer (Pharmacia Biotec) using the primer walking strategy. The 1.1 kb and the 1.7 kb insert contained different receptor isoforms. Sequences are registrated in the GenEMBL database (U27083, U30374). A porcine genomic DNA $EMBL3_{SP6/T7}$ library (Clontech) was screened using a homologous cDNA fragment (*Eco*RI/*Bam*HI) under high stringency conditions. Overlapping insert-fragments were subcloned and sequenced.

Cell culture

Porcine vascular smooth muscle cells were grown in Ham's F12 medium supplemented with 20% DMEM, 15% FCS, penicillin G (100U/ml) and streptomycine (0.1 mg(ml). Porcine aortic endothelial cells (PAE) were grown in DMEM supplemented with FCS and antibiotics as above. COS-7 cells were grown in DMEM-Glutamax, supplemented with 10% FCS and antibiotics as above.

Receptor expression analysis

Total RNA was prepared with TRIZOL (GIBCO-BRL) and used in the RT-PCR reaction. RT-PCR was performed with homologous EP_3-receptor specific primers which bind in the 4. and 7. transmembrane domain of the corresponding protein, resulting in a 452 bp PCR product. PCR was performed at 58°C, 2mM $MgCl_2$, 20 cycles. Heterologous GAPDH primers were added in each reaction for semiquantitative analysis. Detection of the PCR fragments was performed on an ALF*express* sequencer.

RESULTS AND DISCUSSION

Sequence analysis of several cDNA fragments from a porcine heart cDNA library revealed that two different isoforms of the porcine EP_3-receptor had been cloned. A 1.7 kb cDNA fragment contained the complete receptor sequence. A second cDNA insert lacking parts of the 5'translated sequence differed at the 3'end of the translated region from the full length insert (Fig. 1). This results in different carboxyterminal-domains of the predicted receptor proteins. The two different isoforms were named isoform A and isoform B, respectively.

The amino acid sequence of both porcine EP_3-receptor isoforms was deduced from the nucleic acid sequence. The two receptor isoforms consist of 395 amino acids (isoform A) or

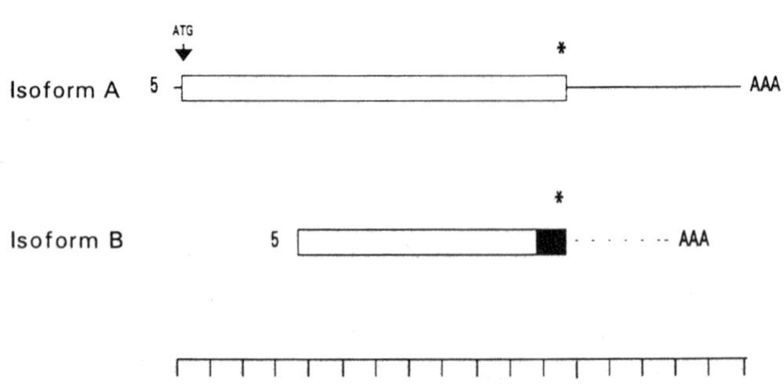

Figure 1. A 1.7 kb cDNA fragment contained the complete receptor sequence (isoform A). The open reading frame is indicated with the ATG and the stop codon (asterisk). A 1.1 kb cDNA insert differed in the 3'part of the translated region from isoform A (black box) and the following untranslated region (isoform B).

```
isoform A:      FCQAVSQKQREEAATLIFTHLSISRTEPGEARVLFSKSKC

isoform B:      FCQVANAVCSCSKNGQKVQTISLSHEITQTEA
human EP3-II:   FCQVANAVSSCSNDGQKGQPISLSNEIIQTEA
murine EP3γ:    FCQVANAVSSCSSDGQKGQAISLSNEVVQPGP
              *******  ***   ***  *  ****  *   *
```

Figure 2. Sequence comparison of the C-termini of the two porcine EP_3-receptor isoforms.

387 amino acids (isoform B), respectively, with an estimated molecular weight of about 43 kd. A sequence comparison revealed strong homology to the human EP_3-receptor (2) (88% identity on protein level) and the bovine EP_3-receptor (3). A strong sequence homology between isoform B, the human $EP_{3\text{-II}}$- and the murine $EP_{3\gamma}$-receptor (4) was detected in a clustal alignment of the C-termini (Fig. 2). It is known that these receptors are coupled to G_i proteins. Isoform A showed no sequence homologies to the C-termini of any hitherto cloned receptor isoform.

A Kyte-Doolittle hydropathy plot analysis revealed 7 putative transmembrane domains (Fig. 3). Two putative phosphorylation sites for cAMP-dependent protein kinases were located in the first intracellular loop (Ser_{81} and Ser_{86}) of the predicted receptor proteins. Isoform A showed additionally a putative protein kinase C-dependent phosphorylation site in the carboxyterminal domain (Ser_{362}).

Sequencing of the genomic region of the porcine EP_3-receptor gene revealed an intron which was located at the end of the sixth transmembrane domain of the corresponding protein (Leu_{298}). This position matches exactly the corresponding intron in the TXA_2- and the EP_1 receptor gene (4, 5).

COS-7 cells expressing the cloned porcine EP_3-receptors showed specific binding of PGE_1. Displacement of $[^3H]$-PGE_1 occured in experiments using PGE_1 and M&B 28.767 but not butaprost (data not shown). Analysis of the receptor expression in different tissues and cells of the pig by means of RT-PCR revealed the presence of EP_3-receptor transcripts in heart, kidney and cultured vascular smooth muscle cells (VSMC) but not in cultured aortic endothelial cells (PAE) (data not shown).

In conclusion we present the complete sequence of the porcine EP_3-receptor isoform A and isoform B. As suggested for previously published EP_3-receptor isoforms, different

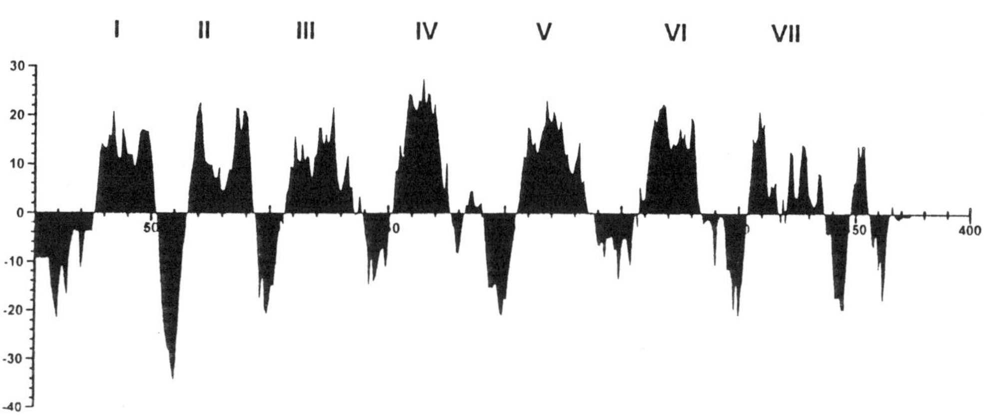

Figure 3. Kyte-Doolittle hydropathy plot analysis of the porcine EP_3-receptor isoform A. The seven hydrophobic domains (I-VII) correspond to the seven transmembrane domains of the receptor protein.

isoforms may allow different modulation of signal transduction by coupling to different G proteins. Coupling of the porcine EP_3-receptor, at least isoform B, to inhibitory G proteins seems very likely due to the sequence homologies. This could explain at least partly the cardioprotective effects of prostaglandins by diminishing detrimental catecholamine effects. A possible regulation of the receptor protein by phosphorylation, as expected due to the putative PKA- and PKC-dependent phosphorylation sites, has to be investigated.

ACKNOWLEDGMENT

This study was supported by the DFG (SFB 242, A9, C5).

REFERENCES

1. Y. Sugimoto, T. Namba, A. Honda, M. Negishi, A. Ichikawa, and S. Narumiya, Cloning and expression of a cDNA for mouse prostaglandin E receptor EP_3 subtype, *J Biol Chem.* 267: 6463 (1992).

2. M. Kotani, I. Tanaka, Y. Ogawa, T. Usui, K. Mori, A. Ichikawa, S. Narumiya, T. Yoshimi, and K. Nakao, Molecular cloning and expression of multiple isoforms of human prostaglandin E receptor EP_3 subtype generated by alternative messenger RNA splicing: multiple second messenger systems and tissue specific distribution, *Mol Pharmacol.* 48: 869 (1995).

3. T. Namba, Y. Sugimoto, M. Negishi, A. Irie, F. Ushikubi, A. Kakizuka, S. Ito, A. Ichikawa, and S. Narumiya, Alternative splicing of C-terminal tail of prostaglandin E receptor subtype EP_3 determines G-protein specificity, *Nature* 365: 166 (1993).

4. B. Båtshake, C. Nilsson, and J. Sundelin, Molecular characterization of the mouse prostanoid EP_1 receptor gene, *Eur J Biochem* 231: 809 (1995).

5. R.M. Nüsing, M. Hirata, A. Kakizuka, T. Eki, K. Ozawa, and S. Narumiya, Characterization and chromosomal mapping of the human thromboxane A_2 receptor gene, *J Biol. Chem.* 268: 25253 (1993).

BISALLYLIC HYDROXYLATION OF LINOLEIC AND ARACHIDONIC ACIDS BY ADULT AND FETAL HUMAN LIVER MICROSOMES AND A COMPARISON WITH HUMAN RECOMBINANT CYTOCHROMES P450

Lena Hörnsten, Johan Bylund and Ernst H Oliw

Department of Pharmaceutical Biosciences,
Uppsala University Biomedical Center,
S-751 24 Uppsala, Sweden

INTRODUCTION

Cytochromes P450 (CYP) of liver microsomes can oxygenate polyunsaturated fatty acids[1,2]. Hydroxylations of carbons at the ω end and epoxidations of the double bonds have attracted greatest attention. ω-Hydroxylation of polyunsaturated fatty acids is efficiently catalyzed by CYP4A, while other enzymes catalyze ω2 and ω3 hydroxylations of arachidonic acid (20:4n-6). Enzymes of the CYP2B and CYP2C subfamily oxygenate 20:4n-6 to epoxides[1,2], but all the constitutive epoxygenases may not have been identified.

Liver microsomes can catalyze at least two additional NADPH-dependent oxygenation reactions: bisallylic hydroxylation and hydroxylation with double bond migration[3]. For example, the bisallylic C-11 of 18:2n-6 can be hydroxylatad by liver microsomes of phenobarbital-treated rats to 11-HODE[3] and the three bisallylic carbons of 20:4n-6 can be hydroxylated to yield three hydroxyeicosatetraenoic acids, 7-HETE, 10-HETE and 13-HETE[4]. Adult human liver microsomes have previously been found to form 13-HETE and 11-HODE[5]. These bisallylic metabolites are chemically unstable at low pH and decompose to *cis-trans* conjugated hydroxy fatty acids[6]. We recently reported that dexamethasone (DEX) treatment potently and selectively induced the bisallylic hydroxylation activity of rat liver microsomes[7].

Our main objective was to investigate adult and fetal human liver microsomes for bisallylic hydroxylation activity with 18:2n-6 and 20:4n-6 as substrates. We also investigated eight recombinant human CYP enzymes for bisallylic hydroxylation activity. Linoleic acid is convenient for this purpose, while arachidonic acid can be oxygenated by some enzymes to a complex mixture of metabolites. Finally, we also examined and compared the effect of treatment with erythromycin (ERY), DEX, starvation and gender on rat liver bisallylic hydroxylation activity.

Recent Advances in Prostaglandin, Thromboxane, and Leukotriene Research
Edited by Sinzinger *et al.*, Plenum Press, New York, 1998

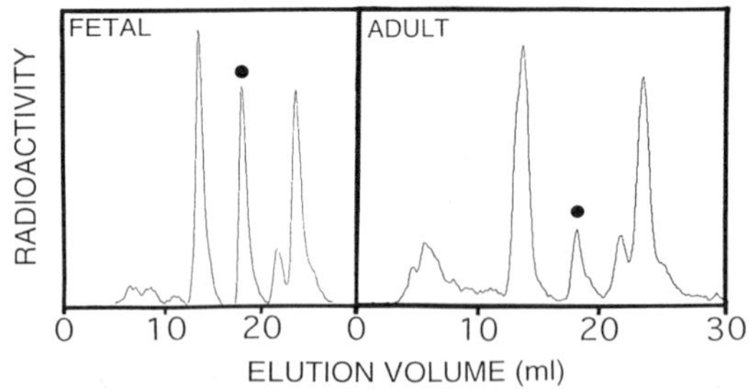

Figure 1. RP-HPLC of metabolites formed from [^{14}C]18:2n-6 by human by fetal and adult human liver microsomes and NADPH. The product with the elution volume of 11-HODE is marked •. 17-HODE and 18-HODE elute after 22 and 24 min, respectively, and diols after 13 min.

EXPERIMENTAL

Adult human liver samples were obtained from the liver bank of the Department of Clinical Pharmacology at Huddinge sjukhus, Karolinska institutet, Stockholm, and human fetal liver samples from the Division of Clinical Pharmacology, Uppsala university. Incubations of fatty acids with liver microsomes were performed and the products analyzed by RP-HPLC and SP-HPLC as described[7]. Microsomes of recombinant human CYP enzymes, which were expressed in lymphoblastoid cells (CYP1A2, CYP2A6, CYP2B6, CYP2D6, CYP2E1) or in insect cells (CYP3A4, CYP2C8, CYP2C9) were purchased from Gentest Corp. (MA, USA) as well as control microsomes. The recombinant enzymes were used as recommended by the supplier.

RESULTS AND DISCUSSION

Two samples of fetal human liver microsomes metabolized 18:2n-6 to the bisallylic hydroxy metabolite 11-HODE as a major product, while adult liver microsomes formed relatively small amounts of 11-HODE (Figure 1). 20:4n-6 was metabolized to 13-HETE, 10-HETE and 7-HETE by both adult and fetal liver microsomes as judged from SP-HPLC.

Our next goal was to identify human recombinant CYP enzymes with significant bisallylic hydroxylation activity using HPLC for analysis of 20:4n-6 and 18:2n-6 metabolites. CYP1A2, CYP2C8, CYP2C9 and CYP3A4 converted [^{14}C]18:2n-6 to [^{14}C]11-HODE as well as to other metabolites (Figure 2), while as series of other enzymes (CYP2A6, CYP2B6, CYP2D6, and CYP2E1) did not form significant amounts of 11-HODE. [^{14}C]20:4n-6 was also oxygenated by CYP1A2, CYP3A4, and to some extent by CYP2C8 to ^{14}C-labeled 13-HETE, 10-HETE and 7-HETE. Both substrates were also epoxidized and hydroxylated at the terminal carbons. A comparison of metabolites formed by CYP3A4 and liver microsomes of DEX treated rats is shown in Figure 3.

CYP2C9 formed 12R-HETE as a major monohydroxy metabolites as judged from straight phase and chiral HPLC (Chiralcel ODH). Biosynthesis of 12-HETE by CYP2C9 is in agreement with a recent study of Rifkind et al[8]. The mechanism of biosynthesis of 12R-HETE has not been determined, but it could be formed by hydroxylation with double bond migration[3].

Bisallylic hydroxylases were also studied in liver microsomes of Fischer 344 rats. Treatment with erythromycin (ER; 734 mg/kg for 5 days) induced bisallylic hydroxylation of 18:2n-6 to 11-HODE about 2- fold, while DEX treatment (150 mg/kg for 4 days) induced it 10-fold[7]. A large number of different agents increase bisallylic hydroxylation activity of rat

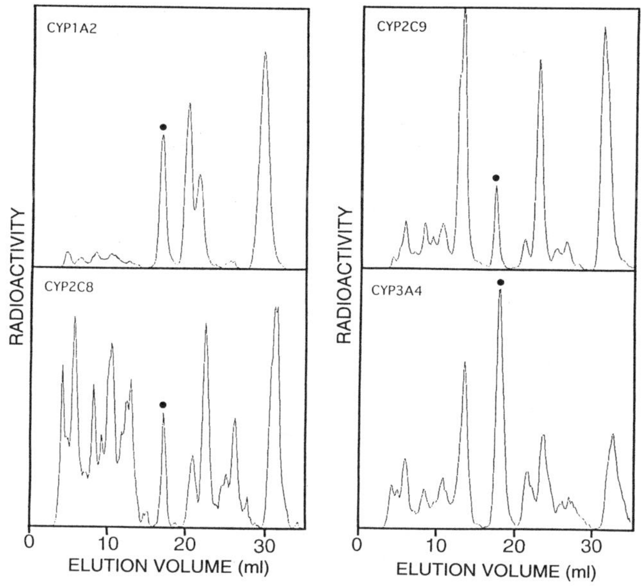

Figure 2. RP-HPLC of metabolites formed from [^{14}C]18:2n-6 by human recombinant CYP1A2, CYP2C8, CYP2C9 and CYP3A4. The peak marked • in the chromatograms had the same elution time as 11-HODE and it was totally absent in microsomes without expressed protein.

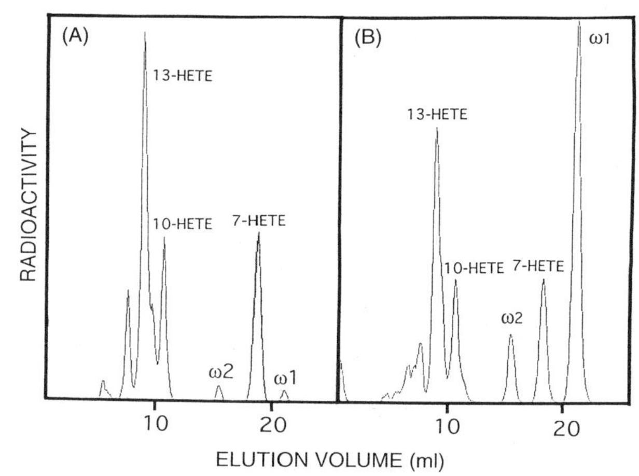

Figure 3. HPLC of metabolites formed from [^{14}C]20:4n-6 by human recombinant CYP3A4 and liver microsomes of DEX treated rats. A, SP-HPLC of HETEs formed by CYP3A4. B, SP-HPLC of HETEs formed by liver microsomes of DEX treated rats. These microsomes generate a mixture of 13-HETE, 10-HETE, 7-HETE as well as 19-HETE and 20-HETE, which has been characterized by GC-MS[7].

liver microsomes, but most of them also increased hydroxylation of the ω side chain[7]. Only DEX and to some extent ERY treatment appeared to selectively increase bisallylic hydroxylases of rat liver microsomes. Both DEX and ERY also increased testosterone 6β-hydroxylase activity, which can be attributed to induction of CYP3A1. ERY appeared to be a poor inducer of enzyme activity in comparison with DEX, possibly related to inhibition of CYP3A1 by ERY. CYP3A1 is also inhibited by troleandomycin (TAO)[7]. The biosynthesis of 11-HODE by microsomes of DEX treated rats was strongly inhibited by TAO with an ED_{50} of

1 μM. Microsomes of DEX treated rats oxidized 20:4n-6 to 13-HETE, 10-HETE and 7-HETE, but the induction of bisallylic hydroxylation activity by DEX was less apparent with 20:4n-6 than with 18:2n-6 as a substrate. There appeared to be no difference between rats of both sexes concerning bisallylic hydroxylation activity after treatment with DEX. The results suggested that CYP3A1 of liver microsomes of DEX treated rats could catalyze bisallylic hydroxylations of 18:2n-6 and 20:4n-6.

TAO neither inhibit the biosynthesis of 11-HODE by human adult liver nor by fetal liver microsomes, indicating that other human CYP enzymes than members of the CYP3A subfamily could be constitutive bisallylic hydroxylases. Studies are now in progress with specific inhibitors to determine whether CYP1A2 or enzymes of the CYP2C subfamily might have this function.

ACKNOWLEDGMENTS

Supported by the Swedish Medical Research Council (6523).

REFERENCES

1. E.H. Oliw, Oxygenation of polyunsaturated fatty acids by cytochrome P450 monooxygenases. *Prog. Lipid Res.* 33:329 (1994).
2. E.H. Oliw, J. Bylund, and L. Hörnsten, Bisallylic hydroxylation and epoxidation of polyunsaturated fatty acids by cytochrome P450. *Lipids* 31:1003 (1996)
3. E.H Oliw, I.D. Brodowsky, L. Hörnsten, and M. Hamberg, *bis*-Allylic hydroxylation of polyunsaturated fatty acids by hepatic monooxygenases and its relation to the enzymatic and non-enzymatic formation of conjugated hydroxy fatty acids. *Arch. Biochem. Biophys.* 300:434 (1993).
4. A.R. Brash, W.E. Boeglin, J.H. Capdevila, S. Yeola, and I.A. Blair, 7-HETE, 10-HETE, and 13-HETE are major products of NADPH-dependent arachidonic acid metabolism in rat liver microsomes: analysis of their stereochemistry and the stereochemistry of their acid-catalyzed rearrangement. *Arch. Biochem. Biophys.* 321:485 (1995).
5. E.H. Oliw, *bis*-Allylic hydroxylation of linoleic acid and arachidonic acid by human hepatic monooxygenases. *Biochim. Biophys. Acta* 1166:258 (1993).
6. M. Hamberg, W.H. Gerwick, and P.A. Åsen, Linoleic acid metabolism in the red alga *Lithothamnion corallioides*: biosynthesis of 11(R)-hydroxy-9(Z), 12(Z)-octadecadioenoic acid. *Lipids* 27:487 (1992).
7. L. Hörnsten, J. Bylund, and E.H. Oliw, Dexamethasone induces bisallylic hydroxylation of polyunsaturated fatty acids by rat liver microsomes. *Arch. Biochem. Biophys.* 332:261 (1996).
8. A.B. Rifkind, C. Lee, T.K.H. Chang, and D.J. Waxman, Arachidonic acid metabolism by human cytochrome P450s 2C8, 2C9, 2E1, and 1A2: regioselective oxygenation and evidence for a role for CYP2C enzymes in arachidonic acid epoxygenatiion in human liver microsomes. *Arch. Biochem. Biophys.* 320:380 (1995).

ORGANIZATION OF THE HUMAN PROSTACYCLIN SYNTHASE GENE AND ASSOCIATION ANALYSIS OF A NOVEL CA REPEAT IN ESSENTIAL HYPERTENSION

Tomohiro Nakayama, Masayoshi Soma, Katsuo Kanmatsuse

Second Department of Internal Medicine, Nihon University School of Medicine, Tokyo, Japan

INTRODUCTION

In 1976, Vane et al discovered prostacyclin (PGI_2) that is a potent inhibitor of platelet secretion and aggregation, smooth muscle cell proliferation and vasoconstriction [1-3]. It is considered to play a key role in vasoprotection and its deficiency may lead to thrombosis and other vascular lesions [4,5]. Biosynthesis of PGI_2 is catalyzed serially by phospholipase A_2 which liberates arachidonic acid from the sn-2 position of membrane phospholipids, prostaglandin H_2 synthase, a bifunctional enzyme which catalyzes the conversion of arachidonic acid to prostaglandin G_2 and subsequently to prostaglandin H_2, and prostacyclin synthase (PGIS, EC 5.3.99.4) which catalyzes the formation of PGI_2 from prostaglandin H_2. PGIS has been shown to be widely distributed, predominantly in vascular endothelial and smooth muscle cells [6,7]. The enzyme is a membrane-bound protein located in the endoplasmic reticulum. It exhibits spectral characteristics of cytochrome P450s but has no mono-oxygenase activity [8,9]. Instead, it catalyzes an isomerization reaction and thus does not require a reductase to initiate the catalytic activity. A description has been given of the cDNA of human PGIS [10] and the deduced amino acid sequence. In the present study, the genomic organization of the human PGIS gene was investigated [11], and the association study of essential hypertension was carried out using a novel CA/TG dinucleotide repeat polymorphism in this gene [12].

MATERIALS AND METHODS

Isolation and Characterization of PGIS Genomic Clones

Based on the information for the cDNA of human PGIS, we were able to design some oligonucleotide primers. Fortunately, a genomic clone (λ hPGIS141) had previously been isolated in the middle of the cDNA of PGIS [10], and certain primers were designed and synthesized using the nucleotide sequence of this clone. The primer at the 5'- upstream

Recent Advances in Prostaglandin, Thromboxane, and Leukotriene Research
Edited by Sinzinger *et al.*, Plenum Press, New York, 1998

region and the primer at the 3'- downstream region of this genomic clone were designed to amplify the intron regions adjoining this clone. The long and accurate polymerase chain reaction (LA PCR)[13] was performed with each of by these primers, and succeeded in amplifying the single PCR products. More than 10 primers were thus designed and synthesized to certain fragments of PCR products covering all genomic sequences from the start codon to the termination codon.

Sequencing

The LA PCR-amplified DNA segments were directly sequenced using a Thermo Sequenase kit (U.S. Amersham LIFE SCIENCE Corp., Cleveland, OH), after purification with a column kit (MICROCON, U.S. Amicon Inc. Beverly, MA). More than 40 nucleotides of the coding regions of the PGIS gene were sequenced together with the exon-intron boundaries.

Isolation of CA/TG repeat microsatellites

CA/TG repeat microsatellites were examined in these PCR products by the hybridization described previously [14]. A CA/TG repeat microsatellite was found at about 9 kb downstream from the exon-intron junction of the exon 6 and was proven to be a sequence tagged site [14]. Primer sequences which can be used to amplify this microsatellite were as follows:
PGIS-F (CA strand) 5'TAAACGGAGGCAAGGCTCCGT3';
PGIS-R (TG strand) 5'CTGCCTGCAGGAAAAAGCTGTGG3'.
The amplification was performed as described earlier [15].

Subjects for association study

The study group consisted of 130 patients (mean age 51.8±9.9) with EH diagnosed according to WHO criteria, which included a sitting systolic blood pressure (SBP) of more than 160 mmHg and/or a diastolic blood pressure (DBP) of more than 95 mmHg on 3 occasions spanning 2 months from the first medical examination and without anti-hypertensive medication. Subjects diagnosed as having secondary hypertension were excluded. One hundred twenty five normotensive healthy subjects (mean age 51.1±6.4) served as controls.The normotensive (NT) subjects had no family history of hypertension, and in all cases their SBP was less than 140 mmHg and, their DBP less than 85 mmHg. A positive family history was defined as hypertension diagnosed in grandparents, uncles, aunts, parents, or siblings.

RESULTS AND DISCUSSION

The structural organization of the human PGIS gene was determined by analysis of the genomic clones and of the total genomic DNA by PCR amplification across introns followed by sequencing of the intron/exon junctions. The cDNA is interrupted by 9 introns of variable size. The PGIS gene is approximately 70 kilobases (kb) in length, which is more than 30 times the length of the cDNA and typical for a mammalian gene (Fig. 1). The size range for the exons is from 74 base pairs (bp) (ATG-exon 1) to 182 bp (exons 6 and 8) and that for the introns is from approximately 1 kbp (intron 7) to 19 kbp (intron 1). All the exon-intron junctions conform to the consensus sequences (AG/GT) [16]. In the present study, we determined the genomic structure of human PGIS by direct sequencing for the PCR products, not by screening of the genomic library. The LA PCR can amplify long lengths of DNA of more than 20 kbp [13]. Confirmation of whether or not the PCR products amplified the PGIS gene was performed by nucleotide sequencing of more than 40 nucleotides within the cDNA region of the 5'-portion and 3'-portion of the exon. Our study demonstrated that the LA PCR and direct sequencing of PCR products were available for determining the structure of the gene.

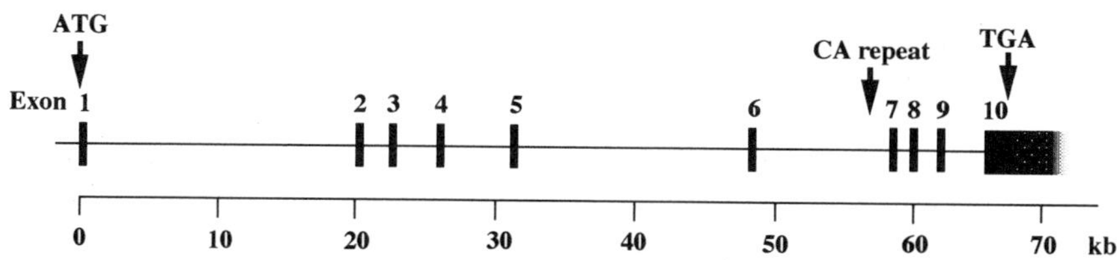

Figure 1. Organization of the human PGIS gene. Top line, exons are indicated by solid bars. The 3' end of exon 10 are not depicted. The arrows indicate the positions of initiation condon, termination condon and CA repeat. Second line, scale in kilobases.

A novel CA/TG dinucleotide repeat polymorphism in intron 6 of the PGIS was identified. It was polymorphic in 245 unrelated Japanese subjects, with the repeat size ranging from 10-14 CAs giving an allele size of 178 to 186 bp. The estimated allele frequencies shows that the twelve CA repeat allele being present more than 90 percent of the time. The overall distribution of the alleles was not significantly different between the EH and NT groups (χ^2= 5.5, df = 3, p = 0.14). Table 1 shows genotype frequencies of the PGIS gene in the 245 subjects. The twelve/twelve homozygote repeat allele was the most frequent, being present 87% of the time in the healthy Japanese subjects. The observed and expected heterozygosities were 0.13 and 0.12, respectively, which were in good agreement with the Hardy Weinberg Equilibrium. The genotypes of 10/10, 10/13, 12/14 were identified only in the EH patients.These patients showed complications of hypertension such as changes in optic fundus and left ventricular hypertrophy.

Table 1. Genotype frequencies of the CA repeat numbers of the PGIS gene

Genotype	EH (n = 130)		NT (n = 125)	
	No.	(Frequency)	No.	(Frequency)
10/10	1	(0.008)	0	(0.000)
10/12	18	(0.138)	10	(0.080)
10/13	1	(0.008)	0	(0.000)
11/12	0	(0.000)	4	(0.032)
12/12	108	(0.831)	109	(0.872)
12/13	1	(0.008)	2	(0.016)
12/14	1	(0.008)	0	(0.000)

In summary, we determined the structural organization of the human PGIS gene and found a novel CA/TG repeat in intron 6. Distributions of allele frequencies in the subjects with essential hypertension and normotensive controls were not different. Thus, the genes detected by examination of this microsatellite polymorphism in the PGIS gene do not associate with essential hypertension.

Acknowledgments

We wish to acknowledge Ms H Tobe for her technical assistance, and Drs A Kubo, M Kunimoto, M Watanabe, N Fukuda, Y Watanabe and Y Izumi for collecting samples.

REFERENCES

1. S. Moncada, R.J. Gryglewski, S. Bunting and J.R. Vane, An enzyme isolated from arteries transforms prostaglandin endoperoxides to an unstable substance that inhibits platelet aggregation, *Nature.* 263:663-665 (1976).
2. J. Svensson, M. Hamberg, and B. Samuelsson, On the formation and effects of thromboxane A_2 in human platelets, *Acta Physiol Scand.* 98:285-294 (1976).
3. J.S. Pober and R.S. Cotran, Physiol rev cytokines and endothelial cell biology. *Physiol Rev.* 70: 427-455 (1990).
4. S, Bunting, S. Moncada, and J.R. Vane, The prostacyclin - thromboxane A_2 balance: pathophysiological and therapeutic implications. *Brit Med Bull.* 39:271-276 (1983).
5. J.A. Oates, G.A. FitzGerald, and R.A. Branch, Clinical implications of prostaglandin and thromboxane A2 formation, *New Eng J Med.* 319:689-717 (1988).
6. C. Ingerman-Wojenski, M.J. Silver, J.B Smith, and E. Macarak, Bovine endothelial cells in culture produce thromboxane as well as prostacyclin, *J Clin Invest* 67: 1292-1296 (1981).
7. W.L. Smith, D.L. DeWitt, and M.L. Allen, Bimodal distribution of the prostaglandin I_2 synthase antigen in smooth muscle cells. *J Biol Chem* 258:5922-5926 (1983) .
8. V. Ullrich, L. Castle, and P. Weber, Spectral evidence for the cytochrome P450 nature of prostacyclin synthetase, *Biochem Pharmacol* 30:2033-2036 (1981).
9. V. Ullrich, M. Hecker, A concept for the mechanism of prostacyclin and thromboxane A2 biosynthesis, *Advances in Prostaglandin, Thromboxane, & Leukotriene Research.* 20:95-101 (1990).
10. A. Miyata, S. Hara, C. Yokoyama, H. Inoue, V. Ullrich, and T. Tanabe, Molecular cloning and expression of human prostacyclin synthase, *Biochem Biophys Res Commun.* 200:1728-1734 (1994).
11. T. Nakayama, M. Soma, Y. Izumi, and K. Kanmatsuse, Organization of the human prostacyclin synthase gene, *Biochem Biophys Res Commun* 221: 803-806 (1996).
12. T. Nakayama, M. Soma, Y. Takahashi, J. Uwabo, Y. Izumi, an K. Kanmatsuse, Novel polymorphic CA/TG repeat identified in the human prostacyclin synthase gene, *Hum Hered* 47:176-177 (1997).
13. S. Cheng, C. Fockler, W.M. Barnes, and R. Higuchi, Effective amplification of long targets from cloned inserts and human genomic DNA, *Proc Natl Acad Sci USA.* 91: 5695-5699 (1994) .
14. M. Iizuka, R. Makino, T. Sekiya, and K. Hayashi, Selective isolation of highly polymorphic (dC-dA)n. (dG-dT)n microsatellites by stringent hybridization. *Genet Anal Tech Appl.* 10:2-5 (1993).
15. T. Nakayama, M. Soma, Y. Izumi, K. Kanmatsuse, and M. Esumi, CA repeat polymorphism of the endothelial nitric oxide synthase gene in Japanese, *Hum Hered.* 45:301-302 (1995).
16. S.M. Mount, A catalogue of splice junction sequences. A catalogue of splice junction sequences. 10: 459-472 (1982).

MECHANISM OF ACTION OF ANTI-INFLAMMATORY DRUGS

John R. Vane and Regina M. Botting

The William Harvey Research Institute,
St Bartholomew's and the Royal London School of Medicine and Dentistry,
Queen Mary and Westfield College, Charterhouse Square,
London, ECIM 6BQ, UK.

INTRODUCTION

25 years ago, Vane proposed that the mechanism of action of the aspirin-like drugs (non-steroid anti-inflammatory drugs; NSAIDs) was through the inhibition of prostaglandin biosynthesis[1] and, there is now a general acceptance of the theory. The inhibition by aspirin is due to the irreversible acetylation of the cyclooxygenase (COX) site of prostaglandin endoperoxide synthase, leaving the peroxidase activity of the enzyme unaffected. In contrast to this unique irreversible action of aspirin, other NSAIDs such as ibuprofen or indomethacin produce reversible or irreversible COX inhibition by competing with the substrate, arachidonic acid, for the active site of the enzyme.

The inhibition of prostaglandin synthesis by NSAIDs has been demonstrated in a wide variety of systems, ranging from microsomal enzyme preparations, cells and tissues to whole animals and man. We now know that there are at least two isoenzymes, named COX-1 (or PGH synthase-1) and COX-2 (or PGH synthase-2).[2] Over the last two decades, several new drugs have reached the market based on COX-1 enzyme screens.

Garavito and his colleagues[3] have determined the three dimensional structure of COX-1, providing a new understanding for the actions of COX inhibitors. Each dimer of COX-1 comprises three independent folding units: an epidermal growth factor-like domain, a membrane binding motif and an enzymatic domain. The sites for peroxidase and cyclooxygenase activity are adjacent but spatially distinct. The conformation of the membrane-binding motif strongly suggests that the enzyme integrates into only a single leaflet of the lipid bilayer and is thus a monotopic membrane protein. Three of the helices of the structure form the entrance to the cyclo-oxygenase channel and their insertion into the membrane could allow arachidonic acid to gain access to the active site from the interior of the bilayer. The cyclooxygenase active site is a long, hydrophobic channel and Garavito et al.[3] provide evidence to suggest that some of the aspirin-like drugs, such as flurbiprofen, inhibit COX-1 by excluding arachidonate from the upper portion of the channel. Tyrosine 385 and serine 530 are at the apex of the long active site. Aspirin irreversibly inhibits COX-1

Recent Advances in Prostaglandin, Thromboxane, and Leukotriene Research
Edited by Sinzinger *et al.*, Plenum Press, New York, 1998

by acetylation of the serine 530, thereby excluding access of arachidonic acid. The S(-) stereoisomer of flurbiprofen interacts via its carboxylate with arginine 120, thereby placing the second phenyl ring within Van der Waal's contact of tyrosine 385. There may be a number of other sub-sites for drug binding in this narrow channel.

The three dimensional structure of COX-2 has also been published.[4] It closely resembles the structure of COX-1, except that the COX-2 active site is slightly larger and can accommodate bigger structures than those which are able to reach the active site of COX-1. A secondary internal pocket of COX-2 contributes significantly to the larger volume of the active site in this enzyme, although the central channel is also bigger by 17%. Selectivity for COX-2 inhibitors is dependent on the presence in the enzyme of valine at position 523 in place of the isoleucine found in COX-1.[5]

TWO ISOENZYMES

The constitutive isoform, COX-1, has clear physiological functions. Its activation leads, for instance, to the production of prostacyclin which when released by the endothelium is anti-thrombogenic[6] and when released by the gastric mucosa is cytoprotective.[7] It is also COX-1 in platelets that leads to thromboxane A_2 production, causing aggregation of the platelets to prevent inappropriate bleeding.[8] The existence of the inducible isoform, COX-2, was first suspected when Needleman and his group showed that cytokines induced the expression of COX protein[9] and that bacterial lipopolysaccharide increased the synthesis of prostaglandins in human monocytes *in vitro*[10] and in mouse peritoneal macrophages *in vivo*.[11] This increase was inhibited by dexamethasone and associated with *de novo* synthesis of new COX protein. A year or so later, COX-2 was identified as a distinct isoform encoded by a different gene from COX-1.[12-14] The human COX-2 gene at 8.3kb is similar to the COX-2 gene of mouse and chicken, but smaller than the 22kb human COX-1 gene. The sequence of its cDNA shows a 60% homology with the sequence of the non-inducible enzyme, with the mRNA for the inducible enzyme being approximately 4.5 kb and that of the constitutive enzyme being 2.8 kb. However, both enzymes have a molecular weight of 71 Kd and similar active sites for the natural substrate and for blockade by NSAIDs. The inhibition by the glucocorticoids of the expression of COX-2 is an additional aspect of the anti-inflammatory action of the corticosteroids. The levels of COX-2, normally very low in cells, are tightly controlled by a number of factors including cytokines, intracellular messengers and availability of substrate.

COX-2 is induced by inflammatory stimuli and by cytokines in migratory and other cells, suggesting that the anti-inflammatory actions of NSAIDs are due to the inhibition of COX-2, whereas the unwanted side effects such as damage to the stomach lining and toxic effects on the kidney are due to inhibition of the constitutive enzyme, COX-1.

FUNCTIONS OF COX-1 AND COX-2

COX-1 performs a "housekeeping" function to synthesize prostaglandins which regulate normal cell activity. The concentration of the enzyme largely remains stable, but small (2-4 fold) increases in expression can occur in response to stimulation with hormones or growth factors.[15,16] Normally, little or no COX-2 is found in resting cells but its expression can be increased dramatically after exposure of cells to bacterial lipopolysaccharide, phorbol esters, cytokines or growth factors. However, "constitutive" levels of COX-2 have been detected in some organs such as the brain and in uterine tissues during gestation.

In the Gastrointestinal Tract

In most species, including man, the bulk of the "cytoprotective" prostaglandins in the stomach are synthesised by COX-1 although small quantities of COX-2 can also be expressed in the stomach of the rat.[17] It might be expected that mice rendered deficient in the gene for COX-1 would develop spontaneous gastric bleeding or erosions but they do not.[18] Perhaps compensatory protective mechanisms, such as the synthesis of nitric oxide or CGRP, take over the protective role of the prostaglandins.

It is interesting that COX-2 is highly expressed in human and animal colon cancer cells as well as in human colorectal adenocarcinomas.[19,20] The consequent increase in synthesis of prostaglandins, (mainly PGE_2) can be prevented with NSAIDs, including some which are selective inhibitors of COX-2. Epidemiological studies have established a strong link between ingestion of aspirin and a reduced risk of developing colon cancer.[21,22] It has also been reported that sulindac causes reduction of prostaglandin synthesis and regression of adenomatous polyps which would otherwise develop into rectal carcinomas unless surgically removed.[23]

The close connection between colon cancer and COX-2 is further emphasized since mice with a mutation of the APC gene develop a condition similar to Familial Adenomatous Polyposis, characterised by the presence of large numbers of intestinal polyps. However, mice carrying the APC mutation made deficient in the COX-2 gene have very few intestinal polyps indicating that COX-2 is involved in polyp formation and in neoplasia of the colon.[24]

In the Kidney

Maintenance of kidney function is dependent on prostaglandins when the function is compromised in both animal models of disease states and in patients with congestive heart failure, liver cirrhosis or renal insufficiency. Patients are therefore at risk of renal ischaemia when prostaglandin synthesis is reduced by NSAIDs.

Those kidney cells which synthesize prostaglandins contain mostly COX-1, but low levels of COX-2 mRNA have also been measured.[25] In addition, cultured rat mesangial cells increase production of PGI_2 and PGE_2 after induction of COX-2 with cytokines or growth factors[26] and up-regulation of COX-2 expression has been observed in the macula densa following salt deprivation.[25]

Mice which lack the gene for production of COX-1 appear healthy and do not show significant signs of kidney pathology. This is in accord with the findings that inhibition of COX-1 by NSAIDs does not alter renal function under normal physiological conditions. However, in COX-2 (-/-) null mice the kidneys failed to continue normal development after birth with the result that the animals died within 8 weeks of birth.[27]

COX-2/COX-1 INHIBITORY RATIOS OF NSAIDS

Individual NSAIDs show different potencies against COX-1 compared with COX-2 and this satisfactorily explains the variations in the side effects of NSAIDs at their anti-inflammatory doses. Drugs with high potency against COX-2 and therefore a lower COX-2/COX-1 activity ratio, have anti-inflammatory activity with fewer side effects in the stomach and kidney. Garcia Rodriguez and Jick,[28] Langman et al.[29] and Henry et al.[30] have published a comparison of epidemiological data of the side effects of NSAIDs. Piroxicam and indomethacin are amongst those with the highest gastrointestinal toxicity. These drugs have a much higher potency against COX-1 than against COX-2.[31] Thus, when epidemiological results are compared with COX-2/COX-1 ratios, there is a parallel relationship between gastrointestinal side effects and COX-2/COX-1 ratios (Figure 1).

Assessment of Selectivity

A number of methods have been described for determining the ED_{50} values for NSAIDs against COX-2 and COX-1, leading to values for the COX-2/COX-1 ratio. Table 1 shows values for indomethacin which vary between 0.07 and 60.[32,33] Clearly, cultures of cells from animals (first group) give relatively high values whereas microsomal human enzyme preparations (middle group) give a different set. Some of the reasons for variation have been identified. For example, Laneuville et al.[34] added drug and substrate to the human microsomal enzyme at the same time but noted that several NSAIDs, including indomethacin, take 10-20 min to produce a full enzyme inhibition. When the drug was incubated for 20 min with the enzyme before adding substrate,[35] indomethacin was far more potent (especially on COX-2) and the ratio was reduced to 3.5. Other conditions that will lead to variation in results are the absence of protein in the medium and the several hours taken to induce COX-2 in cells.

Despite these methodological variations the ratios still fall into three broad groups; high ratios for indomethacin, naproxen and piroxicam, approximately equi-activity on the two enzymes for diclofenac and ibuprofen and selectivity for COX-2 for nimesulide, meloxicam, etodolac, MK-966 and celecoxib. Thus, most of the results published so far support the hypothesis that the unwanted side effects of NSAIDs are due to their ability to inhibit COX-1 whilst their anti-inflammatory (therapeutic effects) are due to inhibition of COX-2.

Selective COX-2 Inhibition

Meloxicam, Nimesulide and Etodolac. These NSAIDs were not specifically designed as COX-2 inhibitors and are already available in some countries for the treatment of inflammation. They preferentially inhibit COX-2 rather than COX-1, with a variation of their COX-2/COX-1 ratios of between 0.1 and 0.01.

Meloxicam, which has a selectivity towards COX-2 of up to 100 fold over COX-1 (depending on the test system)[35] is marketed around the world for use in rheumatoid arthritis and osteoarthritis.[36] In double blind trials in more than 5000 patients with osteoarthritis and rheumatoid arthritis, meloxicam in doses of 7.5mg or 15mg once a day compared in efficacy with standard NSAIDs such as naproxen 750-1000mg, piroxicam 20 mg and diclofenac 100mg. Both doses of meloxicam produced significantly fewer gastrointestinal adverse effects than the standard NSAIDs ($p<0.05$). Discontinuation of treatment due to gastrointestinal side effects was also significantly less frequent with meloxicam.

Nimesulide is currently sold in several European countries and in South America for the relief of pain associated with inflammatory conditions. It is a selective inhibitor of COX-2 with 10 to 50-fold greater potency against this enzyme than against COX-1.[37] In limited clinical trials for its use in acute and chronic inflammation in patients, it was more effective than placebo or had comparable anti-inflammatory activity to established NSAIDs. Epidemiological data suggests that, in long term therapeutic use at anti-inflammatory doses (100mg twice daily) it causes no more serious gastrointestinal symptoms than placebo.

Etodolac is marketed in Europe and North America for the treatment of osteoarthritis and rheumatoid arhtritis. It has a COX-2/COX-1 ratio of 0.09 when tested on recombinant human enzymes and of 0.1 on human whole blood.[38] In healthy human volunteers, etodolac twice a day did not suppress gastric mucosal prostaglandin production and caused less gastric damage than naproxen.[39] Patients with osteoarthritis or rheumatoid arthritis obtained

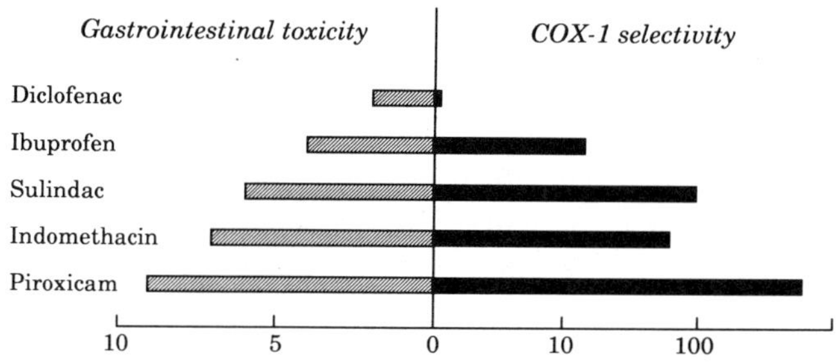

Figure 1. Comparison of gastric damage and cyclooxygenase selectivity of non-steroidal anti-flammatory drugs. The left side represents the ranking of gastrointestinal toxicity for anti-inflammatory doses of the NSAIDs taken from the epidemiological study of Henry et al. 1996 (30). The right side represents log COX-2/COX-1 activity ratios determined in intact cells by measuring IC50 values of the NSAIDs as micromoles/1. (Data taken from ref. 33).

Table 1. Inhibition of COX-1 and COX-2 by indomethacin, determined using different models.

Model	COX-2 IC_{50} (μM)	COX-1 IC_{50} (μM)	COX-2/COX-1 Ratio	Reference
Cultured animal cells	145	6.6	22	Meade et al. 1993
	1.7	0.028	60	Mitchell et al. 1993
	0.006	0.0002	30	Engelhardt et al. 1996
	0.009	0.0015	6	Klein et al. 1994
Human enzyme preparations	0.97	7.4	1.3	Futaki et al. 1994
	>1000	13	>75	Laneuville et al. 1994
	1.4	0.6	2.3	O'Neill et al. 1994
	0.9	0.1	9	Gierse et al. 1995
	0.1	0.35	3.5	Churchill et al. 1996
Human cells	0.36	0.7	0.51	Patrignani et al. 1994
	0.0012	0.017	~0.07	Grossman et al. 1995
	1.7	0.13	12.5	Young et al. 1996

relief from symptoms with etodolac equal to other commonly used NSAIDs but with a lower incidence of serious gastrointestinal toxicity.[40]

Celecoxib and MK-966. These drugs were designed to have high selectivity for COX-2 over COX-1 and are still in clinical development. Monsanto/Searle have made inhibitors, such as SC 58125 which are some 400-fold more potent *in vitro* against COX-2 than against COX-1. One of these, celecoxib (SC 58635), is an effective analgesic in humans for moderate to severe pain following tooth extraction.[41] In Phase II trials, patients with osteoarthritis of the knee were treated with twice daily doses of 40 mg, 100 mg or 200 mg. There was significantly greater improvement on all doses than on placebo and withdrawal for lack of efficacy was greater on placebo than on celecoxib. The incidence of adverse reactions and withdrawal from the study due to adverse reactions was similar in placebo and celecoxib-treated groups. In a study in patients with rheumatoid arthritis, twice daily doses of 40 mg, 200 mg or 400 mg reduced duration of morning stiffness more than placebo and the number of painful joints was significantly less at the two highest doses than on placebo. No serious adverse events or changes in renal parameters were reported. Celecoxib is currently in Phase III trials in arthritic patients.[42]

In whole cell assays, the selective COX-2 inhibitor from Merck-Frosst L-475,337, inhibited COX-2 with an IC_{50} of 20 nM, but was inactive on COX-1 even at doses of 10μM (COX-2/COX-1 ratio of 1/500). A similar highly selective COX-2 inhibitor from Merck-Frosst, MK-966, is currently undergoing Phase II/III clinical trials. In Phase I studies, a single dose of 250 mg daily for seven days produced no adverse effects on the stomach mucosa, as evidenced by gastroscopy. After a single dose of 1g, there was no evidence of COX-1 inhibition in platelets, but activity of COX-2 in LPS-stimulated monocytes was reduced. For dental pain, MK-966 at 25 mg per dose demonstrated equal analgesic activity to ibuprofen and provided relief from symptoms in a six-week study of osteoarthritis.[43]

New selective COX-2 and dual COX-2/5LO inhibitors. Several pharmaceutical companies have selective COX-2 inhibitors under investigation.[42] Parke-Davis are conducting clinical studies on CI-1004, a combined COX-2 and 5-lipoxygenase inhibitor, which has no gastric toxicity in rats or monkeys and is well tolerated in humans in single doses of up to 100mg. This drug inhibits recombinant human COX-2 with an IC_{50} of 0.48μM and has no apparent activity against COX-1. PD 098120-003, another COX-2 inhibitor from Parke-Davis, has no effect on COX-1 and does not cause gastrointestinal toxicity.

RS-57067000, from Roche Bioscience, has a weak effect on COX-1 and does not cause gastric erosions in the rat in doses up to 200mg/kg. The COX-2/COX-1 ratio for this compound on human recombinant enzymes is less than 0.0006.

Other companies are developing dual COX-2/5-lipoxygenase inhibitors as anti-inflammatory drugs. Johnson and Johnson are evaluating RWJ 63556 which inhibits LTB_4 production in human leukocytes with an IC_{50} of 1.02μM and has a COX-2/COX-1 ratio of less than 0.18 determined on human peripheral leukocytes. Procter and Gamble have presented data on PGV 20229 which has no gastrointestinal toxicity in dogs and inhibits leukotriene biosynthesis in RBL-2 cells. The COX-2/COX-1 ratio of PGV 20229 is 0.03 measured on human platelets for COX-1 and on human monocytes for COX-2.

CONCLUSIONS

As well as benefiting arthritic patients, the new highly selective COX-2 inhibitors may demonstrate new important therapeutic benefits, slowing down tumour growth, preventing premature labour and perhaps even retarding the progression of Alzheimer's disease.

Acknowledgments. The William Harvey Research Institute is supported by grants from Schwarz Pharma Ltd., Servier International Research Institute and Boehringer Ingelheim.

REFERENCES

1. Vane JR. Inhibition of prostaglandin synthesis as a mechanism of action for the aspirin-like drugs. *Nature*; 231: 232-235 (1971).
2. Xie W, Robertson DL and Simmons DL. Mitogen-inducible prostaglandin G/H synthase: a new target for nonsteroidal antiinflammatory drugs. *Drug Devel Res*; 25: 249-265 (1992).
3. Picot D, Loll PJ, Garavito RM. The x-ray crystal structure of the membrane protein prostaglandin H_2 synthase-1. *Nature*; 367: 243-249 (1994).
4. Luong, C., Miller, A., Barnett, J., Chow, J., Ramesha, C., and Browner, M.F. Flexibility of the NSAID binding site in the structure of human cyclooxygenase-2. *Nature Structural Biology*; 3: 927-933 (1996).
5. Gierse, J.K., McDonald, J.J., Hauser, S.D., Rangwala, S.H., Koboldt, C.M., and Seibert, K. A single amino acid difference between cyclooxygenase-1 (COX-1) and -2 (COX-2) reverses the selectivity of COX-2 specific inhibitors. *J.Biol.Chem.* 271:15810-15814 (1996).
6. Moncada S, Gryglewski R, Bunting S, Vane JR. An enzyme isolated from arteries transforms prostaglandin endoperoxides to an unstable substance that inhibits platelet aggregation. *Nature*; 263: 663-665 (1976).
7. Whittle BJR, Higgs GA, Eakins KE, Moncada S, Vane JR. Selective inhibition of prostaglandin production in inflammatory exudates and gastric mucosa. *Nature*; 284: 271-273 (1980).
8. Funk, C.D., Funk, L.B., Kennedy, M.E., Pong, A.S., and Fitzgerald, G.A. Human platelet/erythroleukemia cell prostaglandin G/H synthase: cDNA cloning, expression, and gene chromosomal assignment. *FASEB J*; 5:2304-2312 (1991).
9. Raz, A., Wyche, A., and Needleman, P. Temporal and pharmacological division of fibroblast cyclooxygenase expression into transcriptional and translational phases. *Proc Natl Acad Sci USA*; 86: 1657-1661 (1989).
10. Fu JY, Masferrer JL, Seibert K, Raz A and Needleman P. The induction and suppression of prostaglandin H_2 synthase (cyclooxygenase) in human monocytes. *J Biol Chem*; 265: 16737-16740. (1990).
11. Masferrer JL, Zweifel BS, Seibert K and Needleman P. Selective regulation of cellular cyclooxygenase by dexamethasone and endotoxin in mice. *J Clin Invest*; 86: 1375-1379 (1990).
12. Xie W, Chipman JG, Robertson DL, Erikson RL and Simmons DL Expression of a mitogen-responsive gene encoding prostaglandin synthase is regulated by mRNA splicing. *Proc Natl Acad Sci USA*; 88: 2692-2696 (1991).
13. O'Banion MK, Sadowski HB, Winn V and Young DA. A serum- and glucocorticoid-regulated 4-kilobase mRNA encodes a cyclooxygenase-related protein. *J Biol Chem*; 266: 23261-23267 (1991).
14. Kujubu DA, Fletcher BS, Varnum BC, Lim RW and Herschman HR. TIS10, a phorbol ester tumor promoter-inducible mRNA from swiss 3T3 cells, encodes a novel prostaglandin synthase/cyclooxygenase homologue. *J Biol Chem*; 26: 12866-12872 (1991).
15. DeWitt, D.L. Prostaglandin endoperoxide synthase: Regulation of enzyme expression. *Biochim.Biophys.Acta* 1083:121-134 (1991).
16. Wu, K.K., Sanduja, R., Tsai, A.-L., Ferhanoglu, B., and Loose-Mitchell, D.S. Aspirin inhibits interleukin 1-induced prostaglandin H synthase expression in cultured endothelial cells. *Proc Natl Acad Sci USA*; 88: 2384-2387 (1991).
17. Kargman, S., Charleson, S., Cartwright, M., Frank, J., Riendeau, D., Mancini, J., Evans, J., and O'Neill, G.P. Characterization of prostaglandin G/H synthase 1 and 2 in rat, dog,monkey and human gastrointestinal tracts. *Gastroenterology*; 111: 445-454 (1996).
18. Langenbach R, Morham SG, Tiano HF, Loftin CD, Ghanayem BI, Chulada PC, Mahler JF, Lee CA, Goulding EH, Kluckman KD, Kim HS and Smithies O. Prostaglandin synthase 1 gene disruption in mice reduces arachidonic acid-induced inflammation and indomethacin-induced gastric ulceration. *Cell*; 83: 483-492 (1995).
19. Kutchera, W., Jones, D.A., Matsunami, N., Groden, J., McIntyre, T.M., Zimmerman, G.A., White, R.L., and Prescott, S.M. Prostaglandin H synthase 2 is expressed abnormally in human colon cancer: Evidence for a transcriptional effect. *Proc Natl Acad Sci USA*; 93: 4816-4820 (1996).
20. Gustafson-Svärd, C., Lilja, I., Hallböök, O., and Sjödahl, R. Cyclooxygenase-1 and cyclooxygenase-2 gene expression in human colorectal adenocarcinomas and in azoxymethane induced colonic tumours in rats. *Gut*; 38: 79-84 (1996).

21. Thun, M.J., Namboodiri, M.M., and Heath, C.W.J. Aspirin use and reduced risk of fatal colon cancer. *New England Journal of Medicine*; 325: 1593-1596 (1991).
22. Luk, G.D. Prevention of gastrointestinal cancer - the potential role of NSAIDs in colorectal cancer. *Schweiz Med Wochenschr*; 126: 801-812 (1996).
23. Nugent, K.P., Spigelman, A.D., and Phillips, R.K.S. Tissue prostaglandin levels in familial adenomatous polyposis patients treated with sulindac. *Dis Colon Rectum*; 39: 659-662 (1996).
24. Oshima, M., Dinchuk, J.E., Kargman, S.L., Oshima, H., Hancock,B., Kwong, E., Trzaskos, J.M., Evans, J.F., Taketo, M.M. Suppression of intestinal polyposis in *Apc*$^{\Delta 716}$ knockout mice by inhibition of cyclooxygenase 2 (COX-2). *Cell*; 87: 803-809 (1996).
25. Harris, R.C., McKanna, J.A., Akai, Y., Jacobson, H.R., Dubois, R.N., and Breyer, M.D. Cyclooxygenase-2 is associated with the macula densa of rat kidney and increases with salt restriction. *J.Clin.Invest.*; 94: 2504-2510 (1994).
26. Nüsing, R.M., Klein, T., Pfeilschifter, J., and Ullrich, V. Effect of cyclic AMP and prostaglandin E_2 on the induction of nitric oxide- and prostanoid-forming pathways in cultured rat mesangial cells. *Biochem.J.*; 313: 617-623 (1996).
27. Morham, S.G., Langenbach, R., Loftin, C.D., Tiano, H.F., Vouloumanos, N., Jenette, J.C., Mahler, J.F., Kluckman, K.D., Ledford, A., Lee, C.A., and Smithies, O. Prostaglandin synthase 2 gene disruption causes severe renal pathology in the mouse. *Cell*; 83: 473-482 (1995).
28. Garcia Rodriguez LA, Jick H. Risk of upper gastrointestinal bleeding and perforation associated with individual non-steroidal anti-inflammatory drugs. *Lancet*; 343: 769-772 (1994).
29. Langman MJS, Weil J, Wainwright P, Lawson DH, Rawlins MD, Logan RFA, Murphy M, Vessey MP, Colin-Jones DG. Risks of bleeding peptic ulcer associated with individual non-steroidal anti-inflammatory drugs. *Lancet*; 343: 1075-1078 (1994).
30. Henry D, Lim LL-Y, Rodriguez LAG, Gutthann SP, Carson JL, Griffin M, Savage R, Logan R, Moride Y, Hawkey C, Hill S, Fries JT. Variability in risk of gastrointestinal complications with individual non-steroidal anti-inflammatory drugs: results of a collaborative meta-analysis. *Br Med J*; 312: 1563-1566 (1996).
31. Vane JR and Botting RM. New insights into the mode of action of anti-inflammatory drugs. *Inflamm Res*; 44: 1-10 (1995).
32. Grossman CJ, Wiseman J, Lucas FS, Trevethick MA, Birch PJ. Inhibition of constitutive and inducible cyclooxygenase activity in human platelets and mononuclear cells by NSAIDs and Cox 2 inhibitors. *Inflamm Res*; 44: 253-257 (1995).
33. Mitchell JA, Akarasereenont P, Thiemermann C, Flower RJ, Vane JR. Selectivity of nonsteroidal antiinflammatory drugs as inhibitors of constitutive and inducible cyclooxygenase. *Proc Natl Acad Sci USA*; 90: 11693-11697 (1993).
34. Laneuville O, Breuer DK, DeWitt DL, Hla T, Funk CD and Smith WL. Differential inhibition of human prostaglandin endoperoxide H synthases-1 and -2 by nonsteroidal anti-inflammatory drugs. *J Pharmacol Exp Therap*; 271: 927-934 (1994).
35. Churchill L, Graham AG, Shih C-K, Pauletti D, Farina PR, Grob PM. Selective inhibition of human cyclo-oxygenase-2 by meloxicam. *Inflammopharmacology*; 4: 125-135 (1996).
36. Barner A. Review of clinical trials and benefit/risk ratio of meloxicam. *Scand J Rheumatol*; 25 (Suppl 102): 29-37 (1996).
37. Rabasseda, X. Nimesulide: a selective cyclooxygenase 2 inhibitor antiinflammatory drug. *Drugs of Today*; 32 (Suppl D): 1-23 (1996).
38. Glaser, K., Sung, M.-L., O'Neill, K., Belfast, M., Hartman, D., Carlson, R., Kreft, A., Kubrak, D., Hsiao, C.-L. and Weichman, B. Etodolac selectively inhibits human prostaglandin G/H synthase 2 (PGHS-2) versus human PGHS-1. *Eur J Pharmacol*; 281: 107-111 (1995).
39. Laine, L., Sloane, R., Ferretti, M. and Cominelli, F. A randomized double-blind comparison of placebo, etodolac and naproxen on gastrointestinal injury and prostaglandin production. *Gastrointest Endosc*; 42: 428-433 (1995).
40. Cummings, D.M. and Amadio, P. Jr. A review of selected newer nonsteroidal anti-inflammatory drugs. *Am Fam Physician*; 49: 1197-1202 (1994).
41. Hubbard RC, Mehlisch DR, Jasper DR, Nugent MJ, Yu S, Isakson PC. SC-58635, a highly selective inhibitor of COX-2, is an effective analgesic in an acute post-surgical pain model. *J Invest Med*; 44: 293A (1996).
42. Parnham, M.J. COX-2 inhibitors at the 8th International Conference of the Inflammation Research Association. *Exp. Opin. Invest. Drugs*; 6: 79-82 (1997).
43. Ford-Hutchinson, A. New highly selective COX-2 inhibitors. In: *Selective COX-2 Inhibitors: Pharmacology, Clinical Effects and Therapeutic Potential*. Abstracts of the William Harvey Research Conference, Cannes, p 23 (1997).

DIFFERENTIAL EFFECTS OF 2,3,7,8-TETRACHLORODIBENZO-*p*-DIOXIN ON THE EXPRESSION OF PROSTAGLANDIN-H SYNTHASE ISOENZYMES IN MOUSE TISSUES

Christoph Vogel[1], Ulrike S. Schuhmacher[2], Gisela H. Degen[2], Carsten Goebel[3], and Josef Abel[1]

[1] Medical Institute of Environmental Hygiene at the Heinrich-Heine-University Düsseldorf, Department of Toxicology, Auf'm Hennekamp 50, 40225 Düsseldorf, FRG
[2] Institute of Occupational Physiology at the University Dortmund, Ardeystrasse 67, 44139 Dortmund, FRG
[3] Diabetes Research Institute at the Heinrich-Heine-University Düsseldorf, Clinical Department, Auf'm Hennekamp 65, 40225 Düsseldorf, FRG

INTRODUCTION

2,3,7,8-tetrachlorodibenzo-*p*-dioxin (TCDD) and related compounds cause a number of toxic responses including wasting syndrome, thymus atrophy and immunotoxicity, reproductive and developmental toxicity, dermal toxicity, hyperplasias, and carcinogenesis. Most if not all toxic effects of TCDD have been found to be mediated by the arylhydrocarbon receptor (AhR) to which TCDD binds with high affinity. The TCDD:AhR complex is known to translocate into the nucleus where it dimerizes with ARNT, the socalled AhR nuclear translocator protein. This complex can bind to dioxin responsive elements (DREs) located in the regulatory region of several TCDD-sensitive genes and leads to activation of their transcription (rev. in Landers and Bunce, 1991). This has been well shown for several phase 1 and phase 2 enzymes, for example cytochrom P450 1A1 (CYP1A1) and glutathione S-transferase Ya (Gonzalez et al., 1985; Pimental et al., 1993). Recently two DRE motifs were identified in the regulatory region of the mouse prostaglandin endoperoxide H synthase-2 (PGHS-2) gene (Kraemer et al., 1996); this raised the question whether PGHS-2 is induceable by TCDD.

PGHS enzyme is believed to play a role in the metabolic activation of xenobiotics, especially in extrahepatic tissues (Eling et al., 1990). Two PGHS isoenzymes, PGHS-1 and PGHS-2, were described. While PGHS-1 is constitutively expressed in a wide variety of cell types, PGHS-2 is barely expressed in most tissues (Smith and DeWitt, 1995). Its expression can be induced by numerous factors, such as cytokines, growth factors, and phorbolester (Herschman, 1996). Here, we describe tissue-specific changes of PGHS-2 mRNA expression in TCDD treated C57BL/6 mice. Moreover, we analyzed CYP1A1 induction to verify the TCDD responsiveness of the tissues examined.

METHODS

In vitro studies. Peritoneal macrophages (PMΦ) were derived from female 6-8 weeks old C57BL/6J and DBA/2 mice (R. Janvier, Le Genest St Isle, France). Preparation and cultivation of PMΦ were performed as described (Goebel et al., 1995). In brief, the harvested peritoneal cells were washed and suspended in RPMI 1640 medium supplemented with 10% FCS at a density of $1x10^6$ cells/ml. After 20 h, nonadherent cells were removed by vigorous washing and about 90% of adherent cells were identified as macrophages by Pappenheim's stain. Cells were then treated with 10 nM TCDD for 48 h. Vehicle treated cells (0.1% DMSO) were used as controls. As positive control for PGHS-2 induction cells were treated with 1 μg lipopolysaccharide (LPS) and 25 U interferon-γ (INFγ) per ml medium.

In vivo studies. Animals were kept under specific pathogen-free conditions. Three C57BL/6 mice were treated with a single dose (i.p.) of 10 μg TCDD/kg bw. Control animals received an equal amount of vehicle (corn oil, 1 ml/kg bw). The animals were killed by CO_2 asphyxiation after 48 h, and liver, lung, kidney, thymus, and spleen were excised, quickly frozen in liquid nitrogen and stored at -70°C for analysis.

RNA isolation and reverse transcription-polymerase chain reaction (RT-PCR). Isolation of total RNA and the following RT-PCR were carried out as previously described (Döhr et al., 1995). The sources of primer sequences used for amplification of CYP1A1 and β-actin were published by Abel et al. (1996). Primer sequences for PGHS-1 and PGHS-2 were chosen from published cDNA sequences (DeWitt et al., 1990; O'Banion et al., 1992) using a primer selection program. Specific conditions of PCR reaction are given in the legend of figure 1.

RESULTS

Effect of TCDD on PGHS mRNA expression in PMΦ. In figure 1 the effect of 10 nM TCDD on PGHS-2 mRNA in PMΦ is illustrated. In PMΦ derived from C57BL/6 mice TCDD led to 10-fold increase compared to vehicle treated controls, whereas TCDD did not change the mRNA expression of PGHS-2 in PMΦ derived from DBA/2 mice. To compare the responsiveness of PMΦ toward TCDD, the expression of CYP1A1 was monitored. PMΦ obtained from C57BL/6 mice respond with a 60-fold increase of CYP1A1 mRNA level after 10 nM TCDD whereas only a 8-fold increase of CYP1A1 mRNA expression was detected in PMΦ with DBA/2 background. To test the PGHS responsiveness of cultured PMΦ, cells were treated with LPS and INFγ, a well known inducer of PGHS-2. As shown in figure 1, the PGHS-2 mRNA level in PMΦ of both mouse strains is elevated similarly by LPS/INFγ treatment. In contrast, the expression of PGHS-1 was not affected by TCDD or LPS/INFγ (Fig. 1).

In vivo studies. To examine the responsiveness of PGHS-2 toward TCDD *in vivo* the expression was analyzed in different tissues of C57BL/6 mice after TCDD treatment. The results are illustrated in figure 2. In liver and kidney, TCDD had no effect on PGHS-2 expression. An increase of PGHS-2 mRNA expression was detected in lung and spleen, the mRNA content was about 2-fold higher compared to control animals. In thymus, however, TCDD led to a decrease of PGHS-2 mRNA; the mRNA content was reduced about 60% compared to control mice. For comparison, the expression of CYP1A1 in different tissues was analyzed. As seen in figure 2, CYP1A1 mRNA is induced after TCDD treatment in all tissues examined. The most pronounced response was found in liver, lung, and kidney, respectively. Again, TCDD did not affect the expression of PGHS-1 mRNA (Fig. 2).

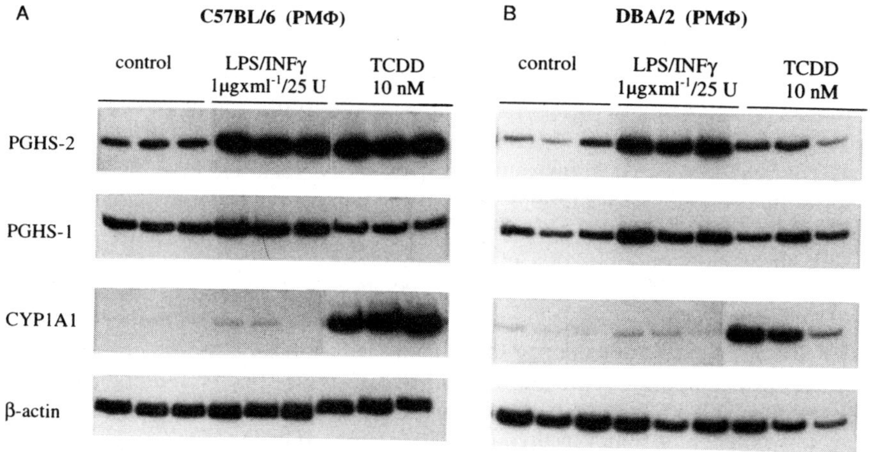

Figure 1. RT-PCR analysis of PGHS-2, PGHS-1 and CYP1A1 mRNA expression in PMΦ derived from C57BL/6 mice (Fig. 1A) or DBA/2 mice (Fig. 1B). PMΦ were treated with 10 nM TCDD or 1 μg LPS/25 U INFγ per ml medium for 48 h. Total RNA was reversely transcribed and aliquots of the resulting cDNA were subjected to 30, 26, 30, and 20 cycles of PCR amplification for mouse PGHS-2, PGHS-1, CYP1A1 and β-actin-specific oligonucleotide primers, respectively; the annealing temperature was 52, 56, 62, and 62 °C for the respective primers. The [α-^{32}P]CTP labeled PCR-products were analyzed on 10% polyacrylamide gels and exposed to X-ray film. The optical density of each band was quantified using an OmniMedia scanner (Millipore, Überlingen, FRG) and was normalized with that of its corresponding β-actin band. Shown are three typical results and means are given as arbitrary units.

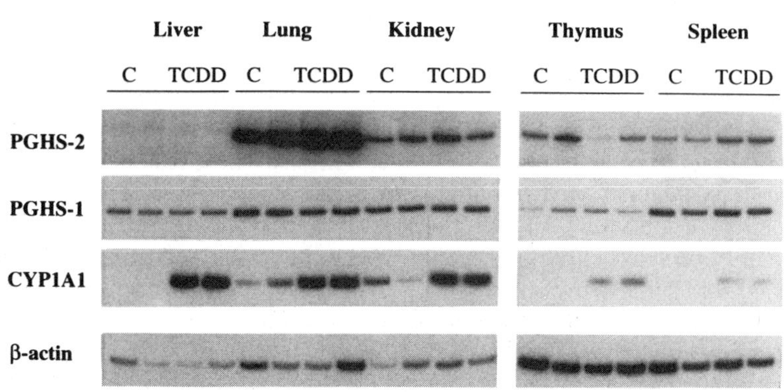

Figure 2. Effect of TCDD on PGHS-2, PGHS-1, and CYP1A1 mRNA levels in liver, lung, kidney, thymus, and spleen. Female C57BL/6 mice were treated with a single dose of 10 μg TCDD/kg for 48 h. The mRNA levels were detected by RT-PCR as described in the legend of figure 1. Shown are representative bands from two animals and means of PGHS-2 mRNA levels are given as arbitrary units.

DISCUSSION

Recently, several studies have shown that TCDD leads to changes of PGHS-2 expression *in vitro* (Kraemer et al., 1996; Olnes et al., 1996; Liu et al., 1997). The present study was aimed to investigate whether TCDD can modulate PGHS-2 mRNA expression under *in vivo* conditions. As shown in figure 1, TCDD treatment of female C57BL/6 mice led to a stimulation of PGHS-2 in lung and spleen, whereas in liver and kidney no change of PGHS-2 expression was observed. To test the TCDD responsiveness of the tissues examined, the induction of CYP1A1 was analyzed. In contrast to CYP1A1, which is induced in liver, lung, kidney, and thymus, PGHS-2 mRNA expression was reduced by TCDD in thymus. Although the relevance of the downregulation of PGHS-2 expression in thymus is not known, it is of interest that TCDD led to tissue-specific changes of PGHS-2 expression, indicating that beside the AhR other cell/tissue-specific factors may be involved in pathways of TCDD-mediated effects on PGHS-2 expression. A decrease of PGHS-2 expression was recently found in thymocytes from rats (Olnes et al., 1996). It is likely that the modulated PGHS-2 expression and an inappropriate generation of prostaglandins, which are important regulators of immune functions, contribute to the thymus toxicity associated with TCDD exposure. So far, the function of PGHS-2 is not completely understood, it has been suggested that this enzyme is involved in cell growth, apoptosis and differentiation (DeWitt and Smith, 1995). Thus, an altered PGHS-2 expression by TCDD may provide new insights into its mode of action.

In addition we were interested whether the AhR is involved in TCDD mediated PGHS-2 mRNA induction. Primary cultured macrophages are a well established *in vitro* system to study the induceability of PGHS-2 expression by exogeneous stimuli like LPS and phorbolester (Reddy and Herschman, 1994). Therefore we analyzed the effect of TCDD on PGHS-2 mRNA expression in macrophages obtained from DBA/2 and from C57BL/6 mice. These mouse strains are known to respond to TCDD with different susceptibility. The different susceptibility has been linked to different allelic forms of the AhR. Bearing the Ah^{d} allele, the DBA/2 mice are less sensitive toward TCDD than C57BL/6 mice, bearing the Ah^{b-1} allele (Poland and Glover, 1990). In agreement with a recent report on TCDD-mediated PGHS-2 induction *in vitro* (Kraemer et al., 1996) we could demonstrate that TCDD stimulates PGHS-2 mRNA expression *in vitro* in PMΦ derived from female C57BL/6 mice. As shown in figure 1, TCDD did not change PGHS-2 mRNA levels in PMΦ from DBA/2 mice under the described conditions. These results indicate that the AhR is involved in TCDD-mediated stimulation of PGHS-2 mRNA expression. This suggestion is confirmed by results of CYP1A1 mRNA analyses which show that CYP1A1 is less induceable in PMΦ from DBA/2 mice compared to PMΦ derived from C57BL/6 mice (Fig. 1). The transcriptional activation of CYP1A1 by TCDD has been demonstrated to be mediated through binding of the activated AhR on DRE motifs which are located in the regulatory region of the CYP1A1 gene. However, until now it is unclear whether DRE motifs are essential for the TCDD-stimulated PGHS-2 induction. Studies in Mardin Darvey canine kidney cells (MDCK) show that DREs seem not to be strongly involved in transcriptional activation of PGHS-2 (Kraemer et al., 1996), whereas other studies suggest that DREs may be involved in transcriptional activity of PGHS-2 (Olnes et al., 1996; Liu et al., 1997). Thus the mechanism by which TCDD mediates changes of PGHS-2 expression remains to be elucidated further.

REFERENCES

Abel, J., Li, W., Döhr, O., Vogel, C., and Donat, S., 1996, Dose-response relationship of cytochrome P4501b1 mRNA induction by 2,3,7,8-tetrachlorodibenzo-*p*-dioxin in livers of C57BL/6J and DBA/2J mice, *Arch. Toxicol.* 70:510.

DeWitt, D.L., El-Harith, E.A., Kraemer, S.A., Andrews, M.J., Yao, E.F., Armstrong, R.L., and Smith, W.L., 1990, The aspirin and heme-binding sites of ovine and murine prostaglandin endoperoxide synthases, *J. Biol. Chem.* 265:5192.

DeWitt, D., and Smith, W.L., 1995, Yes, but do they still get headaches?, Cell 83:345.

Chang, C., Smith, D.R., Prasad, V.S., Sidman, C.L., Nebert, D.W., and Puga, A., 1993, Ten nucleotides differences, five of which cause amino acid changes, are associated with the Ah receptor locus polymorphism of C57BL/6 and DBA/2 mice, *Pharmacogenetics* 3:312.

Döhr, O., Vogel, C., and Abel, J., 1995, Different response of 2,3,7,8-tetrachlorodibenzo-*p*-dioxin (TCDD)-sensitive genes in human breast cancer MCF-7 and MDA-MB 231 cells, *Arch. Biochem. Biophys.* 321:405.

Eling, T.E., Thompson, D.C., Foureman, G.L., Curtis, J.F., and Hughes, M.F., 1990, Prostaglandin H synthase and xenobiotic oxidation, *Annu. Rev. Pharmacol. Toxicol.* 30:1.

Goebel, C., Kubicka-Muranyi, M., Tonn, T., Gonzalez, J., and Gleichmann, E., 1995, Phagocytes render chemicals immunogenic: oxidation of gold (I) to the T cell sensitizing gold (III) metabolite generated by mononuclear phagocytes, *Arch. Toxicol.* 69:450.

Gonzalez, F.J., Kimura, S., and Nebert, D.W., 1985, Comparison of the flanking regions and introns of the mouse 2,3,7,8-tetrachlorodibenzo-*p*-dioxin inducible cytochrome P_1-450 and P_3-450 genes, *J. Biol. Chem.* 260:5040.

Herschman, H.R., 1996, Prostaglandin synthase 2, *Biochim. Biophys. Acta* 1299:125.

Kraemer, S.A., Arthur, K.A., Denison, M.S., Smith, W.L., and DeWitt, D.L., 1996, Regulation of prostaglandin endoperoxide H synthase-2 expression by 2,3,7,8-tetrachlorodibenzo-*p*-dioxin, *Arch. Biochem. Biophys.* 330:319.

Landers, J.P., and Bunce, N.J., 1991, The Ah receptor and mechanism of dioxin toxicity, *Biochem. J.*, 276:273.

Liu, Y., Levy, G.N., and Weber,W.W., 1997, Induction of human prostaglandin endoperoxide H synthase-2 (PHS-2) mRNA by TCDD, *Prostaglandins*, 53:1.

O'Banion, M.K., Winn, V.D., and Young, D.A., 1992, cDNA clonong and functional activity of a glucocorticoid-regulated inflammatory cyclooxygenase, *Proc. Natl. Acad. Sci.* 89:4888.

Olnes, M.J., Verma, M., and Kurl, R.N., 1996, 2,3,7,8-Tetrachlorodibenzo-*p*-dioxin modulates expression of the prostaglandin G/H synthase-2 gene in rat thymocytes, *J. Pharmacol. Exp.Therap.* 279:1566.

Pimental, R.A., Liang, B., Yee, G.K., Wilhelmsson, A., Poellinger, L., and Paulson, K.E., 1993, Dioxin receptor and C/EBP regulate the function of the glutathione S-transferase Ya gene xenobiotic response element, *Mol. Cell. Biol.* 13:4365.

Poland, A., and Glover, E., 1990, Characterization and strain distribution pattern in murine Ah receptor specified by the Ah^{d} and Ah^{b-3} alleles, *Mol. Pharmacol.* 38:306-312.

Reddy, S.T., and Herschman, H.R., 1994, Ligand-induced prostaglandin synthesis requires expression of the TIS/PGS-2 prostaglandin synthase gene in murine fibroblasts and macrophages, *J. Biol. Chem.* 269:15473.

Smith, W.L., and DeWitt, D.L., 1995, Biochemistry of prostaglandin endoperoxide H synthase-1 and synthase-2 and their differential susceptibility to nonsteroidal anti-inflammatory drugs, *Seminars in Nephrology* 15:179.

IN VIVO MEASUREMENT OF INTERSTITIAL TXB_2-CONCENTRATIONS IN SKELETAL MUSCLE DURING ISCHEMIA AND REPERFUSION

Markus Müller

Department of Clinical Pharmacology, Vienna General Hospital (AKH), University of Vienna School of Medicine, Vienna Austria

INTRODUCTION

Acute arterial occlusion of an extremity frequently leads to prolonged periods of ischemia of skeletal muscle[1,2]. Reestablishment of blood flow to previously ischemic skeletal muscle initiates various biochemical cascades which may lead to the development of a reperfusion syndrome[1,2]. This syndrome is characterized by a paradoxical impairment of nutritive blood flow during reperfusion and may further aggravate local tissue damage. Thus, most of the damage to skeletal muscle occurs after oxygenation is restored rather than during ischemia[2]. A sudden release of metabolites from postischemic muscle tissue may have additional profound systemic consequences like acute renal failure or fatal multiple organ failure[2]. It was estimated that one third of deaths from peripheral embolism may result from the metabolic events of local ischemia and reperfusion.

The relative contribution of the various, locally released mediators to the development of a reperfusion syndrom is not settled. Liberation of arachidonic acid from damaged cells and subsequent synthesis of thromboxane-A_2 (TxA_2), however, was observed in ischemia-reperfusion models and has been implicated in causing reperfusion injury[3,4]. Selective inhibition of thromboxane synthetase has been shown to alleviate postischemic tissue damage by reducing TxA_2 effects on nutritive blood flow[5].

In the present study in vivo tissue microdialysis[6,7,8] was employed to characterize the time course of TxA_2 concentrations, measured as TxB_2, in ischemic and postischemic human skeletal muscle. In addition, the effect of oral aspirin, administered at different time-points during the experimental procedure, on interstitial TxB_2 concentrations was assessed.

Adress for correspondence: Markus Müller, MD, Department of Clinical Pharmacology, AKH, Währinger Gürtel 18-20, A-1090 Vienna, Austria, Tel. -43-1-40400-2981, Fax: -43-1-40400-2998
E-mail: Klin-Pharmakologie @ univie.ac.at

MATERIALS AND METHODS

Microdialysis measurements

The principles of microdialysis have been described in detail previously[8]. Briefly, microdialysis is based on sampling of analytes from the interstitial space by means of a semipermeable membrane at the tip of a microdialysis probe. The probe is constantly perfused with a physiological solution (perfusate) at a flow-rate of 1-10 µl/min. Once the probe is implanted into the tissue, substances present in the interstitial fluid at a certain concentration (c_{tissue}) are filtered by diffusion out of the interstitial fluid into the probe, resulting in a concentration ($c_{dialysate}$) in the perfusion medium. In the present experimental procedures a commercially available microdialysis probe (CMA-10®, CMA, Stockholm, Sweden) with a molecular cutoff of 20 kD, an outer diameter of 500 µm and a membrane length of 16 mm was used. The probe was perfused by means of a precision infusion pump (Precidor ®, Ilfors-AG, Basel, Switzerland) at a flow rate of 1,5 µl/min. Ringer's solution was used as perfusion fluid. The dialysis probe was inserted without anesthesia into the medial vastus muscle 10 cm above the knee joint. The intramuscular position of the microdialysis membrane was confirmed by high frequency (7.5MHz) 2D ultrasound scanning. Samples were collected on ice and stored at -25°C before being assayed.

Study population

The study population included 10 healthy, normal weight, drug free, male volunteers, at 20-34 years of age. The study was approved by the local ethics committee. All volunteers were given a detailed description of the study and their written consent was obtained.

Study protocol

The experimental procedures were started at 08.00 a.m. after an overnight fast. Volunteers were resting in a supine position throughout the study period. Room temperature was kept at 25°C. After the insertion of the microdialysis probe, three 15 minutes samples were collected to obtain stable intercellular baseline concentrations of analytes before the onset of ischemia. Subsequently, one thigh was constricted by a blood pressure cuff which served as a tourniquet; the cuff pressure was 150 mmHg above systolic blood pressure. The leg was constricted for 20 minutes. Sampling intervals during and immediately after leg constriction were 10 or 15 minutes. An additional microdialysis probe was inserted into the contralateral, non-ischemic leg to serve as a baseline control. In additional studies 500mg oral acetylsalicylic acid (Aspirin ®, Bayer Austria, Vienna, Austria) was administered 30 minutes before insertion of the microdialysis probe or 5 to 10 minutes after the onset of ischemia.

Chemical analysis

TxB_2 concentrations in the dialysate fluid were measured without prior extraction employing the TxB_2 [^{125} I] RIA Kit with Biomag ® Magnetic Separation (Advanced Magnetics Inc., Cambridge, Massachusetts, USA)

RESULTS

The time course of mean TxB_2 concentrations in the dialysate during the occlusion experiments with (n=4) and without (n=6) administration of acetylsalicylic acid is depicted in figure 1. Mean TxB_2 concentration during occlusion did not differ from baseline concentration. Immediately after a 20 minute occlusion period, mean TxB_2 concentrations in the dialysate increased from baseline values of 312 ±281 pg/ml to 2119 ±925 pg/ml ($p < 0.02$). Administration of 500 mg acetylsalicylic acid 30 minutes before the experiment resulted in TxB_2 concentrations below the detection limit of 20 pg/ml throughout the study (data not shown). Postischemic TxB_2 concentrations following administration of 500 mg acetylsalicylic acid 5 to 10 minutes after onset of a 20 minute occlusion period were not significantly different from baseline values.

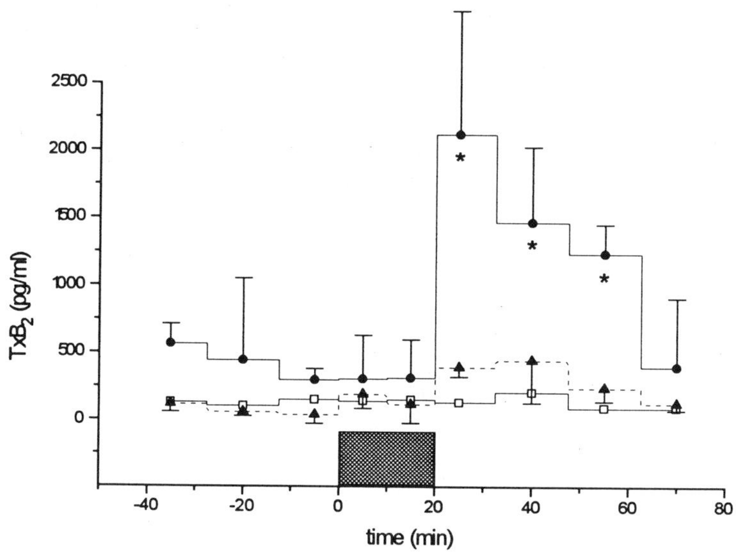

Figure 1. *Interstitial TxB_2 concentration in human skeletal muscle before (-50 min - 0 min) during (0 min - 20 min) and after (20 min - 80 min) a 20 minute constriction period (n=6).* Open squares represent contralateral control leg values. Dashed lines represent experiments with administration of 500mg acetylsalicylic acid (n=4) 5-10 minutes after onset of constriction (closed triangles). Bar indicates time of ischemia. * denotes $p < 0.05$ as compared to baseline. (Müller el al. 1995, *Europ J Clin Invest 25: 601-607)*

DISCUSSION

In the present study the microdialysis technique was applied for the measurement of TxB_2 kinetics in the interstitial space in human skeletal muscle. The rationale for measuring interstitial TxB_2 kinetics in an ischemia reperfusion model was derived from previous studies showing that during reperfusion following tissue ischemia the high concentration of arachidonic acid triggers a cascade of enhanced prostanoid synthesis as soon as the tissue becomes reoxygenized[3,4]. This was shown to be associated with increased TxA_2 levels in postischemic tissue[3,9] and effluent local venous blood[4]. TxA_2, a potent vasoconstrictor may thus aggravate the ischemic tissue damage by platelet activation and direct vasoconstrictor effects. However, the relative contribution of the TxA_2 effect remains to be assessed, since a multitude of factors e.g. free radicals, cell adhesion molecules or cytokines contribute to the development of ischemia reperfusion injury[1,2]. It is also not clear whether prostanoids are synthesized during or after ischemia. Our study revealed a significant increase in extracellular-fluid-dialysate concentration of TxB_2 immediately after but not during ischemia. These observations confirm previous results[3,4,5,7,9]. In addition, a time-course of interstitial TxB_2 concentrations in ischemic and postischemic skeletal muscle was shown for the first time. Since tissue TxB_2 concentration increases rapidly within the first minutes after release of ischemia, we propose to use an increase in TxB_2 tissue concentrations as a marker of reperfusion of skeletal muscle tissue. As expected, 500mg acetylsalicylic acid, given before the onset of ischemia, completely suppressed the postischemic rise in TxB_2. Surprisingly, the same dose of acetylsalicylic acid caused almost complete suppression of the rise in TxB_2, even when given orally 5-10 minutes after application of the tourniquet. Since peak plasma concentrations of acetylsalicylic acid following p.o. administration are achieved in about 14 minutes, this indicates that local activation of fresh inflowing platelets, rather than platelets trapped in the occluded leg or parenchymal cells, accounts for most of the local postischemic TxB_2 production. Notably, these findings suggest that acetylsalicylic acid might be beneficial for reducing reperfusion tissue injury even when administered after the onset of ischemia.

REFERENCES

1. P.A.Grace, Ischemia-reperfusion injury. *Br J Surg. 81, 637-647 (1994).*
2. M. Odeh, The role of reperfusion induced injury in the pathogenesis of the crush syndrome. *New Engl J Med. 324 (20) 1417-1422 (1991)*
3. G. Hotter, J. Rosello-Catafau, D. Closa et al. , Pancreas prostanoid production in ischemia and reperfusion. *Prostaglandins. 43:497-501 (1992).*
4. M. Manner, R. Seelos, K. Roy, N. Runkel, G. Otto, The role of prostaglandins in reperfusion injury after liver transplantation in pigs. *Transplant. Proc. 23 (5): 2413 (1991).*
5. L.C. Pettigrew, J.C. Grotta, H.M. Rhoades, K.K. Wu, Effect of Thromboxane synthetase inhibition on eicosanoid levels and blood flow in ischemic rat brain. *Stroke 20,5:627-632 (1989).*
6. M. Müller, A. Holmäng, O. Anderson, H.G. Eichler and P. Lönnroth, In vivo characterization of interstitial glucose and lactate concentrations in human skeletal muscle by calibrated microdialysis. *Am J Physiol 271 (Endocrinol Metab 34). E1003-E1007 (1996).*
7. M. Müller, R. Schmid, M. Nieszpaur-Los, P. Lönnroth, A. Molnar, P. Fasching and H.G. Eichler, Key metabolite kinetics in human skeletal muscle during ischemia and reperfusion: measurement by microdialysis. *Europ J Clin Invest 25: 601-607 (1995).*
8. M. Müller, R. Schmid, A. Georgopoulos, A. Buxbaum, C. Wasicek and H.G. Eichler, Application of microdialysis to clinical pharmacokinetics in man. *Clin Pharmacol Ther 57: 371-80 (1995).*
9. I. Huk, J. Nanobashvili, H. Lassner, A. Punz, M. Müller, H. Siegl, C. Neumayer, M. Mittlboeck and E. Roth, Influence of a short term acute ischemia and reperfusion on skeletal muscle metabolism and morphology in rabbits. *Int Angiol. 5:59-63 (1996).*

MUTATION OF TYR-383 IN LEUKOTRIENE A_4 HYDROLASE: EFFECTS ON ENZYME ACTIVITIES

Martina Andberg, Anders Wetterholm and Jesper Z. Haeggström

Department of Medical Biochemistry and Biophysics, Karolinska Institutet, S-171 77 Stockholm, SWEDEN

INTRODUCTION

Leukotriene (LT) A_4 hydrolase catalyzes hydrolysis of LTA_4 (5(*S*)-*trans*-5,6-oxido-7,9-*trans*-11,14-*cis*-eicosatetraenoic acid) into the proinflammatory compound LTB_4 (5(*S*),12 (*R*)-6,14-*cis*-8,10-*trans*-dihydroxyeicosatetraenoic acid, in a reaction without any cofactor requirement. The substrate LTA_4 is in turn derived from arachidonic acid by two consecutive reactions catalyzed by 5-lipoxygenase [1].

From sequence comparisons with other zinc metalloenzymes, *e.g.* aminopeptidase M and thermolysin, a zinc-binding motif was identified in LTA_4 hydrolase [2, 3], and the enzyme was shown to contain one catalytic zinc atom complexed to His-295, His-299 and Glu-318 [4-6]. Furthermore, LTA_4 hydrolase was found to possess peptidase activity towards synthetic substrates [5, 7]. Mutagenetic studies identified Glu-296 as a catalytic amino acid in the peptidase reaction, in which it presumably acts as a general base [8]. Such a reaction mechanism postulates a proton transfer to the nitrogen of the peptide bond and further sequence comparisons with aminopeptidase M led to the identification of a proton donor motif in LTA_4 hydrolase and Tyr-383 was suggested as the potential proton donor. To investigate the catalytic role of this amino acid, we replaced the tyrosine residue with a Phe, His or Gln, purified the recombinant proteins, and studied the effects on enzyme activities.

MATERIALS AND METHODS

The mutants were produced by site-directed mutagenesis on the recombinant plasmid pULTA4, an *E. coli* expression vector containing the entire cDNA for mouse LTA_4 hydrolase. The mutagenesis was carried out by a method based on PCR as previously described [9]. For expression, purification and activity determinations see reference [9].

Recent Advances in Prostaglandin, Thromboxane, and Leukotriene Research
Edited by Sinzinger *et al.*, Plenum Press, New York, 1998

RESULTS AND DISCUSSION

Tyr-383 is required for the peptidase activity

From sequence comparisons with other zinc hydrolases, a tyrosine in position 383 in LTA_4 hydrolase has been suggested as a possible catalytic amino acid for the peptidase activity [10]. By site-directed mutagenesis on mouse LTA_4 hydrolase cDNA, Tyr-383 was exchanged for a Phe, His or Gln. The mutated enzymes were expressed in *E. coli* and the resulting recombinant proteins [Y383F], [Y383H] and [Y383Q]LTA_4 hydrolase (denoted in single letter code for the amino acid change) were purified to apparent homogeneity by FPLC to allow assays of both the epoxide hydrolase and the peptidase activities [9]. None of the mutated enzymes exhibited significant peptidase activity towards alanine-4-nitroanilide, all of the mutants showing activities ≤ 0.3 % of that for the wild type enzyme (Table 1).

Table 1. Enzyme activities of recombinant LTA_4 hydrolase and mutants [Y383F], [Y383H], [Y383Q] and [G386A]LTA_4 hydrolase.

Enzyme	Epoxide hydrolase activity (nmol/mg/min)		Peptidase activity (nmol/mg/min)	
Wild type	662±160	100 %	240 ± 74	100 %
[Y383F]	71±30	11 %	0.78 ± 0.46	0.3 %
[Y383H]	107±47	16 %	0.22 ± 0.23	0.1 %
[Y383Q]	112±55	17 %	0.09 ± 0.10	0.04 %
[G386A]	513±85	77 %	46 ± 28	19 %

The epoxide hydrolase activity was reduced but still clearly detectable and corresponded to 11, 16 and 17 % of the activity of the unmutated enzyme. To serve as a control, a Gly was replaced by an Ala at pos 386, and this mutant exhibited an epoxide hydrolase activity of 77 % and a peptidase activity of 19 % relative to that of the wild type enzyme.

To further study the effects of mutation on the epoxide hydrolase activity, [Y383Q]LTA_4 hydrolase was selected for kinetic analysis. For this mutant, the reduced epoxide hydrolase activity was mostly due to an increased K_m value for LTA_4, which was calculated to be 60 μM as compared to 5 μM for the wild type enzyme. At saturating substrate concentrations the difference in specific activity between [Y383Q] and wild type enzyme was less evident, and [Y383Q] displayed 60 % of the catalytic activity of the unmutated enzyme. In contrast, the peptidase activity was abolished even at substrate concentrations of alanine-4-nitroanilide 60 times the normal K_m value. These results show that a tyrosine residue in position 383 is of major importance for the peptidase activity where it may serve as a proton donor. Since all mutants converted LTA_4 into LTB_4, Tyr-383 is not critical for the epoxide hydrolase activity.

Formation of a novel metabolite of LTA_4 by Tyr-383 mutants

LTA_4 hydrolase mutated in position 383 *i.e.* [Y383F], [Y383H] and [Y383Q]LTA_4 hydrolase, were found to catalyze hydrolysis of LTA_4 into a novel enzymatic metabolite in addition to the expected LTB_4. Reverse-phase HPLC analysis of products formed when these mutants were incubated with LTA_4 revealed five peaks (I-V) originating from the substrate (Fig. 1). Peaks I and II corresponded to the epimers at C12 of 5(*S*),12-dihydroxy-6,8,10-*trans*-14-*cis*-eicosatetraenoic acid, non-enzymatic hydrolysis products of LTA_4, whereas peak III eluted with the expected enzymatic product LTB_4. Material eluting under peak V, the novel metabolite, was collected and subjected to UV spectrophotometry, which revealed a conjugated triene spectrum, typical of leukotrienes. The retention time in RP-HPLC was similar to one of the non-enzymatically formed 5,6-dihydroxyeicosatetraenoic acids, which suggested that compound V could be an isomer of 5,6-DHETE. It has been previously reported that soluble epoxide hydrolase (sEH) converts LTA_4 into 5(*S*),6(*R*)-DHETE [11]. Comparison of the retention time of 5(*S*),6(*R*)-DHETE formed by sEH with the retention time of compound V, showed that these two compounds did not coelute. Thus, mutagenetic replacement of Tyr-383 leads to an enzyme capable of forming an isomer of 5,6-DHETE. From these results one can speculate that sEH and LTA_4 hydrolase are functionally and structurally related.

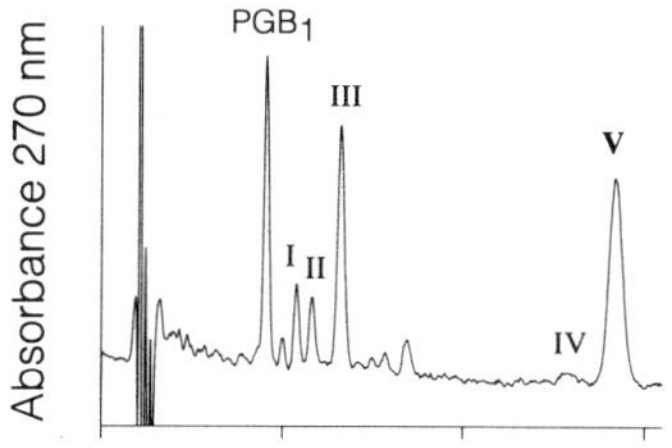

Figure 1. Reverse-phase HPLC analysis of products formed by [Y383Q]LTA_4 hydrolase with LTA_4.

ACKNOWLEDGMENTS

This work was financially supported by the Swedish Medical Research Council (03X-10350), The European Union (BMH4-CT960229), Vårdalstiftelsen, and Konung Gustav V:s 80-årsfond.

REFERENCES

1. B. Samuelsson, Leukotrienes: Mediators of immediate hypersensitivity reactions and inflammation, *Science* 220:568 (1983).
2. B. Malfroy, H. Kado-Fong, C. Gros, B. Giros, J.-C. Schwartz, and R. Hellmiss, Molecular cloning and amino acid sequence of rat kidney aminopeptidase M: A member of a super family of zinc-metallohydrolases, *Biochem. Biophys. Res. Commun.* 161:236 (1989).
3. B.L. Vallee, and D.S. Auld, Zinc coordination, function, and structure of zinc enzymes and other proteins, *Biochemistry* 29:5647 (1990).
4. J.Z. Haeggström, A. Wetterholm, R. Shapiro, B.L. Vallee, and B. Samuelsson, Leukotriene A_4 hydrolase: A zinc metalloenzyme, *Biochem. Biophys. Res. Commun.* 172:965 (1990).

5. M. Minami, N. Ohishi, H. Mutoh, T. Izumi, H. Bito, H. Wada, Y. Seyama, H. Toh, and T. Shimizu, Leukotriene A_4 hydrolase is a zinc-containing aminopeptidase, *Biochem. Biophys. Res. Commun.* 173:620 (1990).
6. J.F. Medina, A. Wetterholm, O. Rådmark, R. Shapiro, J.Z. Haeggström, B.L. Vallee, and B. Samuelsson, Leukotriene A_4 hydrolase: Determination of the three zinc-binding ligands by site-directed mutagenesis and zinc analysis, *Proc. Natl. Acad. Sci. USA* 88:7620 (1991).
7. J.Z. Haeggström, A. Wetterholm, B.L. Vallee, and B. Samuelsson, Leukotriene A_4 hydrolase: An epoxide hydrolase with peptidase activity, *Biochem. Biophys. Res. Commun.* 173:431 (1990).
8. A. Wetterholm, J.F. Medina, O. Rådmark, R. Shapiro, J.Z. Haeggström, B.L. Vallee, and B. Samuelsson, Leukotriene A_4 hydrolase: Abrogation of the peptidase activity by mutation of glutamic acid-296, *Proc. Natl. Acad. Sci. USA* 89:9141 (1992).
9. M. Blomster, A. Wetterholm, M.J. Mueller, and J.Z. Haeggström, Evidence for a catalytic role of tyrosine 383 in the peptidase reaction of leukotriene A_4 hydrolase, *Eur. J. Biochem.* 231:528 (1995).
10. M. Minami, H. Bito, N. Ohishi, H. Tsuge, M. Miyano, M. Mori, H. Wada, H. Mutoh, S. Shimada, T. Izumi, K. Abe, and T. Shimizu, Leukotriene A_4 hydrolase, a bifunctional enzyme: Distinction of leukotriene A_4 hydrolase and aminopeptidase activities by site-directed mutagenesis at Glu-297, *FEBS Lett.* 309:353 (1992).
11. J. Haeggström, J. Meijer, and O. Rådmark, Leukotriene A_4 : Enzymatic conversion into 5,6-dihydroxy-7,9,11,14-eicosatetraenoic acid by mouse liver cytosolic epoxide hydrolase, *J. Biol. Chem.* 261:6332 (1986).

PROSTAGLANDIN RECEPTORS OF THE EP2 AND EP4 SUBTYPES REGULATE B LYMPHOCYTE ACTIVATION AND DIFFERENTIATION TO IGE-SECRETING CELLS

Eric R. Fedyk[1,2], Sarah G. Harris[1,2], Josue Padilla[1,2,5] and Richard P. Phipps[1,2,3,4]

University of Rochester Cancer Center[1] and Departments of Microbiology and Immunology[2], Pediatrics[3], Environmental Medicine[4], and the Eastman Dental Center[5], University of Rochester School of Medicine and Dentistry, 601 Elmwood Avenue, Rochester, New York, 14642

INTRODUCTION

Immunoglobulin E (IgE) and T-helper type 2 (Th2) cytokines initiate physiologic cascades which cause allergic disease and asthma. Understanding how these cascades are triggered is critical for comprehending disease pathogenesis and for developing effective therapies. Novel therapies are needed to curb the increasing global incidence of allergy and asthma. Numerous mechanisms regulate IgE and Th2 cytokine production, however, an emerging concept is that E-series prostaglandins (PGEs), shift the immune response towards allergy by promoting production of IgE and a Th2 profile of cytokines (1). PGEs are lipid molecules which regulate diverse processes throughout the body. Prostaglandins are produced in most tissues (including lymphoid) by "constitutive" cyclooxygenases (COX-1), and in response to inflammatory stimuli, production is increased by expression of "inducible" cyclooxygenases (COX-2), (2). PGEs bind receptors (EP receptors) expressed by hematopoietic cells and promote Th2 and IgE responses at multiple levels. Firstly, PGEs induce development of Th2 cells via modulating cytokine production by antigen presenting cells. Secondly, they inhibit Th1 and promote Th2 profiles of cytokine production from mature differentiated T cells. Lastly, PGEs directly target B lymphocytes and enhance cytokine-directed immunoglobulin isotype class switching.

PGEs INDUCE DEVELOPMENT OF TH2 LYMPHOCYTES

PGEs regulate the development of T lymphocytes into mature cytokine-producing Th cells by modulating the activity of antigen-presenting cells (APC) with which T cells interact. Activation of APCs by antigen induces production of cytokines. The type of cytokines produced depends on the nature of the antigen and the conditions in which it was encountered. These APC-derived cytokines then direct T cell development towards a Th1 (IL-2 and IFN-γ) or Th2 (IL-4, IL-5 and IL-10) phenotype. For example, macrophage-derived IL-12 promotes differentiation to Th1 lymphocytes while simultaneously inhibiting differentiation of Th2 cells (3). Monocytes express EP_2, EP_3 and EP_4 receptors (4, 5) and when stimulated by PGEs, EP_2 receptors elevate cAMP

(5). PGE_2 stimulation and cAMP elevation inhibit production of IL-12 by monocytes (6, 7). Therefore, one method by which PGEs promote Th2 responses is by inhibiting development of Th1 lymphocytes. In contrast, IL-10 reciprocally promotes development of Th2 cells. PGE_2 and cAMP-elevating agents increase IL-10 production by monocytes (8, 9). Increased production of IL-10 is a second mechanism by which PGEs promote development of Th2 lymphocytes. Thus by increasing IL-10 and decreasing IL-12 production, PGE_2 promotes development of Th2 responses (Figure 1).

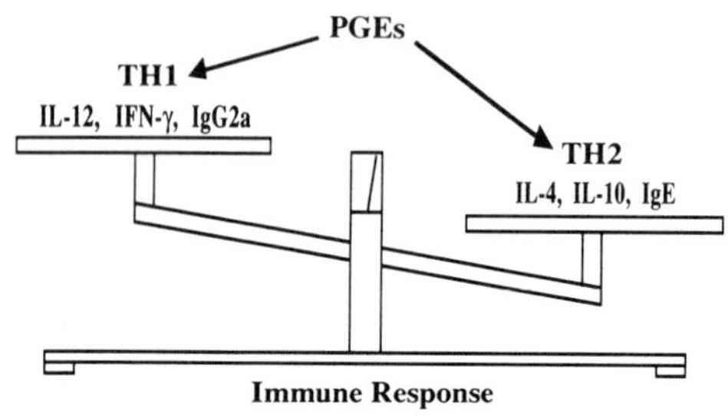

Figure 1. PGEs shift the balance of the immune response towards a Th2 profile of cytokine production and IgE. PGEs promote production of Th2-type cytokines by the immune response which in turn, induce IgE production by B lymphocytes.

PGEs MODULATE CYTOKINE PRODUCTION BY MATURE T LYMPHOCYTES

PGEs also modulate cytokine production from mature differentiated Th lymphocytes. T lymphocytes express EP_2, EP_3 and EP_4 receptors and when stimulated by PGEs, EP_2 and EP_4 receptors elevate cAMP (4, 8, 9). PGEs and cAMP-elevating agents profoundly inhibit production of the Th1 cytokines IL-2 and IFN-γ and inhibit growth of Th1 cells requiring IL-2 (1, 10, 11). In stark contrast, these agents do not inhibit Th2 cytokine production and may even increase production of IL-4 and IL-5 (1, 11). These data indicate that a third mechanism by which PGEs promote a Th2 response is by inhibiting production of Th1-type cytokines from mature Th1 lymphocytes. Therefore, by regulating production of IL-4, IL-10 and IL-12, PGEs shift the balance of the immune response towards a Th2 phenotype (Figure 1), (1).

PGEs ENHANCE IgE PRODUCTION BY B LYMPHOCYTES

IgE production is initiated by immunoglobulin isotype class switching to the ε locus of the Ig heavy chain. Immunoglobulin class switching involves a DNA recombination event which switches expression of immunoglobulin heavy chains from μ and δ to specific γ, ε or α chains. Isotype switching requires at least two signals. The first is stimulation by antigen and/or contact-mediated costimulatory signals (including gp39 interaction with CD40), Epstein-Barr viral infection (human only) or lipopolysaccharide (mouse only), (12). Cytokines provide the second signal and direct the DNA recombinase activity to a specific heavy chain locus downstream from μ and δ (e.g. γ, ε or α). In humans, IL-4 and/or IL-13 direct the recombinase to the immunoglobulin ε heavy chain locus, whereas in mice, only IL-4 specifies switching to IgE (12).

PGEs inhibit activation of B lymphocytes and promote differentiation to Ig secreting cells. Mouse B lymphocytes express EP_1, EP_2 $EP_{3\beta}$ and EP_4 receptors (13). Exogenous PGE_2 elevates cAMP within B lymphocytes (14, 15) and this presumably, is mediated by EP_2 and/or EP_4 receptors (13). PGEs inhibit B cell activation events, such as hyperexpression of class II MHC and FcεRII, induced by LPS, IL-4 and/or anti-IgM antibody, via binding EP_2 and EP_4 receptors and elevating cAMP (13, 14). These signals promote differentiation of B lymphocytes. Addition of PGEs to mouse B lymphocytes cultured in LPS and IL-4 potently enhances IgE production while inhibiting IgM. PGEs do not promote IgE production in the absence of IL-4 suggesting that PGEs operate by enhancing the effect of cytokines on isotype class switching (13, 14, 16). PGE_2-stimulated IgE production is also mimicked by selective agonists of EP_2 and EP_4 receptors (13), the cAMP-elevating agents cholera toxin and dbcAMP and is completely blocked by RpcAMP, a competitive inhibitor of protein kinase A (14). Thus, PGEs promote Th2 and IgE responses by binding EP_2 and EP_4 receptors on B lymphocytes (13), elevating cAMP, and enhancing recombination and transcription of the ε locus (16). This further shifts the balance of the immune response towards IgE (Figure 1). Similar observations have been reported for the effect of PGE_2 on *ex vivo*, antigen-specific IgE responses (17).

Preliminary investigations examining the effect of PGEs on human IgE production yielded conflicting conclusions. However, these data actually agree. It must be acknowledged that the effect of PGEs is dose-dependent and in particular, that supra-optimal concentrations of PGEs exist. Initially, one group reported that PGE_2 inhibited IgE production by human peripheral blood mononuclear cells (PBMC) and indeed it did at the high concentrations that they examined (18). *In vitro*, the optimal concentration of PGE_2 for enhancing IL-4-induced IgE production by human PBMC is $1 \times 10^{-9} - 10^{-8}$ M (19) and is at least a log more for mouse B cells (14). Furthermore, this same group (and others) more recently reported that similar cAMP-elevating agents *enhance* IgE production by human PBMC in this system (20, 21). Lastly, it has also been reported that PGE_2 ($< 1 \times 10^{-7}$ M) inhibits IL-4-induced IgE production from human B lymphocytes stimulated with the CD40-ligand (22). Interaction of gp39 and CD40 is required for isotype switching *in vivo* and hence, important additional studies are underway to further investigate what effect PGEs have on CD40-induced isotype class switching.

ELEVATED LEVELS OF PGEs POSITIVELY CORRELATE WITH PATHOGENIC TH2 AND IgE RESPONSES

Data obtained from *in vivo* studies corroborate that an overproduction of PGE_2 correlates with elevated Th2 and IgE responses in humans. Elevated levels of PGE_2, Th2 cytokines and IgE are also characteristic of patients with hyper-IgE syndrome (1), atopy (1, 12), recipients of autologous bone marrow transplants (1, 23) trauma patients (24), HIV infection (25) and Hodgkin's disease (1). Thus, *in vivo* studies support the model that elevated levels of PGEs, shift the balance of the immune response toward Th2 and IgE responses (Figure 1).

CONCLUSIONS

PGEs are potent immunomodulators which promote Th2 and IgE responses via the elevation of cAMP in target cells. These target cells include APCs, T and B lymphocytes. PGEs induce development of Th2 cells via increasing IL-10 production by macrophages while decreasing IL-12 production. Moreover, PGEs inhibit Th1 (IL-2 and IFN-γ) and promote Th2 (IL-4 and IL-5) profiles of cytokine production from mature differentiated T cells. PGEs inhibit B cell activation events by binding EP_2 and EP_4 receptors and elevating cAMP. These signals promote differentiation of B lymphocytes and potentiate IL-4-directed isotype class switching to IgE.

ACKNOWLEDGMENTS

This work was supported by USPHS Grants DE11390, AI07285, CA11198, DE07061, HL007216.

REFERENCES

1. Phipps, R. P., Stein, S. H. and Roper, R. L., A new view of prostaglandin E regulation of the immune response. *Immunol. Today.* 12:349-52 (1991).
2. Smith, W. L., Meade, E. A. and DeWitt, D. L., Pharmacology of prostaglandin endoperoxide synthase isozymes-1 and -2. *Annals N. Y. Acad. Sci.* 714:136-142 (1994).
3. O'Garra, A. and Murphy, K., Role of cytokines in determining T-lymphocyte function. *Curr. Opin. Immunol.* 6:458-466 (1994).
4. Mori, K., Tanaka, I., Kotani, M., Miyaoka, F., Sando, T., Muro, S., Sasaki, Y., Nakagawa, O., Ogawa, Y., Usui, T., Ozaki, S., Ichikawa, A., Narumiya, S. and Nakao, K., Gene expression of the human prostaglandin E receptor EP4 subtype - differential regulation in monocytoid and lymphoid lineage cells by phorbol ester. *J. Mol. Med.-Imm.* 74(6):333-336 (1996).
5. Zeng, L., An, S. and Goetzl, E.J., Independent downregulation of EP2 and EP3 sutyoes of the prostaglandin E2 receptor on U937 human monocytic cells. *J.Immunol.* 86:620-628 (1995).
6. van der Pouw Kraan, T.C., Boeije, L.C., Snijders, A., Smeenk, R.J., Wijdenes, J. and Aarden, L.A., Regulation of IL-12 production by human monocytes and the influence of prostaglandin E2. *Annals New York Acad. Sci.* 795:147-57 (1996).
7. van der Pouw-Kraan, T. C. T. M., Boeije, C. M., Smeenk, R. J. T., Wijdenes, J. and Aarden, L. A., Prostaglandin-E2 is a potent inhibitor of human interleukin 12 production. *J. Exp. Med.* 181:775-779 (1995).
8. Kambayashi, T., Alexander, H. R., Fong, M., and Strassmann, G., Potential involvement of IL-10 in suppressing tumor-associated macrophages. Colon-26-derived prostaglandin E2 inhibits TNF-alpha release via a mechanism involving IL-10. *J. Immunol.* 154:3383-90 (1995).
9. Strassmann, G., Patil-Koota, V., Finkelman, F., Fong, M. and Kambayashi, T., Evidence for the involvement of interleukin 10 in the differential deactivation of murine peritoneal macrophages by prostaglandin E2. *J. Exp. Med.* 180:2365-2370 (1994).
8. Devries, G.W., Guarino, P., Mclaughlin, A., Chen, J., Andrews, S. and Woodward, D.F., An EP receptor with a novel pharmacological profile in the T cell line Jurkat. *Brit. J. Pharmacol.* 115(7):1231-1234 (1995).
9. Zeng, L., An, S.Z. and Goetzl E.J., Regulation of expression of matix metalloproteinase-9 in early human T cells of the HSB.2 cultured line by the EP3 subtype of prostaglandin E2 receptor. *J. Biol. Chem.* 271(44):27744-27750 (1996).
10. Snijdewint, F. G. M., Kalinski, P., Wierenga, E. A., Bos J. D. and Kapsenberg, M., Prostaglandin E_2 differentially modulates cytokine secretion profiles of human T helper lymphocytes. *J. Immunol.* 150:5321-5329 (1993).
11. Hilkens, C.M.U., Vermeulen, H., Joost van Neerven, R. J., Snijdewint, F.G.M., Wierenga, E.A. and Kapsenberg, M.L., Differential modulation of T helper type 1 (Th1) and T helper type 2 Th2) cytokine secretion by prostaglandin E_2 critically depends on interleukin-2. *Eur. J. Immunol.* 25:59-63 (1995).
12. de Vries, J., Atopic allergy and other hypersensitivities. Curr. Opin. Immunol. 6, 835-837, 1994.
13. Fedyk, E. R. and Phipps, R. P., Prostaglandin E2 receptors of the EP2 and EP4 subtypes regulate activation and differentiation of mouse B lymphocytes to IgE-secreting cells. *Proc. Natl. Acad. Sci. U. S. A.* 93(20):10978-10983 (1996).
14. Roper, R. L. and Phipps, R. P., Prostaglandin E2 and cAMP inhibit B lymphocyte activation and simultaneously promote IgE and IgG1 synthesis. *J. Immunol.* 149:2984-2991 (1992).
15. Brown, D. M. and Phipps, R. P., Characterization and regulation of prostaglandin E2 receptors on normal and malignant murine B lymphocytes. *Cell. Immunol.* 161:79-87 (1995).
16. Roper, R. L., Brown, D. M. and Phipps, R. P., Prostaglandin E2 promotes B lymphocyte Ig isotype switching to IgE. *J. Immunol.* 154:162-70 (1995).
17. Hikida, M., Takai, T. and Ohmori, H., Selective regulation of antigen-specific IgE response by cyclic AMP level in murine lymphocytes. *Immunol. Letters.* 33:301-306 (1992).
18. Pene, J., Rousset, F., Briere, F., Chretien, I., Bonnefoy, J. Y., Spits, H., Yokota, T., Arai, N., Arai, K., Banchereau, J. and de Vries, J. E., IgE production by normal human lymphocytes is induced by interleukin 4 and suppressed by interferons gamma and alpha and prostaglandin E2. *Proc. Natl. Acad. Sci. U. S. A.* 85:6880-6884 (1988).
19. Byron, K.A., Liberatos, S., Varigos, G.A. and Wootton, A.M., Interferon-gamma production in atopic dermatitis: a role for prostaglandins? Intl. Arc.Allergy Immunol. 99(1):50-5 (1992).
20. Paul-Eugene, N., Pene, J., Bousquet, J. and Dugas, B., Role of cyclic nucleotides and nitric oxide in blood mononuclear cell IgE production stimulated by IL-4. *Cytokine.* 7:64-69 (1995).
21. Coqueret, O., Demarquay, D. and Lagente, V., Role of cAMP in the modulation of IgE production by the beta(2)-adrenoceptor agonist, fenoterol. *Euro. Resp. J.* 9(2):220-225 (1996).
22. Garrone, P., Galibert, L., Rousset, F., Fu, S.M. and Banchereau, J., Regulatory effects of prostaglandin E_2 on the growth and differentiation of human B lymphocytes activated through their CD40 antigen. *J. Immunol.* 152:4282-4290 (1996).

23. Sarayan, J. A., Rappeport, J., Leung D. Y. M., Parkman, R. and Geha, R. S., Regulation of human IgE synthesis in acute graft versus host disease. *J. Clin. Invest.* 71:556-564 (1983).
24. DiPiro, J. T., Hamilton, R. G., Howdieshell, T. R., Adkinson, N. F. Jr. and Mansberger, A. R. Jr., Total IgE in plasma is elevated after traumatic injury and is associated with sepsis syndrome. *Annals of Surgery.* 215:460-466 (1992).
25. Mastino, A., Grelli, S., Piacentini, M., Oliverio, S., Favalli, C., Perno, C. F. and Garci, E., Correlation between induction of lymphocyte apoptosis and prostaglandin E2 production by macrophages infected with HIV. *Cell. Immunol.* 152:120-130 (1993).

URINARY 9α, 11β-PGF_2 AS A MARKER OF MAST CELL ACTIVATION IN ALLERGIC AND ASPIRIN-INTOLERANT ASTHMA

Siobhán O'Sullivan[1], Barbro Dahlén[2], Annika Roquet[2], Lilian Larsson[1], Sven-Erik Dahlén[1] and Maria Kumlin[1].

[1] Experimental Asthma & Allergy Research, Institute of Environmental Medicine and
[2] Division of Respiratory Medicine at Karolinska Hospital, Karolinska Institutet, Stockholm, Sweden.

INTRODUCTION

Prostaglandin D_2 (PGD_2), is the major cyclooxygenase product of mast cells (1). It has been shown to be markedly overproduced in disorders involving mast cell activation, e.g. systemic mastocytosis. Few if any other cells are capable of producing PGD_2 in comparable quantities to the mast cell which makes it an attractive candidate for a mast cell marker. In the lung, the NADPH-dependant enzyme 11-ketoreductase converts PGD_2 to 9α,11β-PGF_2, which is excreted intact into the urine (2). In the past, urinary metabolites of PGD_2 have been measured with a stable dilution assay using negative ion chemical ionisation gas chromatography mass spectroscopy (GC NICI-MS) (3). Although this method is highly sensitive and specific, it is also very labour intensive making it impractical for routine analysis. Therefore, we adopted a commercially available enzyme-immunoassay for semi-automated measurements of 9α,11β-PGF_2 in urine.

Once validated we applied the methodology to investigate the occurrence of mast cell activation in allergen-induced and aspirin-induced bronchoconstriction, respectively. It is generally accepted that the early asthmatic response (EAR) to allergen is a result of IgE-mediated mast cell degranulation. However, the mechanism behind the late asthmatic response (LAR) is much less clear, with conflicting reports with respect to mast cell activation. In the case of aspirin-intolerant asthma (AIA), inhibition of the cyclooxygenase enzyme by aspirin or other non-steroidal anti-inflammatory drugs provokes symptoms in sensitive patients. However, the precise mechanism behind the intolerance reaction remains to be established. Convincing evidence has recently accumulated which supports a pivotal role for the leukotrienes (LTs) as mediators of aspirin-induced airway obstruction (4). With respect to the provocation-induced release of the LTs the mast cell is one potential source of their formation.

EXPERIMENTAL PROCEDURES

Bronchoprovocations were performed on 12 well-characterised atopic asthmatics with dual reactions and 8 aspirin-sensitive asthmatics by having the subjects inhale allergen and lysine-aspirin respectively. Bronchoprovocations were always performed in the mornings (between 7.30 and 8.30 am). Pulmonary function was measured by a spirometer as forced expiratory volume in one second, FEV_1, every 15 min during the first hour after the challenges and thereafter at hourly intervals for 12 hours. The aspirin-sensitive subjects were provoked on two occasions separated by at least 6 weeks, on the second occasion, the patients received a dose of lysine-aspirin which was higher (mean four-fold increase). Finally, four of the eight aspirin-sensitive patients who had previously undergone lysine-aspirin challenges were also subjected to challenge with histamine on two occasions separated by at least 6 weeks. Urine was collected prior to the provocations and thereafter at hourly intervals.

In order to identify the immunoreactivity in the samples, urine was purified prior to analysis by enzyme-immunoassay. Samples were subjected to solid-phase extraction on activated Isolute C_{18} columns washed with 2ml of water, 2ml of 50% methanol in water, 9α,11β-PGF_2 was eluted with 100% methanol. The methanol elutes were reconstituted in HPLC mobile phase (26/74/0.01; acetonitrile/water/acetic acid vol/vol/vol). Samples were injected onto a RP-HPLC column and eluted with mobile phase at a flow rate of 1ml/min. HPLC fractions were analysed with EIA. EIA analyses was performed with polyclonal antisera and acetylcholinesterase-linked tracer (Cayman Chemical), as previously described (5). Creatinine was measured in all urine samples by a commercially available colorimetric assay by using an alkaline picrate method (Sigma Chemical Company).

RESULTS

The specificity of the 9α,11β-PGF_2 antibody was assessed in a series of cross-reactivity studies. The antibody proved to be highly specific for 9α,11β-PGF_2 and $PGF_{2\alpha}$ (<1.7%), PGD_2 (<1.7%), 13,14-dihydro-15-keto-$PGF_{2\alpha}$ (<0.1%) and 8-epi-$PGF_{2\alpha}$ (<0.01%) did not cause any significant displacement of the tracer. The accuracy of the EIA for 9α,11β-PGF_2 was assessed in both buffer and urine and was excellent (r^2=0.99). The EIA was linear from 15.6-200pg/ml, samples exceeding this concentration were subsequently diluted to ensure they fell within this region of the standard curve. In purified urine samples from patients with atopic asthma challenged with allergen, the retention time of the immunoreactivity correlated with the retention time of the synthetic standard. Immunoreactivity was not detected in any other HPLC fractions. Recovery after purification, as assessed by tritiated $PGF_{2\alpha}$ was approximately 40% in urine (42 ± 9, n=8).

The airway response after bronchial challenge of asthmatics with inhaled allergen consisted of a fall in FEV_1, which began within 10 min after the start of the challenge and reached its peak after 30min (EAR). The group mean maximum drop during the EAR was 29±2%. This was followed by a more protracted decrease in FEV_1 corresponding to the LAR with the mean maximal drop (28±4%) occurring 6 to 8 hours after allergen challenge. During the EAR, there was a significant increase in the urinary excretion of both 9α,11β-PGF_2 and LTE_4 (Fig.1). This was followed by significant consistent increases in the levels of both metabolites during the LAR (Fig.1). Pulmonary function and urinary levels of either metabolite did not return to pre-challenge values until 24 hours after allergen provocation.

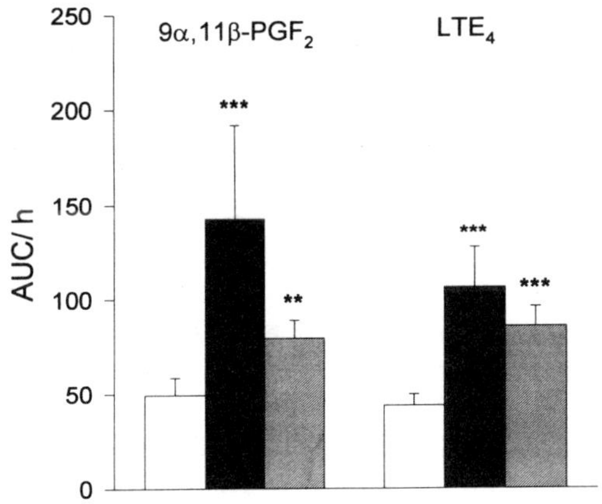

Figure 1. Mean concentrations (ng/mmol creatinine) of 9α,11β-PGF_2 and LTE_4 expressed as AUC/h, before challenge with allergen (open bar), during the EAR (defined as 0-3h), depicted by the filled bar and during the LAR (3-12h), depicted by the hatched bar.

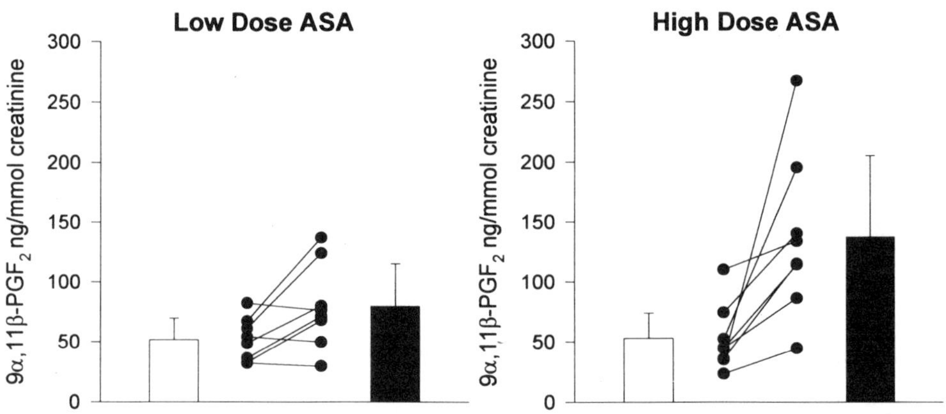

Figure 2. Pre-challenge (open bars) and post-challenge (filled bars) levels of 9α,11β-PGF_2 in urine samples from eight patients with aspirin-sensitive asthma after challenge with lysine-aspirin (low dose inhalation shown in the left panel, high dose inhalation shown in the right panel). The urine sample collected immediately prior to inhalation of lysine-aspirin is defined as the pre-challenge level, and the peak detected after maximal airway obstruction, as the post-challenge level.

Bronchial challenge of the aspirin-sensitive asthmatics with inhaled lysine-aspirin elicited bronchoconstriction without any extra-pulmonary symptoms. This was followed by a significant increase in the urinary excretion of 9α,11β-PGF_2 ($p<0.05$) (Fig. 2), with the peak concentration found in samples collected 1-3 hours after administration of the last dose of aspirin. Challenge with a higher dose of aspirin on another occasion produced an even greater increase ($p<0.01$) in urinary 9α,11β-PGF_2 (Fig.2).

Finally, in contrast to the allergen and aspirin challenges, there was no post-challenge increase in urinary 9α,11β-PGF_2 after histamine provocation, (pre-challenge value, 53.1±26.7 ng/mmol creatinine vs. post-challenge value, 49.0±13.3 ng/mmol creatinine mean ± SE).

DISCUSSION

The specificity of the 9α,11β-PGF_2 antibody was very high with very low cross-reactivities against structurally related compounds such as $PGF_{2\alpha}$ and 8-epi-$PGF_{2\alpha}$

Furthermore, the specificity of the EIA was supported by RP-HPLC analysis of urine samples, where the immunoreactivity measured by EIA was found to be confined to the fractions corresponding to authentic 9α,11β-PGF_2. Therefore, the EIA was deemed sufficiently selective and suitable for use in unpurified urine samples.

Here we present initial findings on increased levels of 9α,11β-PGF_2 in urine after bronchoprovocation with allergen, which represents the first documentation that increased levels of this metabolite may be detected in the urine during a provocation that elicits activation of lung mast cells, and consequently, production of PGD_2. Urinary excretion of 9α,11β-PGF_2 increased during both the EAR and LAR, which supports mast cell activation as a feature of both phases of the asthmatic response to allergen. In addition, there is increased formation of the cysteinyl-leukotrienes as indicated by increased levels of urinary LTE_4 during the EAR and LAR. In combination with the studies of attenuation of late allergen-induced bronchoconstriction by cys-LT receptor antagonists (6) and biosynthesis inhibitors (7), this strongly argues for a role for LTs in the LAR.

This investigation provided data, which strengthened the hypothesis that mast cells resident in the lungs are activated during aspirin-induced bronchoconstriction. We were able to document a clear cut rise in urinary 9α,11β-PGF_2 after lysine aspirin provocation. Moreover, the effect of lysine-aspirin on excretion of the PGD_2 metabolite appears to be dose-dependant since increased doses of inhaled lysine-aspirin resulted in an increased excretion of 9α,11β-PGF_2. None of the reactions were accompanied by extra-pulmonary symptoms, which lends strong support for the pulmonary formation of PGD_2.We considered the possibility that PGD_2 was being formed in response to the bronchoconstriction produced by the LTs, rather than during the initial intolerance reaction which triggers release of inflammatory mediators. However, the finding that histamine-induced bronchoconstriction did not increase urinary excretion of 9α,11β-PGF_2 strongly refutes that PGD_2 is being formed as a consequence of the bronchoconstriction.

In conclusion, measurements of urinary 9α,11β-PGF_2 with EIA appears to qualify as a new, simple and non-invasive method to monitor *in vivo* mast cell activation.

REFERENCES

1. R. Lewis, N Soter, P Diamond, K Austen, J Oates, LJ Roberts II. Prostaglandin D_2 generation after activation of rat and human mast cells with anti-IgE. *J Immunol.* 129:1627-31 (1982).

2. TE Liston, LJ Roberts II. Metabolic fate of radiolabeled prostaglandin D_2 in a normal human male volunteer. *J Biol Chem.* 260:13172-80. (1985).

3. JD Morrow, LJ Roberts II. Mass spectrometry of prostanoids: F_2-isoprostanes produced by non-cyclooxygenase free radical-catalysed mechanism. *Methods in Enzymology.* 233:163-75 (1994).

4. PE Christie, P Tagari, AW Ford-Hutchinson, S Charlesson, P Chee, JP Arm TH Lee. Urinary leukotrieneE_4 concentrations increase after aspirin challenge in aspirin-sensitive asthmatic subjects. *Am Rev Respir Dis.* 143:1025-1029. (1991).

5. S O'Sullivan, B Dahlén, S-E Dahlén, M Kumlin. Increased urinary excretion of the prostaglandin D_2 metabolite 9α,11β-PGF_2 after aspirin challenge supports mast cell activation in aspirin-induced airway obstruction. *J Allergy Clin Immunol.* 98:421-32. (1996).

6. A Roquet, B Dahlén, M Kumlin, E Ihre, G Anstrén, S Binks, S-E Dahlén. Combined antagonism of leukotrienes and histamine produces predominant inhibition of allergen-induced early and late phase airway obstruction in asthmatics. *Am J Respir Crit Care Med.* In press (1997).

7. Z Diamant, MC Timmers, H van der Veen, BS Friedman, et al. The effect of MK-0591, a novel 5-Lipoxygenase activating protein inhibitor, on leukotriene biosynthesis and allergen-induced airway responses in asthmatic subjects *in vivo*. *J Allergy Clin Immunol.* 95:42-51. (1995).

INDUCTION OF NITRIC OXIDE SYNTHASE (NOS) AND VASCULAR ENDOTHELIAL GROWTH FACTOR (VEGF) IN EXPERIMENTAL MODEL OF ANGIOPLASTY AND HEART ISCHEMIA

Aldona Dembinska-Kiec,[1] Jozef Dulak,[1] Lukasz Partyka,[1] Robert Krzesz,[1] Dariusz Dudek,[1] Stanislaw Bartus,[1] Michal Polus,[1] Ibeth Guevara,[1] Iwona Wybranska,[1] Tadeusz Krzeminski[2]

[1]Department of Clinical Biochemistry, Jagiellonian University School of Medicine, Kraków, Poland
[2]Department of Pharmacology, Silesian Medical Academy, Zabrze-Rokitnica, Poland
Department of Clinical Biochemistry,
Jagiellonian University School of Medicine,
Kopernika 15a, 31-501 Kraków,
POLAND
fax: +48 12 214073

INTRODUCTION

Balloon angioplasty (e.g. PTCA) leads to the activation of a number of genes in injured vessels[1]. As a consequence of the endothelium damage several characteristic reactions, such as thrombus formation, vasoconstriction and finally - vascular smooth muscle cell (VSMC) proliferation, lead to restenosis[1]. Among other things, the diminished generation of nitric oxide (NO), due to the lack of constitutively expressed endothelial NO synthase (eNOS) is causatively connected with this process.

The other isoform of NOS - inducible NOS (iNOS), expressed in the presence of cytokines, in VSMC, macrophages, neutrophils and other cells is responsible for the NO generation in much higher amounts[2]. This enzyme can be also induced during inflammatory reactions developing after vascular injury[2,3] and myocardial infarction[2,4].

Either hypoxemia[5] or IL-1β[6] induces the expression of a vascular endothelial growth factor (VEGF) gene in VSMC. VEGF is a mitogenic factor for endothelial cells promoting angiogenesis and induction of collateral blood vessel formation in infarcted tissue.

Since we found that VEGF may directly activate NO biosynthesis in intact endothelium[7], the aim of this study was to investigate the induction of iNOS and VEGF expression in experimental model of angioplasty and myocardial infarction.

MATERIALS AND METHODS

Animals

Male Wistar rats were used for experiments

Experimental models of endothelial injury

Two kinds of endothelial injury were performed: the crush-injury of rat carotid artery and balloon angioplasty with Fogarty catheter of rat carotid artery. The damaged arteries were removed at 3, 6, 12, 16, 24 hours and 3 and 14 days after surgery. The sham-operated vessels served as a control.

Myocardial infarction

Myocardial infarction was induced in anaesthetised rats by complete occlusion of the left coronary artery. Animals were sacrificed at 14 hours or at 11, 21, 28 or 35 days after infarct, and samples of infarcted and healthy myocardium were isolated.

Studies on gene expression

a) RNA isolation: total RNA was isolated from samples of rat arteries or myocardium by guanidine thiocyanate-phenol-chloroform- extraction, as formerly describe[d2].

b) reverse transcription (RT) and PCR was performed with Tth polymerase in Biometra Uno-Thermoblock, with primers for rat iNOS, VEGF and GAPDH, according to the method described previously[2].

RESULTS

Experimental endothelial injury resulted in the iNOS mRNA expression observed already at 3 h after vascular injury (Fig. 1, Table 1) and lasted until two weeks in the balloon injured vessels (Table 1). In the crush-injury model the iNOS expression was observed only during the first 24 hours. At 48 hours, and two or four weeks after injury no iNOS mRNA was detected (data not shown).

The $VEGF_{121}$ isoform was found in non-injured arteries (Fig. 1).

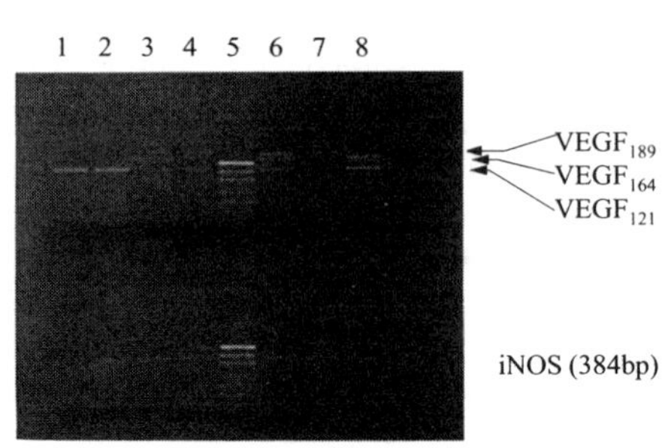

Figure 1. iNOS and VEGF gene expression in rat arteries after balloon angioplasty:
1) control artery; 2-4) 3 h, 6h, 12 h after insult; 5) DNA marker; 6,7) 18 h and 24 h after insult; 8) products of RT-PCR of RNA isolated form VSMCs stimulated in vitro with IL-1B; (2-6 animals studied per group)

Angioplasty caused the expression of two additional isoforms, $VEGF_{164}$ and $VEGF_{189}$, observed for the first time at 6 h after injury (Table 1, Fig. 1). VEGF expression was observed until 3 days after injury, and at two weeks after injury no VEGF mRNA was detected (Table 1). GAPDH mRNA was permanently present in the studied vessels, at a similar level at all time studied (not shown).

In intact rat's heart a weak expression of iNOS but not VEGF mRNA was observed after surgery. iNOS induction was found both in the infarcted muscle and intact part of a myocardium at 14 h but not in the hearts of rats on 11, 21 and 28 days after ischaemia (Fig. 2)[2]. A weak expression of iNOS mRNA was also detected at 35 days after infarct.

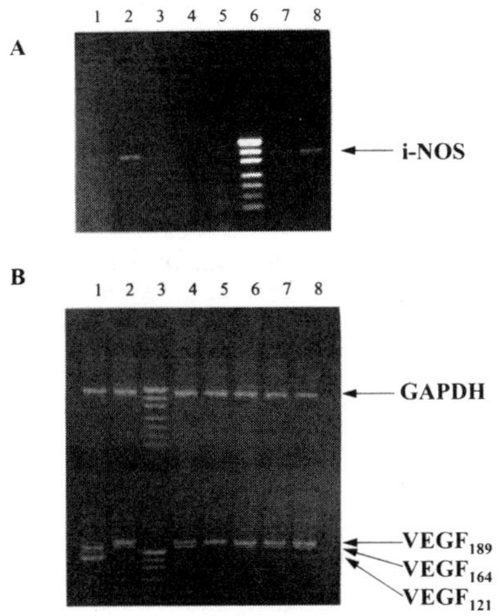

Figure 2. iNOS (A), GAPDH (B) and VEGF (B) expression after myocardial infarction:
A) 1,2 - 14 h after insult; 3,4) 11 days; 5) 28 days; 6) DNA marker; 7) 35 days; 8) iNOS expression in VSMC in vitro stimulated with IL-1b
B) 1) positive control - VEGF expression in VSMC in vitro stimulated with IL-1b; 2) 35 days after myocardial infarction; 3) DNA marker; 4) 28 days after insult; 5,6) 11 days after insult; 7,8) 14 hours after insult

Strong VEGF mRNA expression was observed at all the time after artery occlusion (Fig. 2). The strongest expression was observed for $VEGF_{189}$ isoform, while in VSMC in vitro $VEGF_{121}$ and $VEGF_{164}$ were more abundant (Fig. 1).

DISCUSSION

Our studies demonstrated for the first time the co-induction of iNOS and VEGF mRNA in injured rat vessels at a short time after balloon angioplasty. It seems that iNOS expression was previous to up-regulation of VEGF gene expression. The similar sequence of events is observed in VSMC in vitro after activation with IL-1β (in preparation).

Like other researchers, we observed the induction of iNOS expression during a short time after myocardial infarction[4]. Additional iNOS expression at 14 hours after artery occlusion, but not later (except for a weak expression at 35 days after insult) may result from the induction of iNOS in the inflammatory cells invading the necrotic tissue.

Induction of iNOS after balloon angioplasty was observed by several authors[1,3]. The role of augmented NO release is not clear, however, intimal hyperplasia can be inhibited by NO-donors, chronic NO inhalation, L-arginine treatment, or expression of experimentally

Table 1. iNOS and VEGF genes induction at different time after endothelial injury - a balloon angioplasty model

	Control arteries	3 h	6 h	12 h	24 h	3 days	14 days
iNOS	–	+	+	+	+	+	+
$VEGF_{121}$	+	+	+	+	+	–	–
$VEGF_{164}$	–	–	+	+	+	–	–
$VEGF_{189}$	–	–	+	+	+	–	–

transfected exogenous cNOS and iNOS genes, producing active NOS enzyme (for review, see [8-10]). We can thus suggest that in spite of NOS induction after vascular injury (as observed in our study), the amount of NO may be not sufficient to prevent formation of neointima.

Alternatively, the conditions in the injured vessels may shift the antiproliferative activity of NO to the mitogenic[11]. The observed upregulation of VEGF gene in VSMC may result from inflammatory cytokines or hypoxia[5,6]. VEGF synthesis after myocardial infarction or vascular injury may be necessary for revascularization of ischemic regions. The PTCA experimental studies indicate that VEGF infusion, or VEGF CND transfer to balloon-injured vessels enhance reendothelialization[8].

In summary, the co-induction of NO and VEGF after vascular injury indicates that the VEGF/NO relationship seems to be important for reendothelialisation and control of the neointima formation.

REFERENCES

1. V. Fuster, E. Falk, J.T. Fallon, L. Badimon, J.H. Chesbro, J.J. Badimon, The three processes leading to post PTCA restenosis: dependence on the lesion substrate. Thromb. Haemostasis 74: 552 (1995).
2. J. Dulak, R. Krzesz, M. Polus, D. Dudek, L. Partyka, A. Polus, I. Wybranska, I. Guevara, T. Krzeminski, A. Dembinska-Kiec, Inducible nitric oxide synthase gene expression in rat cells studied by Northern blot hybridization and RT-PCR amplification. Acta Angiol. 2: 57 (1996).
3. G.A. Joly, V.B. Schini, P.M. Vanhoutte, Balloon injury and interleukin-1β induce nitric oxide synthase activity in rat carotid arteries. Circ. Res. 71: 331 (1992).
4. S.M. Wildhirt, R.R. Dudek, H. Suzuku, V. Pinto, K.S. Narayan, R.J. Bing, Immunochemistry in the identification of nitric oxide synthase isoenzymes in myocardial infarction. Cardiovasc. Res. 29: 526 (1995).
5. J. Li, M.A. Perrela, J.C. Tsai, S.F. Yet, C.H. Hsieh, M. Yoshizumi, C. Patterson, W.O. Endege, F. Zhou, M.E. Lee, Induction of vascular endothelial growth factor gene expression by interleukin -1β in rat aortic smooth muscle cells. J. Biol. Chem. 270: 308 (1995).
6. G.T. Stavri, Y. Hong, I.C. Zachary, G. Breier, P.A. Baskerville, S. Yld-Herttuala, W. Risau, J.F. Martin, J.D. Erusalimsky, Hypoxia and platelet-derived growth factor BB synergistically upregulate the expression of vascular endothelial growth factor in vascular smooth muscle cells. FEBS Let. 358: 311 (1995).
7. A. Dembinska-Kiec, I. Huk et al. Vascular endothelial growth factor stimulates human endothelial cells to release nitric oxide. in preparation
8. L.J. Feldman, J.M. Isner, Gene therapy for the vulnerable plaque. J. Am. Coll. Cardiol. 26: 826 (1995).

9. H. Von der Leyen, G.H. Gibbons, R. Morishita, N.P. Lewis, L. Zhang, M. Nakajima, Y. Kaneda, J.P. Cooke, V.J. Dzau, Gene therapy inhibiting neointimal vascular lesion: in vivo transfer of endothelial cell nitric oxide synthase gene. Proc. Natl. Acad. Sci. USA 92: 1137 (1995).
10. E. Tzeng, L.L. Shears II, P.D. Robbins, B.R. Pitt, D.A. Geller, S.C. Watkins, R.L. Simmons, T.R. Billiar, Vascular gene transfer of the human inducible nitric oxide synthase: characterization of activity and effects on myointimal hyperplasia. Mol. Med. 2: 211 (1996).
11. A. Hassid, H. Arabsahani, T. Boucier, G.S. Dhaunsi, C. Matthews, Nitric oxide selectively amplifies FGF-2-induced mitogenesis in primary rat aortic smooth muscle cells. Am. J. Physiol. 267: H1040 (1994).

CGP57698: A STRUCTURALLY SIMPLE, HIGHLY POTENT PEPTIDO-LEUKOTRIENE (PLT) ANTAGONIST OF THE QUINOLINE TYPE

Andreas von Sprecher, Marc Gerspacher, Andreas Beck, Gary P. Anderson, Ulrich Niederhauser, Natarajan Subramanian, Howard A. Ball, Conrad Gentsch, Annick Vassout, Aina Felner, Helmut Bittiger, Kathleen Hauser, Karl Giese, Joseph Kraetz and Michael A. Bray

Novartis Pharma AG
CH-4002 Basel
Switzerland

A number of peptido-leukotriene antagonists coming from a variety of different structural classes have been developed over the last ten years and are in clinical development as anti-asthmatic agents.[1,2,3] Many of these compounds are complex molecules and we have aimed at developing novel compounds which combine structural simplicity with high oral potency. Compounds of the quinoline type were considered to be suitable target structures for the design of new potent and highly bioavailable LTD_4 antagonists.[4,5] The clinically most advanced compound of this type is at present MK-0476 (montelukast sodium).[6] The quinoline antagonists, characterized by a substituent in the 2 position consisting of a spacer with 2 atoms, an additional aromatic moiety and a terminal acidic group, show good *in vitro* and *in vivo* activity including ease of synthesis. The nature of this acidic residue is of key importance for the pLT antagonist potency of quinoline type antagonists. Combination of the required structural elements led to the synthesis of CGP57698 (4-[3-(7-fluoro-2-quinolinyl-methoxy)phenyl-amino]-2,2-diethyl-4-oxo-buta-noic acid) In CGP57698 the acidic group is derived from succinic acid with two geminal ethyl groups, which are essential for the high potency of CGP57698.

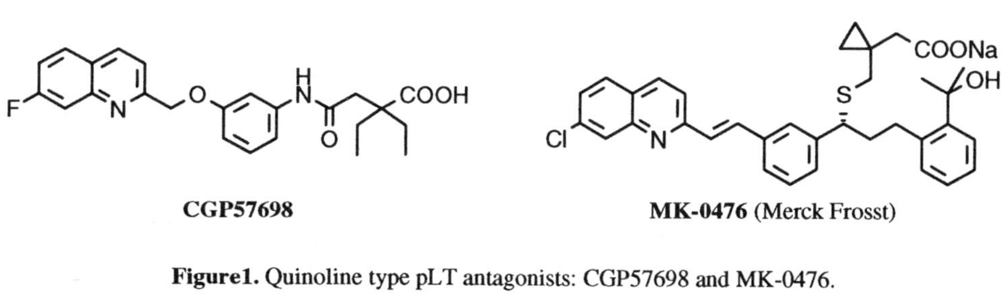

CGP57698 | **MK-0476** (Merck Frosst)

Figure1. Quinoline type pLT antagonists: CGP57698 and MK-0476.

Recent Advances in Prostaglandin, Thromboxane, and Leukotriene Research
Edited by Sinzinger *et al.*, Plenum Press, New York, 1998

PHARMACOLOGY

CGP57698 is a potent antagonist of pLT induced smooth muscle contractions *in vitro* (guinea-pig ileum: IC_{50} = 0.24 nM versus LTD_4 and 0.62 nM versus LTE_4). *In vitro* the compound is a potent inhibitor of [^{3}H]-LTD_4 binding to guinea-pig lung membranes with an IC_{50} of 6nM (K_i=2.5 nM). Other *in vitro* studies indicate that CGP57698 is a specific pLT antagonist, as it is inactive (at 0.24 mM) as an antagonist of histamine, bradykinin, acetylcholine, serotonin, prostaglandin E1 and substance P. Also, CGP57698 does not have intrinsic smooth muscle relaxant activity on the guinea-pig ileum (inactive at 0.24 mM against $BaCl_2$).

In vivo, the compound is potent but short acting as an aerosol when tested against LTD_4-induced bronchoconstriction in the guinea-pig. The IC_{50} data (as % of compound in the aerosol) when given 15 minutes prior to challenge is 0.00065% however when given one hour pre-challenge the compound is incative with a dose of 0.001%. When given orally, CGP57698 is extremely potent as an antagonist of both LTD_4 and LTE_4 induced bronchoconstriction in the anaesthetized guinea-pig with a well defined biological activity of at least 12-24 hours. The ED_{50} data are: 0.07, 0.53, 0.036 and 0.66 mg/kg at 2, 4, 8 and 12 h pretreatment time respectively, versus LTD_4 and 0.046, 0.28, 0.1 and 4.0 at -2, -4, -8 and -24h respectively, versus LTE_4. CGP 57 698 is also a potent long acting inhibitor of antigen induced, pLT dependent, bronchoconstriction in the guinea-pig when given orally (ED_{50} - 2h: 0.16 mg/kg; -4h: 0.5 mg/kg; -8h: 0.29 mg/kg).

A comparison between CGP57698 and ICI204219 (Table 1) confirms the attractive biological profile of CGP57698.

Table1. Comparison of CGP57698 and ICI204219.

	CGP57698	ICI204219[1]
LTD_4 antagonism guinea-pig ileum: IC_{50}(nM)	0.24	5.4
Inhib. of [^{3}H]-LTD_4 binding to guinea-pig lung membranes: IC_{50}(nM)	6	44
Inhib. of LTD_4(aerosol)-induced bronchoconstriction, guinea-pig:		
ED_{50}(mg/kg p.o.) -2h	0.07	0.55
ED_{50}(mg/kg p.o.) -4h	0.53	0.46
ED_{50}(mg/kg p.o.) -8h	0.04	1.30
Inhib. of antigen-induced, pLT mediated bronchoconstriction, guinea-pig:		
ED_{50}(mg/kg p.o.) -2h	0.16	0.38
ED_{50}(mg/kg p.o.) -4h	0.47	1.40
ED_{50}(mg/kg p.o.) -8h	0.29	n.d.

[1]Values determined at Ciba-Geigy, Basel
n.d.: Not determined

It is apparent that the potency of CGP57698 versus LTD_4/LTE_4 when given orally, is not dose related with time. This may be due to different absorption kinetics. This variation is unlikely to be due to an active metabolite and/or compartmentalization in the body, as when the compound is given i.v. there appears to be no difference in potency when the drug is given 2 or 4h prior to challenge. ED_{50} data are as follows: 0.14, 0.22, 0.066 and 0.1 mg/kg at 15 min, 1h, 2h and 4h respectively prior to LTD_4 challenge. Definitive answers to these apparent differences in potency will have to await absorption and metabolism studies.

CGP57698 has also been shown to be a potent antagonist of inhaled LTD_4 in the squirrel monkey when given orally two hours prior to challenge with an ED_{50} value of 2.0 mg/kg.

SAFETY PHARMACOLOGY

Cardiovascular pharmacology: CGP57698 was tested *in vivo* for cardiovascular and respiratory effects in anaesthetized cats at cumulative intravenous doses of 0.01-10 mg/kg and for renal effects in conscious rats at single oral doses of 10-100 mg/kg, and *in vitro* for effects on isolated guinea-pig atria and on isolated rabbit aortic rings preconstricted with either angiotensin II or noradrenaline, at concentrations up to 1 μmol/l. *In vitro*, CGP57698 had no relevant effect in isolated atria, and no effects on isolated aortic rings up to the highest concentration tested. *In vivo*, CGP57698 had no relevant effects on cardiovascular and respiratory functions in cats, or on renal excretory function in rats. These doses are considerably higher than the ED_{50}s for inhibition of LTD_4-induced bronchoconstriction in guinea-pigs. It can therefore be assumed that up to these doses CGP57698 should not give rise to significant effects on either the cardiovascular system and respiration, or on renal excretory function.

Central nervous system pharmacology: CGP57698 (in oral doses between 10 and 100 mg/kg) exhibited no relevant CNS-effects in the following tests: observation test in mice, body temperature in rats, rotarod test in rats, passive avoidance in mice, motility in mice and rats and ethanol induced sleep in mice. In addition, CGP57698 was inactive at a dose of 10μ M in the following neurotransmitter receptor binding assays: $alpha_1$-adrenergic (^{3}H-prazosin), $alpha_2$-adrenergic (^{3}H-clonidine), beta-adrenergic (^{3}H-dihydroalprenolol), 5-HT_1 (^{3}H-5-HT), 5-HT_2 (^{3}H-ketanserin), 5-HT_3 (^{3}H-BRL 43694), glycine (^{3}H-DCKA), histamine-1 (^{3}H-pyrilamine), opiate μ-type (^{3}H-naloxone) and muscarinic agonist (^{3}H-CMD).

TOXICOLOGY

CGP57698 showed no acute toxicity in rats in the limit test with 2000 mg/kg. Neither mortality nor adverse reactions were seen, and there were no gross pathological findings at necropsy after a 14 day observation period.

Oral administration of 100, 300 or 1000 mg/kg to rats for 14 days was well tolerated. However, dose-dependent decreased blood cell parameters, atrophy of the haemopoietic tissue of the bone marrow and decreased thymus weight was seen in females. Males were only minimally affected by the treatment.

In a 3 month oral toxicity study at doses of 10, 80 or 200 mg/kg female rats treated with 200mg/kg and later on also the female rats of the 80 mg/kg dose group had to be killed for humane reasons, as the treatment in females was associated with severely decreased blood cell parameters, decreased cellularity in thymus and bone marrow and centrilobular liver necrosis. Male rats showed increased liver weights from 80 mg/kg.

No significant findings were seen in marmosets either in a rising-dose study at oral doses of 200->500->1000 mg/kg or in a 3 month oral toxicity study at daily oral doses of 100, 300 or 800 mg/kg.

CGP57698 was shown to be Ames test negative with doses up to 5000 μg per plate. No mutagenicity was seen in a bacterial mutagenicity test with strains of S. typhimurium or E. coli WP2 uvrA.

The above results do not preclude further development of the compound.

CONCLUSION

CGP57698 is a highly potent orally active pLT antagonist of the quinoline type with a long duration of action. Due to its simple molecular structure CGP57698 can be synthesized in only 4 steps starting from 3-fluoroaniline. CGP57698 is a potential anti-asthmatic drug with a similar pre-clinical profile as the best competitors of the same class.

REFERENCES

1. C.D.W. Brooks, and J.B. Summers, *Modulators of leukotriene biosynthesis and receptor activation,* J. Med. Chem. 39: 2629-2654 (1996).
2. A. von Sprecher, A. Beck, M. Gerspacher, and M.A. Bray, *Peptidoleukotriene antagonists: state of the art,* Chimia. 46: 304-311 (1992).
3. A. von Sprecher, A. Beck, M. Gerspacher, A. Sallmann, G.P. Anderson, N. Subramanian, U. Niederhauser, and M.A. Bray, *Strategies in the design of peptidoleukotriene antagonists,* J. Lipid Mediators. 6(1-3): 265-73 (1993).
4. R.A.Galemmo, A. Gavai, and F.-C. Huang, *Recent developments in sulphidopeptide leukotriene receptor antagonists,* Curr. Opin. Ther. Patents. 811-833 (1992).
5. J.H. Musser, and A.F. Kreft, *Substituted-[2-quinolinyl(bridged)aryl] compounds: modulators of eicosanoid biosynthesis and action,* Drugs Future. 15: 73-80 (1990).
6. M. Labelle, M. Belley, Y. Gareau, J.Y. Gauthier, D. Guay, R. Gordon, S.G. Grossmann, T.R. Jones, Y. Leblanc, M. McAuliffe, C. McFarlane, P. Masson, K.M. Metters, N. Ouimet, D.H. Patrick, H. Piechuta, C. Rochette, N. Sawyer, Y.B. Xiang, C.B. Pickett, A.W. Ford-Hutchinson, R. Zamboni, R.N. Young, *Discovery of MK-0476, a potent and orally active leukotriene D_4 receptor antagonist devoid of peroxisomal enzyme induction,* Bioorg. Med. Chem. Lett. 5: 283-288 (1995).

S 18886, A NEW THROMBOXANE (TP)-RECEPTOR ANTAGONIST IS THE ACTIVE ISOMER OF S 18204 IN ALL SPECIES, EXCEPT IN THE GUINEA-PIG.

Serge Simonet[1], Jean-Jacques Descombes[1], Marie-Odile Vallez[1], Thierry Dubuffet[2], Gilbert Lavielle[2] and Tony J. Verbeuren[1].

Servier Research Institute,
[1]Division of Angiology,
[2]Division of Chemistry,
11 Rue des Moulineaux,
92150 Suresnes, France.

INTRODUCTION

Thromboxane (TXA_2) causes platelet aggregation, vasoconstriction and cell proliferation through the activation of specific cell membrane TP-receptors[1,2]. TXA_2 is implicated in thrombotic disorders and TP-receptor antagonists have been described as an anti-thrombotic approach[3]. S 18204 (Fig. 1) is a recently developed orally active and long acting TP-receptor antagonist[4]. Since S 18204 is a racemic compound, the aim of this study was to select the active isomer (S 18885 or S 18886) of S 18204 on platelet and contractile responses induced by the TP-receptor agonist U 46619 under *in vitro*, *in vivo* and *ex vivo* conditions.

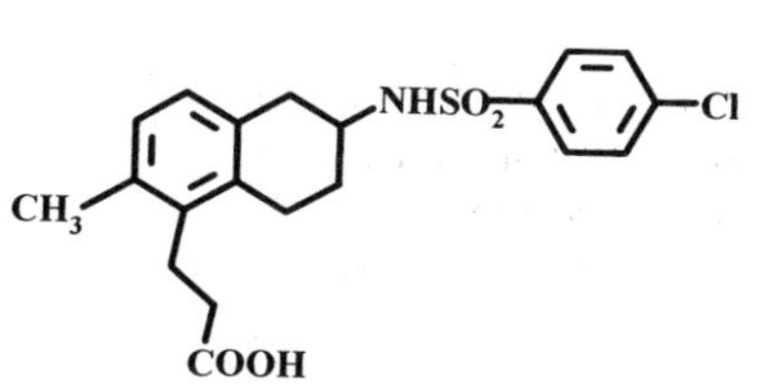

Figure 1. S 18204 (3-[(6-amino-(4-chlorobenzenesulfonyl)-2-methyl-5,6,7,8-tetrahydronapht]-1-yl)propionic acid) is a racemic compound. S 18885 and S 18886 are the (-) and (+) isomers respectively of S 18204.

Recent Advances in Prostaglandin, Thromboxane, and Leukotriene Research
Edited by Sinzinger *et al.*, Plenum Press, New York, 1998

METHODS

In vitro and *ex vivo* platelet aggregation studies

The effects of S 18885 and S 18886 were examined *in vitro* in different species. Citrated whole blood was taken by venipuncture from healthy volunteers or was obtained from carotid or femoral arteries of anesthetized guinea-pigs, rabbits and dogs. Citrated blood was centrifuged to obtain platelet-rich plasma (PRP) and platelet poor plasma (PPP). Platelet aggregation was measured by the turbidometric technique[5] using a four channel Chrono-log aggregometer. PRP was incubated in the absence or in the presence of S 18885 or S 18886 at different concentrations for 3 min before addition of agonist. Aggregation was induced with the stable thromboxane analog U 46619 (9,11-dideoxy-9α -epoxymethano-prostaglandin $F_{2\alpha}$); in canine PRP, aggregation was initiated with the α_2- adrenoceptor agonist adrenaline (10 μM) followed after 30 sec by U 46619. The amplitude of the aggregation was measured as the difference of light transmission between PRP (0 %) and PPP (100 %). The activity of the antagonists was calculated as an IC_{50} value, which is the concentration of the antagonists that permitted to obtain 50 % of the maximal inhibition.

In the *ex vivo* study, S 18885 or S 18886 was administered orally to conscious rats at 1 mg/kg. 48 h after the oral administration, hirudinated whole blood was taken from the anesthetized animals and platelet aggregation was studied in a whole blood aggregometer (Chrono-log) according to Cardinal & Flower[6]. Platelet aggregation was induced with increasing concentrations of U 46619 and the change in impedance (measured in ohm) was taken as an estimate of aggregation.

In vitro contractile studies

In vitro contractile responses of guinea-pig trachea were studied as described previously[7]. Segments were suspended in organ baths filled with physiological salt solution at 37°C. Cumulative concentration-response curves to U 46619 were performed. After washing, the antagonists were added to the organ bath and 30 min later, a second curve to U 46619 was performed. Results were expressed as percentage of the maximal control contraction induced by U 46619. Concentration-response curves obtained were analysed to determine the EC_{50} values and the pA_2 value was calculated following the method described by Arunlakshana and Schild[8].

In vivo studies

Guinea-pigs were anesthetized and the bronchoconstriction measured as described previously[9]. U 46619 (2 μg/kg i.v) was injected 5 min before and after the intravenous injection of the antagonists, administered in a dose-dependent manner. The bronchoconstrictions were compared before and after treatment with the antagonist and the percent inhibition by the antagonist of the U 46619-induced bronchoconstriction was calculated to obtain ID_{50} values, the inhibitory doses that permitted to obtain 50 % of the maximal inhibition.

In anesthetized pigs, increases in pulmonary artery pressure (PAP) were induced by U 46619 (0.05 μg/kg i.v.). Antagonists were administered at the dose of 10 μg/kg by i.v. bolus and the changes in inhibitory effects of antagonists on U 46619 induced pressor responses were compared before and after treatment with the antagonist. The percent inhibition by the antagonist of the U 46619-induced bronchoconstriction was calculated.

Statistical analysis

Data were expressed as mean $\pm$ s.e.m.. Statistical analysis was performed by Student's *t* test. The differences between means were considered to be statistically significant if $p \leq 0.05$.

RESULTS

The data of the *in vitro* platelet aggregation studies in dogs, guinea-pigs, rabbits and humans are summarized in table 1 and show that both isomers of S 18204 inhibit U 46619-induced platelet aggregation. In human, dog and rabbit platelets, S 18886 was significantly more active than S 18885; in guinea-pig platelets, no significant difference was noted between the two compounds.

Table 1. Inhibition of U 46619 (1 μM) -induced platelet aggregation (IC_{50}, μM ; n=5)

	Human	Dog	Guinea-pig	Rabbit
S 18885	0.82 ± 0.27	0.044 ± 0.013	0.40 ± 0.08	0.23 ± 0.06
S 18886	0.24 ± 0.08	0.015 ± 0.005	0.61 ± 0.03	0.13 ± 0.02

Ex vivo, 48 h after oral administration of the compounds at 1 mg/kg to rats, platelet aggregation induced by increasing concentrations of U 46619 was inhibited by S 18886 but not by S 18885 (Fig. 2).

In guinea-pig trachea, S 18885 and S 18886 competitively and similarly antagonized the contractile response to U 46619 with pA_2 values of 9.1 $\pm$ 0.1 (n=6) and 9.0 $\pm$ 0.2 (n=7), respectively.

In the anesthetized guinea-pig, S 18885 and S 18886 similarly inhibited the U 46619-induced bronchoconstriction in a dose-dependent manner (ID_{50} values were 30 ± 7 μg/kg (n=4) and 32 $\pm$ 1 μg/kg (n=6) for S 18885 and S 18886, respectively).

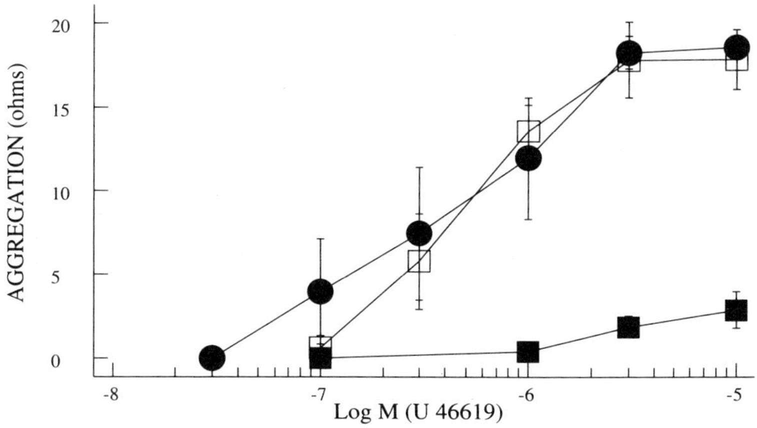

Figure 2. Inhibition of U 46619 induced platelet aggregation 48 h after oral administration of vehicule (●), S 18885 (□) or S 18886 (■) at the dose of 1 mg/kg in the rat.

In the anesthetized pig, S 18886 inhibited the pulmonary pressor responses to U 46619 for more than 6 hours. By contrast, the effect of the (-) isomer S 18885 is weaker and shorter (data not shown)

DISCUSSION

These data illustrate that the two isomers of S 18204 are potent TP-receptor antagonists. The present results also indicate that the effects of the TP-receptor antagonists are species-dependent ; in all species, except in the guinea-pig, S 18886 is the more active isomer of S 18204.

The results obtained *in vitro*, *in vivo* and *ex vivo* demonstrate that S 18886 is a TP-receptor antagonist in vascular and bronchial smooth muscles as well as in platelets. The oral activity and the long duration of action of S 18886 was demonstrated in conscious rats where platelet aggregation to U 46619 was completely inhibited during 48 h. A long lasting anti-platelet effect has also been observed with oral administration of S 18204 *ex vivo* in dogs[4].

A remarkable finding of our study is that both *in vitro* and *in vivo* in guinea-pigs, the (-) isomer, S 18885 shows a similar antagonist activity than that of the (+) isomer, S 18886.

In conclusion, S 18885 and S 18886 are active TP-receptor antagonists in all species. S 18886 is the more active isomer of S 18204, in all species, except in the guinea-pig and shows oral activity. Thus, S 18886 was selected for further evaluation as an anti-platelet drug.

REFERENCES

1. M. Hamberg, J. Svensson and B. Samuelsson. Thromboxanes: a new group of biologically active compounds derived from prostaglandin endoperoxides, *Proc. Natl. Acad. Sci. USA* 72 : 2994-2998 (1975).
2. F.K. Ko, S.M. Yu, Y.F. Kang and C.M. Teng. Characterization of the thromboxane (TP-) receptor subtype involved in proliferation in cultured vascular smooth muscle cells of rats, *Br. J. Pharmacol.* 116 : 1801-1808 (1995).
3. P. Gresele, H. Deckmyn, G.G. Nenci and J. Vermylen. Thromboxane syntase inhibitors, thromboxane receptor antagonists and dual blockers in thrombotic disorders, *TIPS.* 12 : 158-163 (1991).
4. T.J. Verbeuren, S. Simonet, J.J. Descombes, M.O. Vallez, M. Laubie and G. Lavielle. S 18204 is a new powerful TXA_2-receptor antagonist with a long duration of action, *Thromb. Haemostasis.* 73 : 1324 (1995).
5. G.V.R. Born. Aggregation of blood platelets by ADP and its reversal, *Nature.* 194, 927 (1962).
6. D.C. Cardinal and R.J. Flower. The electronic aggregometer : a novel device for assessing platelet behaviour in blood, *J. Pharm. Meth.* 3 : 135-158 (1980).
7. S.R. O'Donnell. and J.C. Wanstall. The importance of agonists in studies designed to predict $beta_2$:$beta_1$ adrenoceptor selectivity of antagonists from pA_2 values on guinea-pig trachea and atria, *Naunyn-Schmiedeber's Arch. Pharmacol.* 308 : 183-190. (1979).
8. O. Arunlakshana and H.O. Schild. Some quantitative uses of drug antagonism, *Br. J. Pharmacol.* 14 : 48-58 (1959).
9. H. Konzett. and R. Rossler. Versuchsanordnung zu untersuchungen an der bromkialmuskulatur, *Arch. Exp. Path. Pharmak.* 195 : 71-74 (1940).

GLIAL AND NEURONAL EXPRESSION OF CYCLOOXYGENASE-2: RELEVANCE TO ALZHEIMER'S DISEASE

M. Kerry O'Banion,[1,2] Julia W. Chang,[2] Mitchell D. Kaplan,[2]
Anna Yermakova,[2] and Paul D. Coleman[2]

[1] Department of Neurology
[2] Department of Neurobiology and Anatomy
University of Rochester School of Medicine and Dentistry
Rochester, NY 14642

INTRODUCTION

Alzheimer's disease (AD) is the leading cause of dementia in the elderly, affecting over 4 million individuals in the United States at an estimated annual cost of $100 billion. AD typically presents with memory loss followed by progressive cognitive deterioration over a period of 5 to 10 years. This clinical pattern is mirrored by neuropathology first in the hippocampus and associated structures with later involvement of cortical regions and some subcortical nuclei. Recent research efforts have lead to significant advances in our understanding of the genetics and pathological hallmarks of AD. However, we still do not understand the cause(s) of this disease, and current therapeutic interventions do not influence the overall progression of AD.

Neuropathological examination of AD brain reveals significant activation of microglia and astrocytes, and increased levels of proinflammatory cytokines, acute phase reactant proteins, complement components, and MHC class II antigens (reviewed by McGeer and McGeer, 1995; Rogers et al., 1996). The possibility that inflammatory processes contribute to AD pathogenesis is supported by over twenty epidemiological studies indicating that non-steroidal anti-inflammatory drugs (NSAID) and other anti-inflammatory therapies offer protection from AD (reviewed by Breitner, 1996). Although most of these studies were retrospective, a placebo-controlled trial of indomethacin showed benefit in 14 AD patients (Rogers et al., 1993). Furthermore, in a recent longitudinal study with over 1600 participants, Stewart et al. (1997) reported a relative risk of 0.40 (95% confidence interval 0.19-0.84) for developing AD in people reporting NSAID use for two or more years. Importantly, no significant effect was observed for aspirin or acetaminophen.

To investigate the apparent benefits of NSAID in AD, we have characterized cyclooxygenase-2 (COX-2; O'Banion et al., 1992) expression in cultures of brain glial cells treated with proinflammatory cytokines and in human AD and control brain tissue.

PROINFLAMMATORY CYTOKINES INDUCE COX-2 IN CULTURED GLIA

Interleukin-1ß (IL-1ß) treatment of primary astrocyte cultures prepared from the cortices of neonatal mice resulted in a dose- and time-dependent induction of prostaglandin production, cyclooxygenase activity, and COX-2 mRNA (O'Banion et al., 1996). By northern blot analysis, COX-2 message reached maximal levels 2 to 4 hours following treatment and declined to control levels by 24 hours. As in many cell types, astrocytic COX-2 was also induced by other inflammatory mediators, including tumor necrosis factor alpha and lipopolysaccharide, growth factors such as basic fibroblast growth factor, and by protein kinase C activation with phorbol ester. As seen in Figure 1, the induction of COX-2 mRNA and cyclooxygenase activity by phorbol ester was strongly augmented by co-treatment with IL-1ß. In other studies, we have investigated the effects of proinflammatory cytokines and growth factors in rat primary astrocyte cultures and find similar inductions of COX-2 mRNA (Chang et al., 1996).

Astrocytes represent the primary support cells in brain. Another glial cell type, the microglia, are believed to function as resident macrophages in the central nervous system. As would be expected, several recent studies demonstrated COX-2 induction and prostaglandin production in cultured microglia responding to lipopolysaccharide and other factors (Fiebich et al., 1996; Minghetti et al., 1996; Bauer et al., 1997). Thus two different glial cell types, which are both activated in AD, have the capacity to express COX-2.

NEURONS EXPRESS COX-2 IN HUMAN BRAIN

Based on evidence for inflammatory changes in AD tissue, the apparent benefits of NSAID in AD, and demonstration of COX-2 induction in glial cells, we hypothesized that COX-2 mRNA levels would be elevated in AD brain. Surprisingly, our initial quantification revealed a nearly three-fold reduction of COX-2 mRNA in affected AD neocortex relative to control neocortex. In contrast, levels of COX-2 were virtually unchanged in putamen, a brain region that is generally spared in Alzheimer's disease (Chang et al., 1996).

To determine the cellular localization of COX-2 in human brain we combined immunocytochemistry to glial and neuronal markers with *in situ* hybridization using a COX-2 riboprobe. These studies revealed prominent expression in select groups of neurons; specifically pyramidal cells of the hippocampus and to a lesser extent pyramidal cells of the cortex and granule cells of the dentate gyrus (Chang et al., 1996). Similar patterns of neuronal expression have been observed in rat (Yamagata et al., 1993; Breder et al., 1995; Kaufmann et al., 1996), mouse and monkey (A.Y. and M.K.O., unpublished observations). Recently, we have confirmed that COX-2 protein is expressed in human hippocampal neurons using immunocytochemistry (Figure 2B).

In contrast to fibroblasts and astroctyes, where COX-2 protein is localized in the vicinity of the nucleus, COX-2 appears in the soma and dendritic processes of neurons (Figure 2B, arrowheads). Indeed, there is some evidence that COX-2 localizes in specialized sites of

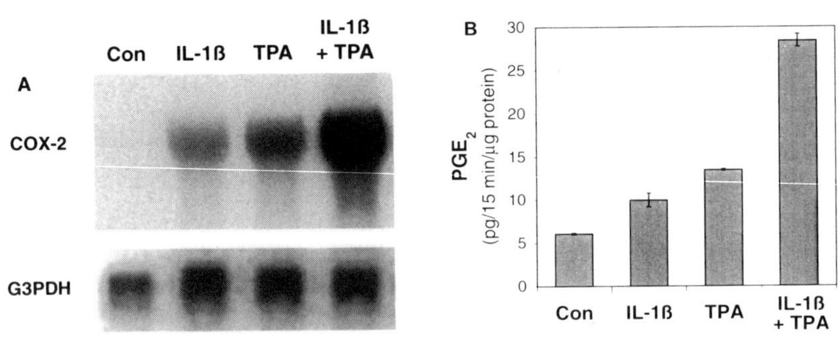

Figure 1. A) COX-2 mRNA levels are elevated in cultured murine astrocytes 2 hours after treatment with IL-1ß (1 ng/ml), phorbol ester (TPA, 100 nM), or both compounds. The relative amount of RNA loaded in each lane is indicated by the hybridization signal with a G3PDH probe. **B)** Increases in COX-2 mRNA abundance correlated with levels of enzyme activity at 4 hours. Cyclooxygenase activity was determined by radioimmunoassay measurement of PGE_2 generated during a 15 min incubation with 30 µM arachidonic acid and is reported as mean +/- SEM, normalized to total cellular protein (n = 4). The cyclooxygenase activity in control astrocytes is due to constitutive expression of COX-1 (O'Banion et al., 1996).

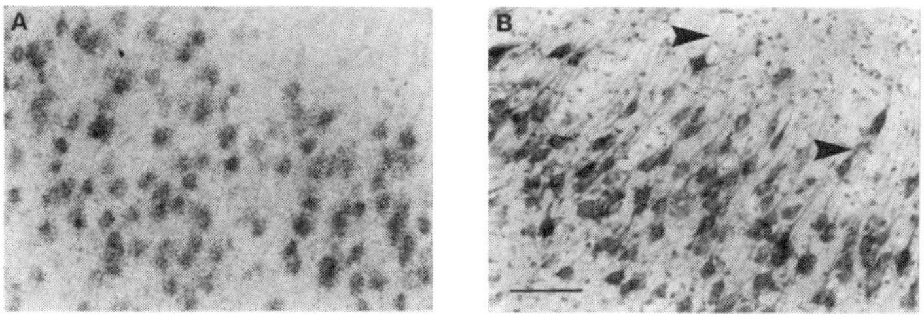

Figure 2. **A)** *In situ* hybridization with a human COX-2 riboprobe reveals significant COX-2 mRNA expression in pyramidal neurons of the human hippocampus. See Chang et al., (1996) for methods. **B)** Hippocampal neurons stain prominently with a monoclonal antibody raised against human COX-2 (Transduction Laboratories, Lexington, KY). Diaminobenzidine was used as the chromagen and sections were counterstained with cresyl violet. Note immunoreactivity localized in soma and dendrites (arrowheads). Bar = 100 μM in both pictures.

synaptic input known as dendritic spines (Kaufmann et al., 1996). Furthermore, neuronal COX-2 is regulated by synaptic activity: seizures dramatically increase COX-2 in hippocampal and cortical neurons; and loss of sensory input leads to a decrease in COX-2 message levels (Yamagata et al., 1993). These findings of prominent COX-2 expression in neurons and its regulation by synaptic activity provide a basis for our observation of decreased COX-2 mRNA in AD brain, which is known to exhibit extensive synaptic and neuronal loss (Terry et al., 1991; O'Banion et al., 1994).

CONCLUSION

Our studies have revealed two potential sites of action for NSAID in AD. One possibility is that NSAID inhibit COX-2 in activated glia, resulting in an attenuation of inflammatory processes in the brain. Another possibility is that COX-2 expression makes neurons vulnerable to other insults, perhaps by increasing free radical formation. Obviously, these mechanisms are not exclusive of one another. COX-1 is also expressed in brain, albeit at lower levels than COX-2. Defining the cellular substrates and pathological mechanisms involved represent critical issues for the rationale application of NSAID therapy in AD.

ACKNOWLEDGMENTS

The authors thank L. Callahan for assistance with *in situ* hybridization, J. Olschowka for advice on immunocytochemistry, and J. Rogers (Sun Health Research Institute, Sun City, Arizona) and the Rochester Alzheimer's Disease Center Brain Bank for human tissue. This work was supported by grants from the NIH to M.K.O. (R29 NS33553), P.D.C. (LEAD AG09016), and the Rochester Alzheimer's Disease Center (R30 AG08665), and a grant from the Lucille P. Markey Charitable Trust. M.D.K. is supported by a Dental Scientist Training Fellowship (K16 DE00159).

REFERENCES

Bauer, M.K.A., Lieb, K., Schulzeostohoff, K., Berger, M., Gebicke-Haerter, P.J., Bauer, J., and Fiebich, B.L., 1997, Expression and regulation of cyclooxygenase-2 in rat microglia, *Eur. J. Biochem.* 243:726.

Breder, C.D., DeWitt, D., and Kraig, R.P., 1995, Characterization of inducible cyclooxygenase in rat brain, *J. Comp. Neurol.* 355:296.

Breitner, J.C.S., 1996, Inflammatory processes and antiinflammatory drugs in Alzheimer's disease: a current appraisal, *Neurobiol. Aging* 17:789.

Chang, J.W., Coleman, P.D., and O'Banion, M.K., 1996, Prostaglandin G/H synthase-2 (cyclooxygenase-2) mRNA expression is decreased in Alzheimer's disease, *Neurobiol. Aging* 17:801.

Fiebich, B.L., Biber, K., Lieb, K., van Calker, D., Berger, M., Bauer, J., and Gebicke-Haerter, P.J., 1996, Cyclooxygenase-2 expression in rat microglia is induced by adenosine A_{2a}-receptors, *Glia* 18:152.

Kaufmann, W.E., Worley, P.F., Pegg, J., Bremer, M., and Isakson, P., 1996, Cox-2, a synaptically-induced enzyme, is expressed by excitatory neurons at postsynaptic sites in rat cerebral cortex, *Proc. Natl. Acad. Sci. USA* 93:2317.

McGeer, P.L., and McGeer, E.G., 1995, Chronic inflammatory mechanisms of the brain: implications for the treatment of Alzheimer disease, *Dementia* 9:111.

Minghetti, L., Polazzi, E., Nicolini, A., Créminon, C., and Levi, G., 1996, Interferon gamma and nitric oxide down-regulate lipopolysaccharide-induced prostanoid production in cultured rat microglial cells by inhibiting cyclooxygenase-2 expression, *J. Neurochem.* 66:1963.

O'Banion, M.K., Coleman, P.D., and Callahan, L.M., 1994, Regional neuronal loss in aging and Alzheimer's disease: a brief review, *Sem. Neurosci.* 6:307.

O'Banion, M.K., Dusel, J.C., Chang, J.W., Kaplan, M.D., and Coleman, P.D., 1996, Interleukin-1ß induces prostaglandin G/H synthase-2 (cyclooxygenase-2) in primary murine astrocyte cultures, *J. Neurochem.* 66:2532.

O'Banion, M.K., Winn, V.D., and Young, D.A., 1992, cDNA cloning and functional activity of a glucocorticoid-regulated inflammatory cyclooxygenase, *Proc. Natl. Acad. Sci. USA* 89:4888.

Rogers, J., Kirby, L.C., Hempelman, S.R., Berry, D.L., McGeer, P.L., Kasznaik, A.W., Zalinski, J., Cofield, M., Mansukhani, L., Willson, P., and Kogan, F., 1993, Clinical trial of indomethacin in Alzheimer's disease, *Neurology* 43:1609.

Rogers, J., Webster, S., Lue, L.-F., Brachova, L., Civin, W.H., Emmerling, M., Shivers, B., Walker, D., and McGeer, P., 1996, Inflammation and Alzheimer's disease pathogenesis, *Neurobiol. Aging* 17:681.

Stewart, W.F., Kawas, C., Corrada, M., and Metter, E.J., 1997, Risk of Alzheimer's disease and duration of NSAID use, *Neurology* 48:626.

Terry, R.D., Masliah, E., Salmon, D.P., Butters, N., DeTeresa, R., Hill, R., Hansen, L.A., and Katzman, R., 1991, Physical basis of cognitive alterations in Alzheimer's disease: synapse loss is the major correlate of cognitive impairment, *Ann. Neurol.* 30:572.

Yamagata, K., Andreasson, K.I., Kaufmann, W.I., Barnes, C.A., and Worley, P.F., 1993, Expression of a mitogen-inducible cyclooxygenase in brain neurons: regulation by synaptic activity and glucocorticoids, *Neuron* 11:371.

PROSTAGLANDIN E_2 DOWNREGULATES INDUCIBLE NITRIC OXIDE SYNTHASE EXPRESSION IN MICROGLIA BY INCREASING cAMP LEVELS

Luisa Minghetti[1], Alessia Nicolini[1], Elisabetta Polazzi[1], Christophe Créminon[2], Jacques Maclouf[3] and Giulio Levi[1].

[1]Laboratory of Pathophysiology, Istituto Superiore di Sanità, Rome Italy;
[2]CEA, Service de Pharmacologie et d'Immunologie, DRM, CE / Saclay, France
[3]I.F.R. Biologie de la Circulation-Lariboisière, INSERM U348,
Hôpital Lariboisière, Paris, France

INTRODUCTION

Microglial cells are ontogenetically related to the cells of the mononuclear phagocyte lineage, unlike all the other cells types in the CNS, and are widely spread throughout the brain. In normal brain, resting microglia show a downregulated phenotype adapted to the specialised microenvironment of the CNS. However, they can be rapidly activated to become important effector cells in most brain pathologies. Depending on the type or intensity of the stimulus and the concurrence of other local factors, they may foster neuroprotective and repair processes as well as contribute to the establishment and/or the amplification of tissue damage[1].

Among the numerous substances released by activated microglial cells, nitric oxide (NO) and prostanoids (including prostaglandins and thromboxanes), are likely to be important mediators of microglial functions, as they are known to regulate inflammation, immune response, blood vessel dilatation, neurotransmission and neural cell survival[2,3].

NO is derived from the oxidation of L-arginine by the NADPH-dependent enzyme NO synthase (NOS), while prostanoids are synthesised from arachidonic acid through the cyclooxygenase (COX) pathway. Both NOS and COX exist as constitutive forms (cNOS and COX-1), and as inducible forms (iNOS and COX-2), which are rapidly expressed upon stimulation with mitogens, cytokines and endotoxin[3,4].

Increasing evidence indicates that the NOS and COX pathways are mutually linked, but the type of interactions existing between the two pathways appears to dependent on the cell type and on the level of the end products released (for review, see Di Rosa et al., 1996[5]). For example, in the RAW 264.7 macrophage cell line and in primary astrocyte cultures, NO up-regulates COX activity[6,7], whereas in other cells, including the J774.2 macrophage cell line and microglial cells, NO inhibits prostanoid production[8,9]. Similarly, exogenous prostaglandin E_2 (PGE_2) was shown to be a positive or a negative modulator of NOS activity in distinct macrophage models, such as bone marrow-derived macrophages or established cell lines[10,11,12,13]. Therefore, in the light of the diversity of the observations reported, we have studied the role of PGE_2 in the regulation of NO synthesis in microglia.

Recent Advances in Prostaglandin, Thromboxane, and Leukotriene Research
Edited by Sinzinger *et al.*, Plenum Press, New York, 1998

RESULTS AND DISCUSSION

By using purified microglial cultures, obtained from rat neonatal brain, we have previously shown that COX-2 and iNOS are rapidly expressed upon stimulation with bacterial endotoxin (LPS), and that prostanoids and NO are synthesised with a comparable time- and dose-dependence[9]. Moreover we showed that, at variance with other macrophage cell models, COX-2 expression and prostanoid synthesis in microglial cultures are down regulated by endogenously produced or exogenously added NO[9].

PGE_2 inhibits microglial NO production and iNOS expression

To study the role of PGE_2, one of the major COX products in microglial cells[14,15], in the regulation of LPS-induced NO production, we stimulated microglial cultures with LPS (10 ng/ml) for 24 h in the presence of increasing amounts of PGE_2. We found that PGE_2 caused a dose-dependent reduction of LPS-induced nitrite accumulation, without affecting the basal release (Figure 1A). The inhibitory effect of PGE_2 was more pronounced in the presence than in the absence of 1 μM indomethacin (INDO), a specific inhibitor of COX activity which, at this concentration, completely abrogated the endogenous production of prostanoids[14]. Indeed, INDO, as well as other COX inhibitors such as acetyl-salicylic acid and 6-methoxy naphthalene acetic acid, caused *per se* a moderate inhibition of LPS-induced NO synthesis, suggesting that endogenous prostanoids could counteract the inhibitory effect of exogenous PGE_2 (see also Minghetti et al., 1997[16]).

In order to determine whether PGE_2 exerted its inhibitory effect by regulating iNOS expression, we analysed microglial proteins by western blot using a specific polyclonal antibody raised against a 18-residue peptide corresponding to the murine macrophage iNOS amino acid sequence 55-71, deduced from the cDNA sequence[16]. In LPS-treated samples, the antibody recognised a major band of approximately 130 kDa, co-migrating with purified macrophage iNOS, while no immunoreactivity was detected in unstimulated cultures.

Similar to what observed for nitrite accumulation, the expression of iNOS induced by LPS was markedly reduced by the presence of increasing amounts of PGE_2 (Fig. 1B). The inhibitory effect of 100nM PGE_2 on LPS-induced iNOS expression and nitrite accumulation was maximal when

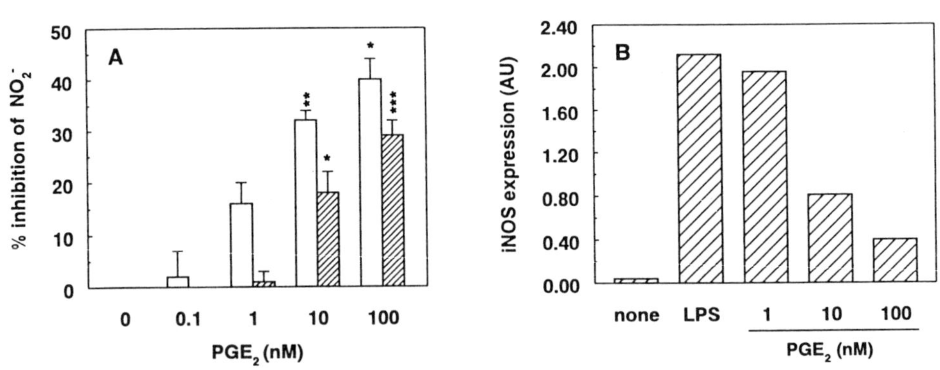

Figure 1. Microglial cells were obtained and cultured as previously described[14]. Cells were stimulated with 10 ng/ml LPS, with or without 1μM INDO, in the presence or in the absence of increasing concentrations of PGE_2. (A) NO production was measured as nitrite (NO_2^-) accumulation in culture media by a modified procedure based on the Griess reaction (Minghetti et al., 1996). Data are expressed as inhibition (%) of nitrite accumulation in the presence of LPS (13.99 ± 1.16 μM, shaded bars) or of LPS+INDO (10.30 ± 0.73 μM, open bars) and are means ± SEM of three to nine independent experiments, run in duplicate. * $p<0.025$, ** $p<0.01$, *** $p<0.0025$ compared with cultures stimulated by LPS or LPS+INDO, Student's "t" test for paired samples. (B) Cell lysates from stimulated and control cultures were prepared and equal amounts of proteins (10 μg) were analysed by western blot, using anti-iNOS specific antibodies (1:5000). Protein bands were visualised using horseradish peroxidase-conjugated secondary antibodies and ECL detection system and iNOS expression assessed by densitometric analysis. Data are means of two independent experiments.

microglial cells were treated simultaneously with the two agents, while the addition of PGE_2 2 h after LPS, when iNOS protein was already detectable, had no effect. Although our observations based on western blot analysis do not allow to clarify at which level PGE_2 is acting, the observed time-dependence of the PGE_2 inhibitory activity is suggestive of a reduction of the transcription rate rather than of the stability of iNOS mRNA.

PGE_2 inhibitory activity on iNOS is mediated by cAMP

Four types of specific EP receptors, associated to different signal transduction systems, have been so far identified[17], but data on microglial EP receptors are still lacking. However, the detection of EP3 mRNA in microglial cells[18] and their ability to respond to PGE_2 with a substantial increase in cAMP level[19,20], suggest that microglia could express at least these two EP receptor subtypes.

We then sought to determine whether the inhibitory effect of PGE_2 on NO production, required the activation of the EP2 receptor subtype, which is coupled to adenylyl cyclase through a Gs protein. We observed that 11-deoxy-16,16 dm-PGE_2, a specific EP2 receptor agonist, mimicked the inhibitory effect of PGE_2 in a concentration range (0.01-1 μM) including that effective with the natural agonist PGE_2. On the contrary, sulprostone, a potent agonist at EP3 and EP1 receptors, associated with an inhibition of adenylyl cyclase activity and with the generation of IP3 and Ca^{2+} mobilisation, respectively, caused a statistically significant inhibition of nitrite accumulation only at the highest concentration used (1 μM). Thus, the activation of EP2 receptors may be involved in the PGE_2-induced inhibition of NO production.

Other two obervations supported the involvement of cAMP in the down-regulation of NO synthesis by PGE_2. First, the inhibitory effect of PGE_2 on LPS-induced NO production was substantially reduced, in a dose-dependent manner, by SQ 22,536, a cell permeable adenylyl cyclase inhibitor which has been successfully employed as a probe for the involvement of adenylyl cyclase activation in other systems[21]. Second, the β-adrenergic agonist isoproterenol (100 nM), the adenylyl cyclase activator forskolin (100μM) and the cAMP analogue dibutyryl cAMP (10 μM), all effectively decreased the accumulation of nitrite and the expression of iNOS induced by LPS. At these concentrations, isoproterenol, forskolin and dbcAMP had no significant effect on the basal iNOS expression and NO production, although at higher concentrations the three agents induced a moderate NO production.

CONCLUSION

In recent years, increased attention has been focused on the role played by prostanoids and NO in the establishment and/or prevention of brain damage. In addition to the well known proinflammatory and immunodepressive activities of prostanoids, several authors have described a neuroprotective role for PGE_2 and/or prostacyclin against hypoxia, glutamate-induced neurotoxicity, microglial toxic products and actinomycin C toxicity[22,23,24,25,26]. Furthermore, PGE_2 has been shown to protect embryonic neural retinal cells against and may depress local immune responses by downregulating the expression of the costimulatory molecule B7-2 in microglia[27].

On the contrary, microglia-produced NO and reactive nitrogen oxides have been suggested to mediate neuronal cell death in ischemic and neurodegenerative disorders[28,29] and have been implicated in the establishment of oligodendrocyte damage in multiple sclerosis[30].

Our study indicates that PGE_2 downregulates iNOS expression and NO synthesis in LPS-activated microglial cultures by a mechanism involving EP2 receptor activation and cAMP generation and suggests that the inhibition of NO synthesis by PGE_2 may represent an additional mechanism by which prostanoids affect inflammatory, immune and degenerative processes in the CNS. In particular, by this mechanism, PGE_2 may help to prevent the neuronal and oligodendroglial damage that can be induced by an excessive production of NO.

REFERENCES

1. G.W. Kreutzberg, Microglia: a sensor for pathological events in the CNS, *Trends Neurosci.*, 19:312 (1996)
2. T. Shimizu and L.S. Wolfe, Arachidonic acid cascade and signal transduction, *J.Neurochem.*, 55: 1 (1990).
3. C. Nathan, Nitric oxide as a secretory product of mammalian cells, *FASEB J.*, 6: 3051(1992).
4. M. Goppelt-Struebe, Regulation of prostaglandin endoperoxide (cyclooxygenase) isozyme expression, *Prostaglandins Leukot. Essent. Fatty Acids*, 52: 213 (1995).
5. M. Di Rosa, A. Ialenti, A. Ianaro and L. Sautebin, Interaction between nitric oxide and cyclooxygenase pathways, *Prostaglandins LeuKot. Essent. Fatty Acids*, 54: 229 (1996).
6. D. Salvemini, T.P. Misko, J.L. Masferrer, K. Seibert, M.G. Currie and P. Needleman, Nitric oxide activates cyclooxygenase enzymes, *Proc.Natl.Acad.Sci.USA*, 90: 7240 (1993).
7. F. Molina-Holgado, A. Lledó and C. Guaza, Evidence for cyclooxygenase activation by nitric oxide in astrocytes, *Glia*, 15: 167 (1995).
8. T.A. Swierkosz, J.A. Mitchell, T.D. Warner, R.M. Botting and J.R. Vane, Co-induction of nitric oxide synthase and cyclo-oxygenase: interactions between nitric oxide and prostanoids, *Br.J.Pharmacol.*, 114: 1335 (1995).
9. L. Minghetti, E. Polazzi, A. Nicolini, C. Créminon and G. Levi, Interferon-γ and nitric oxide down-regulate lipopolysaccharide-induced prostanoid production in cultured rat microglial cells by inhibiting cyclooxygenase-2 expression, *J. Neurochem.*, 66: 1963 (1996).
10. S. Betz Corradin, N. Fasel, Y. Buchmüller-Rouiller, A. Ransijn, J. Smith and J. Mauël, Induction of macrophage nitric oxide production by interferon-γ and tumor necrosis factor-α is enhanced by interleukin-10, *Eur. J. Immunol.*, 23: 2045 (1993).
11. J. Mauël, A. Ransijn, S. Betz Corradin and Y. Buchmüller-Rouiller, Effect of PGE_2 and of agents that raise cAMP levels on macrophage activation induced by IFN-γ and TNF-α, *J. Leukocyte Biol.*, 58: 217 (1995).
12. P. Marotta, L. Sautebim and M. Di Rosa, Modulation of the induction of nitric oxide synthase by eicosanoids in the murine macrophages cell line J774, *Br. J. Pharmacol.*, 107: 640(1992).
13. S. Milano, F. Arcoleo, M. Dieli, R. D'Agostino, P, D'Agostino, G. De Nucci, and E. Cillari, Prostaglandin E_2 regulates inducible nitric oxide synthase in the murine macrophage cell line J774, *Prostaglandins*, 49: 105 (1995).
14. L. Minghetti and G. Levi, Induction of prostanoid biosynthesis by bacterial lipopolysaccharide and isoproterenol in rat microglial cultures, *J. Neurochem.*, 65: 2690 (1995).
15. N. Slepko, L. Minghetti, E. Polazzi, A. Nicolini and G. Levi, Re-orientation of prostanoid production accompanies "activation" of adult microglial cells in culture, *J Neurosci. Res.*, in press (1997)
16. L. Minghetti, A. Nicolini, E. Polazzi, C: Créminon , J. Maclouf, and G. Levi, Inducible nitric oxide synthase expression id downregulated by exogenous prostaglandin E2 and by cyclooxygenase inhibitors, *Glia* 19:152 (1997).
17. M. Negishi, Y. Sugimoto, and A. Ichikawa, Molecular mechanisms of diverse actions of prostanoid receptors, *Biochim. Biophys. Acta*, 1259: 109 (1995).
18. J. Kitanaka, H. Hashimoto, M. Gotoh, K. Kondo, K. Sakata, Y. Hirasawa, M. Sawada, A. Suzumura, T. Marunouchi, T Matsuda, and A. Baba, Expression pattern of messenger RNAs for prostanoid receptors in glial cell cultures, *Brain Res.*, 707: 282 (1996).
19. M. Patrizio, T. Costa, and G. Levi, Interferon-γ and lipopolysaccharide reduce cAMP responses in cultured glial cells: reversal by a type IV phosphodiesterase inhibitor, *Glia*, 14: 94 (1995).
20. M. Patrizio, D. Riitano, T. Costa, and G. Levi Selective enhancement by serum factors of cyclic AMP accumulation in microglia cultures, *Neurochem. Int.*, 29: 89 (1996).
21. E. Talpain, R.A. Armstrong, R.A Coleman and C.J. Vardey, Characterization of the PGE receptor subtype mediating inhibition of superoxide production in human neurotrophils, *Br. J. PharMacol.*, 114: 1459 (1995).
22. C. Cazevieille, A. Muller and C. Bonne, Prostacyclin (PGI_2) protects rat cortical neurons in culture against hypoxia/reoxygenation and glutamate-induced injury, *Neurosci.Lett.*, 160: 106 (1993).
23. C. Cazevieille, A.Muller, F. Meynier, N. Dutroit, and C. Bonne, Protection by prostaglandins from glutamate toxicity in cortical neurons, *Neurochem. Int.*, 24: 395 (1994).
24. A. Akaike, S. Kaneko, Y., Tamura, N. Nakata, H. Shiomi, F. Ushikubi, and S. Narumiya, Prostaglandin E_2 protects cultured cortical neurons against *N*-methyl-D-aspartate receptor-mediated glutamate cytotoxicity, *Brain Res.*, 663: 237 (1994).
25. C. Théry, A. Dobbertin and M. Mallat, Downregulation of in vitro neurotoxicity of brain macrophages by prostaglandin E_2 and a β-adrenergic agonist, *Glia*, 11: 383 (1994).
26. J.B. Dymond and G.W. Kalmus, The cytoprotective properties of prostaglandin E_2 against the toxic effects of actinomycin C on embryonic neural retinal cells, *Prostaglandins*, 44: 129 (1992).
27. B. Menendez-Iglesias, J. Cerase, C. Ceracchini, G. Levi and F. Aloisi, Analysis of B7-1 and B7-2 costimulatory ligands in cultured mouse microglia: upregulation by interferon-γ and lipopolysaccharide and downregulation by interleukin-10, prostaglandin E_2 and cyclic AMP-elevating agents, *J. Neuroimmunol.*, 72:83 (1997)
28. K.M. Boje and P.K. Arora, Microglial-produced nitric oxide and reactive nitrogen oxides mediate neuronal cell death, *Brain Res.*, 587: 250 (1992).
29. C.C. Chao, S. Hu, T.W. Molitor, E.G. Shaskan and P.K. Peterson, Activated microglia mediate neuronal cell death injury via a nitric oxide mechanism, *J. Immunol.*, 149: 2736 (1992).
30. J.E. Merrill, L.J. Ignarro, M.P. Sherman, J. Melinek and T.E. Lane, Microglial cell cytotoxicity of oligodendrocytes is mediated through nitric oxide, *J. Immunol.*, 151: 2132 (1993).

FORMATION OF A NOVEL CLASS OF F_3-ISOPROSTANES DURING PEROXIDATION OF EICOSAPENTAENOIC ACID (EPA)

Jaffar Nourooz-Zadeh [1], Barry Halliwell [2] and Erik. E. Änggård [3]

1) Centre for Clinical Pharmacology and Therapeutic Toxicology, Department of Medicine, UCL, 5 University Street, WC1E 6JJ, London, England.

2) Neurodegerative Disease Research Centre, King's College, Chelsea Campus, Manresa Road, SW3, London, England.

3) The William Harvey Research Institute, EC1M 6BQ, London, England.

INTRODUCTION

Accumulating data suggest that dietary supplementation of fish oils rich in eicosapentaenoic acid (EPA) and eicosapentaenoic acid (DHA) increases susceptibility of isolated low density lipoprotein (LDL) to peroxidation [1-4]. Oxidative modication of LDL may contribute to tissue damage through cytotoxic reactions with endothelial cells or through further reactions to generate "modified" LDL which is selectively accummulated by "scavenger" receptors [5]. Therefore, concern has been raised with respect to the high intake of EPA and DHA in numerous trials in the prevention of disease as well as encouragement to self medication by the health food industry.

The most widely used assay for the assessment of lipid peroxidation in model systems or biological materials is based on the measurement of thiobarbituric acid reactive substances (TBARs). Neither the TBA-test nor various methods for peroxide determination identify the specific fatty acids which have been peroxidized [6]. Recently, measurement of one of the major F_2-isoprostanes, 8-epi $PGF_{2\alpha}$, has been shown to be a specific and reliable index of lipid peroxidation in humans and animal experimental models [7-10]. F_2-isoprostanes constitutes subfamilies of PGF_2-like compounds, formed during free radical-mediated peroxidation of arachidonic acid, independent of the cyclooxgenase pathway. Each F_2-isoprostane subfamily would comprise a cyclopentane ring containing 1,3-diol moiety and two alkyl moities containing

Recent Advances in Prostaglandin, Thromboxane, and Leukotriene Research
Edited by Sinzinger *et al.*, Plenum Press, New York, 1998

the original carboxylic and methyl terminus. Therefore, we aim to develop a similar biomarker of peroxidation derived from EPA.

RESULTS

EPA (C20:5ω3) has 4 carbon centres (C7, C10, C13 and C16) located between adjacent double bonds. Free radical attack at these positions theoretically yields six subfamilies of F_3-isoprostanes as the end products (Figure 1). Type I and type VI are derived from initial free radical attack at positions C7 and C16. Type II and type III are derived from initial attack at positions C10 whereas type IV and Type V are derived from attack at position C13. Theoretically, 96 F_3-isoprostane isomers could be formed during peroxidation of eicosapentaenoic acid. Three hydroxyl groups on the prostanoid skeleton can be arranged in 2^3 configurations. Regio-isomer IV represents the 8-epi PGF_3 stereoisomers.

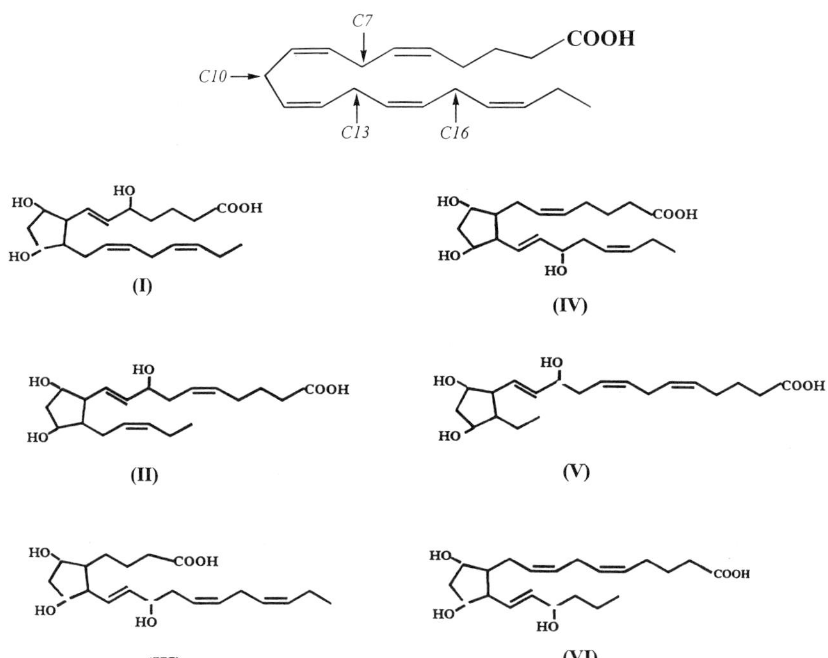

Figure 1

To explore this hypothesis, we exposed EPA liposomes (1mg/ ml phosphate buffered saline; PBS) to the peroxyl radical generator 2,2-Azobis-(2-amidinopropane) dichloride (AAPH; 1mM, final concentration) at 37 °C for 24 h. F_3-isoprostanes were isolated by chromatography on a C_{18}- and an NH_2 cartridges as described previously [9]. Final determination of the F_3-isoprostanes as pentafluorobenzyl/trimethylsylil ether (PFB/TMS) derivatives by gas chromatography-mass spectrometry/negative ion chemical ionisation (GC-MS/NICI) using the deuterated $PGF_{2\alpha}$ as the internal standard [8].

Figure 2 shows GC-MS/NICI chromatogram of the PGF_3-like compounds (PFB ester /TMS ether) derivative formed during peroxidation of EPA liposomes with AAPH (1 mM) at 37 °C for 24 h. The signal monitored at m/Z 567 and 573 represent the carboxylate anion $[M-180]^-$ for the PGF_3-like compounds and the tetradeutrated $PGF_{2\alpha}$, respectively. As seen in Figure 2, components (I and II) were the predominant oxidation products whereas the 8-epi $PGF_{3\alpha}$ was a minor component.

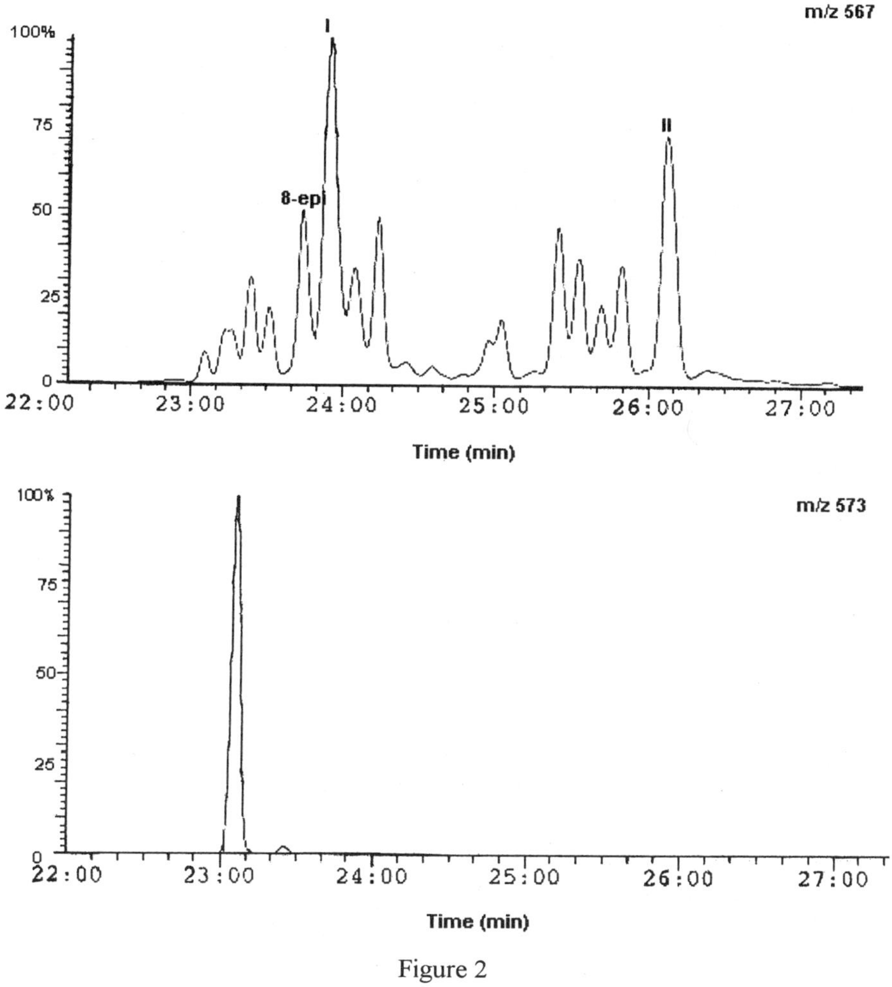

Figure 2

Components I and II were tentatively identified as F_3-isoprostanes by forming PBF/$[^2H_9]$ TMS and PBF/cyclic boronate/TMS derivatives. Freshly prepared EPA liposomes already contained small but quantifiable amounts of component I and component II (1.68 nmol/mol EPA; for each) as well as 8-epi $PGF_{3\alpha}$ (0.61 nmol/mol EPA. At 24 h, AAPH oxidised EPA samples contained 48 and 67.5 nmol/mol EPA of components I and II, respectively.

In conclusion, our study indicates that it may be feasible to analyse specific isoprostanes derived from EPA. The possibility of studying lipid peroxidation products from this and other (eg. DHA) individual fatty acids provide new opportunities for nutritional research.

ACKNOWLEDGMENTS

We thank the British Heart Foundation and the Ministry of Agriculture, Fisheries, and Foods for research support.

REFERENCES

1) Harats, D., Dabach, Y., Hollander, G., Ben-Naim, M., Schwartz, R., Berry, E.M. , Stein, O., and Stein, Y., Fish oil ingestion in smokers and nonsmoskers enhances peroxidation of plasma lipoproteins, Atheroscler. 90:127 (1991).

2) Lussier-Cancan, S., Dubreuil-Quidoz, S., Roederer, G., Leboeuf, N., Boulet, L., de Langavant, G.C., Davignon, J., and Naruszewicz, M., Influence of probucol on enhanced LDL oxidation after fish oil treatment of hypertriglyceridemic patients, Atheroscler. Thromb. 13:1790 (1993).

3) Oostenbrug, G.S., Mensink, R.P., and Honstra, G., Effect of fish oil and vitamin E supplementation on copper catalysed oxidation of human LDL in vitro, Eur. J. Clin. Nutr. 48:895 (1994).

4) Suzukawa, M., Abbey, M., Howe, P.R.C., Nestel, P.J., Effects of fish oil fatty acids on LDL size, oxidizability, and uptake by macrophages, J. Lipid Res. 36:473 (1995).

5) Steinberg, D., Parthasarathy, S., Carew, T.E., Khoo, J.C., and Witztum, J.L., Beyond cholesterol: modification of LDL that increases its atherogenicity, New Engl. J. Med. 320:915 (1989).

6) Halliwell, B., Oxidative stress, nutrition and health: experimental strategies for optimization of nutritional antioxidant intake in humans, Free Radical. Res. 25:57 (1996).

7) Morrow, J.D., Hill, K.E., Burk, R.F., Tarek, M., Nammour, T.M., Badr, K.F., Robert II, J.L., A series of prostaglandin F_2-like compounds are produces in vivo in humans by a non-cyclooxgenase, free radical-catalyzed mechanism, Proc. Natl. Acad. Sci. USA 87:9383 (1990).

8) Morrow J.D., Roberts II, J.L., The isoprostanes: current knowledge and directions for future research, Biochem. Pharmcol. 51:1 (1996).

9) Nourooz-Zadeh, J., Gopaul, N.K., Barrow, S., Mallet, A.I. and Änggård, E.E., Analysis of F_2-isoprostanes as indicators of non-enzymatic lipid peroxidation in vivo by GC-MS: develoment of a solid phase extraction procedure, J. Chromatogr. Biomed. Appl. 667:199 (1995).

10) Gopaul, N.K., Änggård, E.E., Betteridge, D.J., Wolff S.P. and Nourooz-Zadeh, J., Plasma 8-epi $PGF_{2\alpha}$ levels are elevated in individuals with NIDDM, FEBS Letters 368:225 (1995).

PLASMA 8-EPI $PGF_{2\alpha}$ LEVELS IN HEALTH AND DISEASE

Jaffar Nourooz-Zadeh[1] and Erik E. Änggård [2]

1) Department of Medicine, Division of Clinical Pharmacology and Toxicology, Department of Medicine, UCL, 5 University Street, London WC1E 6JJ, England.
2) The William Harvey Research Institute, London EC1M 6BQ, England.

INTRODUCTION

Lipid peroxidation has been proposed to play an important role in the development of several diseases including atherosclerosis, chronic renal failure, diabetes, inflamation and neurodegenerative disorders. However, this issue has been difficult to address since specific and reproducible methods for the measurement of oxidative stress have been unavailable or unreliable [1].

A major step forward came with the discovery of F_2-isoprostanes. F_2-isoprostanes comprises a unique family of PGF_2-like compounds produced by a free radical-mediated from arachidonic acid (AA), independent of the cyclooxygenase pathway [2]. Non-cyclooxgenase peroxidation of AA yields four regio-isomers of PGF_2 having side chains and the hydroxyl groups on the cyclopentane ring *cis*-oriented, respectively. Each of these PGF_2-like compounds is theoretically composed of a mixture of eight racemic diastereomers [2]. Measurement of one of the major F_2-isoprostanes, 8-epi $PGF_{2\alpha}$, has been shown to represent a sensitive and reliable marker of oxidative stress [2]. In addition, the 8-epi $PGF_{2\alpha}$ has been shown to be biologically active [2].

Using a specific and reliable assay [3], we report plasma concentration of total (sum of free and esterified) 8-epi $PGF_{2\alpha}$ from non-insulin dependent diabetes (n=39), chronic renal failure (n=38), primary hyperlipidaemic (n=14) and control subjects (n=15).

RESULTS AND DISCUSSION

Plasma F_2-isoprostanes were analyzed using an imporved assay developed in this laboratory [3]. The Technique involves chromatography on a octadecylisane (C_{18}) followed by an aminopropyl (NH_2) cartridge followed by gas chromatography- mass spectrometry (GC-MS). The main advantage of our assay lies in improved recovery, speed and gas chromatographic separation as compared to the established techinque [4].

Establish isoprostane assay involves solid-phase extraction followed by one or two steps of thin layer chromatography prior to final determination by (GC-MS).

Plasma levels of total (sum of free and esterified) 8-epi $PGF_{2\alpha}$ in 39 individuals with non-insulin dependent diabetes mellitus (NIDDM) varied from 0.32 to 2.19 nM with a mean of 0.93 ± 0.46 nM. In the patient with chronic renal failure (CRF) (n=38), the levels were between 0.21 and 2.47 nM with a mean of 0.68 ± 0.41 nM whereas those with primary hyperlipidaemic (PH) patients varied from 0.13 to 0.8 nM with a mean of 0.43 ± 0.22 nM. Plasma 8-epi $PGF_{2\alpha}$ levels in NIDDM, CRF and PH were markedly higher than that recorded for the controls 0.28 ± 0.02; range 0.02-0.63 nM (at the 95% confidence level). The highest difference was recorded for individuals with NIDDM and CRF, respectively ($p<0.0005$, Mann Whitney test). Data for NIDDM, CRR, PH and controls subjects is shown in Figure 1.

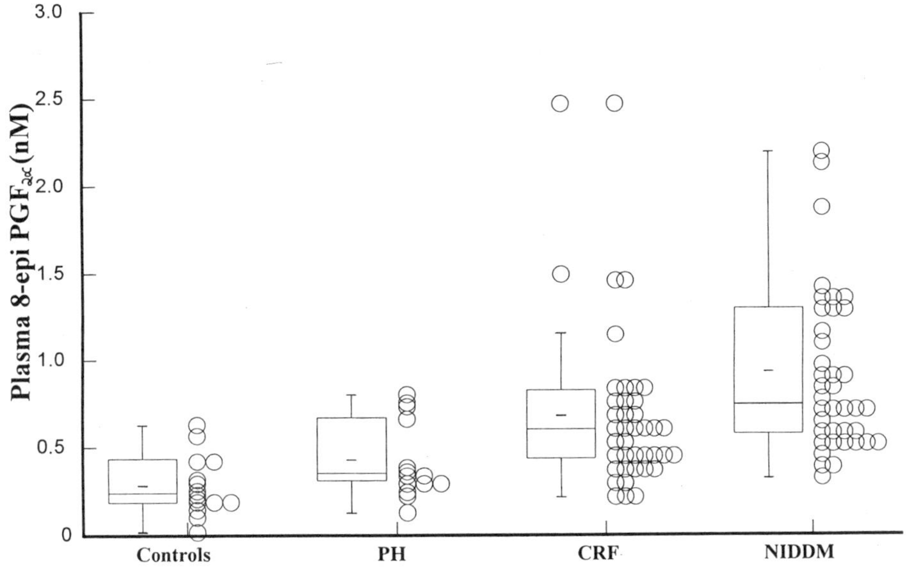

Figure 1

It has been shown that F_2-isoprostanes may be generated in plasma during prolonged or improper storage. Morrow et al. [2] demonstrated that addition of 0.002% butylated hydroxytoluene suppressed the *in vitro* formation of F_2-isoprostanes in plasma by 90%, but did not inhibit the F_2-isoprostanes originally presented in plasma. In this study, F_2-isoprostanes analysis were carried out after 2 weeks of sample collection to minimise artefact formation during the storage of plasma at -70 °C. In addition, the absence of free in the whole plasma and isolated low density lipoproteins of control confirm the reliability of our SPE/GC-MS assay [3,4,5].

Morrow et al. [2] reported the presence of to 0.05 ± 0.02 nM (range 0 to 0.14 nM; n=5) of presumably free (unesterified) 8-epi $PGF_{2\alpha}$ in plasma from healthy individuals. In a recent study, the presence of 0.09 ± 0.05 and 0.29 ± 0.09 nM of free and esterified 8-epi $PGF_{2\alpha}$, respectively, in plasma samples from healthy donors (n=8) [7]. Lynch et al. [8] reported the presence of 0.02 and 0.09 nM of free and total (sum of free and esterified) 8-

epi $PGF_{2\alpha}$, respectively, in a plasma sample from normolipidaemic individual. On the contrary, we did not find any quantifiable levels of free 8-epi $PGF_{2\alpha}$ in plasma samples from either controls or diabetic at the detection limit of 0.03 nM [4,9].

Enhanced levels of free as well as esterified 8-epi $PGF_{2\alpha}$ has been reported in plasma from smokers as compared to non-smokers 0.62 ± 0.21 vs 0.25 ± 0.15 nM [7]. The levels of free and esterifed 8-epi $PGF_{2\alpha}$ were reduced by 40 and 25%, respectively, after two weeks of abstinence from smoking. Reilly et al. [10], have reported elevated level of urinary of unmetabolized 8-epi $PGF_{2\alpha}$ in smokers compared to controls. The levels of unmetabolized urinary 8-epi $PGF_{2\alpha}$ were markedly reduced after two weeks abstinence from smoking or following dietary supplementation of a combination of vitamin C and vitamin E. These data together provides further evidences for the ramification of the measurement of 8-epi $PGF_{2\alpha}$ as an index of oxidative stress *in vivo*.

In conclusion, we have shown that GC-MS/NICI analysis of plasma 8-epi $PGF_{2\alpha}$ is a specific and reliable marker of non-cyclooxgenase peroxidation of arachidonic acid. In addition, we have established that plasma 8-epi $PGF_{2\alpha}$ levels are elevated in three clinical situations associated with oxidative stress. F_2-isoprostane measurement may prove to be useful in studying the relationships between lipid peroxidation and complications in disease as well as the impact of various therapeutic interventions.

ACKNOWLEDGMENT

The British Heart Foundation for financial support.

REFERENCES

1) Halliwell, B., Oxidative stress, nutrition and health: experimental strategies for optimization of nutritional antioxidant intake in humans, Free Radical Res. 25:57 (1996).

2) Morrow, J.D., Hill, K.E., Burk, R.F., Mammour, T.M., Badr. K.F., and Roberts II, L.J., A series of prostaglandin F_2-like compounds are produces in vivo in humans by a non-cyclooxgenase, free radical-catalyzed mechanism, Proc. Natl. Acad. Sci. USA 87: 9383 (1990).

3) Nourooz-Zadeh, J., Gopaul, N. K., Barrow, S., Mallet, A.I. and Änggård, E.E., Analysis of F_2-isoprostanes as indicators of non-enzymatic lipid peroxidation in vivo by GC-MS:develoment of a solid phase extraction procedure, J. Chromatogr. Biomed. Appl. 667:199 (1995).

4) Morrow, J.D., Harris, T.M., and Roberts II, L.J., Non-cyclooxgenase oxidative formation of a series of novel prostaglandins: analytical ramifications for measurement of eicosanoids, Anal Biochem. 184:1 (1990).

5) Gopaul, N.K., Nourooz-Zadeh, J., Mallet, A.I., and Änggård, E.E., Formation of PGF_2-isopratanes during the oxidative modication of LDL, Biochem. Biophysc. Res. Comm. 200:338 (1994).

6) Gopaul, N.K., Nourooz-Zadeh, J., Mallet, A.I., and Änggård, E.E., Formation of F_2-

isoprostanes during aortic endothelial cell-mediated oxidation of LDL, FEBS Letters 348:297 (1994).

7) Morrow, J.D., Frei, B., Longmire, A.W., Gaziano, J.M., Lynch, S.M., Shyr, Y., Strauss, W.E., Oates, J.A., and Roberts II, L.J., Increase in circulating products of lipid peroxidation (F_2-isoprostanes) in smokers, New Engl. J. Med. 332:1198-1203 (1995).

8) Lynch, S.M., Morrow, J.D., Roberts II, L.J., and Frei, B., Formation of non-cyclooxgenase-derived prostanoid (F_2-isoprostanes) in plasma and LDL exposed to oxidative stress in vitro, J. Clin. Invest. 93:998 (1994).

9) Gopaul, N.K., Änggård, E.E., Mallet, A.I., Bettridge, D.J., Wolff, S.P., and Nourooz-Zadeh, J., Plasma 8-epi $PGF_{2\alpha}$ levels are elevated in individuals with NIDDM, FEBS Letters 368:225 (1995).

10) Reilly, M., Delanty, N., Lawson, J. A., and Fitzgerald, G. A., Modulation of oxidant stress in vivo in chronic cigarette smokers, Circulation 94:19 (1996).

REGULATION OF MAP-KINASE ACTIVATION BY 8-ISO-PROSTAGLANDIN F_{2a} IN CULTURED RAT AORTIC SMOOTH MUSCLE CELLS

Megumu Fukunaga[1], Takafumi Yura[2], Kihito Takahashi[3], and Kamal F. Badr[4]

[1]School of Allied Health Sciences, Faculty of Medicine, Osaka University, Suita-shi, 565, Japan, [2]Yao City Hospital, Yao-shi, 581, Japan, [3]Merk, Sharp, & Dohme Research Laboratories, Tokyo, 107, Japan, and [4]Renal Division, Department of Medicine, Emory University and Veterans Affairs Medical Center, Atlanta, Georgia, 30033, USA.

INTRODUCTION

8-iso-prostaglandin $F_{2\alpha}$ (8-iso-$PGF_{2\alpha}$), an F_2-isoprostane, is produced in vivo by a cyclooxygenase-independent, free radical-catalyzed mechanism involving peroxidation of arachidonic acid[1]. This isoprostane is a highly potent renal vasoconstrictor[2] and stimulates proliferation of cultured rat aortic smooth muscle cells (AoSMC) through enhancement of phosphoinositide turnover[3]. AoSMC have distinct F_2-isoprostane binding sites, although these putative receptors bear homology to the thromboxane A_2 (TxA_2) receptor[4]. In this study, we report the activation of mitogen activated protein kinases (MAP-kinases)[5] by 8-iso-$PGF_{2\alpha}$ in AoSMC and its regulating mechanism involving protein kinase C (PKC) and pertussis toxin (PT)-sensitive GTP binding proteins.

MATERIALS AND METHODS

In time-control studies, quiescent AoSMC were incubated with 100nM 8-iso-$PGF_{2\alpha}$ or vehicle for different incubation times and collected in extraction buffer (20mM Tris-HCl, pH 7.5, 60mM β-glycerophosphate, 10mM EGTA, 10mM $MgCl_2$, 0.1mM NaF, 2mM dithiothreitol, 1mM Na_3VO_4, 20μg/ml aprotinin, and 1mM phenylmethylsulfonyl fluoride) after being washed, followed by homogenization. Supernatants of homogenate centrifuged

Recent Advances in Prostaglandin, Thromboxane, and Leukotriene Research
Edited by Sinzinger *et al.*, Plenum Press, New York, 1998

at 14,000×g for 15min at 4℃ were used for MAP-kinase assay using MAP-kinase assay system (Amersham International, Buckinghamshire, UK) with [^{32}P]-γ-ATP in 30min reaction time at 30℃ after measurement of protein content with DC protein assay system (Bio-Rad Laboratories, Inc., Hercules, CA). In studies of dose-dependency, different concentrations of 8-iso-$PGF_{2\alpha}$ were added to quiescent AoSMC and, after 5min incubation time, cells were treated and MAP-kinase activity was measured at 5min as described above. In some experiments, 500nM phorbol 12-myristate 13-acetate (PMA) or 100ng/ml PT was added to quiescent AoSMC 24h and 16h prior to experiments, respectively, and MAP-kinase activation at 5min by 100nM 8-iso-$PGF_{2\alpha}$ was examined as described above.

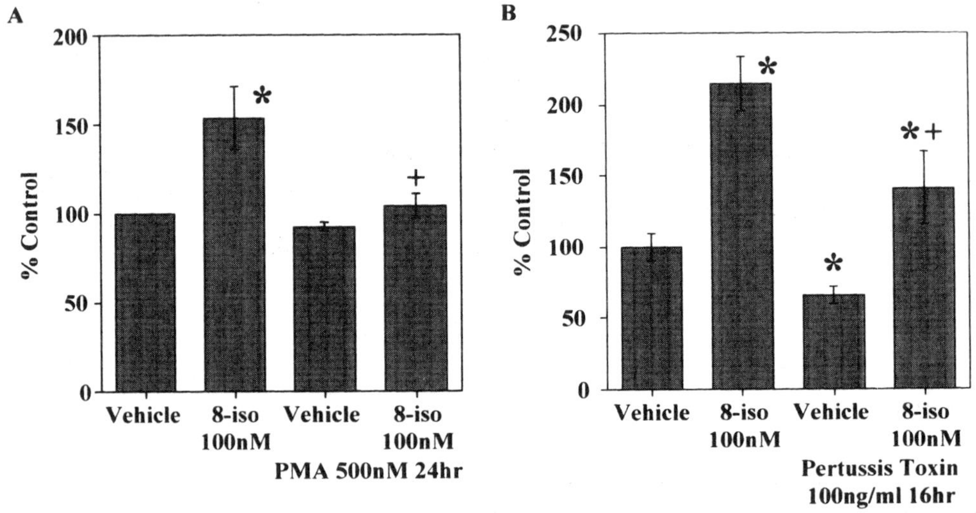

Figure MAP-kinase activity stimulated by 100 nM 8-iso-$PGF_{2\alpha}$ in AoSMC.
A. Effect of exposure to 500 nM PMA for 24 h. B. Effect of exposure to 100 ng/ml PT for 16 h. Data are expressed as a percentile of the value obtained at time 0 (% Control) and shown as means±S.E.M. of four to six experiments. *: $P<0.05$ vs. Vehicle, +: $P<0.05$ vs. 8-iso-$PGF_{2\alpha}$ by Scheffe's multiple comparison analysis. 8-iso: 8-iso-$PGF_{2\alpha}$.

RESULTS AND DISCUSSION

One hundred nanomolar 8-iso-$PGF_{2\alpha}$ stimulated MAP-kinase activity time-dependently, reaching its peak value (250% of time 0) at 5 to 10min followed by gradual decline remaining above its basal value by 60min. Stimulatory effect of this isoprostane at 5min was dose-dependent with significant increases at concentrations higher than 100pM. Since MAP-kinases are known to be involved in several cellular events including proliferation[5], it is strongly suggested that MAP-kinases may play a role in regulating muscle cell proliferation.

Both of PKC and GTP binding proteins are thought to take parts in regulating MAP-kinase cascade[5]. Since, as we reported, 8-iso-$PGF_{2\alpha}$ increases production of Ins 1,4,5-P_3[3], a hydrolyzed product of PtdIns 4,5-P_2[6] and because it was speculated that the other hydrolyzed product of PtdIns 4,5-P_2, diacylglycerol[6], might be increased followed by the activation of PKC induced by itself, we next examined the involvement of PKC in regulating MAP-kinases activated by 8-iso-$PGF_{2\alpha}$ in AoSMC.

Under PKC down-regulation of AoSMC achieved by 24 h exposure to 500nM PMA, which was confirmed both by a significant decrease of PKC activity by about 30% and by decreased expressions of PKC-α and PKC-ε in immunoblotting, stimulation of MAP-kinase activity at 5min by 100nM 8-iso-$PGF_{2\alpha}$ was almost completely abolished with no effect on basal conditions, suggesting an essential role for PKC in MAP-kinase activation by 8-iso-$PGF_{2\alpha}$ (Figure A), while exposure to 100ng/ml PT for 16h significantly reduced the potency of this isoprostane to stimulate MAP-kinase activity with a significant suppression of the basal activity of MAP kinases (Figure B), suggesting that MAP-kinase activity is partially regulated by PT-sensitive GTP-binding protein(s). Thus, in this study, we elucidated the involvement of PKC and GTP-binding protein(s) in regulating mechanism of MAP-kinase activation induced by 8-iso-$PGF_{2\alpha}$.

These our findings, together with recent findings of the increased plasma level of isoprostanes in hepatorenal syndrome[7] and smokers[8], provide further support for the potential pathophysiologic relevance of isoprostane in mediating vascular injury during oxidant stress.

ACKNOWLEDGMENTS

This work was supported by National Institutes of Health Grants, 2P50 DK39261-06.

REFERENCES

1. J.D. Morrow, K.E. Hill, R.F. Burk, T.M. Nammour, K.F. Badr, and L.J. Roberts II. A series of prostaglandin F_2-like compounds are produced in vivo in humans by a non-cyclooxygenase, free radical-catalyzed mechanism. *Proc. Natl. Acad. Sci. USA.* 87:9383 (1992).

2. K. Takahashi, T.M. Nammour, M. Fukunaga, J. Ebert, J.D. Morrow, L.J. Roberts II, R.L. Hoover, and K.F. Badr. Glomerular actions of a free radical-generated novel prostaglandin, 8-epi-prostaglandin $F_{2\alpha}$, in the rat: Evidence for interaction with thromboxane A_2 receptors. *J. Clin. Invest.* 90:136 (1992).

3. M. Fukunaga, N. Makita, L.J. Roberts II, J.D. Morrow, K. Takahashi, and K.F. Badr. Evidence for the existence of F_2-isoprostane receptors on rat vascular smooth muscle cells. *Am. J. Physiol. (Cell Physiol. 33)* 264:C1619 (1993).

4. T. Yura, M. Fukunaga, R. Grygorczyk, N. Makita, K. Takahashi, and K.F. Badr. Molecular and functional evidence for the distinct nature of F_2-isoprostane receptors from those of thromboxane A_2. *Adv. Prostagl. Thrombox. Leukotr. Res.* 23:237 (1995).

5. R. Seger, and E.G. Krebs. The MAPK signaling cascade. *FASEB J.* 9:726 (1995).

6. M.J. Berridge, and R.F. Irvine. Inositol trisphosphate, a novel second messenger in cellular signal transduction. *Nature Lond.* 1984:312 (1984).

7. J.D. Morrow, K.P. Moore, J.A. Awad, M.D. Ravenscraft, G. Marini, K.F. Badr, R. Williams, and L.J. Roberts II. Marked overproduction of non-cyclooxygenase derived prostanoids (F_2-isoprostanes) in the hepatorenal syndrome. *J. Lipid Mediators* 6:417 (1993).

8. J.D. Morrow, B. Frei, A.W. Longmire, J.M. Gaziano, S.M. Lynch, Y. Shyr, W.E. Strauss, J.A. Oates, L.J. Roberts II. Increase in circulating products of lipid peroxidation (F_2-isoprostane) in smokers.: Smoking as a cause of oxidative damage. *N. Engl. J. Med.* 332:1198 (1995).

THE EFFECT OF NON-PROSTANOID PROSTACYCLIN MIMETICS ON CYCLIC AMP PRODUCTION BY NEURONAL SK-N-SH CELLS

Helen Wise and Kevin B.S. Chow

Department of Pharmacology
The Chinese University of Hong Kong
Shatin, N.T.
Hong Kong

INTRODUCTION

Although prostacyclin (PGI_2) is more commonly known for its important role as an antithrombotic agent, there is accumulating evidence suggesting that PGI_2 has additional neuronal activity. For example, the PGI_2 analogue cicaprost can stimulate the release of tachykinins from guinea-pig ileum (Jones and Lawrence, 1993), and can inhibit the spontaneous contractile activity of rat colon by stimulating the release of factors involved in nonadrenergic noncholinergic neurotransmission (Qian and Jones, 1995). Furthermore, the activation of IP-receptors in the rat enteric nervous system appears to be mediated by a novel IP-receptor subtype, distinct from the IP-receptor associated with inhibition of platelet aggregation (Wise et al., 1995). This novel IP-receptor was identified by testing some non-prostanoid PGI_2 mimetics which, while active in inhibiting rat neutrophil aggregation, were without effect in the rat colon.

To date, only one IP-receptor has been cloned and this appears to be the classical platelet IP-receptor coupled to a G_s protein (Hirata et al., 1995). So far we have been unable to identify the signalling pathway(s) used by the IP-receptor in rat enteric neurones, therefore we were interested in examining other neuronal preparations with easier access to second messenger information. Therefore the aim of the present study was to characterise the IP-receptor present in the human neuroblastoma cell line SK-N-SH, by examining the ability of the non-prostanoid PGI_2 mimetics BMY 22389 (octimibate), BMY 42393 and BMY 45778 (Meanwell et al., 1994) to stimulate adenylate cyclase activity.

MATERIALS AND METHODS

SK-N-SH cells were cultured at 37°C in RPMI-1640 medium supplemented with 10%-FCS, 100 U/ml-penicillin and 100 μg/ml-streptomycin. Confluent monolayers of cells in 24-well plates were labelled with [^{3}H]-adenine (1 μCi/well) 24 hr before assay of [^{3}H]-cyclic AMP production. Cells were incubated for 10 min at 37°C in Hepes-buffered RPMI-1640 containing 10%-FCS, 3 μM-indomethacin and 100 μM-rolipram, followed by 10 min with the addition of test compound. The reaction was stopped by replacing the medium with ice-

Recent Advances in Prostaglandin, Thromboxane, and Leukotriene Research
Edited by Sinzinger *et al.*, Plenum Press, New York, 1998

cold 7.5%-trichloroacetic acid and intracellular [^{3}H]-cyclic AMP was separated from [^{3}H]-ATP by column chromatography (Wise and Jones, 1994). Results are shown as [^{3}H]-cyclic AMP production relative to basal adenylate cyclase activity in the absence of exogenous stimuli, and are expressed as mean $\pm$ SEM of 4 experiments, each performed in quadruplicate. pEC_{50} values (negative logarithm of concentration of agonist producing half-maximum response) were calculated using the logistic curve fitting program in PRISM (GraphPad).

RESULTS AND DISCUSSION

In the presence of the phosphodiesterase inhibitor rolipram, the standard IP-receptor agonists cicaprost and iloprost produced a concentration-dependent increase in [^{3}H]-cyclic AMP production in SK-N-SH cells with pEC_{50} values of 7.91 $\pm$ 0.06 and 7.79 $\pm$ 0.09 respectively (Figure 1A). The potency with which these compounds stimulated adenylate cyclase activity in SK-N-SH cells was very similar to that seen in rat neutrophils (Wise, 1996) despite the marked difference in absolute levels of adenylate cyclase activity. For example, [^{3}H]-cyclic AMP production was stimulated approximately 30-fold by cicaprost in SK-N-SH cells, yet only produced a 4-fold increase in rat neutrophils. When adenylate cyclase activity was measured in membranes prepared from other neuroblastoma cell lines such as NG108-15 and NCB-20, iloprost showed similar activity with pEC_{50} values of 7.62 and 7.55 respectively (Kelly et al., 1990), however the extent of stimulation was slightly lower at 10-fold.

Recently, Takechi et al. (1996) described a neuronal IP-receptor from autoradiographic studies of [^{3}H]-isocarbacyclin binding to rat brain. Although [^{3}H]-isocarbacyclin bound with similar affinity in the nucleus tractus solitarius (NTS) and thalamus (3.9 and 7.8 nM respectively), [^{3}H]-iloprost had much lower affinity binding in the thalamus compared with the NTS (159 and 6.8 nM respectively). Furthermore, the different order of potency of IP-agonists in competition for [^{3}H]-isocarbacyclin binding in NTS and thalamus suggested that the binding sites in the NTS were similar to the classical platelet IP-receptor, while the binding sites in the thalamus represent a novel neuronal IP-receptor subtype. Iloprost appears to be about 50-fold more potent than cicaprost at this neuronal IP-receptor, therefore the equipotent nature of iloprost and cicaprost in SK-N-SH cells argues against this being a similar 'neuronal' receptor.

The three non-prostanoid PGI_2 mimetics also produced a concentration-dependent increase in [^{3}H]-cyclic AMP production in SK-N-SH cells, with pEC_{50} values of 7.74 $\pm$ 0.24, 5.83 $\pm$ 0.07 and 5.72 $\pm$ 0.29 for BMY 45778, BMY 22389 and BMY 42393 respectively (Figure 2B). As seen previously in human platelets (Meanwell et al., 1994) and rat neutrophils (Wise, 1996), all three BMY compounds appeared to be partial agonists for stimulation of adenylate cyclase activity in SK-N-SH cells, despite behaving as full agonists in functional assays of platelet and neutrophil aggregation. For example in SK-N-SH cells, BMY 22389, BMY 42393 and BMY 45778, at a concentration of 10 μM, produced only 40, 75, and 50% respectively of the response to cicaprost and iloprost.

BMY 45778 is the most potent of this series of compounds reported by Bristol-Myers Squibb, being approximately 30-fold less active than iloprost at inhibiting human platelet aggregation in platelet-rich plasma (Meanwell et al., 1994). However BMY 45778 exhibits a high degree of protein binding relative to the hydrophilic prostacyclin analogues and this IC_{50} value underestimates the affinity of BMY 45778 for the IP-receptor. For example, in competition for [^{3}H]-iloprost binding to human platelet membranes, i.e. in protein-free conditions, BMY 45778 has a K_i value of 10 nM compared with 12 nM for iloprost (Jones et al., 1997). Therefore the low EC_{50} value for BMY 45778 in stimulating adenylate

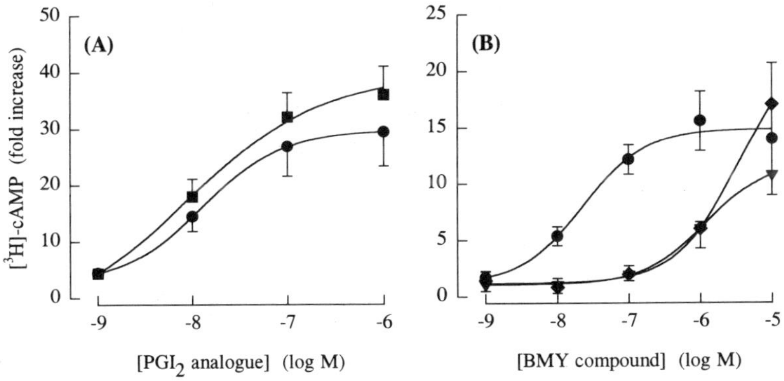

Figure 1. The effects of prostacyclin mimetics on [^{3}H]-cyclic AMP accumulation. Data represent the mean ± SEM of 4 experiments performed in quadruplicate: (A) cicaprost (●), iloprost (○); (B) BMY 22389 (▼), BMY 42393 (◆), BMY 45778 (●).

cyclase activity in SK-N-SH cells (18 nM compared with 16 nM for iloprost) is somewhat surprising given that these experiments were performed in serum-containing medium. Unfortunately we have been unable to obtain a more accurate estimate of the affinity of BMY 45778 for the IP-receptor in SK-N-SH cells due to the low level of [^{3}H]-iloprost binding sites in these cells (unpublished observations). However, the high potency of BMY 45778 in SK-N-SH cells suggest that this IP-receptor subtype is clearly different from the IP-receptor subtype present in rat enteric neurones.

One difficulty in comparing the IP-receptor subtype present in SK-N-SH cells with the neuronal IP-receptor of Takechi et al. (1996) is that the latter appears unable to couple to adenylate cyclase, and at the present time we cannot be certain whether this is due to technical limitations or indeed indicates that the neuronal IP-receptor utilizes a distinct signalling pathway. Certainly there is no need to ascribe these differences to species effects since cicaprost and iloprost are equipotent in competing for [^{3}H]-iloprost binding to recombinant platelet-like rat IP-receptor (Sasaki et al., 1994) and human IP-receptors (Katsuyama et al., 1994; Nakagawa et al., 1994). However, in addition to the high potency of cicaprost in SK-N-SH cells (Figure 1A) compared with the low potency of cicaprost in rat CNS (Takechi et al., 1996), there are other indications that the IP-receptor in neuroblastoma cell lines differ from the 'neuronal' IP-receptor. For example, K_D values for [^{3}H]-iloprost binding to membranes of NG108-15 and NCB-20 neuroblastoma cells have been estimated at 5 - 15 nM (Kelly et al., 1990), which is similar to the affinity of [^{3}H]-iloprost binding in the NTS rather than in the thalamus (Takechi et al., 1996).

[^{3}H]-Iloprost binding sites in rat NTS appear to represent IP-receptors present on the central terminals of the primary sensory afferents, and relatively high levels of [^{3}H]-iloprost binding sites are also found in the rat dorsal horn (Matsumura et al., 1995). In addition, *in situ* hybridization studies using a probe derived from the cloned mouse IP-receptor showed approximately 40% of neurones in mouse dorsal root ganglia contain mRNA for this IP-receptor (Oida et al., 1995). Taken together, these results suggest that the NTS and dorsal root ganglia contain the classical platelet IP-receptor. It is therefore not surprising that the IP-receptor subtype in neuroblastoma cell lines looks more like the platelet than the neuronal IP-receptors, because neuroblastoma cell lines derive from neural crest cells which ordinarily give rise to spinal ganglionic cells (Ross et al., 1983).

Therefore in conclusion, the IP-receptor in the human neuroblastoma cell line SK-N-SH has the characteristics of the platelet IP-receptor and appears to differ from the IP-receptor subtype present in rat CNS and in rat enteric neurones.

ACKNOWLEDGMENTS

The gifts of BMY 22389, BMY 42393 and BMY 45778 (Bristol-Myers Squibb USA) and cicaprost, iloprost and rolipram (Schering AG Germany) are gratefully acknowledged. This work was supported by the Research Grants Council of Hong Kong.

REFERENCES

Hirata, M., Ushikubi, F. and Narumiya, S., 1995, *J. Lipid Mediators Cell Signalling,* 12:393.
Jones, R.L. and Lawrence, R.A., 1993, *Pharmacol. Commun.* 3:147.
Jones, R.L., Qian, Y., Wise, H., Wong, H.N.C., Lam, W.L., Chan, H.W., Yim, A.P.C. and Ho, J.K.S.,1997, *J. Cardiovasc. Pharmacol.* in press.
Katsuyama, M., Sugimoto, Y., Namba, T., Irie, A., Negishi, M., Narumiya, S. and Ichikawa, A., 1994, *FEBS Lett.* 344:74.
Kelly, E., Keen, M., Nobbs, P. and MacDermot, J., 1990, *Br. J. Pharmacol.* 99:309.
Matsumura, K., Watanabe, Y., Onoe, H. and Watanabe, Y., 1995, *Neurosci.* 65:493.
Meanwell, N. A., Romine, J. L., and Seiler, S. M., 1994, *Drugs of the Future*, 19:361.
Nakagawa, O., Tanaka, I., Usui, T., Harada, M., Sasaki, Y., Itoh, H., Yoshimasa, T., Namba, T., Narumiya, S. and Nakao, K., 1994, *Circ.* 90:1643.
Oida, H., Namba, T., Sugimoto, Y., Ushikubi, F., Ohishi, H., Ichikawa, A. and Narumiya, S., 1995, *Br. J.Pharmacol.* 116:2828.
Qian, Y. and Jones, R.L., 1995, *Br. J. Pharmacol.* 115:163.
Ross, R.A., Spengler, B.A. and Biedler, J.L., 1983, *J Natl. Cancer Inst.* 71:741.
Sasaki, Y., Usui, T., Tanaka, I., Nakagawa, O., Sando, T., Takahashi, T., Namba, T., Narumiya, S. and Nakao, K., 1994, *Biochim. Biophys. Acta* 1224:601.
Takechi, H., Matsumura, K., Watanabe, Y., Kato, K., Noyori, R. and Suzuki, M., 1996, *J. Biol. Chem.* 271:5901.
Wise, H., 1996, *Prostaglandins Leuk. Essen. Fatty Acids* 54:351.
Wise, H., and Jones, R. L., 1994, *Br J Pharmacol.* 113:581.
Wise, H., Qian, Y. & Jones, R.L., 1995, *Eur. Pharmacol.* 278, 265.

BIOSYNTHESIS OF 2-ARACHIDONOYL-GLYCEROL, A NOVEL CANNABIMIMETIC EICOSANOID, IN MOUSE NEUROBLASTOMA CELLS

Tiziana Bisogno[1], Nunzio Sepe[1], Luciano De Petrocellis[2], and Vincenzo Di Marzo[1*]

[1]Istituto per la Chimica di Molecole di Interesse Biologico and [2]Istituto di Cibernetica C.N.R., Via Toiano 6, 80072, Arco Felice, Napoli, Italy

INTRODUCTION

Following the original isolation from porcine brain of the first endogenous ligand of cannabinoid receptors, anandamide (arachidonoyl-ethanolamide)[1], another arachidonic acid (AA)-derived metabolite, 2-arachidonoyl-glycerol (2-AG), was found in canine gut[2] and reported to share with anandamide and plant cannabinoids several pharmacological actions. 2-AG binds to either brain (CB1) or spleen (CB2) cannabinoid receptor subtypes with Ki values in the high nmolar-low μmolar range and inhibits forskolin-induced adenylate cyclase activation; *in vivo*, 2-AG reproduces the typical cannabimimetic effects of hypothermy, hypomotility and analgesia[2]. Moreover, 2-AG inhibits cell proliferation in lymphocytes that do not express the CB1 receptor subtype[3], and, due to the inability of anandamide to functionally activate CB2 receptors[4], has been suggested to act as the possible endogenous cannabinoid agonist in the immune system[3], also on the basis of its recent finding in the spleen[5]. The isolation of mono-arachidonoyl-glycerol isomers from rat brain in levels 50-100 fold higher than those of anandamide[6] and the recent report of a CB1 receptor-mediated stimulatory action by 2-AG on intracellular free Ca^{2+} in neuroblastoma x glioma hybrid cells[7], however, raised the possibility that the monoglyceride might function as a possible 'endocannabinoid' also in the CNS. In this study we provide biochemical grounds to the hypothesis that 2-AG may be a neuronal mediator by describing: a) its Ca^{2+}-dependent biosynthesis in mouse neuroblastoma N18TG2 cells, and b) the occurrence in the same cells of enzymatic activities responsible for 2-AG formation from diacylglycerols and (lyso)phosphatidylcholine.

MATERIALS AND METHODS

N18TG2 cells, prelabelled by overnight incubation with [^{3}H]AA (NEN Dupont, 230 Ci/mmol), were grown and stimulated with the agents shown in Table 1 (purchased from Sigma) as described previously[8]. After the incubation cells or cells plus medium were extracted, and the metabolites shown in Table 1 purified as described elsewhere[8,9]. The 1- or 2-AG isomers were separated by means of thin layer chromatography (TLC) carried out on borate-impregnated cells as described elsewhere[9]. Conclusive identification of 2-AG produced by cells was obtained by means of gas chromatography-electron impact mass spectrometry of the di-acetoxy derivative[9]. N18TG2 cell homogenates were prepared in 20 mM HEPES buffer pH 7.4 containing 2 mM $CaCl_2$ and 2 mM dithiothreitol, and incubated at 37 °C with various

* Author to whom all correspondence should be addressed.
Fax: +39-81-8041770. E-mail: +VDM@TRINC.ICMIB.NA.CNR.IT

[^{3}H]-labelled precursors synthesized as described elsewhere[10]. Incubation conditions are described in Table 2. Incubates were extracted with chloroform methanol 2:1 (v:v) and the organic phase purified by TLC as described elsewhere[10]. Radioactivity incorporated into biosynthetic products was measured by a one-dimensional radioactivity TLC scanner and by scraping the bands off the silica and counting their β-emission by a liquid scintillation scanner.

RESULTS AND DISCUSSION

The effect of various agents on the release of [^{3}H]-2-AG, [^{3}H]AA and 1-acyl-2-[^{3}H]arachidonoyl-glycerols (AArGs) from radiolabelled N18TG2 cells is shown in Table 1. Formation of [^{3}H]-2-AG was always observed concomitantly to [^{3}H]AArG release. Stimulation with ionomycin led to a 7-fold increase of [^{3}H]-2-AG levels and resulted also in the production of [^{3}H]AA and [^{3}H]AArG. This effect was dependent on extracellular calcium since it was significantly reduced by co-incubation with the Ca^{2+} chelating agent EGTA. Stimulation of cells with exogenous phospholipases caused a striking release of both [^{3}H]-2-AG and [^{3}H]AArG only when using phospholipase C. Exogenous phospholipase A_2 (PLA_2) alone induced release of [^{3}H]AA but not of [^{3}H]-2-AG or [^{3}H]AArG. However, pre-treatment of cells with PLA_2, which causes a strong depletion of the amounts of *sn*-2-arachidonate in membrane phosphoglycerides, resulted in a sensible potentiation of ionomycin-evoked [^{3}H]-2-AG and [^{3}H]AArG formation. Pre-incubation of cells with the phospholipase C inhibitor neomycin sulphate did not prevent but slightly potentiated ionomycin-induced [^{3}H]-2-AG and [^{3}H]AArG biosynthesis.

Table 1. Effect of various substances on [^{3}H]-2-AG, [^{3}H]AArG and [^{3}H]AA biosynthesis in intact N18TG2 mouse neuroblastoma cells

Substance (concentration)	2-AG (cpm)	AArG (cpm)	AA (cpm)
None	66±68	36±15	23±13
Ionomycin (5 μM)	437±40*	162±7*	65±6*
Ionomycin (5 μM)+ EGTA (1mM)	205±15**	91±20**	34±6**
Ionomycin + Neomycin (1mM)	491±59*	235±50*	61±22
Phospholipase D (10 U/ml)	139±10	49±2	34±5
Phospholipase C (10 U/ml)	14228±2955*	1718±578*	572±50*
Phospholipase A2 (10 U/ml)	30±30	90±32	3820+75*
Ionomycin + Phospholipase A2	1053±82**	532±50**	6821±61**

*=Statistically significant from None; **=statistically significant from Ionomycin only. $P<0.01$, unpaired Student's T test, means±S.E.M., n=3.

These data suggest that ionomycin stimulation of N18TG2 cells triggers 2-AG biosynthesis: a) by causing Ca^{2+} entry into the cells, b) through the formation of AArGs but not necessarily following phospholipase C-catalyzed hydrolysis of *sn*-2-arachidonate-containing phosphoglycerides, and c) *via* a pathway that is also, but not only, sensible to AA release from membrane phospholipids.

In agreement with a role for AArGs as 2-AG precursors in N18TG2 cells, we found that high levels of [^{3}H]-2-AG could be produced upon incubation of whole cell homogenates with synthetic 1-acyl-2-[^{3}H]arachidonoyl-glycerols (Table 2). This conversion was negligible when using boiled homogenates and was not sensitive to the presence of EGTA. Interestingly, we found that significant albeit little amounts of [^{3}H]-2-AG could be obtained also from radiolabelled 1,2-*sn*-di-arachidonoyl-phosphatidylcholine (AAPC). Accordingly, [^{3}H]AAPC was enzymatically converted, in an EGTA-sensitive fashion, into 2-AG precusor [^{3}H]-1,2-*sn*-di-arachidonoyl-glycerol (DArG) at a rate higher than that observed for its conversion into [^{3}H]-2-AG (Table 2). AAPC serves also as acyl donor for the regioselective transfer of *sn*-1 arachidonate to phosphatidylethanolamine in the biosynthesis of anandamide precursor, *N*-arachidonoyl-phosphatidylethanolamine[11,14] (N-ArPE, Table 2). Finally, enzymatic formation of [^{3}H]-2-AG was also observed when using as a precursor 1-lyso-2-[^{3}H]arachidonoyl-phosphatidylcholine, a by-product of N-ArPE biosynthesis[11]. This conversion was not inhibited, but, on the contrary, was enhanced by EGTA (Table 2), thus suggesting that the phospholipase C-like enzyme involved is different from that catalyzing [^{3}H]AAPC hydrolysis to [^{3}H]DArG.

Table 2. Conversion (pmol/hour/mg protein) of various [^{3}H]-labelled precursors into various products by N18TG2 mouse neuroblastoma cell whole homogenates

PRECURSORS	PRODUCTS		
	[^{3}H]-2-AG	[^{3}H]-DArG	[^{3}H]-N-ArPE
1-arachidonoyl-2-[^{3}H]-arachidonoyl-phosphatidylcholine (18.2 nmol/0.4ml)	6.1±0.7^a (N.D.)	46.0±1.4^a (15.4±0.8)	3.3±0.3^a (N.D.)
1,2-di-[^{3}H]-arachidonoyl-phosphatidylcholine (18.2 nmol/0.4ml)	9.5±2.1^a (N.D.)	70.2±0.3^a (41.8±2.9)	21.1±5.4^a (4.0±0.3)
1-lyso-2-[^{3}H]-arachidonoyl-phosphatidylcholine (18.2 nmol/0.5ml)	61.6±8.4^b (91.3+4.9)	N.D.	N.D.
1-arachidonoyl-2-[^{3}H]-arachidonoyl-glycerol (9.1 nmol/1.0 ml)	6960±204^b (8520±492)	-	N.D.
1-stearoyl-2-[^{3}H]-arachidonoyl-glycerol (1.3 pmol/1.0 ml)	0.756±0.036^c (N.D.)	-	N.D.

a, b, c=respectively, 2 hour, 30 min and 15 min incubations. N.D.= not determined. Data between brackets are in the presence of 2 mM EGTA. Conversion was <1 pmol/hour/mg protein with heat-inactivated homogenates.

From these findings it is possible to suggest that 2-AG may be released also from (lyso)phospholipid precursors provided that neuronal cells are stimulated with agonists capable of up-regulating phospholipase C enzymes that may not be sensitive only to ionomycin stimulation (Fig. 1)[12]. On the other hand, when cells are stimulated with agonists that cause Ca^{2+} influx without altering phospholipase C activity, it is possible that 2-AG be synthesized from AArGs derived from *de novo* synthesis and not from the hydrolysis of phosphoglycerides (Fig. 1). It is worthwhile mentioning that ionomycin, which probably does not significantly induce the activation of phospholipases other than PLA_2 in N18TG2 cells, is, in this neuronal cell type, a poor stimulant of anandamide formation[9], which requires the activation of a phospholipase D-like enzyme in rat central neurons[14].

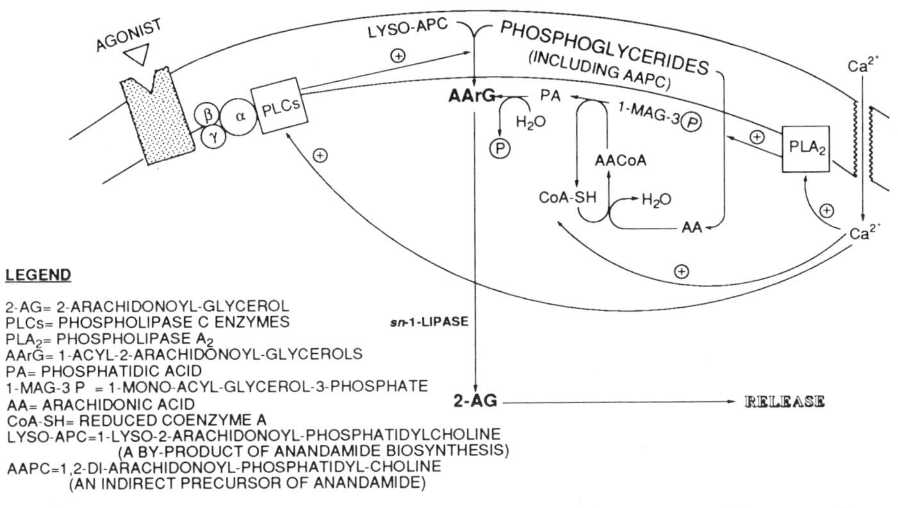

Figure 1. Potential biosynthetic routes for 2-AG biosynthesis in mouse neuroblastoma cells

From the above data it is also possible to hypothesize, for 2-AG and anandamide formation in neurons, the existence of concurring and/or competing biosynthetic pathways. On the one hand, AAPC, which is one of the di-arachidonoyl-phosphoglyceride species utilized in the biosynthesis of anandamide precursor N-ArPE[11], may serve as a potential precursor also for 2-AG; on the other hand, physiological stimuli leading to N-ArPE (and, therefore, anandamide) biosynthesis[13], with subsequent formation of 1-lyso-2-arachidonoyl-phosphatidylcholine, may also lead to 2-AG formation *via* the hydrolysis of this lysophosphoglyceride.

In conclusion, we have shown that N18TG2 cells, a neuronal model widely used for studies on the endogenous cannabinoid system, a) synthesize 2-AG in response to ionomycin stimulation and subsequent Ca^{2+} influx into cells, and b) express an *sn*-1 diacylglycerol lipase activity responsible for the formation of 2-AG from diacylglycerols. Moreover we have shown that these cells contain two distinct phospholipase C-like enzymes which, if activated by adequate agonists, may potentially relate the biosynthesis of 2-AG to that of anandamide. Further studies will be required in order to characterize the enzymatic activities described here as well as to investigate the possibility of a joint regulation of the biosynthesis of the two AA-derived 'endocannabinoids'.

Acknowledgments

The authors wish to thank Dr. T. Sugiura, Teikyo University, Japan, for the generous gift of synthetic [^{3}H]AAPC and 1-lyso-2-[^{3}H]arachidonoyl-phosphatidylcholine, and the Human Frontiers in Science Organization for financial support (grant RG26/95 to VDM).

REFERENCES

1. Devane WA, Hanus L, Breuer A, Pertwee RG, Stevenson LA, Griffin G, Gibson D, Mandelbaum A, Etinger A and Mechoulam R, Isolation and structure of a brain costituent that binds to the cannabinoid receptor, *Science* 258:1946 (1992).
2. Mechoulam R, Ben-Shabat S, Hanus L, Ligumsky M, Kaminski NE, Schatz AR, Gopher A, Almog S, Martin BR, Compton DR, Pertwee RG, Griffin G, Bayewitch M, Barg J and Vogel Z, Identification of an endogenous 2-monoglyceride, present in canine gut, that binds to cannabinoid receptors, *Biochem Pharmacol* 50:83 (1995) .
3. Lee M, Yang KH and Kaminski NE, Effects of putative cannabinoid receptor ligands, anandamide and 2-arachidonoyl-glycerol on immune function in B6C3F1 mouse splenocytes, *J Pharmacol Exp Ther* 275:529 (1995).
4. Bayewitch M, Avidor-Reiss T, Levy R, Barg J, Mechoulam R and Vogel Z, The peripheral cannabinoid receptor: adenylate cyclase inhibition and G protein coupling. *FEBS Lett* 375:143 (1995).
5. Ben-Shabat S, Fride E, Vogel Z, Bisogno T, De Petrocellis L, Di Marzo V and Mechoulam R, An entourage effect: endogenous, inactive acyl-glycerol esters enhance 2-arachidonoyl-glycerol activity, *Mol. Pharmacol,* submitted.
6. Sugiura T, Kondo S, Sukagawa A, Nakane S, Shinoda A, Itoh K, Yamashita A and Waku K, 2-Arachidonoyl-glycerol: a possible endogenous cannabinoid receptor ligand in brain. *Biochem Biophys Res Comm* 215:89 (1995).
7. Sugiura T, Kodaka T, Kondo S, Tonegawa T, Nakane S, Kishimoto S, Yamashita A and Waku K, 2-arachidonoyl-glycerol, a putative endogenous cannabinoid receptor ligand, induces rapid, transient elevation of intracellular free Ca^{2+} in neuroblastoma X glioma hybrid NG108-15 cells, *Biochem Biophys Res Commun* 229:58 (1996).
8. Di Marzo V, De Petrocellis L, Sepe N and Buono A, Biosynthesis of anandamide and related acylethanolamides in mouse J774 macrophages and N18 neuroblastoma cells, *Biochem J* 316:977 (1996).
9. Bisogno T, Sepe N, Melck D, Maurelli S, De Petrocellis L and Di Marzo V, Biosynthesis, release and degradation of the novel endogenous cannabinomimetic metabolite 2-arachidonoyl-glycerol in mouse neuroblastoma cells, *Biochem J* 322:671 (1997).
10. Di Marzo V, De Petrocellis L, Sugiura T and Waku K, Potential biosynthetic connections between the two cannabimimetic eicosanoids, anandamide and 2-arachidonoyl-glycerol, in mouse neuroblastoma cells, *Biochem Biophys Res Comm* 227:281 (1996).
11. Sugiura T, Kondo S, Sukagawa A, Tonegawa T, Nakane S, Yamashita A, Ishima Y and Waku K, Transacylase-mediated and phosphodiesterase-mediated synthesis of N-arachidonoylethanolamine, an endogenous cannabinoid-receptor ligand, in rat brain microsomes. Comparison with the synthesis from free arachidonic acid and ethanolamine, *Eur J Biochem* 240:53 (1996).
12. Allen AC, Gammon CM, Ousley AH, McCarthy KD and Morell P, Bradykinin stimulates arachidonic acid release through the sequential action of an *sn*-1 diacylglycerol lipase and a monoacylglycerol lipase, *J Neurochem* 58:1130 (1992).
13. Cadas H , di Tomaso E and Piomelli D, Occurrence and biosynthesis of endogenous cannabinoid precursor, *N*-arachidonoyl-phosphatidylethanolamine, in rat brain, *J Neurosci* 17:1226 (1997).
14. Di Marzo V, Fontana A, Cadas H, Schinelli S, Cimino G, Schwartz JC and Piomelli D, Formation and inactivation of endogenous cannabinoid anandamide in central neurons, *Nature* 372:686 (1994).

HYPERALGESIA-ASSOCIATED SPINAL SYNTHESIS AND RELEASE OF PROSTAGLANDINS

David M. Dirig‡ and Tony L. Yaksh†‡

Departments of Anesthesiology† and Pharmacology‡
University of California, San Diego
La Jolla, CA 92093

INTRODUCTION

Following tissue injury, an animal will display spontaneous pain behavior and an exaggerated response to a noxious stimulus, a state otherwise referred to as hyperalgesia. Consistent with the behavioral effects, injury or inflammation produces a left shift in the relationship between discharge rates of primary afferents and stimulus intensity (Reeh et al., 1986). This peripheral hypersensitivity can be explained, in part, by a local release of pro-inflammatory substances, such bradykinin or prostaglandins (Baccaglini and Hogan, 1983). Vane (1971) and Smith and Willis (1971) revealed that agents such as acetylsalicylic acid and indomethacin, which were known to alter the hyperalgesia which occurred secondary to inflammation (e.g. tissue injury), inhibited cyclooxygenase (COX), the enzyme responsible for prostaglandin synthesis. This observation provided a unifying link in explaining the ability of a diverse class of agents called non-steroidal anti-inflammatory drugs (NSAIDs), to normalize the otherwise lowered thresholds (i.e. hyperalgesia) observed in the face of local tissue injury (Ferreira et al., 1971).

Despite the ability of this hypothesis to provide a unifying explanation for tissue injury-associated hyperalgesia and the efficacy of NSAIDs, agents such as acetaminophen were poor anti-inflammatory drugs, but were considered active anti-hyperalgesic agents. Beyond the peripheral effects of NSAIDs, it became apparent that there were non-peripheral sites of NSAID action. Evidence from a number of investigators described below supports the hypothesis that prostaglandins play a key role in modulation of centrally-mediated hypersensitivity that is observed after a peripheral injury.

SPINAL FACILITATION AND PROSTAGLANDINS

It is appreciated that increased sensory afferent activity after nociceptor sensitization after tissue injury may change the spinal processing of sensory input (For review, see Yaksh, 1993). Thus, peripheral injury or inflammation can generate a state of facilitation in which a moderate stimulus will evoke a profound discharge in dorsal horn nociceptive neurons (Dickenson and Sullivan, 1987), and repetitive C fiber activation will increase spinal cord neuron excitability (Woolf, 1983). The hypothesis that spinally-derived prostanoids may be essential for a component of this central facilitation and hyperalgesia is supported by studies demonstrating i) reversal of tissue-injury associated hyperalgesia by spinal administration of COX inhibitors; ii) spinal prostanoid release after tissue injury, and iii) increase in nocifensive behavior after spinal administration of prostanoids.

SPINALLY DELIVERED COX INHIBITORS

Yaksh (1982) demonstrated that intrathecal (IT) administration of the COX inhibitors, zomepirac and aspirin, reduced acetic acid-evoked writhing in mice, and IT pre-treatment with S(+)-ibuprofen blocked the thermal hyperalgesia induced in rats by IT Substance P (SP) or the excitatory amino acid (EAA) agonist, NMDA (Malmberg and Yaksh, 1992b). Additionally, IT administration of a series of NSAIDs reversed the hyperalgesia evoked by hindpaw injection of formalin with relative potency ratios that mirrored their structure-activity relationships for COX inhibition (Malmberg and Yaksh, 1992a). While these findings argue for involvement of spinal prostaglandins in central facilitation, the fact that NSAIDs reverse hyperalgesia, but do not increase thresholds in control animals suggests that prostaglandins are important for a component of central facilitation not involved in acute nociception.

SPINAL PROSTAGLANDIN RELEASE

The effects of cyclooxygenase inhibitors suggest a local release of prostaglandins from the site of NSAID action. Prostanoids of the E series were reported in spinal perfusates following high-intensity electrical stimulation in frogs over thirty years ago (Ramwell et al., 1966). This hypothesized prostaglandin synthesis and release from spinal parenchyma associated with protracted nociception and hyperalgesia is substantiated by studies employing spinal microdialysis catheters (Marsala et al., 1995). By combining such methods with the hyperalgesic models described above, relative changes in spinal cerebrospinal fluid transmitter levels and behavioral measures of nociception can be correlated in un-anesthetized, un-restrained rodents. Using such methods, Malmberg and Yaksh (1995a) demonstrated that spinal PGE2 microdialysate levels increase after paw injection of the irritant, formalin. These increases were reversed by intrathecal NSAIDs (Malmberg and Yaksh, 1995a) and systemic morphine treatment (Malmberg and Yaksh, 1995b). These data support the hypothesis that spinal prostanoid synthesis and release are necessary for the hyperalgesia associated with tissue injury.

Using an experimental model of arthritis, Yang et al. (1996) reported similar findings within 24 hours after intra-articular injection of the irritant mixture, kaolin/carrageenan. This model is associated with peripheral inflammation and sensitization of second order spinal neurons, and this central sensitization has shown to be sensitive to both central and peripheral NSAIDs (Neugebauer et al., 1995). After intra-articular injection of irritant, there was an increase in spinal microdialysate PGE2 content immediately and 24 hours post-injection.

SPINAL PROSTAGLANDIN ADMINISTRATION

The hypothesis of spinally-derived prostanoids affecting neuronal electrophysiology and sensory processing has been driven by studies utilizing spinally delivered COX inhibitors and the demonstration of prostaglandin release from spinal parenchyma under conditions associated with hyperalgesia. While the synthetic source of spinal prostanoids is unclear, if spinal prostanoids are involved in central facilitation, then administration to the spinal parenchyma should induce a state of hypersensitivity. Administration of prostanoids to the intrathecal space decreases thermal nociceptive thresholds in rats (Yaksh, 1982) and mice (Uda et al., 1990). Hayaishi and colleagues have also described the effects of a number of prostanoids injected intrathecally in mice, suggesting that prostanoids may act in concert with EAA receptor activation to generate a spinally-mediated hypersensitive state, behaviorally assessed as allodynia or hyperalgesia (Uda et al., 1990). Suggestive of an interaction between spinal EAA receptor systems, intrathecal injection of PGE2 or PGF2α in mice produced a touch-evoked allodynia which was inhibited by EAA receptor antagonists. This touch-evoked allodynia was abrogated in an EAA receptor subtype-specific manner (Minami et al., 1994). Such studies suggest that spinal prostanoids and EAA's may work in concert to generate a hypersensitive state. However, there may be different EAA receptor-mediated sensory systems depending on the individual prostanoid cascades involved.

CONCLUSION

Given the above relationships, what is the mechanism of prostaglandin action in spinal facilitation? PGE2 increases tetrodotoxin-resistant sodium influx in cultured rat sensory neurons (Gold et al., 1996) which is sensitive to the μ-opiate receptor agonist, DAMGO (Gold and Levine, 1996). The opiate, morphine, and ketorolac, an NSAID, were synergistic in reducing hyperalgesia (Malmberg and Yaksh, 1993), suggesting that prostaglandins may decrease the activation threshold of opiate receptor-expressing, primary afferent terminals.

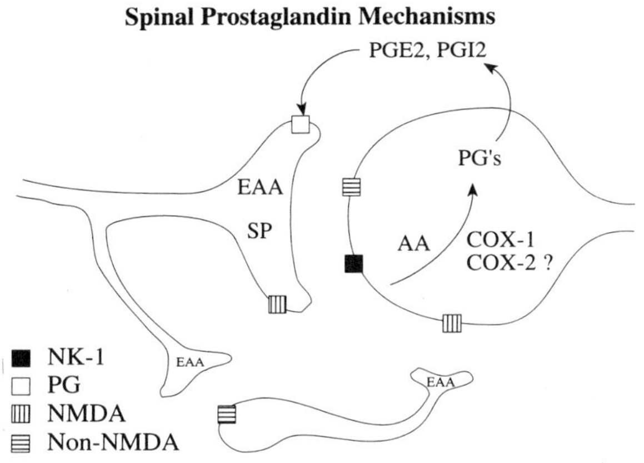

Figure 1. Cartoon depicts possible receptors (squares) associated with prostaglandin mediated spinal facilitation. After tissue injury, increased primary afferent activity leads to prolonged terminal release of SP and EAA's. SP, acting at NK-1 receptors, and EAA's, acting at the NMDA and non-NMDA receptors, will activate interneurons and higher-order neuronal pathways, leading to liberation of arachidonate (AA) from membrane phospholipids and conversion to one of several prostanoids (PG's) by COX-1 or COX-2. Prostanoids can feedback to potentiate depolarization-evoked SP/EAA release, increasing NK/EAA receptor activation and spinal neuronal responses relative to peripheral input.

Also suggestive of an interplay between sensory system transmitters and prostanoids, PGE2 or PGI2 exposure potentiates the capsaicin-evoked release of SP from cultured rat sensory neurons (Hingtgen et al., 1995). SP can evoke release of PGE2 from spinal cord tissue, and both the basal and SP-evoked in vitro spinal PGE2 release are elevated after 24 hours of knee joint inflammation (Dirig and Yaksh, unpublished data). A positive feedback loop (depicted in Figure 1) is suggested from these studies, whereby increased primary afferent activity after peripheral tissue injury sensitizes spinal neurons and increases synaptic glutamate and SP release from primary afferent terminals. These transmitters act on post-synaptic cells to evoke synthesis and release of prostanoids. Given that prostaglandins can potentiate the evoked release of neuropeptides, prostanoids may feed back on the primary afferent terminal to increase quantal transmitter release from pre-synaptic terminals. In this fashion, any subsequent action potential that reaches the sensory afferent terminal would result in an increased neuropeptide release, a greater excitation of post-synaptic neurons, and increased ascending neuronal activity relative to the peripheral stimulus (i.e. an increased input/output ratio), and thus spinal prostaglandin-mediated central facilitation of peripheral sensory input.

Clinical pain syndromes are rarely examples of states mediated by acute afferent input. Treatment of tissue injury or post-surgical trauma cases have more in common with protracted pain models described above than studies of acute nociception, in that the protracted afferent barrage associated with tissue injury and inflammation can produce tonic changes in sensory processing at spinal and supra-spinal levels. One essential aspect of such protracted nociceptive states is the development of central facilitation or centrally-mediated hyperalgesia. The convergent insights described herein emphasize the probable role played by spinal cyclooxygenase products in modulating nociceptive processing. These insights suggest the development of centrally acting agents which alter the receptor mediated actions of prostanoids and the investigation of enzyme inhibitor effects which can specifically target prostaglandin synthetic cascades that are implicated in the spinal processing of peripheral input associated with tissue injury and hyperalgesia.

REFERENCES

Baccaglini, P.I. and Hogan, P.G., 1983, Some rat sensory neurons in culture express characteristics of differentiated pain sensory cells, *Proc Natl Acad Sci U S A*. 80:594.

Dickenson, A.H. and Sullivan, A.F., 1987, Subcutaneous formalin-induced activity of dorsal horn neurones in the rat: differential response to an intrathecal opiate administered pre or post formalin, *Pain*. 30:349.

Ferreira, S.H., Moncada, S., and Vane, J.R., 1971, Indomethacin and aspirin abolish prostaglandin release from the spleen, *Nature New Biol*. 231:237.

Gold, M.S. and Levine, J.D., 1996, DAMGO inhibits prostaglandin E2-induced potentiation of a TTX-resistant Na+ current in rat sensory neurons in vitro, *Soc. for Neurosci*. 22:541.

Gold, M.S., Reichling, D.B., Shuster, M.J., and Levine, J.D., 1996, Hyperalgesic agents increase a tetrodotoxin-resistant Na+ current in nociceptors, *Proc Natl Acad Sci U S A*. 93:1108.

Hingtgen, C.M., Waite, K.J., and Vasko, M.R., 1995, Prostaglandins facilitate peptide release from rat sensory neurons by activating the adenosine 3',5'-cyclic monophosphate transduction cascade, *J Neurosci*. 15:5411.

Malmberg, A.B. and Yaksh, T.L., 1992a, Antinociceptive actions of spinal nonsteroidal anti-inflammatory agents on the formalin test in the rat, *J Pharmacol Exp Ther*. 263:136.

Malmberg, A.B. and Yaksh, T.L., 1992b, Hyperalgesia mediated by spinal glutamate or substance P receptor blocked by spinal cyclooxygenase inhibition, *Science*. 257:1276.

Malmberg, A.B. and Yaksh, T.L., 1993, Pharmacology of the spinal action of ketorolac, morphine, ST-91, U50488H, and L-PIA on the formalin test and an isobolographic analysis of the NSAID interaction, *Anesthesiology*. 79:270.

Malmberg, A.B. and Yaksh, T.L., 1995a, Cyclooxygenase inhibition and the spinal release of prostaglandin E2 and amino acids evoked by paw formalin injection: a microdialysis study in unanesthetized rats, *J Neurosci*. 15:2768.

Malmberg, A.B. and Yaksh, T.L., 1995b, The effect of morphine on formalin-evoked behaviour and spinal release of excitatory amino acids and prostaglandin E2 using microdialysis in conscious rats, *Br J Pharmacol*. 114:1069.

Marsala, M., Malmberg, A.B., and Yaksh, T.L., 1995, A chronic spinal dialysis catheter for use in the unanesthetized rat: Methodology and application, *J. Neurosci. Meth*. 62:43.

Minami, T., Nishihara, I., Uda, R., Ito, S., Hyodo, M., and Hayaishi, O., 1994, Involvement of glutamate receptors in allodynia induced by prostaglandins E2 and F2 alpha injected into conscious mice, *Pain*. 57:225.

Neugebauer, V., Geisslinger, G., Rumenapp, P., Weiretter, F., Szelenyi, I., Brune, K., and Schaible, H.G., 1995, Antinociceptive effects of R(-)- and S(+)-flurbiprofen on rat spinal dorsal horn neurons rendered hyperexcitable by an acute knee joint inflammation, *J Pharmacol Exp Ther*. 275:618.

Ramwell, P.W., Shaw, J.E., and Jessup, R., 1966, Spontaneous and evoked release of prostaglandins from frog spinal cord, *Am J Physiol*. 211:998.

Reeh, P.W., Kocher, L., and Jung, S., 1986, Does neurogenic inflammation alter the sensitivity of unmyelinated nociceptors in the rat?, *Brain Res*. 384:42.

Smith, J.B. and Willis, A.L., 1971, Aspirin selectively inhibits prostaglandin production in human platelets, *Nature New Biol*. 231:235.

Uda, R., Horiguchi, S., Ito, S., Hyodo, M., and Hayaishi, O., 1990, Nociceptive effects induced by intrathecal administration of prostaglandin D2, E2, or F2 alpha to conscious mice, *Brain Res*. 510:26.

Vane, J.R., 1971, Inhibition of prostaglandin synthesis as a mechanism of action for aspirin-like drugs, *Nature New Biol*. 231:232.

Woolf, C.J., 1983, Evidence for a central component of post-injury pain hypersensitivity, *Nature*. 306:686.

Yaksh, T.L., 1982, Central and peripheral mechanism for the antialgesic action of acetylsalicylic acid, in: *Acetylsalicylic Acid: New Uses for an Old Drug*, J.M. Barnet, J. Hirsh, and J.F. Mustard, ed., Raven Press, New York.

Yaksh, T.L., 1993, The spinal pharmacology of facilitation of afferent processing evoked by high-threshold afferent input of the postinjury pain state, *Curr Opin Neurol Neurosurg*. 6:250.

Yang, L., Marsala, M., and Yaksh, T.L., 1996, Characterization of time course of spinal amino acids, citrulline, and PGE2 release after carrageenan / kaolin-induced knee joint inflammation: a chronic microdialysis study, *Pain*. 67:345.

FUNCTIONAL CONSEQUENCES OF THE IDENTIFICATION AND LOCALISATION OF CYCLOOXYGENASE ISOFORMS IN DORSAL HORN OF RAT SPINAL CORD

Blair D. Grubb, Hilary L. Willingale, Natalie J. Gardiner and Susan Giblett

Department of Cell Physiology and Pharmacology
University of Leicester
PO Box 138
Leicester LE1 9HN
United Kingdom

INTRODUCTION

In recent years several studies have examined whether eicosanoids might facilitate neurotransmission in pain pathways. Discrepancies between the doses of some non-steroidal antiinflammatory drugs (NSAIDs) required to produce antiinflammatory and analgesic effects suggest that the analgesic actions may arise through a central action of NSAIDs (McCormack and Urquart, 1995). Evidence supporting this hypothesis has come from behavioural experiments which have shown that intrathecally applied NSAIDs reduce pain-related behaviour in the rat formalin model of inflammation (Malmberg and Yaksh, 1992). Furthermore, NSAIDs applied intravenously reduce the responses of rat spinal cord neurones, rendered hyperexcitable by an acute arthritic lesion, to innocuous and noxious stimuli applied to the inflamed knee (Neugebauer et al., 1995).

Several studies have suggested that NSAIDs are only active in models of inflammatory pain since behavioural and electrophysiological responses to acutely applied noxious stimuli are not reduced by NSAIDs in the normal animal (Malmberg and Yaksh, 1992; Chapman and Dickenson, 1992). The implication of these studies is that prostaglandins may modulate spinal nociceptive processing when there has been a prolonged activation of nociceptive C- and A-δ afferents but that they do not modulate responses to acute noxious stimuli in the non-inflamed situation. The aim of this study was to investigate whether the enzymes responsible for prostanoid synthesis were present in the spinal cord of normal spinal cord and in a location consistent with an involvement in spinal nociceptive processing. Furthermore, we examined the effect of non-selective NSAIDs and a selective cox-2 inhibitor on the responses of spinal cord neurones to acute noxious stimulation of the ankle joint in normal rats and the wind-up of the flexor withdrawal reflex in normal rats.

RESULTS

Identification of cyclooxygenase isoforms cord

Spinal cord samples were prepared for Western blotting by homogenisation in ice-cold lysis buffer (NaCl 150mM, Tris pH 7.4 10mM, EDTA 1mM, EGTA 1mM, Triton-X-100 1%, PMSF 0.2mM, aprotinin 20μg/ml) and centrifugation (20,000rpm, 4°C, 10min). Denatured samples were separated by SDS-PAGE and transferred to nitrocellulose membranes. Membranes were incubated in primary antibody overnight and then for 2h in peroxidase-conjugated secondary antibody. Protein bands were visualised using an ECL-detection kit (Amersham). Western blots revealed the presence of both cox-1 and cox-2 in the protein samples (figure 1).

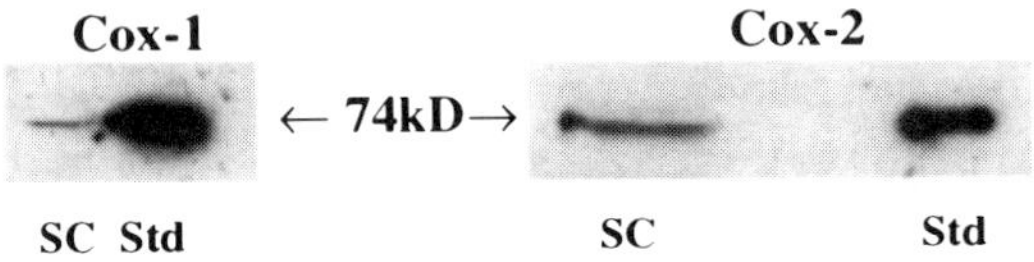

Figure 1. Western blot showing the presence of cox-1 and cox-2 in protein extracts of spinal cord. SC=spinal cord, Std =standard which was purified ovine protein.

Immunocytochemistry of spinal cord tissue for was performed using fresh frozen tissue sections (10μm) which were post-fixed in 2% paraformaldehyde. Sections were incubated in a rabbit anti-mouse cox-2 primary antibody (Cayman) or a mouse anti-ovine cox-1 primary antibody (Cayman) overnight at room temperature. Secondary incubation in FITC-conjugated secondary antibody was carried out for 2 hours before mounting sections for examination using a laser confocal microscope (Biorad). Figure 2 shows the dorsal horn of the spinal cord of a normal rat. The tissue section was incubated with a cox-2 antibody and specific cox-2-like immunoreactivity (cox-2-li) can be observed in the superficial dorsal horn (laminae I and II), and to a lesser extent in medial aspect of the deep dorsal horn. Cox-2-li was also observed around the central canal and around some motoneurones. Using the cox-1 antiserum, we were unable to show cox-1 protein in any discrete regions of the spinal cord.

Electrophysiology

Extracellular recordings from spinal dorsal horn neurones receiving input from the ankle joint of the normal rat have been characterised in the anaesthetised rat (Grubb et al, 1993) and consist of a slowly adapting response. The responses of these neurones to noxious mechanical stimulation of the ankle joint was unaffected by intravenous or intrathecal administration of the NSAIDs, indomethacin or meclofenamic acid, even at highest doses (data not shown). Since peripheral inflammatory stimuli elicit repetitive C-fibre activation and produce hyperexcitability in spinal neurones, which is sensitive to NSAIDs, an experimental paradigm was chosen in the normal rat which mimicked this situation. Neuronal wind-up, which is dependent on C-fibre activation can be defined as an increased neuronal response to a standard electrical stimulus applied to a peripheral nerve.

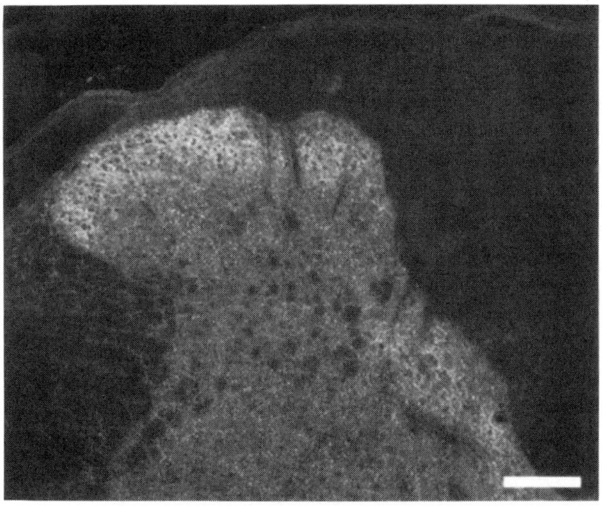

Figure 2. A confocal micrograph of the dorsal horn of normal rat spinal cord. The tissue section has been incubated with a cox-2 primary antibody and an FITC-conjugated secondary antibody. Labelled regions show as white. Cox-2-li can be observed in the superficial dorsal horn (laminae I/ II). Scale bar = 200μm.

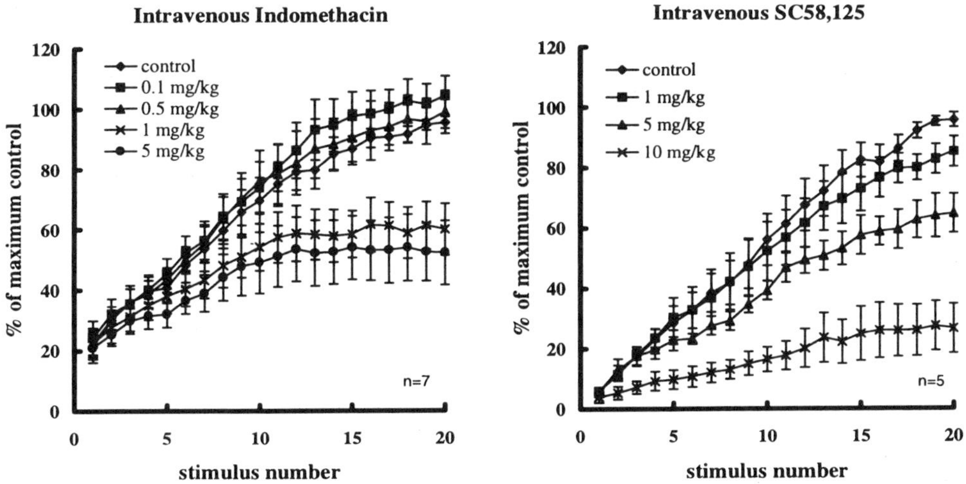

Figure 3. Graph showing the wind-up of a spinal nociceptive reflex in response electrical stimulation of the sural nerve (1.5 times C-fibre threshold, 0.6 Hz, 20 stimuli per train). Both drugs reduced the magnitude of the reflex wind-up in a dose dependent manner.

In barbiturate anaesthetised rats, the sural nerve was dissected free and stimulated at C-fibre strength whilst recordings were made from the biceps femoris muscle using needle electrodes (the reflex activity in the muscle is correlated with neuronal activity in the spinal cord). During a train of 20 stimuli (0.6 Hz, repeated every 5 minutes), the C-fibre component of the emg response increased markedly but then recovered prior to the next train (figure 3). Following intravenous (0.1-5.0 $mgkg^{-1}$) or spinal administration of indomethacin (0.01-1.0mM in 40μl saline pH,7.2; data not shown) or intravenous administration of the selective cox-2 inhibitor SC58,125 (1-10 $mgkg^{-1}$), wind-up was significantly reduced (Figure 4).

CONCLUSION

In normal rats, both cox-1 and cox-2 isoforms are present in the spinal cord in a location, the superficial dorsal horn, which is commensurate with their involvement in spinal nociceptive processing. NSAIDs are unable to modulate the responses of single dorsal horn neurones to noxious mechanical stimuli applied to the ankle joint in the normal rat although the wind-up of a spinal nociceptive reflex is inhibited by both non-selective cox inhibitors and by a selective cox-2 inhibitor. These data suggest that repetitive C-fibre activation, either in models of inflammatory pain or following electrical activation of C- fibres in normal rats, results in activation of a process(es) in the spinal cord which induce prostaglandin synthesis and/or release, and that these prostaglandins facilitate nociceptive processing in the spinal cord.

REFERENCES

Chapman, V. and Dickenson, A.H., 1992, The spinal and peripheral roles of bradykinin and prostaglandins in nociceptive processing in the rat. *Eur. J. Pharmacol.* 219: 427-433.

Grubb, B.D., Stiller, R.U. and Schaible, H.-G., 1993, Dynamic changes in the receptive field properties of spinal cord neurones with ankle input in rats with chronic unilateral inflammation in the ankle region. Exp. Brain Res. 92:441-452.

McCormack, K. and Urquart, E., 1995, Correlation between non-steroidal anti-inflammatory drug efficacy in a clinical pain model and the dissociation of their anti-inflammatory and analgesic properties in animal models. *Clin. Drug Invest.* 9: 88-97.

Malmberg, A.B. and Yaksh, T.L., 1992, Antinociceptive actions of spinal non-steroidal anti-inflammatory agents on the formalin test in the rat. *J. Pharm.Exp. Ther.* 263: 136-146.

Neugebauer, V. Geisslinger, G., Ruemenapp, P., Weiretter, F., Szelenyi, I. Brune, K. and Schaible, H.-G., 1995, Antinociceptive effects of R(-) and S(+)-flurbiprofen in rat spinal dorsal horn neurones rendered hyperexcitable by an acute knee joint inflammation. *J. Pharm.Exp. Ther.* 275: 618-628.

CYCLOOXYGENASE-2 IN THE SPINAL CORD: LOCALIZATION AND REGULATION AFTER A PERIPHERAL INFLAMMATORY STIMULUS

Margarete Goppelt-Struebe and Flora Beiche

Department of Medicine IV, University of Erlangen-Nürnberg, Loschgestr. 8, D-91054 Erlangen, Germany

Summary

Prostaglandins are known as mediators in spinal nociceptive processing after peripheral inflammation. Cyclooxygenase isozymes Cox-1 and Cox-2 as enzymes essential for prostaglandin biosynthesis were therefore investigated in rat spinal cord. mRNA and protein of both enzymes was detected in cervical and lumbar sections of the spinal cord of normal animals. Enzyme activity of Cox-1 and Cox-2 could be determined by the use of specific inhibitors. Cox-2-immunoreactivity (IR) was found in the spinal cord of untreated rats in neurons of laminae II-III, motoneurons of lamina IX and in glial cells. Most prominent staining was observed in the nuclear envelope.

Under an acute peripheral inflammatory stimulus a transient 2-fold increase in Cox-2 mRNA was noticed bilaterally in the lumbar spinal cord by reverse transcription-polymerase chain reaction. This increase was reflected on a smaller scale at protein levels in Western blot analysis of spinal membrane preparations. In the chronic state of adjuvant-induced arthritis, Cox-2 levels were raised again significantly.

Cox-2 might therefore be regarded as the Cox isozyme responsible for increased spinal prostanoid release in nociceptive processing under peripheral stimulation.

Introduction

In the central nervous system, prostaglandins are thought to play a role in a variety of neurological functions including pain control[1] and nociceptive processing[2]. Hyperalgesic effects were observed after intrathecal administration of prostaglandins[3] and a spinal site of action was proposed for the analgesic actions of non-steroidal anti-inflammatory drugs (NSAIDs)[4]. As activation of prostaglandin biosynthesis thus seems to be involved in neuronal pain transmission, we studied expression and activity of Cox isoforms in rat spinal cord. Involvement of cyclooxygenase isozymes in nociceptive processing was investigated in the rat model of adjuvant-induced inflammation.

Recent Advances in Prostaglandin, Thromboxane, and Leukotriene Research
Edited by Sinzinger *et al.*, Plenum Press, New York, 1998

Results

Peripheral inflammation was induced by injection of complete Freund's type adjuvant (CFA) into one hindpaw of male Sprague-Dawley rats as previously described[5]. Development of arthritis was monitored by plethysmographic measurement of paw volumes which is regarded as a sensitive method for quantifying the inflammatory response.

Total RNA was isolated and levels of Cox-1 and Cox-2 mRNA expression were analyzed by reverse transcription polymerase chain reaction (RT-PCR; for details see Beiche et al.[5]). Cox-1 mRNA expression remained fairly unchanged throughout the first week after CFA injection and was not raised, either, in the chronic inflammation at day 22. Cox mRNA levels were compared within each group of animals as well as between groups at different timepoints. A transient but significant increase in spinal Cox-2 mRNA levels could be detected 6 h after induction of peripheral inflammation which returned to baseline within 3 days (Fig. 1). In the chronic phase of arthritis at day 22 spinal Cox-2 mRNA levels were raised again significantly as compared with untreated animals.

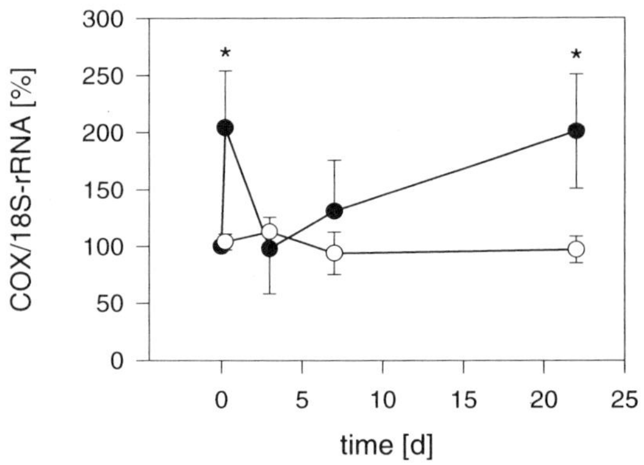

Figure 1. Expression of Cox-2 mRNA in the lumbar spinal cord in the acute and chronic phase of adjuvant-induced inflammation. Levels of mRNA expression were determined using RT-PCR technique. Cox-2 mRNA expression (n=3 per timepoint, triplicate values) is corrected for 18S-rRNA levels, expression in untreated animals is set to 100% in each screen.* = $p<0.05$.

Cox protein expression was detected by Western blotting using a combined microsomal and nuclear membrane preparation from spinal cord homogenates. A single IR band for Cox-1 at approximately 70 kDa was detected in homogenates of all animals investigated during the course of peripheral inflammation. Cox-2 protein expression maintained a similar profile as was found for mRNA levels. A slight increase in Cox-2 protein was seen in the acute phase 6 h after CFA-inoculation as well as in the chronic phase of arthritis at day 22 (Fig. 2). As interindividual differences in Cox-2 protein expression were quite varying, both increases did not reach statistical significance ($0.05<P<0.2$; Fig. 2a).

Biological activity of both isoforms was shown by specific inhibitors: In vitro conversion of arachidonic acid into PGE_2 was almost completely inhibited by diclofenac, a non selective NSAID, whereas NS398, a Cox-2 specific inhibitor reduced Cox activity by about 30 % (Fig. 2b).

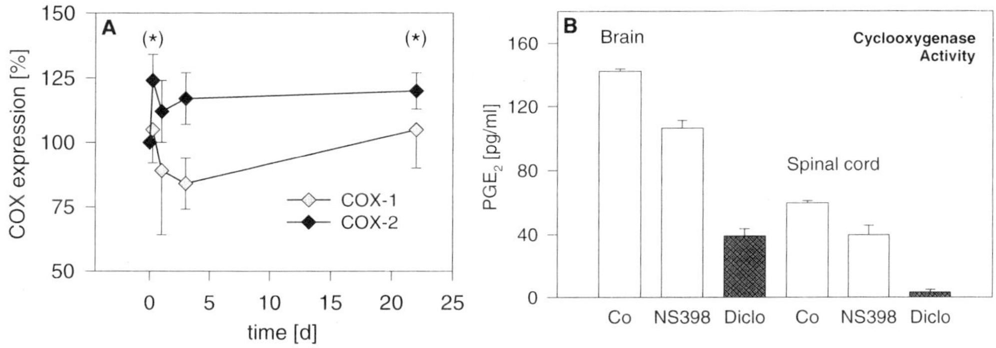

Figure 2. Cox protein expression in the lumbar spinal cord.
Panel (a): Timecourse of spinal Cox-2 protein expression in the acute inflammatory phase and chronic state of arthritis (n=3 per timepoint, duplicate values); expression in untreated animals is set to 100% in each screen (*) = $0.05<p<0.2$. The specific Cox-1 primary antibody was kindly provided by D. DeWitt (Michigan State University, East Lansing, MI), the anti-rat Cox-2 antibody (M-19) directed against a unique carboxy-terminal region of Cox-2 was obtained from Santa Cruz Biotechnology, Inc.
Panel (b): Cox activity (conversion of arachidonic acid into PGE_2) was determined with protein of brain and spinal cord of untreated animals in the presence of diclofenac (10 μM) or NS398 (10 μM) as indicated. Data represent a typical experiment.

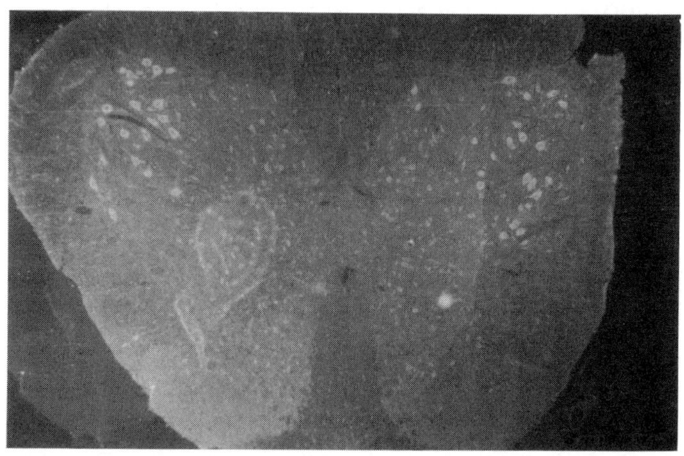

Figure 3. Immunocytochemical staining with anti-Cox-2 antibody in the L4 segment of the spinal cord. The photograph was taken from the lumbar section L4 of a CFA-treated animal 6 h after induction of peripheral inflammation.

Immunocytochemical analysis of Cox-2 expression was performed on 10 μm slices of the spinal cord. Cox-2 immunoreactive neuronal populations were distributed in the superficial layers, laminae I-II, the deeper layers, laminae III and IV, as well as motoneurons of lamina IX (Fig. 3). Cox-2 IR was mapped focally to the nuclear envelope and to the Nissl granules in the perikarya. In laminae II-III, Cox-2 IR seemed to include larger areas of the stained neurons apart from nucleus and Nissl substance. Investigation of the white matter of the spinal cord showed positive IR for the nuclei of a large number of glial cells. Specificity of the primary antibody was determined by preadsorbing the antibody to the carboxy-terminal Cox-2 peptide against which the antibody was raised. Control experiments were also performed with non-specific IgG and omission of primary antibody.

L4 sections from untreated animals or animals killed after any of the indicated time points after CFA inoculation showed essentially a similar pattern of staining: a significant increase in Cox-2 IR could not be detected in number, distribution or intensity of Cox-2 immunoreactive cells in any area of the spinal cord sections investigated.

Conclusion

In this study, we could show the presence and localization of Cox-2-IR in neurons and glial cells in the spinal cord of untreated rats. The prominent expression of Cox-2 in untreated animals may seem surprising, but other than cultured cells or peripheral tissues, animals taken into these kinds of experiment were never completely deprived of all sensory input. The observation that Cox-2 is produced in the ventral, deep dorsal and superficial dorsal horn makes an involvement of prostaglandins in various physiological aspects of spinal function and sensory processing very likely.

Under a peripheral inflammatory stimulus an increase in Cox-2 mRNA and protein was noticed bilaterally in the lumbar spinal cord by RT-PCR and Western blot analysis. Possible progression of sensory information may result in an ascending increase of Cox-2 through the spinal cord as cervical sections also revealed enhanced Cox-2 expression, yet to a smaller extent. Cox-2 induction following peripheral inflammation is thus not restricted to a defined area of the spinal cord. Communication between the two sides of the spinal cord may work via possible polysynaptic pathways important in pain adaptation.

Regulation of Cox-2 expression in neuronal structures is in the centre of a very complex modulatory system[6-8]. As the two phases of adjuvant-induced arthritis are different in their pathophysiology, it is more than likely that different types of mediators may be involved in the occurring induction of Cox-2 expression. Further studies on the regulation of Cox-2 expression in the spinal cord will help to clarify the role of this enzyme in various disease states but also in normal physiology.

References

1. Z.M. Poddubiuk, A comparison of the central actions of prostaglandins A_1, E_1, E_2, F_{1alpha}, and F_{2alpha} in the rat, *Psychopharmacology* 50:89 (1976).

2. S.H. Ferreira, B.B. Lorenzetti, and F.M.A. Correa, Central and peripheral antialgesic actions of aspirin-like drugs, *Eur. J. Pharmacol.* 53:39 (1978).

3. A.B. Malmberg, and T.L. Yaksh, Antinociceptive actions of spinal non-steroidal anti-inflammatory agents on the formalin test in the rat, *J. Pharmacol. Exp. Ther.* 263:136 (1992).

4. V. Neugebauer, G. Geisslinger, P. Ruemenapp, F. Weiretter, I. Szelenyi, K. Brune, and H.-G. Schaible, Antinociceptive effects of R(-) and S(+)-flurbiprofen on rat spinal dorsal horn neurons rendered hyperexitable by an acute knee joint inflammation, *J. Pharmacol. Exp. Ther.* 275:618 (1995).

5. F. Beiche, S. Scheuerer, K. Brune, G. Geisslinger, and M. Goppelt-Struebe, Up-regulation of cyclooxygenase-2 mRNA in the rat spinal cord following peripheral inflammation, *FEBS Lett.* 390:165 (1996).

6. K. Yamagata, K.I. Andreasson, W.E. Kaufmann, C.A. Barnes, and P.F. Worley, Expression of a mitogen-inducible cyclooxygenase in brain neurons: regulation by synaptic activity and glucocorticoids, *Neuron*, 11:371 (1993).

7. Adams, J., Y. Collaco-Moraes, and J. de Belleroche, Cyclooxygenase-2 induction in cerebral cortex: An intracellular response to synaptic excitation, *J. Neurochem.* 66:6-13 (1996).

8. V.L. Marcheselli, and N.G. Bazan, Sustained induction of prostaglandin endoperoxide synthase-2 by seizures in hippocampus, *J. Biol. Chem.* 271:24794 (1996).

THE PRODUCTION OF BIOLOGICALLY ACTIVE EICOSANOIDS BY YEASTS

J. Lodewyk F. Kock,[1] Pierre Venter,[1] Alfred Botha,[1] Dennis J. Coetzee,[1] Piet J. Botes,[1] and Santosh Nigam [2]

[1]Department of Microbiology and Biochemistry
University of the Free State
P.O. Box 339, Bloemfontein, 9300
South Africa
[2]Eicosanoid Research Division
Department of Gynecology
Free University of Berlin, Germany.

INTRODUCTION

The lipid composition of fungi has been well studied. It is generally accepted that C16 and C18 chain length fatty acids (FAs) are the major FAs in fungi. Fungi are also characterised by the presence of both the ω3 and ω6 series of polyunsaturated FAs (PUFAs). The ω3 series include the PUFAs i.e. α-linolenic acid (18:3, ω3) while the ω6 series include mainly linoleic acid (18:2, ω6), γ-linolenic acid (18:3, ω6), di homo γ-linolenic acid (20:3, ω6) and arachidonic acid (20:4, ω6). Both the ω3 and ω6 PUFAs are derived from 18:2 ω6 by the participation of various desaturase and elongase enzymes (Smit, 1993; Van der Westhuizen, 1994).

Interesting patterns are observed when the distribution of the ω3 and ω6 PUFAs in fungi and the phylogenetic development (Kendrick, 1992) of these organisms are compared (Figure 1). In this scheme, the aquatic protoctistan fungi which are characterised by the production of zoospores that swim by means of flagella, are considered "ancestors" of the true fungi. From these fungi the terrestrial true fungi evolved which include the zygomycotan and dikaryomycotan fungi which produce no motile cells. On the basis of hyphal structure, the zygomycotan fungi closely resembles the protoctistan fungi (i.e. both produce sporangia and coenocytic hyphae characterised by the absence of septa) while the dikaryomycotan fungi do not produce these types of sporangia and have septate hyphae. When comparing the distribution of PUFAs in fungi and this phylogenetic scheme, it can be concluded that the more "primitive" protoctistan fungi are capable of producing the ω6 series of PUFAs comprising of eighteen carbons (C18) and twenty carbons (C20). The evolved zygomycotan fungi contain mainly the C18 ω6 series of PUFAs and in only a few cases the C20 ω6 PUFAs. In general, representatives of the Dikaryomycota as well as affiliated anamorphs do not produce the ω6 series of PUFAs. Some members of this phylum i.e. wine yeasts, are considered reduced fungi, and do not produce the ω3 or ω6 PUFAs. In fact, these fungi are incapable of producing unsaturated FAs greater than C18 monoenoic FAs *ab initio.*

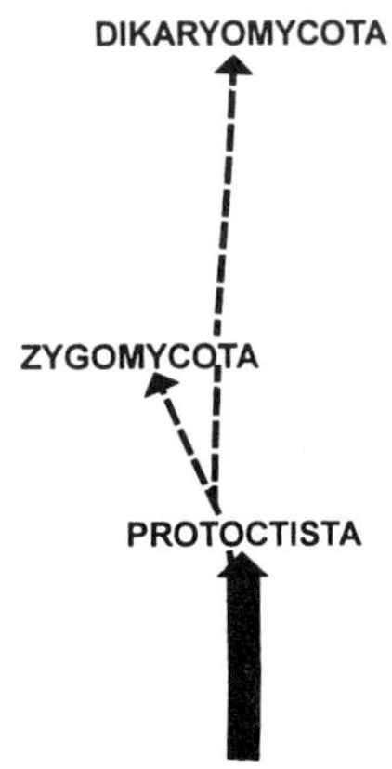

Figure 1. A schematic representation indicating the phylogenetic development and classification of the fungi.

THE PRODUCTION OF 3-HYDROXY FATTY ACIDS BY YEAST

Although the dikaryomycotan fungi do not produce C20 PUFAs and hence eicosanoids *ab initio,* our group has demonstrated the production of a prostacyclin metabolite as well as a novel biologically active 3- hydroxy 5,8,11,14-eicosatetraenoic acid (3-HETE) by the dikaryomycotan fungus (yeast) *Dipodascopsis uninucleata* when fed with arachidonic acid (Van Dyk et al., 1991; Kock et al., 1991; Kock et al., 1997). The accumulation of 3-HETE by this yeast has been shown to be suppressed by aspirin and salicilate.

By feeding this yeast the PUFAs octadecadienoic- [18:2 (9,12)], eicosatrienoic- [20:3 (5,8,11)] and [20:3 (11,14,17)], eicosatetraenoic- [20:4 (5,8,11,14)] and eicosapentaenoic [20:5 (5,8,11,14,17)], 3-hydroxy FAs were produced (Venter et al., 1997). In the case of 18:2 (9, 12) and 20:3 (11,14,17), 3-hydroxy tetradecadienoic acid [14:2 (5,8)] and 3-hydroxy tetradecatrienoic acid [14:3 (5,8,11)] were produced respectively. When PUFAs with double bonds on position C5 i.e. 20:3 (5,8,11), 20:4 (5,8,11,14) and 20:5 (5,8,11,14,17) were fed to this yeast, 3-hydroxy C20 FAs were produced. No 3-hydroxy FAs were observed when octadecatrienoic- [18:3 (6,9,12)], eicosanoic (20:0) or C22 PUFAs were fed to this yeast. The latter were toxic, while 18:3 and 20:0 were almost completely utilised. These results suggest that the presence of a *cis* double bond at the C5 position (either directly or upon corresponding shortening of the molecule by β-oxidation - see Figure 2) may be the signal for the accumulation of the 3-hydroxy poly-enoic FAs by this yeast. Studies are in progress to elucidate the metabolic pathway of the formation of these compounds (Venter et al., 1997).

Thus, the identification of the stereo-specificity of the 3-hydroxy group will be decisive. The L-enantiomere would occur as intermediate step of β-oxidation, whereas the D-enantiomere would exclude this pathway and could at the same time afford a possible explanation why the 3-hydroxy FA escape the β-oxidation and therefore accumulate. In any case, the involvement of the lipoxygenase or a related enzyme in the biosynthesis of this novel arachidonic acid metabolite can be excluded since the methylene group at C3 is not *bis*-allelic which is a prerequisite for the action of such an enzyme.

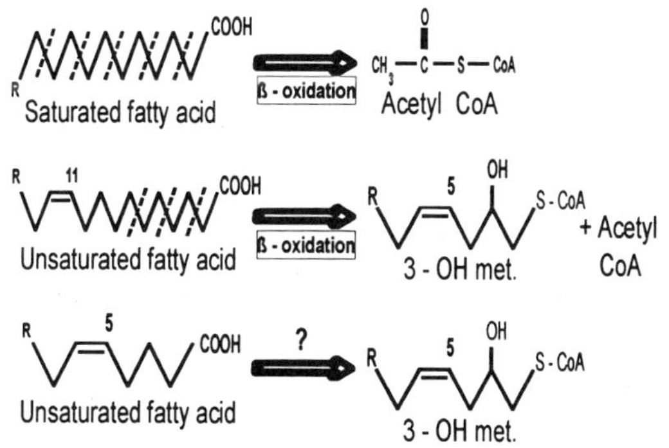

Figure 2. The formation of unsaturated 3-hydroxy FAs by *Dipodascopsis uninucleata.* (met.=metabolite).

ACKNOWLEDGMENTS

The authors wish to thank the South African Foundation for Research Development, Foreign Office of the Free University Berlin and in part the Deutsche Forschungsgemeinschaft, Bonn (Ni 242/9-1) for supporting this study.

REFERENCES

Kendrick, B., 1992, *The Fifth Kingdom,* Mycologue Publications, Ontario.

Kock, J.L.F., Coetzee, D.J., Van Dyk, M.S., Truscott, M., Cloete, F.G., Van Wyk, V., and Augustyn, O.P.H., 1991, Evidence for pharmacologically active prostaglandins in yeasts. *S. Afr. J. Sci.* 87: 73-76.

Kock, J.L.F., Jansen van Vuuren, D., Botha, A., Van Dyk, M.S., Coetzee, D.J., Botes, P.J., Shaw, N., Friend, J., Ratledge, C., Roberts, A.D., and Nigam, S., 1997, The production of biologically active 3-hydroxy- 5,8,11,14-eiosatetraenoic acid (3-HETE) and linoleic acid metabolites by *Dipodascopsis. System. Appl. Microbiol.* 20:39-49.

Smit, M.S., 1993, The turnover of exogenous arachidonic acid by the yeast *Dipodascopsis uninucleata.* Ph.D. Thesis. Department of Microbiology and Biochemistry, University of the Orange Free State, Bloemfontein, South Africa.

Van der Westhuizen, J.P.J., 1994, The distribution of the ω3 and ω6 series of cellular long-chain fatty acids in fungi associated with disease. Ph.D. Thesis. Department of Microbiology and Biochemistry, University of the Orange Free State, Bloemfontein, South Africa.

Van Dyk, M.S., Kock, J.L.F.,Coetzee, D.J., Augustyn, O.P.H., and Nigam, S., 1991, Isolation of a novel arachidonic acid metabolite 3-hydroxy 5,8,11,14-eicosatetraenoic acid (3-HETE) from the yeast *Dipodascopsis uninucleata* UOFS-Y128. *FEBS Lett.* 283:195-198.

Venter, P., Kock, J.L.F., Botha, A., Coetzee, D.J., P.J. Botes, and Nigam, S., 1997, The production of 3-hydroxy fatty acids by the yeast *Dipodascopsis uninucleata.* (manuscript in preparation).

IS LIPOXYGENATION OF PATHOGEN-DERIVED ARACHIDONIC ACID INVOLVED IN PLANT PROTECTION?

Tankred Schewe and Santosh Nigam

Eicosanoid Research Division, Dept. of Gynecology, University Clinics Benjamin Franklin, Free University of Berlin, D-12200 Berlin, Germany

INTRODUCTION

In the plant kingdom the occurrence of arachidonic acid and, therefore, of eicosanoids appears to be restricted to fungi and related organisms, algae, mosses and ferns. In higher plants, however, lipoxygenases capable of oxygenating arachidonic acid are widely distributed. At first sight, the reactions of higher plant lipoxygenases with arachidonic acid seem to be only of academic interest. However, under special circumstances a higher plant lipoxygenase is capable of converting arachidonic acid of fungal origin. Such a situation is given by the infection of the plant by a fungal pathogen. Some aspects of the putative role of these interactions will be discussed in this article. For a detailed survey on the present knowledge of the role of lipoxygenase in plant resistance to infection the reader is referred to the recent paper by Slusarenko.[1]

HIGHER PLANT LIPOXYGENASES AND FUNGAL ARACHIDONIC ACID

In Table 1 some arguments in favour of a possible role of an interaction between higher plant lipoxygenases and arachidonic acid (AA) are compiled. First, AA is present in a number of lower fungi and related organisms such as oomycetes which are parasites of higher plants including those having tremendous agricultural and horticultural importance. The AA is there esterified to membrane lipids in high amounts and represents a potential source for eicosanoids (see *e.g.* the other papers of this Section). Second, AA and eicosapentaenoic acid (EPA) are elicitors of the defence reactions which are induced upon infection of the plant by the fungus.[2] These defence reactions include a variety of events such as oxygen burst and oxidative stress, induction of enzymes of the secondary plant metabolism, among others lipoxygenases, production of antifungal compounds - the so-called phytoalexins, and finally necrosis of the infected tissue. Some of these defence reactions do not only occur in the infected cells but also in the unaffected parts of the host plant and are collectively designated as "systemic aquired resistance", which suggests the involvement of infection-induced signal molecules - the elecitors. Compounds of different chemical classes such as proteins, carbohydrates, lipids and certain polyenoic fatty acids have been identified as elicitors.

Table 1. Implications for a putative role of eicosanoids in fungus-host plant interactions

A number of fungi that are pathogens for higher plants contain large amounts of arachidonic acid (AA) esterified to membrane lipids
AA and EPA are elicitors of pathogen-induced host defence reaction
Elicitation of defence reactions includes induction of lipoxygenase genes
Higher plants contain lipoxygenases (constitutive or infection-induced) capable of oxygenating AA and EPA

The third argument for a putative role of eicosanoids in this process is the fact that the pathogen-induced defence reactions also include the induction of lipoxygenases. Selected examples are shown in Table 2. It can be seen, that the induction of lipoxygenase in the host plant is related to a number of important plant diseases challenged by fungal pathogens. It should be stressed that in many cases a certain lipoxygenase activity is already present in plants before the infection, but both total activity and the number of expressed lipoxygenase genes are enhanced upon infection.

Table 2. Induction of lipoxygenases by fungal pathogen infection

Host plant	Pathogen/Disease	Authors
Wheat	Puccinia graminis tritici	Ocampo et al., 1986[3]
Tomato	Powdery mildew	Kato et al., 1992[4]
Potato	Rhizoctonia solani	Reddy et al., 1992[5]
Tobacco	Phytophthora parasitica	Fournier et al., 1993[6]
Tobacco	Phytophthora cryptogea	Suty et al., 1996[7]
Rice	Magnaporthe grisea	Ohta et al., 1991[8]

Thus, there are two components for an eicosanoid-mediated elicitor action - the pathogen-derived AA and the host-derived lipoxygenases. Nevertheless, the pathogen infection is the only situation in which these two components can interact with each other. In Table 3 the reaction products from AA are listed for a number of plant lipoxygenases. Three characteristics can be recognized hereto:

1. All lipoxygenases from higher plants so far tested are capable of oxygenating AA.

2. In some cases such as in potato the infection-induced enzyme differs from the constitutive isoenzyme with respect to the reaction specificity towards AA.

3. Unlike most of the mammalian lipoxygenases, many plant enzymes exhibit a multiple positional specificity of the oxygenation which leads to a more complex product pattern. Thus, a triple specificity has been observed with the enzymes from tomato fruits, with the elicitor-induced enzyme from hairy roots of potato and with the enzyme from lipid bodies of cucumber seedlings. The product pattern of the tomato enzyme is shown in Figure 1. Three products, 11-HpETE, 8-HpETE and 5-HpETE, are formed in comparable proportions. Mechanistically these oxygenation products can be explained by alternative hydrogen abstractions from all three *bis*-allylic methylene groups present in the AA molecule. Despite this peculiarity, the reaction is specific as expected for a lipoxygenase catalysis, since the analysis of the stereoisomers of the products in Figure 1 revealed exclusively the *S*-enantiomers.

Table 3. Arachidonic acid products of higher plant lipoxygenases

Lipoxygenase	AA products	Authors
Soybean-I	15*S*-HpETE	Hamberg & Samuelsson, 1967[9]
Pea-I	HpETE isomer mixture	Regdel et al., 1995[10]
Pea-II	15*S*-HpETE (5*S*-HpETE)	Regdel et al., 1995[10]
Lupin	15*S*-HpETE (5*S*-HpETE, 8*S*-HpETE)	Najid et al., 1988[11]
Wheat	5*S*-HpETE (8*S*-HpETE)	Heydeck et al., 1990[12]
Barley	5*S*-HpETE	van Aarle & Vliegenthart' 1991[13]
Tomato	5*S*-HpETE 8*S*-HpETE 11*S*-HpETE	Regdel et al., 1993[14]
Potato (constitutive enzyme)	5*S*-HpETE (other HpETE's)	Mulliez et al, 1987[15]
Potato (elicitor-induced enzyme)	5*S*-HpETE 8*S*-HpETE 11*S*-HpETE	Reddy et al., 1992[5]
Cucumber (lipid bodies)	15S-HpETE 8*S*-HpETE 12*S*-HpETE	Feussner & Kühn, 1995[16]

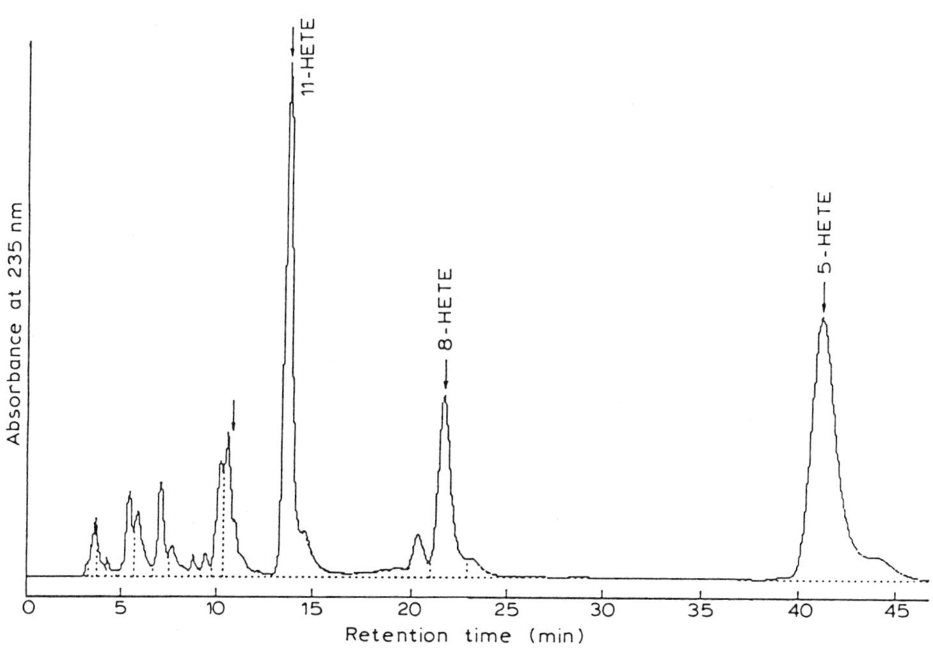

Figure 1. Straight-phase HPLC separation of the reduced reaction products formed from AA by tomato fruit lipoxygenase (from Regdel et al.[14], with permission).

LIPOXYGENASE-DERIVED EICOSANOIDS AND PLANT RESISTANCE

The four arguments discussed above thus strongly support a putative role of eicosanoids in the interaction of pathogenic fungi with the host plant. Moreover, they implicate that AA is a precursor of a lipoxygenase-derived eicosanoid that serves as signal molecule for the elicitation of the host defence reactions.

In order to investigate which eicosanoid is the elicitor *per se*, Pena-Cortez and coworkers from the Institute for Gene Biological Research of the Free University Berlin tested a number of compounds in a suitable elicitation system using pieces of leaves of a special transgenic potato plant (Table 4). In accordance with earlier reports in the literature,[2] among the free fatty acids only AA and EPA exhibited elicitor activity. All 12 conceivable lipoxygenase-derived monohydroxyeicosatetraenoic acids were inactive and may therefore be excluded as elicitors. The corresponding hydroperoxy derivatives were inactive as well. However, the hydroperoxy fatty acids are metabolically labile and, if exogenously added, may undergo rapid reduction by hydroperoxidases to the corresponding hydroxy derivatives prior to reaching the biological target. Surprisingly, a number of prostaglandins exhibited elicitor activity at concentrations as low as 1 micromolar, the highest activity being observed with PGE_2. Although these results are preliminary, they indicate a prospective role of prostaglandins or prostaglandin-like compounds in the elicitation process.

Table 4. Eliciting activity of selected compounds on potato leaves*

Arachidonic acid	+
Eicosapentaenoic acid	+
C_{18}-Fatty acids	–
HETE's (all possible isomers)	–
HpETE's (all possible isomers)	–
Leukotrienes	–
PGE_2	+

*Pena-Cortez & coworkers, unpublished, cited with permission

Are higher plants capable of producing prostaglandin-like compounds? The answer is an affirmative one. It is well known that the synthesis of the plant hormone jasmonic acid includes the conversion of α-linolenic acid to oxo-phytodienoic acid, which contains a cyclopentane ring. The synthesis of oxo-phytodienoic acid involves a lipoxygenase-catalysed formation of a hydroperoxy fatty acid, followed by an allene oxide synthase reaction with subsequent cyclization of the allene oxide intermediate.[17] Substitution of AA for α-linolenic acid in the above cascade resulted in the production of prostaglandin-like compounds should arise. A lipoxygenase-allene oxide synthase pathway is known to be involved in the prostaglandin synthesis in corals[18]. Similar pathways in algae lead to a number of cyclopentane-containing AA- or EPA-derived oxylipins[19]. For fungus-infected plants the existence of such a metabolic route has yet to be established.

OTHER FUNCTIONS OF LIPOXYGENASES

The foregoing considerations revealed a crucial role of lipoxygenases in the defence of higher plants against infections by fungal pathogens. This role is, however, not restricted to a lipoxygenation of free arachidonic acid. Table 5 summarizes the possible functions of lipoxygenases in this process. The role of lipoxygenase as a key enzyme in the synthesis of jasmonic acid has been well established. Jasmonic acid is not only a plant senescence hormone, but is also apparently involved in the pathogen defence. Another route of the α-linolenic acid cascade leads to the antimicrobial phytoalexin *trans*-2-hexenal by a hydroperoxide lyase reaction.[20]

A quite different role of lipoxygenases could be related to their capability of oxygenating phospholipids including those present in biological membranes. It has been shown earlier by our group, that a mammalian 15-lipoxygenase is involved by virtue of this property in the maturational breakdown of mitochondrial membranes in reticulocytes[21]. Moreover, a lipoxygenation of complex lipids has also been demonstrated for some plant lipoxygenases such as for the lipoxygenase from lipid bodies of cucumber seedlings[22]. Hence, it can be conjectured, that upon pathogen infection a lipoxygenase is induced and unmasked, which attacks the membranes either of the host cell leading to cell necrosis, which is a typical part of the defence reactions, or of the pathogen which is killed by this process. The reactions of lipoxygenases with biological membranes may also contribute to the oxygen burst which accompanies the defence reactions.

In summary, both lipoxygenases and lipoxygenase-derived prostaglandin-like eicosanoids appear to play an important role in the defence reactions of higher plants against the infection by pathogenic fungi. The present knowledge on this subject is, however, still in infancy. The further elucidation of these processes should open new approaches to the enhancement of plant resistance.

Table 5. Possible functions of plant lipoxygenases in pathogen defence

Synthesis of jasmonic acid via α-linolenic acid cascade
Syntheses of phytoalexins (e.g. *trans*-2-hexenal) via α-linolenic acid cascade
Synthesis of eicosanoids from pathogen-derived AA (?)
Involvement in pathogen-induced cell necrosis and in killing of pathogens via direct oxygenation of membranes (?)

REFERENCES

1. A.J. Slusarenko, The role of lipoxygenase in plant resistance to infection, in: *Lipoxygenase and Lipoxygenase Pathway Enzymes*, G.I. Piazza, ed., AOCS Press, Champaigne, Illinois (1996) pp 176-197.
2. R.M. Bostock, J.M. Kuc, and R.A. Laine, Eicosapentaenoic and arachidonic acids from Phytophthora infestans elicit fungitoxic sesquiterpenes in the potato, *Science* 212:67-69 (1981).
3. C.A. Ocampo, B. Moerschbacher, and H.J. Grambo, Increased lipoxygenase activity is involved in the hypersensitive response of wheat leaf cells infected with avirulent rust fungi or treated with fungal elicitor, *Z. Naturforsch.* 41C:559-563 (1986).
4. T. Kato, Y. Maeda, T. Hirukawa, T. Namai, and N. Yoshioka, Lipoxygenase activity increment in infected tomato leaves and oxidation product of linolenic acid by its in vitro enzyme reaction, *Biosci. Biotech. Biochem.* 56:373-375 (1992).
5. G.R. Reddy, P. Reddanna, C.C. Reddy, and W.R. Curtis, 11-Hydroperoxyeicosatetraenoic acid is the major dioxygenation product of lipoxygenase from hairy root cultures of Solanum tuberosum, *Biochem. Biophys. Res. Commun.* 189:1349-1352 (1992).
6. J. Fournier, M.-L. Pouénat, M. Rickauer, H. Rabinovitch-Chable, M. Rigaud, and M.-T. Esquerré-Tugayé, Purification and characterization of elicitor-induced lipoxygenase in tobacco cells, *Plant J.* 3:63-70 (1993).
7. L. Suty, A.S. Petitot, D. Lecourieux, J.P. Blein, and A. Pugin, Isolation of partial length cDNAs corresponding to early differentially expressed genes during elicitation of tobacco cells by cryptogein: use of differential mRNA display, *Plant Physiol. Biochem.* 34:443-451 (1996).
8. H. Ohta, K. Shida, Y.-L. Peng, I. Furusawa, J. Shishiyama, S. Aibara, and Y. Morita, A lipoxygenase pathway is activated in rice after infection with the rice blast fungus Magnaporthe grisea, *Plant Physiol.* 97:94-98 (1991).

9. M. Hamberg and B. Samuelsson, On the specificity of the oxygenation of unsaturated fatty acids catalyzed by soybean lipoxidase, *J. Biol. Chem.* 242:5329-5335 (1967).
10. D. Regdel, T. Schewe, and H. Kühn, Comparative characteristics of lipoxygenase isoenzymes from green pea seeds, *Biochemistry (Moscow)* 60:715-721 (1995).
11. A. Najid, J.L. Beneytout, J.P. Leblanc, M. Tixier, and M. Rigaud, Evidence for arachidonic acid C-5, C-8 and C-15 lipoxygenase in Lupinus albus seeds, *Biochim. Biophys. Acta* 960:26-34 (1988).
12. D. Heydeck, R. Wiesner, H. Kühn, and T. Schewe, On the reaction of wheat lipoxygenase with arachidonic acid and its oxygenated derivatives, *Biomed. Biochim. Acta* 50:11-15 (1991).
13. P.G.M. van Aarle, M.M.J. de Barse, G.A. Veldink, and J.F.G. Vliegenthart, Purification of a lipoxygenase from ungerminated barley. Characterization and product formation, *FEBS Lett.* 280:159-162 (1991).
14. D. Regdel, H. Kühn, and T. Schewe, On the reaction specificity of the lipoxygenase from tomato fruits, *Biochim. Biophys. Acta* 1210:297-302 (1994).
15. E. Mulliez, J.P. Leblanc, J.J, Girard, M. Rigaud, and J.C. Chottard, 5-Lipoxygenase from potato tubers. Improved purification and physiological characteristics, *Biochim. Biophys. Acta* 916:13-23 (1987).
16. I. Feussner and H. Kühn, The lipid body lipoxygenase from cucumber seedlings exhibits unusual reaction specificity, *FEBS Lett.* 367:12-14 (1995).
17. H.W. Gardner, Recent investigations into the lipoxygenase pathway of plants, *Biochim. Biophys. Acta* 1084:221-239 (1991).
18. A.R. Brash, S.W. Baertschi and T.M. Harris, Formation of prostaglandin A analogues via an allene oxide, *J. Biol. Chem.* 265(12):6705-6712 (1990)
19. W.H. Gerwick, M.A. Roberts, A. Vulpanovici and D.L. Ballentine, Biogenesis and biological functions of marine algal oxylipins, *Int. Conference on Lipoxygenases and Their Products: Biological Functions,* Malta, May 21-24, 1997
20. K.P.C. Croft, F. Jüttner, and A.J. Slusarenko, Volatile products of the lipoxygenase pathway evolved from Phaseolus vulgaris (L.) leaves inoculated with Pseudomonas syringae pv phaseolicola, *Plant Physiol.* 101:13-24 (1993).
21. T. Schewe and H. Kühn, Do 15-lipoxygenases have a common biological role? *Trends Biochem. Sci.* 16: 369-372 (1991).
22. H. Kühn and I. Feussner, A lipoxygenase is implicated in lipid mobilisation during germination of cucumber seeds, *Prostaglandins Leukotrienes Essential Fatty Acids* 55(Suppl.1):58 (1996).

THE PRODUCTION OF EICOSANOID PRECURSORS BY MUCORALEAN FUNGI

Alfred Botha [1], J. Lodewyk F. Kock [1], and Santosh Nigam [2]

[1] Dept. of Microbiology and Biochemistry, University of the Free State, PO Box 339, Bloemfontein 9300, South Africa
[2] Eicosanoid Research Div., Dept. of Gynaecology, Univ.- Klinikum Benjamin Franklin, Free University Berlin, D-12200 Berlin, Germany

INTRODUCTION

Eicosanoid precursors for commercial utilization are currently obtained predominantly from animal sources (Ratledge, 1994). However, the mucoralean fungi turned out to be an alternative source (Shimizu and Yamada, 1989; Ratledge, 1994). These fungi are more suitable for the production of large quantities of eicosanoid precursors, such as arachidonic acid [20:4(ω6)], dihomo-gamma-linolenic acid [20:3(ω6)], eicosapentaenoic acid [20:5(ω3)] and gamma-linolenic acid [18:3(ω6)]. In order to develop commercial biotechnological processes, it is necessary that favourable fungal strains are selected and that these are screened for the production of eicosanoid precursors from more affordable substrates. The present study reviews the pivotal role of mucoralean fungi in the production of eicosanoid precursors.

ISOLATION OF MUCORALEAN FUNGI

It was found that most mucoralean fungi are saprophytic and occur readily in soil, leaf litter and dung (Alexander, 1961; Hesseltine and Ellis, 1973). In the past, these fungi were isolated from their respective habitats utilising relative non-selective media which allows for growth of many non-mucoralean fungi (Embree, 1963; Eicker, 1974; Steiman et al., 1995). However, more selective media containing anti-fungal agents such as ketoconazole and benzimidazole fungicides have recently been employed (Bartshi et al., 1991; Botha et al., 1995). A complex medium, containing benzimidazole and sodium acetate as carbon source, was used to isolate members of the genus *Mucor*, known for the production of significant quantities of 18:3(ω6), from soil and leaf litter (Botha et al., 1995).

PRODUCTION OF EICOSANOID PRECURSORS

In order to produce biomass containing eicosanoid precursors, purified favorable fungal strains were selected and cultivated in liquid media as described (Tsuchiura and Sakura, 1988;

Shimizu and Yamada 1989; Sajbidor et al., 1990; Bajpai and Bajpai, 1993; Du Preez et al., 1995; Du Preez et al., 1997). Since it is known that a high carbon:nitrogen ratio (C:N ratio) favours lipid production, including that of eicosanoid precursors (Hansson and Dostalek, 1988; Sajbidor et al., 1990), the C:N ratio in these liquid media were chosen in the range from 10:1 to 80:1. Moreover, the media contained a sulphur source, minerals and growth factors. It was found that low incubation temperatures of *circa* 13°C stimulate the production of eicosanoid precursors such as 20:5(ω3) in *Mortierella* (Sajbidor et al., 1990). However, to produce 20:4(ω6), 18:3(ω6) and 20:3(ω6) from representatives of *Mortierella* and *Mucor*, sufficiently aerated cultures were preferably incubated at temperatures ranging from 20°C to 30°C (Hansson and Dostalek, 1988; Ratledge, 1994; Du Preez et al., 1995; Du Preez et al., 1997).

Using the above mentioned strategies, substantial quantities of eicosanoid precursors have been obtained by various authors using mostly glucose as carbon source (Table 1). Recently, however, it was discovered that mucoralean fungi can grow and produce eicosanoid precursors such as 18:3(ω6) on effluents from the petrochemical industry containing acetic acid as carbon source (Kock and Botha, 1993). This was achieved by continuous feeding of the effluent at subtoxic levels to the fungus. The concentration of 18:3(ω6) obtained by this process was 0.45 $g.l^{-1}$, which was comparable to that obtained with the more expensive carbon source, glucose (0.57 $g.l^{-1}$) (Kock and Botha, 1995).

Table 1. Production of eicosanoid precursors by mucoralean fungi[1]

Species	Carbon source used	Eicosanoid precursor	Percentage in the lipid fraction
Mucor circinelloides	Glucose	18:3(ω6)	18
Mortierella isabellina	Glucose	18:3(ω6)	8
Mortierella alpina	Glucose (+ sesame oil)[2]	20:3(ω6)	23
	Glucose	20:4(ω6)	79
	Glucose	20:5(ω3)	20

[1] Data taken from a review by Ratledge (1991)
[2] The non-oil fraction of sesame oil, which is added to the medium, suppresses the conversion of 20:3(ω6) to 20:4(ω6), thus resulting in an accumulation of 20:3(ω6) in the fungal lipids (Shimizu and Yamada, 1989).

The prospects to develop a commercially exploitable process for the production of 18:3(ω6) by the use of mucoralean fungi, were tremendously enhanced by the observation, that in the presence of 10 $g.l^{-1}$ sodium acetate, *Mucor circinelloides* produced 18:3(ω6) in a medium which contained sunflower oil as sole carbon source (Jeffery et al., 1997). With this process the content of 18:3(ω6) in relation to other C-16 and C-18 fatty acids, was raised from 0.4% in the oil fed to the fungus, to 4.6% in the neutral lipids of the fungus.

FUTURE WORK

Inasmuch as mucoralean fungi are strong, prolific producers of eicosanoid precursors, and that eicosanoids have been found in other fungi (Kock et al., 1991; Van Dyk et al., 1991), it appeared reasonable to us to screen for eicosanoids in the genus *Mucor*.

REFERENCES

Alexander, M., 1961, Fungi, in: *Introduction to Soil Microbiology*, M. Alexander, ed., John Wiley and sons, Inc. New York.

Bajpai, P., and Bajpai, P.K., 1993, Eicosapentaenoic acid (EPA) production from microorganisms: a review, *J. Biotechnol.* 30:161.

Bartshi, C., Berthier, J., Guiguettaz, C., and Valla, G., 1991, A selective medium for the isolation and enumeration of *Mucor* species, *Mycol. Res.* 95:373.

Botha, A., Kock, J.L.F., Roux, C., Coetzee, D.J., and Botes, P.J., 1995, An isolation medium for gamma-linolenic acid producing mucoralean fungi, *System. Appl. Microbiol.* 18:448.

Du Preez, J.C., Immelman, M., Kock, J.L.F., and Killian, S.G., 1995, Production of gamma-linolenic acid by *Mucor circinelloides* and *Mucor rouxii* with acetic acid as carbon substrate, *Biotechnol. Lett.* 17: 933.

Du Preez, J.C., Immelman, M., Kock, J.L.F,. and Killian, S.G., 1997, The effect of acetic acid concentration on the growth and production of gamma-linolenic acid by *Mucor circinelloides* CBS203.28 in fed-batch culture, *World J. Microbiol. Biotechnol.* 13:81.

Eicker, A., 1974, The mycoflora of an alkaline soil of the Open-Savannah of the Transvaal, *Trans. Brit. Mycol. Soc.* 63:281.

Embree, R.W., 1963, Observations on *Mortierella vesiculosa, Trans. Brit. Mycol. Soc.* 46:560.

Hansson, L., and Dostalek, M., 1988, Effect of culture conditions on mycelial growth and production of gamma-linolenic acid by the fungus *Mortierella ramanniana, Appl. Microbiol. Biotechnol.* 28:240.

Hesseltine, C.W., and Ellis, J.J., 1973, Chapter 11, Mucorales, in: *The Fungi, an Advanced Treatise* vol. 4B, G.C. Ainsworth, F.K. Sparrow and A.S. Sussman, eds., Academic Press, London.

Jeffery, J., Kock, J.L.F., Botha, A., Coetzee, D.J., Botes, P.J., and Nigam, S., 1997, Enhanced sunflower oil utilisation and gamma-linolenic acid production by *Mucor circinelloides* f. *circinelloides* CBS108.16 in the presence of acetate, *World J. Microbiol. Biotechnol.* (Accepted for publication).

Kock, J.L.F., and Botha, A., 1993, Acetic acid - a novel source for the production of gamma-linolenic acid and cocoa butter equivalents, *S.Afr. J. Sci.* 89:465.

Kock, J.L.F., and Botha, A., 1995, Biological treatment and cultivation of microorganisms, *United States Patent* No. 5,429,942.

Kock, J.L.F., Coetzee, D.J., Van Dyk, M.S., Truscott, M. Cloete, F.G., Van Wyk, V., and Augustyn, O.P.H., 1991, Evidence for pharmacologically active prostaglandins in yeasts, *S.Afr. J. Sci.* 87:73.

Ratledge, C., 1991, Microorganisms for lipids, *Acta Biotechnol.* 11:429.

Ratledge, C., 1994, Yeasts, molds, algae and bacteria as sources of lipids, in: *Technological Advances in Improved and Alternative Sources of Lipids*, B.S. Kamel and Y. Kakuda, eds., Blackie academic and professional, London.

Sajbidor, J., Dobronova, S., and Certik, M., 1990, Arachidonic acid production by *Mortierella* sp. S-17, influence of C/N ratio, *Biotechnol. Lett.* 12:455.

Shimizu, S., and Yamada, H., 1989, Microbial production of polyunsaturated fatty acids (vitamin - F group), in: *Biotechnology of Vitamins, Pigments and Growth Factors*, E.J. Vandamme, ed., Elsevier Applied Science, London.

Steiman, R., Guiraud, P., Sage, L., Seigle-Murandi, F., and Lafond, J-L., 1995, Mycoflora of soil around the Dead Sea, I - Ascomycetes (including *Aspergillus* and *Penicillium*), Basidiomycetes, Zygomycetes, *System. Appl. Microbiol.* 18:310.

Tsuchiura, O.S.,and Sakura, T.Y., 1988, Method for the preparation of a fungal body and a lipid rich in gamma-linolenic acid therefrom, *United States Patent* No. 4,783,408.

Van Dyk, M.S., Kock, J.L.F., Coetzee, D.J., Augustyn, O.P.H., and Nigam, S., 1991, Isolation of a novel arachidonic acid metabolite 3-hydroxy 5,8,11,14-eicosatetraenoic acid (3-HETE) from the yeast *Dipodascopsis uninucleata* UOFS-Y128, *FEBS Lett.* 283:195.

BIOTRANSFORMATION OF ARACHIDONIC ACID AND OTHER EICOSANOIDS BY THE YEAST *DIPODASCOPSIS UNINUCLEATA*, THE OOMYCETE FUNGI *SAPROLEGNIA DICLINA* AND *LEPTOMITUS LACTEUS* AND THE ZYGOMYCETE FUNGUS *MORTIERELLA ISABELLINA*

Arzu Akpinar, Simon R. Fox, Colin Ratledge and John Friend

Department of Biological Sciences,
University of Hull
Hull, HU6 7RX, UK

INTRODUCTION

Dipodascopsis uninucleata, a lipomycetaceous mycelial yeast with a well-defined life cycle, synthesises small amounts of arachidonic acid (ARA) and also produces the novel pharmacologically active metabolite, 3-hydroxyarachidonic acid (3-hydroxyeicosatetraenoic acid, 3-HETE) from exogenous ARA during its sexual phase (van Dyk *et al*, 1991; Kock *et al*, 1997). The production of 3-HETE from exogenous ARA and both ascosporogenesis and ascospore production in *D. uninucleata* were all inhibited by 1 mM aspirin (van Dyk *et al*, 1991; Botha *et al,* 1992; Coetzee *et al*, 1992).

We have developed an improved assay for the bioconversion of exogenous ARA to 3-HETE by *D. uninucleata* and have used it to show 3-hydroxylation of other C20 acids.

Three other fungi we have examined converted exogenous ARA to a range of mono-, di-, and tri-hydroxy derivatives of arachidonic acid, apparently using lipoxygenase (LOX) or cytochrome P450 to catalyse the reactions; in each case no 3-HETE was formed.

Cell-free extracts of *D. uninucleata* have no detectable PGHS activity, a possible catalyst for bioconversion of ARA to 3-HETE, and only low levels of LOX. There was no evidence for a specific 3-hydroxylase system.

We now have evidence to support the hypothesis that 3-HETE is formed by a modified β-oxidation process akin to that found in rat liver mitochondria (Jin *et al*, 1992), but with the chirality of *Saccharomyces* β-oxidation (Hiltunen *et al*, 1992). The β-oxidation system in *D. uninucleata* is constitutive; we are presently attempting to determine whether it is mitochondrial or peroxisomal.

LARGE-SCALE PRODUCTION OF 3-HETE BY *D. UNINUCLEATA*

"Scale-up" of the production of 3-HETE was successfully achieved in 10 litre stirred-tank fermenters with batch cultivation of 4 litres of medium for 46 hours. However, as the down-stream processing of the harvested cells posed technical problems of large-scale lipid extraction, we have preferred to rely on shake-flask cultures using 400 ml of medium in 1-litre flasks.

We have now developed a more reliable method for estimating 3-HETE biosynthesis, by incubating cultures grown for 46 hours with [1-^{14}C]ARA for 1 hour and determining the radioactivity incorporated into 3-HETE. A maximal yield of 3-HETE from 200 μM ARA of 1.9% in 1 hour, using medium buffered to pH 7 has been obtained (Fox *et al*, 1997).

BIOTRANSFORMATION OF OTHER C_{20} ACIDS BY *D. UNINUCLEATA*

Yeast cells were grown for 40-48 hours, washed, resuspended in buffer and incubated for 6 or 8 hours with ^{14}-C-labelled substrate, extracted and analysed for incorporation of ^{14}C by either densitometry of TLC plates or counting collected HPLC fractions. Structures of products were determined by GC-MS of incubation products from cold substrate. The other C_{20} acids examined, *cis* 5,8,11,14,17-20:4 (EPA), *cis* 8, 11, 14-20:3 and *cis* 11, 14, 17-20:3 were also converted to their respective 3-hydroxy derivatives: 3-hydroxy-*cis* 5,8,11,14,17-20:4 (3-HEPE), 3-hydroxy-*cis* 8, 11, 14-20:3 (3-HETrE, 8,11,14) and 3-hydroxy-*cis* 11, 14, 17-20:3 (3-HETrE, 11, 14, 17). Our results therefore differ from those of Kock *et al* (1996) who reported that *D. uninucleata* only 3-hydroxylated substrates containing a *cis*-5 double bond.

Using identical conditions to the optimal ones described above for 3-HETE production, the yield of 3-HEPE from [1- ^{14}C]EPA was 1.6% in 1 hour.

BIOTRANSFORMATION OF ARA BY OTHER FUNGI

We have also determined the products of ARA bioconversion by fungi that are known to contain larger amounts of arachidonic acid in their lipids than *D. uninucleata*. Two of them, *Saprolegnia diclina* and *Leptomitus lacteus*, were Oomycetes and the third, *Mortierella isabellina*, a Zygomycete. The results are shown in Table 1.

The three trihydroxy-eicosatrienoic acids produced by *S. diclina* have been previously reported as products of a lipoxygenase with hydroperoxide isomerase activity (LOX-HPI) from *S. parasitica* (Hamberg, 1986; Hamberg *et al*, 1986). The tested inhibitors are all well-known LOX inhibitors.

The three major HETEs from the *L. lacteus* incubation, which are ω-oxidation products of ARA, have previously been found to be formed from exogenous ARA by both mycelium and a cytosolic fraction of *Gaeumannomyces graminis* (Brodowsky & Oliw, 1992). Cytochrome P_{450}, the usual catalyst for ω-oxidation, is inhibited by carbon monoxide.

POSSIBLE MECHANISMS OF 3-HETE PRODUCTION BY INTACT CELLS OF *D. UNINUCLEATA*

Cytochrome P_{450}, LOX, COX and β-oxidation are four enzyme systems known to

Table 1 Products of ARA bioconversion by 3 fungi, grown for 48 hours in a medium containing 1% glucose and defatted yeast extract. Mycelium was resuspended in pH 7.5 buffer and incubated with [1-^{14}C]ARA for 1 hour; the products extracted from the medium were identified by TLC, HPLC and GC-MS

FUNGUS & EFFECTS OF INHIBITORS	PRODUCTS
Saprolegnia diclina Formation of all products inhibited by NDGA, gossypol and aesculin, each at 1 mM and by ETYA at 60 mM	11,12,15-trihydoxy-5,8,13-20:3 (11,12,15-THETrE, 5,8,13) 11,14,15-trihydoxy-5,8,12-20:3 (11,14,15-THETrE, 5,8,12) 13,14,15-trihydoxy-5,8,11-20:3 (13,14,15-THETrE, 5,8,11)
Leptomitus lacteus Formation of all products inhibited by carbon monoxide	Major products:- 17-hydroxy-arachidonic acid (17-HETE) 18-hydroxy-arachidonic acid (18-HETE) 19-hydroxy-arachidonic acid (19-HETE) Trace amounts of:- 9-hydroxy-arachidonic acid (9-HETE)
Mortierella isabellina 1% glucose in resuspension medium Formation of all products inhibited by carbon monoxide (10 min)	11,12,17-trihydroxy-5,8,14-20:3 (11,12,17-THETrE, 5,8,14) 11,12,18-trihydroxy-5,8,14-20:3 (11,12,18-THETrE, 5,8,14) 11,12,19-trihydroxy-5,8,14-20:3 (11,12,19-THETrE, 5,8,14)
Mortierella isabellina Glucose-free resuspension medium Carbon monoxide did not inhibit formation of these products	15-hydroxy-5,8,11,13-20:4 (15-HETE, 5,8,11,13) 11,12-dihydroxy-5,8,14-20:3 (11,12-diHETrE, 5,8,14) 13,14-dihydroxy-5,8,11-20:3 (13,14-diHETrE, 5,8,14)

introduce hydroxyl groups into polyunsaturated fatty acids and which thus might be involved in the bioconversion of ARA to 3-HETE.

We have tested the effects of possible inhibitory compounds or reaction conditions. Anaerobiosis gave complete inhibition, indicating a requirement for O_2. Carbon monoxide (10 min) had no effect, indicating that a cytochrome P450 was not involved. Both aspirin and indomethacin were inhibitory at 1 mM, indicating the possible involvement of LOX or COX (Sircar *et al*, 1983). LOX and COX products normally have rearrangements to their double bond systems which are not found in 3-HETE, although two cases are known where COX or dioxygenase products retain their methylene-interrupted double bonds; the formation of 13-hydroxy-5,8,11-eicosatrienoic acid (13-HETrE) from Mead acid by seminal vesicle COX + Ca (Elliot *et al*, 1986) and of 8-hydroxy-9,12-octadecadienoic acid (8-HODE) from linoleic acid by the linoleate 8-dioxygenase plus peroxidase from *G. graminis* (Brodowsky, Hamberg & Oliw, 1992) We have been unable to detect any putative LOX intermediate such as 3-hydroperoxy-ETE, possibly due to hydroperoxide lability. Alternatively, a non-LOX free radical intermediate could be involved in the formation of 3-HETE since both aspirin and indomethacin are free radical scavengers (Ellis *et al*, 1993).

We now believe that a partial β-oxidation somewhat similar to that found in rat liver (Jin *et al*, 1992) causes 3-hydroxylation of C20 acids by *D. uninucleata*. The chirality of 3-HETE was determined by Dr M. Hamberg as 95%R and 5%S. *Saccharomyces* forms the *R*-enantiomer of 3-hydroxy-acyl CoA during peroxisomal β-oxidation (Hiltunen *et al*, 1992). After supplementation with CoA, ATP, NAD and Mg, cell-free extracts of *D. uninucleata* catalysed both formation of 3-HETE and "classical" β-oxidation of ARA but not the "reductive" β-oxidation of ARA (Luthria *et al*, 1995), found only in whole cells. The "classical" β-oxidation system is constitutive; we are now determining whether it is mitochondrial or peroxisomal.

Acknowledgments

We thank the BBSRC for the award of a grant and the University of Uludag, Bursa, Turkey for the award of a scholarship to A.A.

REFERENCES

Botha A, Kock, JLF, Coetzee DJ, Linde NA & van Dyk MS (1992) The influence of NSAIDs on the life cycle of *Dipodascopsis uninucleata*. *System Appl Microbiol*, 15: 155-160

Brodowsky ID & Oliw EH (1992) Metabolism of 18:2, 18:3, 20:4 and 20:5 by the fungus *Gaeumannomyces graminis*. *Biochim Biophys Acta*, 1124, 59-65

Brodowsky ID, Hamberg M & Oliw EH (1992) A linoleic (8*R*) dioxygenase and hydroperoxide isomerase of the fungus *Gaeumannomyces graminis*. *J Biol Chem*, 267: 14738-14745

Coetzee DJ, Kock JLF, Botha A, van Dyk MS, Smit EJ, Botes PJ & Augustyn OPH (1992) Yeast Eicosanoids III. The distribution of arachidonic acid metabolites in the life-cycle of *Dipodascopsis uninucleata*. *System Appl Microbiol*, 15: 311-318

Elliot WJ, Morrison AR, Sprecher H & Needleman P (1986) Calcium-dependent oxidation of 5,8,11-icosatrienoic acid by the cyclooxygenase enzyme system. *J Biol Chem*, 261: 6719-6724

Ellis JS, Keenan PJ, Rathmell WG & Friend J (1993) Inhibition of phytoalexin accumulation in potato tuber discs by superoxide scavengers. *Phytochemistry*, 34: 649-655

Fox SR, Ratledge C & Friend J (1997) Optimisation of 3-hydroxyeicosanoid biosynthesis by the yeast *Dipodascopsis uninucleata*. *Biotechnology Letters*, 19, 155-158

Hamberg, M (1986) Isolation and structures of lipoxygenase products from *Saprolegnia parasitica*. *Biochim Biophys Acta*, 876: 688-692

Hamberg M, Herman CA & Herman P (1986) Novel biological transformations of 15- *L*-hydroperoxy-5,8,11,13-eicosatetraenoic acid. *Biochim Biophys Acta*, 877: 447-457

Hiltunen JK, Wenzel B, Beyer A, Erdmann R, Fossa A & Kunau W-H (1992) Peroxisomal multifunctional β-oxidation protein of *Saccharomyces cerevisiae*. *J Biol Chem*, 267: 6646-6653

Jin S-J, Hoppel CL & Tserng K-Y (1992) Incomplete fatty oxidation - the production and epimerisation of 3-hydroxy fatty acids. *J Biol Chem*, 267: 119-125

Kock JLF, van Vuuren J, Botha A, van Dyk MS, Coetzee DJ, Botes PJ, Shaw N, Friend J, Ratledge C, Roberts AD & Nigam S (1997) The production of biologically active 3-HETE and linoleic acid metabolites by *Dipodascopsis*. *System Appl Microbiol*, 20: 39-49.

Luthria DL, Baykousheva SP & Sprecher H (1995) Double bond removal from odd-numbered carbons during peroxisomal β-oxidation of arachidonic acid. *J Biol Chem*, 270: 13771-13778

Sircar JC, Schwender CF & Johnson EA (1983) Soybean lipoxygenase inhibition by nonsteroidal antiinflammatory drugs. *Prostaglandins*, 25: 393-396

van Dyk MS, Kock JLF, Coetzee DJ, Augustyn OPH & Nigam S (1991) Isolation of a new aspirin-sensitive arachidonic acid metabolite 3-HETE from the yeast *Dipodascopsis uninucleata*. *FEBS Lett*, 283: 195-198

HUMAN PROSTACYCLIN RECEPTOR: CLONING AND CO-EXPRESSION WITH EP3 PROSTAGLANDIN RECEPTOR

Sara Ortiz-Vega and Barrie Ashby

Department of Pharmacology
Temple University School of Medicine
Philadelphia, PA 19140

INTRODUCTION

A variety of agonists are able to activate platelets, including thromboxane A_2 and thrombin. Other agents, such as prostaglandins, inhibit platelet aggregation and secretion. All of these responses are mediated through specific receptors on the platelet surface. The primary event coupled with the action of stimulatory agonists is the activation of phospholipase C, whereas activation of adenyl cyclase is the primary event leading to inhibition of platelet activation. The interest of our laboratory is prostaglandin regulation of cAMP metabolism.

Prostaglandins that stimulate platelet adenyl cyclase include PGE_1, PGD_2, and PGI_2 (prostacyclin). Prostacyclin is the most potent inhibitor of platelet aggregation, and is the most physiologically relevant because it is synthesized by endothelial cells.

Based on kinetic studies of cAMP formation, we have postulated that platelet cAMP metabolism is controlled through separate stimulatory and inhibitory prostaglandin receptors[1,2]. In platelets, low concentrations of Iloprost, a stable analog of prostacyclin, stimulated formation of cAMP with no desensitization. Higher concentrations caused an increase in rate of cAMP formation and an increase in time-dependent inhibition (desensitization) with a different concentration dependence, indicating that stimulation and inhibition of cAMP formation are mediated through distinct receptors. Plotting the level of cAMP attained once desensitization was complete (after 5 minutes) as a function of Iloprost concentration gives a bell-shaped dose-response curve.

There are different receptors specific for each of the naturally occurring prostaglandins, which can be pharmacologically divided into subtypes that differ in their potency toward prostaglandins and in their signal transduction pathways[3]. Some of the receptors (IP, EP2, EP4, DP receptors) stimulate adenyl cyclase, some of them mobilize calcium (EP1, FP, TP), and some inhibit adenyl cyclase (EP3 and its isoforms). In order to provide evidence for a model involving co-localized receptors, we cloned the cDNA for the IP receptor which couples to stimulation of adenyl cyclase and co-expressed it with the EP3 receptor which couples to inhibition of the enzyme.

METHODS

A polymerase chain reaction product from human erythroleukemia (HEL) cell total RNA was prepared using primers based on the sequence of the mouse IP receptor[4] and used to screen a HEL cell cDNA library. We isolated two identical clones which were 100% homologous to the human IP receptor[5,6]. The full length clone was subcloned into the vector pcDNA3 (Invitrogen) and transfected into CHO cells. Clones expressing the IP receptor were selected using Geneticin (GIBCO BRL).

The EP3 receptor was cloned as previously described[7]. The receptor corresponded to isoform I of the EP3 receptor[8]. For some of the work, we used isoform II[8], prepared by reverse transcriptase PCR of HEL cell total RNA to obtain the C-terminus of isoform II which was exchanged for the corresponding region of isoform I.

The EP3 receptor was subcloned into the vector pCEP4 (Invitrogen), which contains a hygromycin B selection marker and transfected into cells containing the IP receptor. Clones expressing IP and EP3 receptors were selected with Geneticin and hygromycin B.

In parallel studies with platelets, the presence of the EP3 receptor was characterized by use of the EP3 specific agonist sulprostone. Platelets were isolated as Platelet Rich Plasma from fresh human blood collected in 0.11 M citrate by centrifugation at 300 x g for 15 min.

For measurement of cAMP formation, cells were labeled with [^{3}H]adenine for 2h, washed and challenged with agonists for 30 min in the presence of the phosphodiesterase inhibitor, isobutyl methyl xanthine. cAMP was determined by the method of Salomon[9].

RESULTS

It is clear that platelets possess the IP receptor because prostacyclin causes a robust increase in cAMP leading to inhibition of platelet aggregation and secretion. To demonstrate the presence of a distinct inhibitory receptor we examined[10] the effect of PGE_2, which can bind to EP2, EP4 and EP3 receptors, and the EP3 specific agonist sulprostone on platelets . Iloprost, a potent agonist at the IP receptor, maximally stimulates adenyl cyclase but acts weakly at the EP3 receptor, whereas sulprostone acts exclusively at the EP3 receptor and PGE_2 acts preferentially at the EP3 receptor. Under these conditions PGE_2 can be observed to inhibit Iloprost-stimulated cAMP formation with a similar potency to sulprostone (Fig. 1A). We also examined the effect of these agents on CHO-K1 cells stably transfected with EP3 receptor (isoform II). Both PGE_2 and sulprostone were potent inhibitors of cAMP formation (Fig. 1B).

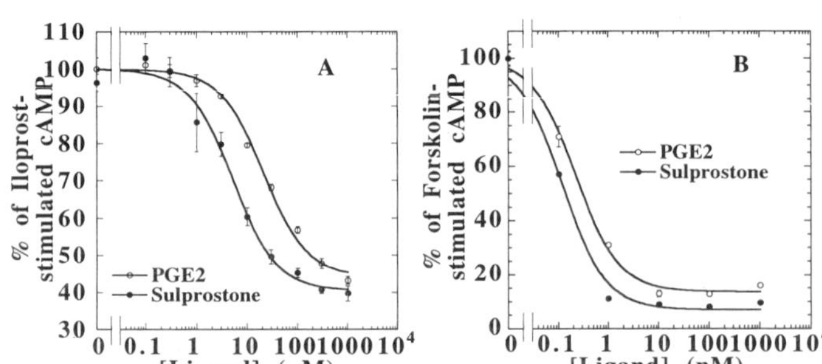

Figure 1. A. Effect of sulprostone (●) and PGE_2 (○) on cAMP formation in intact platelets. Platelets were isolated as Platelet Rich Plasma from fresh human blood. **B.** Effect of sulprostone (●) and PGE_2 (○) on cAMP formation in CHO-K1 cells expressing the EP3 receptor. Reprinted from reference 10 by permission of the publisher. Copyright 1996 by Elsevier Science Inc.

Iloprost stimulated cAMP production in CHO cells stably expressing IP receptor with an EC50 value of 0.1 nM (Fig. 2A). The solid represents a curve fit to a single class of receptor represented by the equation in the figure. In cells expressing both IP and EP3 (isoform I) receptors, Iloprost stimulated cAMP formation in a concentration dependent manner up to 1 nM where it reached a peak, and then cAMP production started to decrease until it reached a plateau around 100 nM (Fig. 2B). These results confirm that Iloprost can interact with a stimulatory and an inhibitory receptor. The solid line represents a curve fit to separate stimulatory and inhibitory receptors according to the equation in the figure.

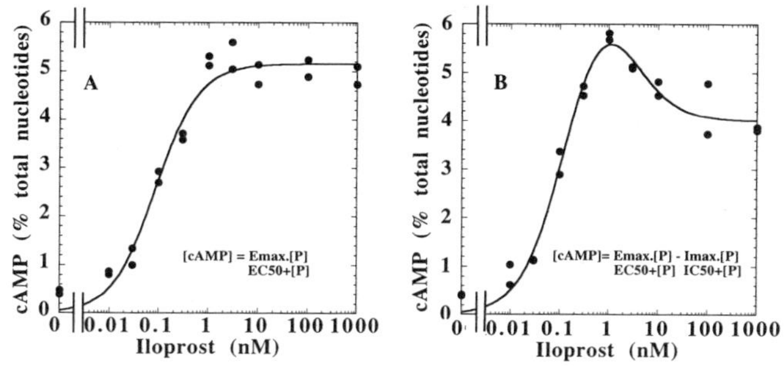

Figure 2. A . Iloprost-stimulated cAMP formation in CHO cells expressing human IP receptor. Cells expressing IP receptor alone gave a hyperbolic dose-response curve that was fit to the equation given in the figure where Emax = maximum level of cAMP formed; EC_{50} = concentration of prostaglandin, P, that gives half-maximal cAMP. **B.** Iloprost-stimulated cAMP formation in CHO cells expressing human IP receptor and EP3 (isoform I) receptor. Cells containing both receptors gave a biphasic, bell-shaped dose response curve that was fit to the equation given in the figure where Imax = maximum extent of inhibition of cAMP formation; IC_{50} = concentration of prostaglandin, P, that gives half-maximal inhibition of cAMP formation.

When CHO cells expressing both receptors were challenged with Iloprost in the presence of sulprostone, an agonist for the EP3 receptor, we obtained a hyperbolic curve, in contrast to the bell-shaped curve that is obtained with Iloprost alone (Fig. 3A). The change from bell-shaped to hyperbolic response is apparent because sulprostone does not cause complete inhibition of adenyl cyclase and Iloprost can stimulate to the inhibited level. Similar curves were obtained with intact platelets (Fig. 3B).

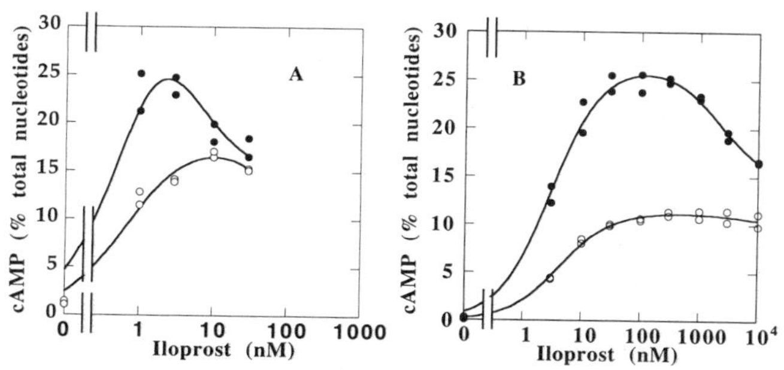

Figure 3. A. Iloprost dose-response curve of cAMP formation in CHO cells expressing both the IP and EP3 (isoform I) receptors in the absence (●) and presence (○) of 10 nM sulprostone. **B.** Iloprost dose-response curve of cAMP formation in platelets in the absence (●) and presence (○) of 10 nM sulprostone.

CONCLUSIONS

We have previously presented kinetic evidence for control of platelet cAMP metabolism through separate stimulatory and inhibitory prostaglandin receptors[1,2]. The current work demonstrates that co-expressed stimulatory and inhibitory receptors can function together to control cAMP.

Co-existence of stimulatory and inhibitory receptors for the same agonist may represent a general mechanism of autocrine or paracrine regulation, buffering cellular response to transient extremes of concentration when agonists are produced close to their sites of action. The mechanism would apply to circulating platelets responding to localized increases in prostaglandin synthesis caused by inflammatory agents. While circulating prostaglandins may be important in maintaining platelets in a non-thrombogenic state, the ability to buffer against excessive prostaglandin-stimulated cAMP formation would prevent platelets from becoming refractory to appropriate stimulatory challenges.

Acknowledgments

This work was supported by National Institutes of Health grant HL48114

REFERENCES

1. B. Ashby. Model of prostaglandin regulated cyclic AMP metabolism in platelets: examination of time-dependent effects on adenylate cyclase and phosphodiesterase activities. *Mol. Pharmacol.* 36: 866 (1989).
2. B. Ashby. Novel mechanism of heterologous desensitization of adenylate cyclase: Prostaglandins bind with different affinities to both stimulatory and inhibitory receptors on platelets. *Mol. Pharmacol.* 38:46 (1990).
3. R.A. Coleman, W.L. Smith, and S. Narumiya. *Pharmacol. Rev.* 46: 205 (1994).
4. T. Namba, H. Oida, Y. Sugimoto, A. Kakizuka, M. Negishi, A. Ichikawa, and S. Narumiya. cDNA cloning of a mouse prostacyclin receptor. *J. Biol. Chem.* 269: 9986 (1994).
5. M. Katsuyama, Y. Sugimoto, T.Namba, A. Irie, M. Negishi, S. Narumiya, and A. Ichikawa. Cloning and expression of a cDNA for the human prostacyclin receptor. *FEBS Letters* 344: 74 (1994).
6. Y. Boie, T.H. Rushmore, A. Darmon-Goodwin, R. Grygorcyzk, D.M.Slipetz, K.M. Metters, M. Abramovitz. Cloning and expression of a cDNA for the human prostanoid IP receptor. *J. Biol. Chem.* 269: 12173 (1994).
7. S.P. Kunapuli, G.F. Mao, M. Bastepe, L. Liu-Chen, S. Li, P.P. Cheung, J.K. DeRiel, and B. Ashby. Cloning and expression of a prostaglandin E receptor EP3 subtype from human erythroleukemia cells. *Biochem. J.* 298: 263 (1994).
8. S. An, J. Yang, S.W. So., L. Zheng, and E.J. Goetzl. Isoforms of the EP3 subtype of human prostaglandin E2 receptor transduce both intracellular calcium and cAMP signals. *Biochemistry* 33: 14496 (1994).
9. Y. Salomon. Adenylate cyclase assay. *Adv. Cyc. Nucl. Res.* 10: 35 (1979).
10. G.F. Mao, J.G. Jin, M. Bastepe, S. Ortiz-Vega , and B. Ashby. Prostaglandin E2 both Stimulates and Inhibits Adenyl Cyclase on Platelets: Comparison of Effects on Cloned EP4 and EP3 Prostaglandin Receptor Subtypes. *Prostaglandins*. 52: 175 (1996).

THREE ISOFORMS OF THE PROSTAGLANDIN E RECEPTOR EP3 SUBTYPE DIFFERENT IN AGONIST-INDEPENDENT CONSTITUTIVE Gi ACTIVITY AND AGONIST-DEPENDENT Gs ACTIVITY

Atsushi Ichikawa, Manabu Negishi and Hiroshi Hasegawa

Department of Physiological Chemistry, Faculty of Pharmaceutical Sciences, Kyoto University, Sakyo-ku, Kyoto 606, Japan

Prostaglandin (PG) E_2 produces a broad range of physiological and pharmacological actions in diverse tissues through specific receptors on plasma membranes for maintenance of local homeostasis in the body[1]. PGE receptors are pharmacologically divided into four subtypes, EP1, EP2, EP3 and EP4, on the basis of their responses to various agonists and antagonists[2]. Among these subtypes, the EP3 receptor has been most well characterized and suggested to be involved in the diverse actions of PGE_2, such as contraction of the uterus, and inhibition of gastric acid secretion, neurotransmitter release, lipolysis in adipose tissue, sodium and water reabsorption in kidney tubules, and modulation of catecholamine release from adrenal chromaffin cells[3]. Although EP3 receptor-mediated actions are believed to be mediated by inhibition of adenylate cyclase, the dose-response curve and potency of PGE_2 vary with the tissues, implying heterogeneity of EP3 receptors. Furthermore, some of the PGE_2 actions appears to be mediated by another second messenger pathway. These different properties of the EP3 imply heterogeneity of this type receptor. We have recently cloned three isoforms of the mouse EP3 receptor with different carboxy-terminal tails, which are produced through alternative splicing[4]. To assess the role of the carboxy-terminal tails of the EP3 receptor in coupling to G proteins, we constructed a mutated EP3 receptor, T-335, in which the carboxy-terminal tail was truncated at the alternative splicing site. We report here that the three isoforms differ in agonist-independent constitutive Gi activity and agonist-dependent Gs activity.

RESULTS AND DISCUSSION

We first examined Gi activities of three isoforms of mouse EP3 receptor and the truncated receptor, T-335, in CHO cell lines stably expressing each receptor[5,6]. As shown in Fig. 1A, forskolin increased the intracellular cAMP levels in the absence of agonist in the order of

Recent Advances in Prostaglandin, Thromboxane, and Leukotriene Research
Edited by Sinzinger *et al.*, Plenum Press, New York, 1998

EP3β > EP3α > EP3γ > T-335. An EP3 agonist, sulprostone, concentration-dependently inhibited the forskolin-stimulated cAMP formation in the three EP3 receptor isoforms. The IC_{50} values for the inhibition and the maximally reduced cAMP levels decreased in the order of the EP3γ (0.03 nM) > EP3α (0.1 nM) > EP3β (30 nM). On the other hand, the agonist-induced inhibition was not observed with T-335. Pertussis toxin (PT) treatment increased

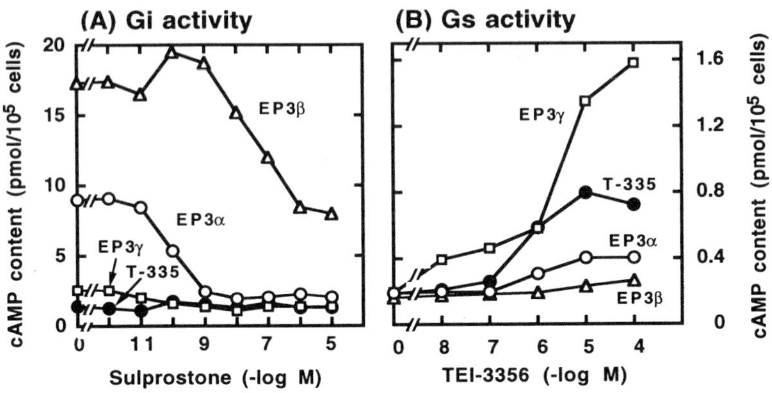

Figure 1. Gi and Gs activities of three EP3 receptor isoforms and the truncated receptor. A, CHO cells expressing EP3α (O), EP3β (Δ), EP3γ (□), or T-335 receptor (●) were incubated at 37°C for 10 min with 10 μM forskolin in the presence of the indicated concentrations of sulprostone. B, after cells expressing each receptor had been treated with 10 ng/ml PT for 5 h, they were incubated at 37°C for 10 min with the indicated concentrations of TEI-3356. cAMP contents were determined, and the values are means of triplicate experiments, which varied by less than 5%.

the forskolin-stimulated cAMP formation in the absence of the agonist in the order of T-335 > EP3γ > EP3α, but this treatment did not increase the cAMP level in the EP3β receptor. These results indicate that three isoforms differ in constitutive Gi activity; the EP3α receptor showed marked constitutive activity, the EP3γ receptor showed mostly full constitutive activity, and the truncated receptor was full constitutive, but the EP3β receptor had no constitutive activity. The different constitutive activities of these receptors on the stimulation of GTPase activity of Gi were also observed.

Lefkowitz and co-workers have proposed a two-state model in which receptors are in equilibrium between the inactive conformation and the active conformation that can associate and activate G protein, and classical agonists increase the concentration of the latter conformation of the receptors[7]. The three EP3 receptor isoforms differ only in the carboxy-terminal tail, showing different constitutive activity, and the truncated receptor exhibits full agonist-independent constitutive activity. This suggests that the carboxy-terminal tails after the alternative splicing site suppress activation of Gi by the EP3 receptor to different extents dependent on the structure of the carboxy-terminal tail; suppression by the tail of the EP3β receptor is complete, that by the tail of the EP3α receptor is partial, and the tail of the EP3γ receptor very weakly suppresses the Gi activation. In addition, the three isoforms differed in IC_{50} value for the agonist-mediated inhibition and the maximal inhibition response (Fig. 1A). According to the two-state model of receptor activation, the two parameters, IC_{50} and maximal response, are dependent on the level of constitutive activity of receptor. As the percentage of constitutive activity increases, the value of IC_{50} decreases and the level of maximal response increases[8]. Variation of the two parameters of three EP3 receptor isoforms are consistent with the diversity of constitutive Gi activities of the receptors. The EP3 receptor isoforms with different constitutive activity produce diverse agonist sensitivity.

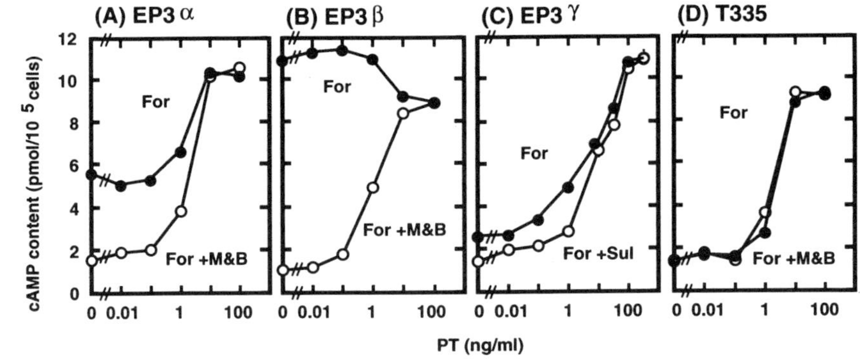

Figure 2. Effect of PT treatment on agonist-dependent or independent inhibition of adenylate cyclase by EP3 receptor isoforms or T-335. After cells expressing EP3α (A), EP3β (B), EP3γ (C), or T-335 (D) had been treated with the indicated concentrations of PT for 6 h, they were incubated at 37°C for 10 min with 10 μM forskolin in the presence (O) or absence (●) of 1 μM M&B 28767 (sulprostone for EP3γ), then cAMP contents were determined, and the values are means of triplicate experiments, which varied by less than 5%.

To assess the role of the carboxy-terminal tails in the coupling of the EP3 receptor to Gs, we next examined the Gs activities of the three EP3 receptor isoforms and T-335 receptor. To ablate the EP3 receptors coupling to Gi, the CHO cells were treated with PT, and we examined the Gs activities of the receptors. As shown in Fig. 1B, TEI-3356, a selective EP3 agonist, concentration-dependently increased the cAMP level in the cells expressing the three isoforms and the truncated receptor, the maximal levels increasing in the order of EP3γ > T-335 > EP3α > EP3β. The cAMP levels in the absence of the agonist in the cells expressing the receptors were the same as that in the mock-transfected cells (0.204 pmol/10^5 cells), indicating that the three isoforms and T-335 receptor have no constitutive Gs activity. The T-335 receptor has agonist-dependent Gs activity without the constitutive activity. Therefore, Gs activation is an intrinsic ability of the EP3 receptor, but the carboxy-terminal tail is not the domain for Gs activation. The Gs activity of the EP3γ receptor was higher than that of the truncated receptor, while the activities of the EP3α and EP3β receptors were lower than that of the T-335 receptor. Thus, the carboxy-terminal tail of the EP3γ receptor may enhance the Gs activity of the EP3 receptor, but the tails of the EP3α and EP3β receptors suppress the activity, suggesting that the carboxy-terminal tails modulate the Gs activity of the EP3 receptor.

In summary, we present here that EP3 receptor isoforms differ in agonist-independent constitutive Gi activity and in modulation of agonist-dependent Gs activity. This study will contribute not only to the understanding heterogeneity of PGE_2 actions but will also help to elucidate the molecular mechanism of G protein activation induced by receptors.

REFERENCES

1. B. Samuelsson, M. Goldyne, E. Granström, M. Hamberg, S. Hammarström, and C. Malmsten, Prostaglandins and thromboxanes. *Ann Rev Biochem.* 47:997-1029 (1978).
2. R.A. Coleman, I. Kennedy, P.P.A. Humphrey, K. Bunce, and P. Lumley, Prostanoids and their receptors, in: *Comprehensive Medicinal Chemistry,* C. Hansch, P.G. Sammes, J.B. Taylor, J.C. Emmett, ed., Pergamon Press, Oxford, 3:643-714 (1990).
3. M. Negishi, Y. Sugimoto, and A. Ichikawa, Molecular mechanisms of diverse actions of prostanoid receptors. *Biochim Biophys Acta.* 1259:109-120 (1995).
4. Y. Sugimoto, M. Negishi, Y. Hayashi, T. Namba, A. Honda, A. Watabe, M. Hirata, S. Narumiya, and A. Ichikawa, Two isoforms of the EP_3 receptor with different carboxyl-terminal domains. *J Biol Chem.* 268:2712-2718 (1993).
5. H. Hasegawa, M. Negishi, and A. Ichikawa, Two isoforms of the prostaglandin E receptor EP3 subtype different in agonist-independent constitutive activity. *J Biol Chem.* 271:1857-1860 (1996).
6. M. Negishi, H. Hasegawa, and A. Ichikawa, Prostaglandin E receptor EP3γ isoform, with mostly full constitutive Gi activity and agonist-dependent Gs activity. *FEBS Lett.* 386:165-168 (1996).
7. R.J. Lefkowitz, S. Cotecchia, P. Samama, and T. Costa, Constitutive activity of receptors coupled to guanine nucleotide regulatory proteins. *Trends Pharmacol Sci.* 14:303-307 (1993).
8. P. Leff, The two-state model of receptor activation. *Trends Pharmacol Sci.* 16:89-97 (1995).

NEUROGENIC INHIBITORY ACTION THROUGH PLATELETS OF PLATELET ACTIVATING FACTOR (PAF) ON VASOCONSTRICTIVE RESPONSIVENESS TO PRESSOR AGENTS

Tsutomu Michibayashi

Department of Laboratory Diagnosis, Faculty of Medicine,
Sapporo Medical University, Sapporo 060, Japan

INTRODUCTION

Platelet activating factor (PAF) is a ubiquitous bioactive phospholipid that is generated by different cell types including platelets (1) , basophils (2) , and endothelial cells (3, 4) .

PAF, as is well known, has a wide variety of biological activities, among which are activation of platelets (5) , hypotensive action (6–9) , and increase in vascular permeability (10, 11) . PAF is also involved in diverse pathological processes such as inflammatory disorders (10, 12) and hypoxia-induced tissue damage (4, 13) . In the present study, to elucidate the participation of PAF in the interaction between autologous platelets and a blood vessel wall, the influences of PAF on the vasoconstrictive responsiveness to pressor agents were examined in the absence and presence of platelets.

MATERIALS AND METHODS

A detailed description of materials and methods is described in references 14 and 15. Briefly, male Japanese white rabbits weighing 2.5 to 3.0 kg were anesthetized with sodium pentobarbital (30 to 60 mg / kg , i.p.) . Following injection of heparin (1000 U/ kg , i.v.) into the marginal ear vein, whole blood was obtained from the femoral artery with needle cannulation and collected into a beaker containing sodium citrate (3.8% w/v, 1 part + 9 parts of blood) . Platelet rich plasma (PRP) was obtained by centrifuging the whole blood at 1000 rpm (182 g) for 10 min. Subsequently, preparation of the perfused arterial segment was performed. This arterial segment was then perfused with a modified Krebs solution (14) . As regards the vasocontractile response to noradrenaline (NA−R) , a suitable concentration, 0.1 mℓ NA, was injected as a bolus into a rubber tube connected to the central arterial cannula.

PAF was then dissolved in 0.9% NaCl containing 0.25% bovine serum albumin (16) . In the experiment examining the effect of PRP with or without PAF or

Recent Advances in Prostaglandin, Thromboxane, and Leukotriene Research
Edited by Sinzinger *et al.*, Plenum Press, New York, 1998

collagen on NA−R, autologous PRP was infused into the perfusion circuit at the rate of 1 to 3 mℓ /hr with a micro−infusion pump (SP−5, Nipro, Japan) .

RESULTS AND DISCUSSION

First, the actions of PAF on NA−R in the absence or presence of platelets were compared (Fig.1) . As shown in Fig. 1−A, PAF was infused into the perfusion circuit at a constant flow rate of 3 mℓ /h by a micro−infusion pump. In this experiment, 10 ng of NA was repetitively injected as a bolus at intervals of 10 min. NA−R remained constant and was observered to be minimally influenced by PAF−infusion.

On the other hand, as seen in Fig. 1−B, PRP with PAF was found to cause biphasic responses to NA, accompanied by unstable elevation of basal perfusion

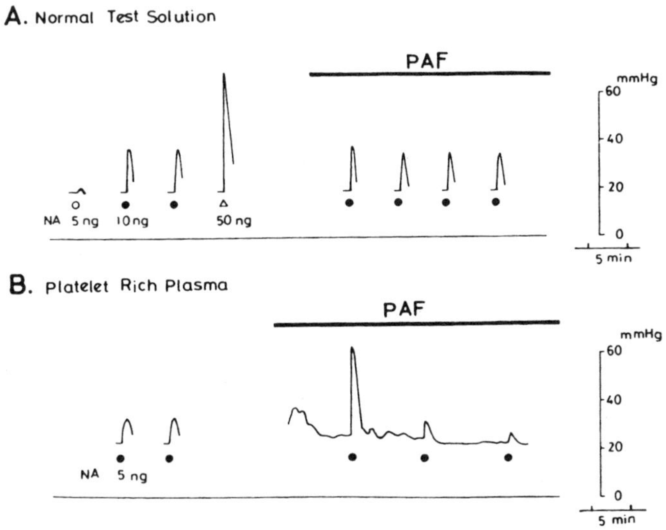

Figure 1. Comparison between actions of PAF on NA−R in the absence (A) and the presence (B) of platelets. In Fig. 1−A. PAF, 0.036 to 10 ng/ mℓ , was applied to the perfusion circuit. In Fig. 1−B. PAF, 50 ng/ mℓ , was added to the syringe containing PRP, and PRP with PAF was infused into the perfusion circuit at a constant flow rate of 3 mℓ /h.

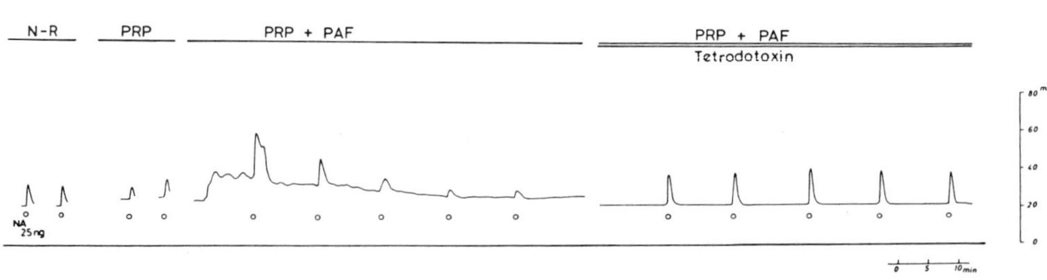

Figure 2. Reversed effect of TTX on the gradually attenuated responses to NA during infusion of PRP with PAF. After the gradually attenuated responses to NA were elicited by the infusion of PRP with PAF (50 ng/ mℓ) , TTX, 10^{-7} g/ mℓ , was applied to the perfusion circuit. Then the effect of PRP with PAF on NA−R was similarly examined and compared at 30 min after TTX perfusion.

pressure, an initially augmented response followed by gradually attenuated responses. In this preparation, platelet aggregation was induced by the addition of PAF, 50 ng/ mℓ , into the syringe containing PRP. PAF−treated PRP was infused into the perfusion system immediately after adding PAF into this syringe. Then 5 ng of NA was repeatedly applied.

Following this experiment, whether the action, through platelets, PAF is affected by neurogenic components of the present arterial preparation was further evaluated (Fig. 2) . Tetrodotoxin (TTX) was used for the purpose of excluding the neurogenic participation from the action of PAF on NA−R. TTX is well known to be a very specific blocker for Na spikes in nerve axon, perivascular autonomic nerves (17) , or skeletal muscle. Under the conditions in which PRP is infused with PAF (50 ng/ mℓ) , the response to 25 ng of NA shows initial elevation, accompanied by unstable elevation of basal perfusion pressure in comparison with the control response, followed by gradually attenuated responses, matching the results shown in Fig. 1−B. Subsequently, TTX, 10^{-7} g/ mℓ , was infused into the perfusion circuit in the presence of platelets with PAF. At 30 min after the biginning of TTX perfusion, 25 ng of NA was again injected as a bolus via connecting rubber tubing. Surprisingly, NA−R was found to be reversed to the level of the initial raised response. In addition, a previous report described that TTX does not show any restrictive action on NA−R during perfusion of normal Krebs' solution, although in an experiment using a perfused central artery of a rabbit ear, TTX completely blocked vasocontractile response to electrical stimulation of perivascular sympathetic nerves (17) . On the basis of these results, it seems reasonable to speculate that neurogenic inhibitory action due to some substances liberated from PAF−activated platelets is involved in the vasodepressive action through platelets of PAF.

Also, a few inhibitors were applied to the perfusion circuit with the aim evaluating that vasoactive substances liberated from activated platelets contribute to the mechanism of NA−R in the presence of platelets stimulated by PAF. Pretreatment with ketanserin (45 to 450 ng/ mℓ , a 5−HT_2 receptor antagonist) elicited a partial reduction of the intially−raised respose to NA (result not shown) , while the gradually−attenuated responses to NA were affected neither by indomethacin (3 μg / mℓ , a cyclo−oxygenase inhibitor) nor N^G −Nitro−L−Arginine (2.2 × 10^{-5} g/ mℓ , a NO synthetic inhibitor, Fig.3) .

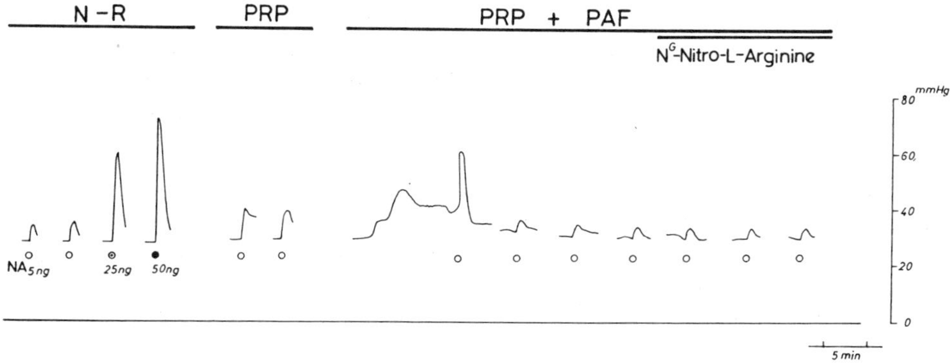

Figure 3. Effect on N^G −Nitro−L−Arginine on the gradually attenuated responses to NA during infusion of PRP with PAF. After the gradually attenuated responses to NA were observed, N^G − Nitro−L−Arginine, 2.2 × 10^{-5} g/ mℓ , was applied to the perfusion circuit.

Finally, to elucidate the participation of the adenylate cyclase−cyclic AMP system in the interaction between platelets activated by PAF and the blood vessel wall, dibutyryl cyclic AMP, 10^{-8} g/ mℓ , was applied to PRP in the syringe approximately 30 min in advance of its infusion into the perfusion circuit. Unexpectedly, dibutyryl cyclic AMP had no any influence on NA−R during infusion of PRP with PAF, in contrast with the marked suppression by dibutyryl cyclic AMP in the presence of PRP with collagen (result not shown) .

CONCLUSION

With the aim of elucidating the interaction between autologous platelets and blood vessel wall, the action of PAF on NA−R was examined using the perfused arterial segment in the absence and the presence of PRP with PAF. Under the condition of infusing platelets activated by PAF, NA−R was found to show the biphasic responses accompanied by unstable elevation of basal perfusion pressure, i.e., an initially augmented response followed by gradually attenuated responses. Since the former was partially reduced by ketanserin, a selective S2−serotonergic antagonist, 5−hydroxytrypramine, liberated from platelets activated by PAF, was estimated to be involved in this initial response. In addition, since the attenuated responses were reversed to the initial raised level by pretreated TTX, it seems likely that the neurogenic inhibitory action by some substances released from the PAF−stimulated Platelets may be involved in these attenuated responses.

Thus, it is possible to draw the conclusion that the inhibitory action through platelets of PAF on NA−R may be due to vasodepressive agents originating from perivascular autonomic nerves stimulated by unknown substances liberated from platelets activated by PAF.

REFERENCES

1. D.J. Hanahan, Platelet-activating factor isolation, identification, and assay. Meth. *Biochem. Analysis.* 31 : 195 (1985) .
2. R.N. Pinckard, R.S. Farr, and D.J. Hanahan, Physicochemical and functional identity of rabbit platelet−activating factor (PAF) released in vivo during IgE anaphylaxis with PAF released in vitro from IgE sensitized basophils, *J. Immunol.* 123 : 1847 (1979) .
3. G.A. Zimmerman, T. M. McIntyre, and S.M. Precott, Production of platelet−activating factor by human vascular endothelial cells : Evidence for a requirement for specific agonists and modulation by prostacyclin, *Circulation* 72 : 718 (1985)
4. M.S. Caplan, L. Adler, A. Kelly, and W. Hsueh, Hypoxia increases stimulus-induced PAF production and release from human umbilical vein endothelial cells, *Biochim. Biophys. Acta* 1128 : 205 (1992) .
5. B.B. Vargaftig, M. Chignard, and J. Benveniste, Commentary, Present concepts on the mechanisms of platelet aggregation, *Biochem. Pharmacol.* 30 : 263 (1981) .
6. M.L. Blank, F. Snyder, L.M. Byers, B. Brooks, and E.E. Muirhead, Antihyper−tensive activity of alkyl ether analog of phosphoryl-choline, *Biochem. Biophys. Res. Commun.* 90 : 1194 (1979) .
7. A. Otsuka, F. Masugi, T. Ogihara, S. Saeki, S.M. Nagano, Y. Koyama, Y. Tabuchi, and Y. Kumahara, Hypotensive mechanism of acetyl glyceryl ether phos-phorylcholine (AGEPC) in dogs. Effects on hemodynamics and humoral factors, *Prostaglandins Leukotrienes Med.* 19 : 25 (1985) .
8. P. Bessin, J. Bonnet, D. Apffel, C. Soulard, L. Desgroux, I. Pelas, and J. Benveniste, Acute circulatory collapse caused by platelet-activating factor (PAF−acether) in dogs, *Eur. J. Pharmacol.* 86 : 403 (1983) .

9. V. Lagente, Z.B. Fortes, J. Garcia–Leme, and B.B. Vargaftig, PAF–acether and endotoxin display similar effects on rat mesenteric microvesseles : inhibition by specific antagonists, *J. Pharmacol. Exper. Ther.* 247 : 254 (1988) .
10. S. Jancar, M.G. Sirois, J. Carrier, P. Braquet, and P. Sirois, PAF induces rat plasma extravasation and releases eicosanoids during anaphylaxis, Inflammation 15 : 347 (1991) .
11. N.F. Voelkel, S. Worthen, J.T. Reeves, P.M. Henson, and R.C. Marphy, Non–immunological production of leukotrienes induced by platelet–activating factor, *Science.* 218 : 286 (1982) .
12. P.V. Peplow, and D.P. Mikkailidis, Platelet-activating factor (PAF) and its relation to prostaglandins, leukotrienes and other aspects of arachidonate metabolism, *Prostaglandins Leukotrienes and Essential Fatty Acids* 41 : 71 (1990) .
13. A. Rizzo, A.E. Taylar, M.I. Townsley, P. Logsdon, P.L. Khimenko, and J.R. Fletcher, Platelet-activating factor increases vascular resistance in rat hind–quarters by thromboxane A_2 , *Am. J. Physiol.* 268 : H1954 (1995) .
14. T. Michibayashi, Vasocontractile responsiveness of perfused arterial segments to platelet–derived thromboxane A_2 , *J. Smooth Muscle Res.* 28 : 25 (1992) .
15. T. Michibayashi, Effect of the adenylate cyclase–cyclic AMP system in platelets on prostanoids metabolism and vasocontractile response, *J. Jpn. Coll. Angiol.* 34 : 535 (1994) .
16. G. Grigoriadis, and A.G. Stewart, Albumin inhibits platelet–activating factor (PAF) –induced responses in platelets and macrophages : implications for the biologically active form of PAF, *Br J. Pharmacol.* 107 : 73 (1992) .
17. T. Michibayashi, Inhibitory action of prostaglandin E on angiotensin II –induced enhancement of vasoconstrictor response to sympathetic nerve stimulation, *Jpn. Circul. J.* 47 : 1065 (1983) .

HIGH AND LOW AFFINITY MUTANTS OF PLATELET-ACTIVATING FACTOR RECEPTOR

Isao Ishii,[1,2] Takashi Izumi,[1] Michio Ui,[2] and Takao Shimizu[1]

[1]Department of Biochemistry and Molecular Biology, Faculty of Medicine, The University of Tokyo, Hongo 7-3-1, Bunkyo, Tokyo 113, Japan
[2]Ui Laboratory, The Institute of Physical and Chemical Research (RIKEN), Hirosawa 2-1, Wako, Saitama 351-01, Japan

INTRODUCTION

Platelet-activating factor (1-*O*-alkyl-2-acetyl-*sn*-glycero-3-phosphocholine; PAF) is a potent phospholipid mediator with diverse physiological actions in a wide variety of cells and tissues (1). PAF as well as histamine, prostaglandines and leukotrienes, is thought to play pivotal roles in allergic and inflammatory diseases. Despite the highly hydrophobic structure, PAF binds to a cell-surface, G protein-coupled, seven-transmembrane (TM)-type receptor, which was first cloned in our laboratory (2). So far, PAF receptor homologs in four mammalian species (guinea pig, human, rat, and mouse) have been cloned. Various PAF receptor antagonists have been developed as antiallergic and antiinflammatory drugs, but little is known of ligand binding sites in the cloned PAF receptors. Most G protein-coupled receptor agonists are considered to bind to a hydrophobic core surrounded by seven-TM α-helices with electrostatic and hydrophobic force. We performed alanine-scanning mutagenesis to replace all 23 polar amino acids individually in the putative TM domains of a guinea pig PAF receptor to search for amino acids involved in the recognition of PAF.

METHODS

Construction of Mutant Receptor cDNAs and Expression in COS-7 Cells

Twenty-three individual point mutations were introduced into pBluescript SK(-) (Stratagene) containing the entire coding region of a guinea pig PAF receptor cDNA (2). The Transformer Site-Directed Mutagenesis Kit (Clontech) was used to exchange polar amino acids with alanine. For construction of epitope-tagged receptors, an 8-amino acid peptide (FLAG) sequence (DYKDDDDK) was inserted between the N-terminal initiator methionine and the second amino acid of the wild type and mutant receptors by polymerase chain reaction, as described (3). The coding regions of all constructs were sequenced on both strands, and were subcloned into a mammalian expression vector pcDNA I (Invitrogen). COS-7 was maintained as a monolayer culture in Dulbecco's modified Eagle's medium supplemented with 10% fetal bovine serum (FBS) and antibiotics. Cells were plated, transfected with the constructs using TRANSFECTAM (BioSepra Inc., MA) according to the manufacturer's protocol, and used for assay after 72 hours.

Table 1. Ligand binding properties observed in COS-7 cells transiently expressing mutant PAF receptors.

TM	Receptor	Relative Specific Binding		Binding Ratio
		PAF	WEB 2086	PAF Binding / WEB 2086 Binding
		% of WT	% of WT	
	WT	100.0 ± 4.9	100.0 ±10.6	1.00
1st	S23A	163.6 ± 6.2	164.4 ± 3.0	1.00
	N33A	39.3 ±11.4	46.6 ± 2.9	0.84
2nd	N58A	28.8 ± 2.4	18.0 ± 1.4	1.60
	T60A	172.7 ±11.9	232.6 ±13.3	0.74
	D63A	94.9 ± 5.1	48.3 ± 2.9	1.96
	T69A	117.7 ± 7.6	28.5 ± 1.5	4.14
3rd	N100A	307.3 ±14.1	168.7 ±11.2	1.82
	T101A	240.0 ± 5.8	168.8 ± 7.4	1.42
	S104A	235.5 ±13.2	113.8 ± 6.2	2.07
	T112A	292.8 ± 5.2	265.1 ±11.2	1.10
4th	S138A	211.9 ±11.7	189.4 ± 7.2	1.12
	S150A	249.6 ±13.7	230.3 ±11.5	1.08
5th	H188A	1.3 ± 2.1	150.3 ± 9.0	0.01
6th	T236A	197.8 ±12.4	313.8 ±22.0	0.63
	H248A	-4.9 ± 1.6	101.0 ± 2.7	< 0
	H249A	12.4 ± 3.4	48.7 ± 2.0	0.26
	Q252A	61.4 ± 5.4	215.7 ± 8.3	0.29
7th	Q276A	1.8 ± 3.7	142.5 ± 4.0	0.01
	T278A	14.2 ± 3.7	258.5 ±10.6	0.06
	S283A	63.6 ± 2.7	224.2 ±10.6	0.28
	T284A	286.2 ±17.6	335.0 ±11.0	0.85
	N285A	273.5 ±14.7	244.2 ±10.2	1.12
	D289A	268.8 ±13.3	189.6 ± 7.0	1.42

Cells were incubated for 1 h at 25 ^{o}C with 3 nM [^{3}H]PAF C-16 and 20 nM [^{3}H]WEB 2086 in the presence (nonspecific) or absence (total) of 10 μM unlabeled WEB 2086. The specific binding (total minus nonspecific) was converted into percentage with that of the wild-type receptor as 100. The binding ratio (PAF binding per WEB 2086 binding) was also calculated. Data are the means or the means ± S.E. of triplicate samples, and are representative of two individual experiments.

Radioligand Binding Assay

Transfected COS-7 cells were washed twice with Hepes-Tyrode's buffer (140 mM NaCl, 2.7 mM KCl, 1.8 mM $CaCl_2$, 0.49 mM $MgCl_2$, 12 mM $NaHCO_3$, 5.6 mM D-glucose, 0.37 mM NaH_2PO_4, 10 mM Hepes pH 7.4) containing 0.1% (w/v) fatty acids-free bovine serum albumin, incubated for 1 hour at 25 ^{o}C with 20 nM [^{3}H]WEB 2086 or 8 nM [^{3}H]PAF C-16 in the presence (for the nonspecific binding) or absence (for the total binding) of 10 μM unlabeled WEB 2086 in the same buffer. The cells were then washed twice with the buffer and solubilized with 1% Triton X-100. The radioactivity associated with the cells was measured by liquid scintillation counting. The balance is expressed as the specific binding with that of the wild type as 100.

Quantitation of Expression Levels of FLAG-tagged PAF Receptors

Immunochemical detection of the FLAG-tagged PAF receptors was done as described (4). Transfected COS-7 cells in 12-well dishes were washed twice with Ca^{2+}, Mg^{2+}-free phosphate-buffered saline (PBS(-)), fixed for 10 min in 2% freshly-prepared paraformaldehyde in PBS(-), washed twice with PBS(-), incubated for 1 hour with 20% FBS in the Hepes-Tyrode's buffer (FBS-Hepes-Tyrode's buffer) to reduce the nonspecific binding. The cells were incubated for 20 min with/without 10 μg/ml anti-FLAG antibody in the FBS-Hepes-Tyrode's buffer. They were washed three times with PBS(-), incubated for 30 min with 1 μCi/ml ^{125}I-labeled sheep anti-mouse IgG F(ab')$_2$ fragment in the FBS-Hepes-Tyrode's buffer, washed three times with PBS(-), solubilized with 0.1 N NaOH, and subjected to gamma counting. The anti-FLAG antibody-dependent binding was employed to show the expression levels of the receptors.

RESULTS AND DISCUSSION

From previous extensive studies of structure-activity relationships using various PAF analogs, the polarity in a PAF molecule seems to be crucial for its biological activities, as reviewed (5). In its glycerol-backboned structure, PAF has an alkyl-ether linkage at the *sn*-1 position, and the acetyl moiety at *sn*-2 and the phosphocholine residue with both positive and negative charges at *sn*-3, which are all important for its biological activities and

Table 2. Ligand and antibody binding properties observed in COS-7 cells transiently expressing FLAG-tagged mutant PAF receptors.

TM	Receptor	Relative Specific PAF Binding	FLAG Expression	Specific PAF Binding per FLAG Expression
		% of FLAG-WT	% of FLAG-WT	% of FLAG-WT
	WT	70.0 ± 4.2	2.1 ± 1.3	-
	FLAG-WT	100.0 ± 3.9	100.0 ± 4.1	100.0
1st	FLAG-S23A	62.0 ± 3.0	93.4 ±10.1	66.4
	FLAG-N33A	19.8 ± 4.9	28.9 ± 0.8	68.5
2nd	FLAG-N58A	29.3 ± 1.7	25.3 ± 1.9	115.8
	FLAG-T60A	139.0 ± 6.0	270.2 ± 7.9	51.4
	FLAG-D63A	53.2 ± 2.7	24.7 ± 0.8	215.4
	FLAG-T69A	54.1 ± 3.6	55.8 ± 1.0	97.0
3rd	FLAG-N100A	254.8 ±30.6	129.9 ± 9.9	196.2
	FLAG-T101A	171.1 ± 8.2	146.8 ± 9.7	116.6
	FLAG-S104A	207.1 ±11.8	78.4 ± 9.8	264.2
	FLAG-T112A	15.1 ± 3.3	7.2 ± 2.3	209.7
4th	FLAG-S138A	94.5 ±10.9	115.3 ± 7.4	82.0
	FLAG-S150A	186.9 ± 8.3	207.8 ±16.8	89.9
5th	FLAG-H188A	7.2 ± 1.4	84.9 ± 3.7	0.1
6th	FLAG-T236A	131.9 ± 3.6	245.8 ± 4.2	53.7
	FLAG-H248A	3.1 ± 1.5	107.9 ± 1.6	2.9
	FLAG-H249A	9.6 ± 0.7	97.7 ± 3.1	9.8
	FLAG-Q252A	21.5 ± 1.6	86.4 ± 3.4	24.9
7th	FLAG-Q276A	2.0 ± 1.1	111.2 ± 2.3	1.8
	FLAG-T278A	8.0 ± 1.2	106.9 ± 1.3	7.5
	FLAG-S283A	38.4 ± 1.5	113.8 ± 3.2	33.7
	FLAG-T284A	118.8 ± 4.1	143.5 ± 4.8	82.8
	FLAG-N285A	201.6 ± 6.9	255.6 ±13.1	78.9
	FLAG-D289A	184.0 ± 7.5	170.6 ± 4.2	107.9

Cells were incubated for 1 h at 25 °C with 3 nM [^{3}H]PAF C-16 in the presence or absence of 10 μM unlabeled WEB 2086. The specific binding was converted into percentage with that of the FLAG-tagged wild-type receptor as 100. Receptor expression was quantified using an anti-FLAG antibody. The FLAG expression was also converted into percentage with that of the FLAG-tagged wild-type receptor as 100. Further, the specific PAF binding per FLAG expression was calculated. Data are the means or the means ± S.E. of triplicate samples, and are representative of two experiments.

receptor binding. PAF receptor homologs of four mammalian species show a high identity (72.5%) in the deduced amino acid sequences. The identity in the putative TM I, II, III, VI and VII domains are especially high (90.5%) compared with that of other portions. All twenty-three polar amino acids clustered in the putative seven-TM regions are perfectly conserved. We reasoned that electrostatical interactions between PAF and these polar residues are likely to be important for binding, and, therefore, selected these residues as targets of alanine substitution.

The wild type and mutant guinea pig PAF receptors were transiently expressed in COS-7 cells and ligand binding properties were investigated (Table 1). The wild type receptor displayed a dose-dependent binding of agonist [^{3}H]PAF C-16 and antagonist [^{3}H]WEB 2086 (data not shown). There was no detectable specific binding of both ligands in untransfected or vector (pcDNAI)-transfected cells (data not shown). To determine if these mutations alter PAF binding affinity, a concentration of 3 nM near the Kd value (3.6 nM in the wild type receptor) was chosen for [^{3}H]PAF binding. On the other hand, a 20 nM concentration near 7-fold of the Kd value (2.9 nM in the wild type) was selected for [^{3}H]WEB 2086 binding, since this was thought to reflect the expression level of the receptor. In each case, the relative specific binding was expressed as percentages of the specific binding, with that of the wild type as 100. The ratio between PAF and WEB 2086 binding was also calculated to represent the relative PAF binding affinity, compared with that observed with the wild type receptor. The binding ratios in Table 1 indicate that some mutations resulted in an apparent gain or loss of PAF binding. PAF preferentially bound to some mutants such as N58A, D63A, T69A, N100A, T101A, S104A, and D289A (with ratios over 1.40) in which substituted residues mainly cluster in TM II and III. In contrast, PAF scarcely bound to other mutants such as H188A, H248A, H249A, Q252A, Q276A, T278A, and S283A (with ratios under 0.30) among TM V, VI and VII domain mutants.

Against our assumption, relative specific binding of WEB 2086 at 20 nM concentration differed between mutants. To accurately estimate the expression levels, N-terminal FLAG-tagged PAF receptors were constructed and analyzed (Table 2). It was previously demonstrated that the epitope-tagged PAF receptor functions comparably to the non-tagged receptor in COS-7 cells (6, 7). The receptor expression levels were quantitated using an anti-FLAG monoclonal antibody. The specific PAF binding per FLAG expression revealed that FLAG-tagged mutants showed similar binding characteristics to those of the corresponding non-tagged mutants; those of N58A, D63A, N100A, T101A, S104A, T112A, and D289A mutants were higher than that of the tagged wild type, whereas those of H188A, H248A, H249A, Q252A, Q276A, and T278A were lower. The T69A mutant showed the highest binding ratio (4.14) in Table 1, but this was due to a decrease of binding affinity to WEB 2086 (Table 2). Consequently, experiments with both types of the receptors identified the six mutants (N58A, D63A, N100A, T101A, S104A and D289A) as high affinity mutants and the other six mutants (H188A, H248A, H249A, Q252A, Q276A and T278A) as low affinity ones.

In conclusion, alanine exchanges of polar amino acids in the transmembrane domains of a guinea pig PAF receptor generate both high and low affinity mutants. According to the two-state model of G protein-coupled receptor activation, receptors exist in an equilibrum between the two conformational states, the inactive (R) and the active (R*) (8). Agonists bind to R* with a higher affinity than to R, and shift the equilibrium to R* by stabilizing it, which in turn activates G proteins and transmits the intracellular signals. In our experiments, the equilibrium might be shifted to R* in high affinity mutants and to R in low affinity mutants, compared with that in the wild type. High affinity mutation may cause a constitutive activation of the receptor in the absence of the agonists. On the other hand, low affinity mutation may cause a loss of the receptor function. Polar residues, the alanine exchanges of which generated low affinity mutants, may be involved in direct or indirect interactions with PAF. Natural mutations of human PAF receptor have not been reported so far, but this study suggests only one-point mutation in the receptor may cause a gain or a loss of the PAF receptor-mediated signaling.

ACKNOWLEDGMENTS

This work was supported in part by grants-in-aid from the Ministry of Education, Science, Sports and Culture and the Ministry of Health and Welfare of Japan, and by grants from Sankyo Foundation of Life Science, the Yamanouchi Foundation for Metabolic Disorders, Human Science Foundation and Senri Life Science Foundation.

REFERENCES

1. Izumi, T. and Shimizu, T. Platelet-activating factor receptor: gene expression and signal transduction. *Biochim. Biophys. Acta* 1259:317-333 (1995)
2. Honda, Z., Nakamura, M., Miki, I., Minami, M., Watanabe, T., Seyama, Y., Okado, H., Toh, H., Ito, K., Miyamoto, T. and Shimizu, T. Cloning by functional expression of platelet-activating factor from guinea-pig lung. *Nature* 349:342-346 (1991)
3. Ali, H., Richardson, R. M., Tomhave, E. D., DuBose, R. A., Haribabu, B. and Snyderman, R. Regulation of stably transfected platelet-activating factor receptor in RBL-2H3 cells. *J. Biol. Chem.* 269:24557-24563 (1994)
4. Ishii, I., Izumi, T., Tsukamoto, H., Umeyama, H., Ui, M. and Shimizu, T. Alanine exchaneges of polar amino acids in the transmembrane domains of a platelet-activating factor receptor generate both constitutively active and inactive mutants. *J. Biol. Chem.* 272:7846-7854 (1997)
5. Snyder, F. *Platelet-activating Factor and Related Lipid Mediators*, Plenum Publishing Corp., New York (1987)
6. Kunz, D., Gerard, N. P. and Gerard, C. The human leukocyte platelet-activating factor receptor. *J. Biol. Chem.* 267:9101-9106 (1992)
7. Gerard, N. P. and Gerard, C. Receptor-dependent internalization of platelet-activating factor. *J. Immunol.* 152:793-800 (1994)
8. Samama, P., Cotecchia, S., Costa, T. and Lefkowitz, R. J. A mutation-induced activated state of the ß2-adrenergic receptor: extending the ternary complex model. *J. Biol. Chem.* 268:4625-4636 (1993)

DIFFERENTIAL EFFECT OF CORTICOTROPIN RELEASING FACTOR ON INTERLEUKIN-1α AND INTERLEUKIN-1β-INDUCED PROSTAGLANDIN SYNTHESIS IN ENDOTHELIAL CELLS AND FIBROBLASTS

Sigal Fleisher-Berkovich and Abraham Danon

Department of Clinical Pharmacology,
Corob Center of Health Sciences,
Ben-Gurion University and Soroka Medical Center,
Beer-Sheva 84105, Israel

INTRODUCTION

The inflammatory response involves two major pathways: one comprises the immune response, predominantly mediated by cytokines such as interleukin 1 (IL-1). The other pathway involves neuroendocrine mechanisms, the most significant of which consists of stimulation of the hypothalamic-pituitary-adrenal axis (HPA). Activation of the neuro-endocrine axis results in corticotropin releasing factor (CRF) release, followed by enhancement of plasma levels of adrenocorticotropic hormone (ACTH) and finally increased production of glucocorticoids by the adrenal cortex. There is increasing evidence for tight connections between both pathways. Thus, it has become evident that IL-1 stimulates the HPA axis by inducing CRF release from the hypothalamus.[1] Cytokines are very potent modulators of cell function, and IL-1 is known to induce prostaglandin (PG) production by many cells, including endothelium and fibroblasts.[2] Additional evidence suggests that not only the end product of the HPA axis, the glucocorticoids, but also other hormones, primarily CRF, might be involved in the regulation of the inflammatory response. CRF has been identified in a number of peripheral sites such as the heart,[3] spleen,[4] adrenal[5] and lung.[6] Recently we discovered a novel neuroimmune mechanism whereby CRF regulates IL-1α-induced PG synthesis in fibroblasts and endothelial cells.[7] Presently we compared the effect of CRF on IL-1α and IL-1β-induced PG synthesis.

MATERIALS AND METHODS

Cell Cultures

Cultures of bovine aortic endothelial cells (BAEC) and human dermal fibroblasts were maintained as monolayer cultures in Dulbecco's modified Eagle's medium (DMEM), supplemented with 10% calf serum. Experiments were initiated by washing twice with serum-free DMEM (SFM). The cells were incubated in 1 ml SFM for 24 hours and then test agents were added for the time indicated for each experiment. Aprotinin was added in each experiment in order to inhibit proteinases.

Determination of 6-keto $PGF_{1\alpha}$ (6-K) and PGE_2

PGE_2 and 6-K, the stable dehydration product of PGI_2, that accumulated in the media were measured in unextracted samples of DMEM, by single antibody radioimmunoassays with dextran-coated charcoal precipitation. Rabbit antibodies to 6-K and to PGE_2 were obtained from Bio-Makor. The sensitivities of both assays were 0.15 ng/ml.

Arachidonic Acid (AA) Release

AA [5,6,8,9,11,12,14,15-^{3}H(N)] (200 Ci/mmol) was added (0.2μCi/ml) to the cell medium for 20-24 hours in order to label cellular lipids. Unincorporated AA was removed by four times washing with SFM. Test agents were added for 24 h and the radioactivity released into the medium was determined.

Prostaglandin H Synthase (PGHS) Activity

Cells were incubated with test agents, then AA (50 μM, unlabeled) was added 30 min before the end of incubation. This concentration of exogenous substrate overrides the release of the endogenous substrate by PLA_2. Thus, the observed changes in 6-K output can be ascribed to changes in PGHS activity.

Data Analysis

Results are expressed as means $\pm$ S.E.M for each experiment. Statistical analysis of the results was performed using two tailed Student's t test. $P \leq 0.05$ was considered significant.

RESULTS AND DISCUSSION

Fig 1 shows that treatment (24 h) of BAEC with IL-1α enhanced 6-K synthesis, and this was completely blocked by CRF at concentrations of 10^{-10}M and higher. By contrast, although IL-1α also increased PGI_2 levels, CRF failed to alter the IL-1β- induced 6-K production. Similar results were observed with IL-1α and IL-1β treated human fibroblasts (data not shown). Because prostacyclin synthesis is dependent on both PGHS and PLA_2, we assessed the activities of the two enzymes. Both IL-1α and IL-1β increased AA release, indicating enhanced PLA_2 activity. CRF restored the IL-1α-induced AA release to control levels but did not affect the IL-1β-enhanced PLA_2 activity (Table 1). Likewise, both types of IL-1 increased PGHS activity. The IL-1α-stimulated PGHS activity was completely blocked by CRF, but the IL-1β- induced enzyme activity was only partialy blocked (Table 1). The observation that CRF blocked the IL-1β-induced PGHS activity without altering the IL-1β - induced prostaglandin production, suggests that in the present system PLA_2 is probably the rate limiting step in prostaglandin synthesis.

IL-1α and IL-1β have similar, although not completely overlapping, biological activities. In fact, significant differences in activity between IL-1α and IL-1β have been described. Thus, a CRF antagonist markedly inhibited fever and thermogenesis that was induced by IL-1β but did not affect the responses to IL-1α.[8] In another report IL-1β showed immunostimulatory activity in vivo, as it could enhance the number of specific antibody producing cells in the spleen of mice that were immunized with antigen, this was inhibited by

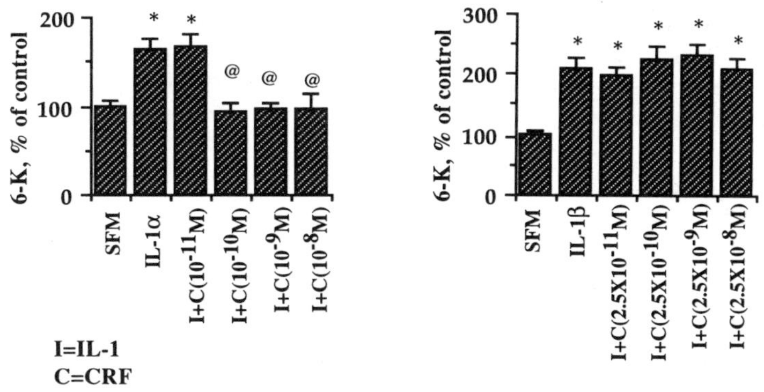

Fig 1. Effects of CRF and IL-1α/IL-1β on 6-K production by endothelial cells. Cells were incubated with IL-1α (left) or IL-1β (right) (100U/ml) and CRF at the concentrations indicated, for 24h. Thereafter, the media were collected and assayed for 6-K. Results are means±SEM of 6 determinations each. *, $P<0.05$ compared with SFM control; @, $P<0.05$ vs. IL-1α.

Table 1. Effects of IL-1α or IL-1β and CRF on AA release and PGHS activity in bovine aortic endothelial cells.

	SFM	CRF	IL-1		CRF + IL-1	
A.			α		α	
AA release (DPM)	299 ± 8.1	289 ± 6.8	391 ± 27.7[@]		291 ± 9.5*	
PGHS activity % of control	100 ± 12.2	88 ± 9.9	162 ± 17.7[@]		100 ± 11.9*	
B.				β		β
AA release (DPM)	359 ± 9.7	346 ± 17.3		450 ± 16.9[@]		431 ± 17.0[@]
PGHS activity % of control	100 ± 11.2	88 ± 9.9		224 ± 22.6[@]		132 ± 7.2*[@]

BAEC were incubated for 24h with IL-1α (A) or IL-1β (B) (100U/ml) alone or in combination with CRF ($2.5X10^{-8}$M). Results are the means ± SEM of 6 determinations each.*$P< 0.05$ compared with IL-1, [@]$P< 0.05$ vs. SFM .

injection of IL-1α.[9] Also IL-1β was more effective than IL-1α in inducing ACTH release both in vitro and in vivo in the rat.[10]

CONCLUSIONS

The results described here suggest that CRF modulates the effect of IL-1α, but not of IL-1βon PG synthesis. Such distinction may have far-reaching implications on neuroimmune regulation of inflammation. Further investigation is required to elucidate site and mechanism of regulation by CRF.

REFERENCES

1. C. Thiemermann, Biosynthesis and interaction of endothelium-derived vasoactive mediators, *Eicosanoids* 4:187-202 (1991).
2. A. Ristimaci A, and L. Viinikka, Modulation of prostacyclin production by cytokines in vascular endothelium cells, *Prost. Leuk. Essent. Fatty acids* 47:93-99 (1992).
3. M.H. Perrin, C. Donaldson, R. Chen, A. Blount, T. Berggren, L. Bilezikjian, P. Sawchenko, and W. Vale, Identification of a second CRF receptor gene and characterization of a cDNA expressed in heart, *Proc. Natl. Acad. Sci. USA* 92:2969-2973 (1995).
4. E.L. Webster, and E.B. Desousa, CRF receptors in mouse spleen: identification, localization and regulation by divalent cations and guanine nucleotides, *Endocrinology* 122:609-617 (1988).
5. J.R Dave, L.E. Eiden, and R.L. Eskay, Corticotropin-releasing factor binding to peripheral tissue and activation of the adenylate cyclase- adenosine 3', 5'-monophosphate system, *Endocrinology* 116:2152-2159 (1985).
6. K.D. Dieterich, D.E Grigoriadis, and E.B. Desousa, Corticotropin-releasing factor receptors in human small lung carcinoma cells: radioligand binding, second messenger and northern blot analysis data, *Endocrinology* 135:1551-1558 (1994).
7. S. Fleisher-Berkovich, and A. Danon, Effect of Corticotropin-releasing factor on prostaglandin synthesis in endothelial cells and fibroblasts, *Endocrinology* 136:4068-4072 (1995).
8. N.J. Busbridge, Dascombe, F.J.H Tilders, J.W.A.M van Oers, E.A. Linton, and N.J. Rothwell, Central activation of thermogenesis and fever by interleukin-1β and interleukin-1α involves different mechanisms, *Biochem. Biophys. Res. Commun.* 162:591-596 (1989).
9. D. Boraschi, V. Luigi, and A. Tagliabue, Differential activity of interleukin-1β and interleukin-1α in the stimulation of the immune response in vivo, *Eur. J. Immunol.* 20:317-321 (1990).
10. A. Uehara, S. Gillis, and A. Arimura, Effect of interleukin-1 on hormone release from normal rat pituitary cells in primary culture, *Neuroendocrinology* 45:343-7 (1987).

BRAIN AND PERIPHERAL ANANDAMIDE AMIDOHYDROLASE AND ITS INHIBITION BY SYNTHETIC ARACHIDONATE ANALOGUES

Luciano De Petrocellis[1], Dominique Melck[2], Natsuo Ueda[3], Yuko Kurahashi[3], Tiziana Bisogno[2], Shozo Yamamoto[3] and Vincenzo Di Marzo[2*]

[1]Istituto di Cibernetica and [2]Istituto per la Chimica di Molecole di Interesse Biologico CNR, Via Toiano 6, 80072, Arco Felice, (NA), Italy, and [3]Biochemistry Department, Tokushima University School of Medicine, Kuramoto-cho, Tokushima 770, Japan

INTRODUCTION

Anandamide (arachidonoyl-ethanolamide, AnNH) was the first brain endogenous substance to be proposed as a physiological agonist at central cannabinoid receptors[1]. Apart from its cannabimimetic properties (recently reviewed in ref. 2), a role as a neuronal chemical signal for AnNH was suggested also by the finding of molecular mechanisms for its inactivation and Ca^{2+}-dependent biosynthesis by rat central neurons[3]. The degradation of AnNH by neurons involves the enzymatic hydrolysis of its amide bond with formation of arachidonic acid (AA) and ethanolamine. The enzyme catalyzing this hydrolysis, originally named 'anandamide amidohydrolase', was characterized from rat[4,5] and porcine[6] brain membranes and was found in high levels also in the liver[4]. The enzyme was inhibited by typical serine and cysteine protease inhibitors, and displayed a peculiar pH dependency profile with optimal activity at $8.5<pH<10.0$. The interest towards AnNH amidohydrolase was increased further by the suggestion[2,8] that the same enzyme may also catalyze the hydrolysis of other bioactive acylethanolamides, such as the anti-inflammatory palmitoylethanolamide (PEA)[7], and the primary fatty acid amide, *cis*-9-octadecenoamide (oleamide), a novel sleep-inducing factor in mammals[9]. The recent cloning and expression of rat liver 'oleamide hydrolase' cDNA[10] confirmed that the same enzyme can hydrolyze both oleamide and, at a higher rate, AnNH[8], leading to propose for this enzyme the name 'fatty acid amide hydrolase' (FAAH)[8,10]. In this paper we report the preliminary characterization of FAAH-like enzymatic activities from a blood cell type, the rat basophilic leukemia (RBL-1) cell line, and from other peripheral mammalian and invertebrate cells, and describe the effect of five synthetic AA-derived inhibitors on these enzymes as well as on the previously described amidohydrolases from mouse neuroblastoma (N18TG2) cells[8] and porcine brain[6].

MATERIALS AND METHODS

RBL-1, N18TG2, human breast carcinoma (EFM-19), and rat pheochromocytoma (PC-12) cells were purchased from DSM (Germany) and grown as advised by the manufacturer. Sea urchin ovaries were dissected from *Paracentrotus lividus* fished in the bay

* Author to whom all correspondence should be addressed.
Fax: +39-81-8041770. E-mail: +VDM@TRINC.ICMIB.NA.CNR.IT

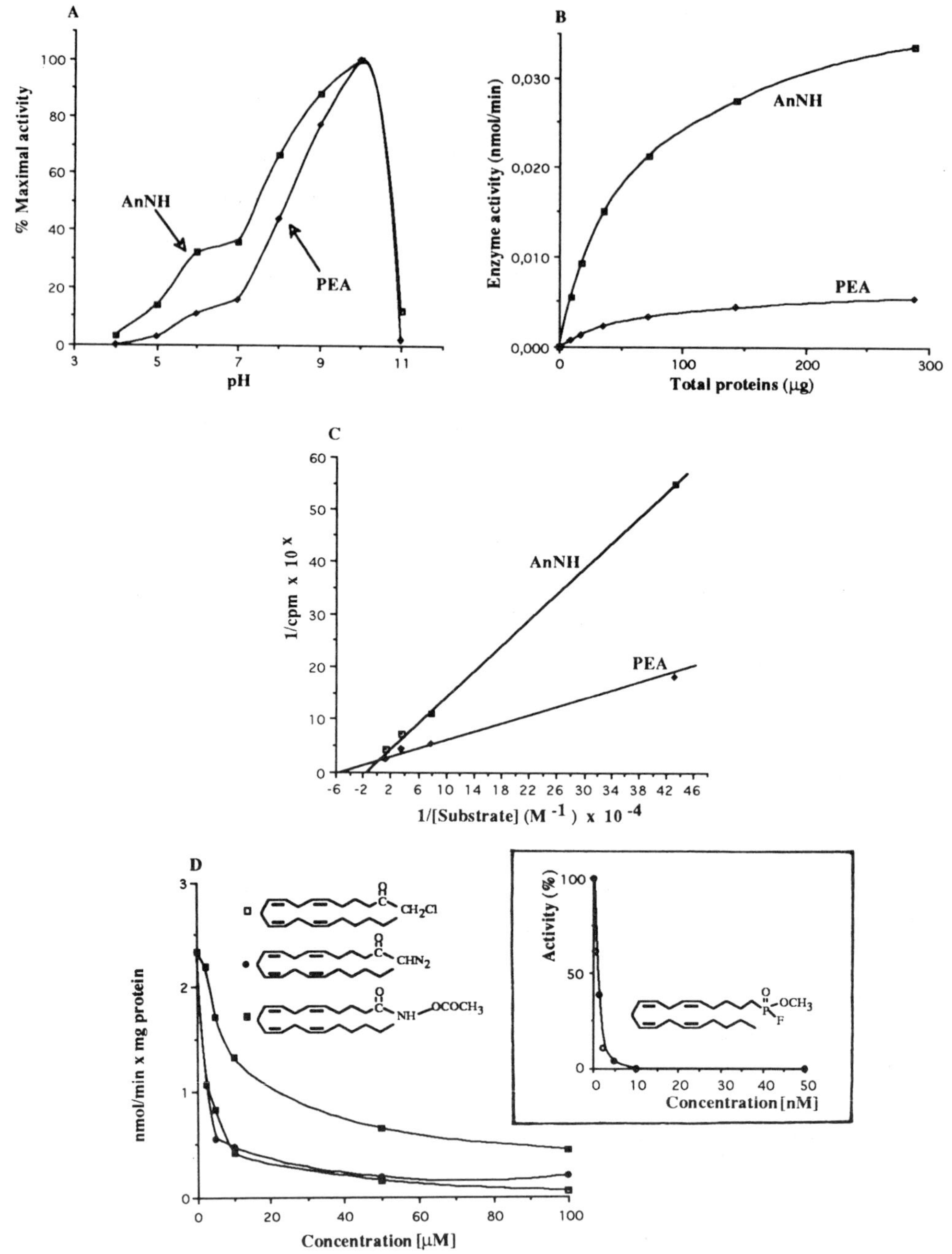

Figure 1. Preliminary characterization of AnNH amidohydrolase from RBL-1 cells.
(A) pH dependency profile; (B) dependency on protein content; (C) dependency on substrate concentration [activity expressed as cpm of ethanolamine released, x=4 or 5, respectively for PEA and AnNH]; (D) inhibition by ADMK (full circles), ACMK (empty squares), AcAHA (full squares) and MAFP (inset).

of Naples. Specimens of *Mytilus galloprovincialis* were purchased from a local mariculture center. Homogenates from porcine brain, cell cultures and invertebrate tissues were prepared and assayed for the presence of AnNH amidohydrolase activity as described previously[6,8,11]. Characterization of this activity was carried out on 10,000 or 100,000 x *g* pellet fractions by studying the effect of different protein or substrate concentrations or of different pH values of the assay buffer. The IC_{50}s for the inhibition of AnNH amidohydrolase by the three novel AA-derivatives (Fig. 1D), arachidonoyl-diazo-methyl-ketone (ADMK), arachidonoyl-chloro-methyl-ketone (ACMK), *N*-acetyl-arachidonoyl-hydroxamate (AcAHA), synthesized as described elsewhere[12], as well as by the previously described methyl-arachidonoyl-fluoro-phosphonate (MAFP)[13] and arachidonoyl-trifluoro-methyl-ketone[14] (ATFMK), were studied using partially purified fractions of the enzyme from porcine brain, RBL-1 and N18TG2 cells, prepared by hydrophobic and/or anion exchange chromatography as described elsewhere[6,12]. A fixed concentration of the inhibitors was then tested also on FAAH-like AnNH hydrolyzing activity from EFM-19, PC-12, sea urchin ovaries and *M. galloprovincialis* particulate fractions. ADMK was also tested on cycloxygenase and lipoxygenase enzymes from various sources[15].

RESULTS AND DISCUSSION

As previously reported[12] for the cognate cell line RBL-2H3, particulate fractions from RBL-1 cells were found to contain an AnNH amidohydrolase activity depending on total protein as well as AnNH concentration (V_{max}=2.4 nmol min^{-1} mg protein^{-1}, K_m=67 μM) and displaying a pH dependency profile typical of FAAH[10], with apparent active site pK_a values of about 6 and 10 (Fig. 1A-C)[16]. Identical pH profiles and similar protein dependency curves (not shown) were found also for AnNH hydrolyzing activities from EFM-19 cells (V_{max}=1.20 nmol min^{-1} mg protein^{-1}, K_m=25.0 μM) or PC-12 cells (V_{max}=0.47 nmol min^{-1} mg protein^{-1}, K_m=46.5 μM) as well as from *P. lividus* ovaries (V_{max}=0.35 nmol min^{-1} mg protein^{-1}, K_m=98.0 μM) and shell-free *M. galloprovincialis* specimens (V_{max}=0.073 nmol min^{-1} mg protein^{-1}, K_m=29.6 μM). Unlike the previously reported AnNH amidohydrolase from N18TG2 cells[8], the enzymes from RBL cells and *P. lividus* ovaries recognized also PEA as a substrate (K_m=20 and 112 μM, respectively), although at a slower rate than AnNH (V_{max}=0.3 and 0.09 nmol min^{-1} mg protein^{-1}), in support of a role for PEA as a local peripheral mediator[7].

'Basophilic' AnNH amidohydrolase, however, was inhibited by the substrate analogues ADMK, ACMK, AcAHA, ATFMK and MAFP (Fig. 1D and inset, Table 1), thus suggesting a certain substrate specificity. As with the enzyme from porcine brain and N18TG2 cells (Table 1), the most active compound was MAFP and the weakest inhibitor AcAHA. Lineweaver-Burk profiles (not shown) indicated that, while the inhibitory action of ADMK, ACMK, ATFMK and AcAHA was competitive and could be reversed by increasing AnNH concentrations, MAFP appeared to act as an irreversible inhibitor. Accordingly, enzymatic activity of MAFP-treated AnNH amidohydrolase could not be recovered upon chromatographic purification of the enzyme (not shown).

Table 1. IC_{50} values (μM) for the inhibition of porcine brain, RBL-1 and N18TG2 cell AnNH amidohydrolase by synthetic substrate analogues.

	ADMK	ACMK	AcAHA	ATFMK	MAFP
RBL-1	2	3	15	2	0.001
N18TG2	3	10	34	3	0.003
Porcine brain	6	23	25	1	N.D.

Data are means of three separate experiments. N.D.= not determined.

Thus, MAFP, a compound previously reported to covalently inhibit phospholipase A_2 with IC_{50}s in the low μM range[13], is the most potent irreversible inhibitor so far reported for

AnNH amidohydrolase (and, therefore, FAAH). ADMK, on the other hand, is one of the most potent competitive inhibitors of the enzyme described to date (Table 1), and is relatively selective as it did not significantly activate sheep cycloxygenase-1 (up to 100 μM) and inhibited porcine leukocyte 12-lipoxygenase and rabbit reticulocyte 15-lipoxygenase only at high concentrations (IC_{50}=27 and 100 μM, respectively). ADMK, however, was found to inhibit 5-lipoxygenase from human leukocytes (IC_{50}<3 μM).

Next, we wanted to probe the effect of the most potent AA analogues mentioned above on the other novel FAAH-like AnNH hydrolyzing activities found in EFM-19 and PC-12 cells or in invertebrate tissues. We found that these synthetic compounds inactivate peripheral mammalian AnNH amidohydrolases to an extent similar to that observed with the corresponding enzymes from porcine brain and N18TG2 cells. Conversely, the invertebrate enzymes appeared to be much less sensitive to inhibition by substrate analogues (Table 2), although they were sensibly inhibited by typical esterase inhibitors (ref. 11 and data not shown). This might suggest that AnNH amidohydrolases from invertebrate organisms may be less specific towards AnNH. On the other hand, all the activities tested were inhibited to some extent by oleamide (Table 2), thus suggesting that the corresponding enzymes also recognize this bioactive fatty acid amide, and confirming their similarity to FAAH[10].

Table 2. Effect of synthetic AA analogues and of oleamide on AnNH hydrolyzing activities from particulate fractions of various central and peripheral cell types and tissues.

	P. brain	N18TG2	EFM-19	PC-12	Sea urchin	Mussel
No inhib.	100	100	100	100	100	100
ADMK	<0.1[d]	12.5±0.9[d]	25.0±2.2[c]	10.0±4.1[c]	N.D.	70.4±2.5[e]
ACMK	17.3±0.7[d]	<0.1[d]	11.9+2.1[c]	<0.1[c]	N.D.	68.8±1.2[e]
ATFMK	<10[d]	15.2±2.2[b]	<0.1[c]	<10[c]	45.5±0.3[d]	70.1±1.6[e]
MAFP	N.D.	<0.1[f]	8.8±1.9[f]	<0.1[f]	N.D.	21.6±0.1[a]
Oleamide	*	27.2±13.1[d]	55.0±2.0[d]	71.0±3.0[d]	N.D.	68.4±3.2[d]

Data are expressed as percent of controls (No inhib.) and are means±S.D. of n=3. [a,b,c,d,e]=1, 10, 50, 100, 200 μM, respectively. [f]= 50 nM. N.D.= not determined. P. brain= porcine brain. *When the partially purified enzyme from P. brain was allowed to react with 100 μM oleamide, the hydrolysis rate was 58% of that observed for AnNH.

In conclusion, we have provided here evidence suggesting that FAAH-like AnNH amidohydrolase, by being present in peripheral cell types and invertebrate tissues, is a widespread enzyme in animal organisms capable of catalyzing the hydrolysis also of bioactive fatty acid amides different from AnNH. Moreover, we have shown that inhibitors obtained from the derivatization of AA can be used as selective probes for the study of this enzyme, and, therefore, for pharmacological and biochemical investigations on its physiological substrates, in both the CNS and periphery. Further studies will be now required in order to determine possible differences in the amino acid sequences of AnNH amidohydrolase enzymes, especially those from lower invertebrates, from that recently reported for the rat liver enzyme[10].

Acknowledgments

This work was partly funded by the Human Frontiers in Science Program (grant RG26/95 to VDM and SY).

REFERENCES

1. Devane WA, Hanus L, Breuer A, Pertwee RG, Stevenson LA, Griffin G, Gibson D, Mandelbaum A, Etinger A and Mechoulam R, Isolation and structure of a brain costituent that binds to the cannabinoid receptor, *Science* 258:1946 (1992).
2. Di Marzo V, De Petrocellis L, Bisogno T and Maurelli S, Pharmacology and physiology of the endogenous cannabimimetic mediator anandamide, *J Drug Dev Clin Pract* 7:199 (1995).

3. Di Marzo V, Fontana A, Cadas H, Schinelli S, Cimino G, Schwartz JC and Piomelli D, Formation and inactivation of endogenous cannabinoid anandamide in central neurons, *Nature* 372:686 (1994).
4. Desarnaud F, Cadas H and Piomelli D, Anandamide amidohydrolase activity in rat brain microsomes. Identification and partial characterization, *J Biol Chem* 270:6030 (1995).
5. Hillard CJ, Wilkison DM, Edgemond WS and Campbell WB, Characterization of the kinetics and distribution of N-arachidonylethanolamine (anandamide) hydrolysis by rat brain, *Biochim Biophys Acta* 1257:249 (1995).
6. Ueda N, Kurahashi Y, Yamamoto S and Tokunaga T, Partial purification and characterization of the porcine brain enzyme hydrolizing and synthesizing anandamide, *J Biol Chem* 270:23823 (1995).
7. Facci L, Dal Toso R, Romanello S, Buriani A, Skaper SD and Leon A, Mast cells express a peripheral cannabinoid receptor with differential sensitivity to anandamide and palmitoylethanolamide, *Proc Natl Acad Sci USA* 92:3376 (1995).
8. Maurelli S, Bisogno T, De Petrocellis L, Di Luccia A, Marino G and Di Marzo V, Two novel class of neuroactive fatty acid amides are substates for mouse neuroblastoma anandamide amidohydrolase, *FEBS Lett* 377:82 (1995).
9. Cravatt BF, Prospero-Garcia O, Siuzdak G, Gilula NB, Henriksen SJ, Boger DL and Lerner RA, Chemical characterization of a family of brain-lipids that induce sleep, *Science* 268:1506 (1995).
10. Cravatt BF, Giang DK, Mayfield SP, Boger DL, Lerner RA and Gilula NB, Molecular characterization of an enzyme that degrades neuromodulatory fatty acid amides, *Nature* 384:83 (1996).
11. Bisogno T, Ventriglia M, Mosca M, Milone A, Cimino G and Di Marzo V, Occurrence and metabolism of anandamide and related acylethanolamides in ovaries of the sea urchin *Paracentrotus lividus*, *Biochim Biophys Acta* (1997) in press.
12. De Petrocellis L, Melck D, Ueda N, Maurelli S, Kurahashi Y, Yamamoto S, Marino G and Di Marzo V, Novel inhibitors of brain, neuronal and basophilic anandamide aminohydrolase, *Biochem Biophys Res Comm* 231:82 (1997).
13. Balsinde J and Dennis EA, Distinct roles in signal transduction for each of the phospholipase A2 enzymes present in P388D1 macrophages, *J Biol Chem* 271:6758 (1996).
14. Koutek B, Prestwich GD, Howlett AC, Chin SA, Salehani D, Akhavan N and Deutsch DG, Inhibitors of arachidonyl ethanolamide hydrolisis, *J Biol Chem* 269:22937 (1994).
15. Alanko J, Kurahashi Y, Yoshimoto T, Yamamoto S and Baba K, Panaxynol, a polyacetylene compound isolated from oriental medicine, inhibits mammalian lipoxygenases, *Biochem Pharmacol* 48:1979 (1994).
16. Patterson JE, Ollmann IR, Cravatt BF, Boger DL, Wong C-H and Lerner RA, Inhibition of oleamide hydrolase catalyzed hydrolysis of the endogenous sleep-inducing lipid *cis*-9-octadecenoamide, *J Am Chem Soc* 118:5938 (1996).

EFFECTS OF 13-HODE AND OTHER MOMOHYDROXIDES ON INTEGRIN/LIGAND BINDING: IMPLICATIONS FOR CELL CELL INTERACTIONS

M. R. Buchanan, [1] and S J. Brister, [2]

Departments of [1] Pathology and [2] Surgery,
McMaster University,
Hamilton, Ontario, Canada,
L8L 8E7

INTRODUCTION

In 1985, we demonstrated that 13-hydroxyoctadecadienoic acid (13-HODE) is synthesized from *endogenous* linoleic acid triglyceride stores by vascular wall cells via the lipoxygenase pathway [1-2]. Specifically, we demonstrated that endothelial cells synthesize more 13-HODE than smooth muscle cells, which in turn, synthesize more 13-HODE than fibroblasts. Since then, we and other investigators have demonstrated that both human and animal vascular wall cells, leukocytes and other white cells, tumour cells and epithelial cells all synthesize 13-HODE[2-5]. In addition, Larrue *et al*[6] reported that anucleated platelets contain *acylated* 13-HODE. The levels of endogenous 13-HODE in all of these cells are highest under basal (unstimulated) conditions, and decrease following cell cell interactions; i.e. at a time when the cells respond to injury, pertubation or stimulation. These observations led us to hypothesize that 13-HODE contributes significantly to blood cell vessel wall biocompatability; i.e. 13-HODE appears to downregulate vessel wall thromboresistance, thereby attenuating thrombogenesis and downregulates extravascular invasion during inflammation and metatasis[7].

Unresolved Issues

Other investigators have demonstrated that 13-HODE is also derived from linoleic acid via the cyclooxygenase pathway [8-9]. It has been suggested that such studies indicate that 13-HODE is synthesized following cell injury/stimulation, particularly since the cyclooxygenase enzyme, COX 2, only becomes active following cell injury or stimulation. While we concur with the results of such studies, namely that many cell types are capable of metabolizing *exogenous* linoleic acid to13-HODE via the cyclooxygenase enzyme, neither we nor anyone else have demonstrated cyclooxygenase-dependent synthesis of 13-HODE by cells from their *endogenous* linoleic acid stores. For example, when endothelial cells are pretreated with aspirin (in a concentration which abolishes PGI_2 production) 13-HODE synthesis is not inhibited, whereas pretreating the endothelial cells with salicylate or the Burroughs Wellcome compound, BW

Recent Advances in Prostaglandin, Thromboxane, and Leukotriene Research
Edited by Sinzinger *et al.*, Plenum Press, New York, 1998

755, abolishes both PGI_2 and 13-HODE production. Under such conditions, both platelet- and tumour cell-endothelial cell adhesion increase[10]. In other studies, increasing the basal level of cyclic AMP also increases the intracellular level of 13-HODE[2]. [High intracellualr levels of cyclic AMP are the norm in cells under basal or unstimulated conditions.] These latter studies are more consistent with the hypothesis that 13-HODE is synthesized in cells under basal conditions, and somehow contributes to cell cell biocompatibility[7].

It is also well recognized that when cells are stimulated, their intracellular cyclic AMP levels decrease, membrane-associated phosphodiesterases become activated, and arachidonic acid is liberated from their phospholipid stores. The free arachidonic acid is rapidly metabolized into a number of prostanoids (*via the cyclooxygenase pathway*) which are released into the ambient surroundings, and also into a number of lipoxygenase-derived monohydroxides such as 5-, 12-, and 15-eicosatetraenoic acids (5-, 12-, 15-HETE) in leukocytes, platelets and tumour cells respectively[7]. While these constituents are eventually released from these cells, it is important to note that their release does not parallel the release of the prostanoids, but rather occurs sometime after the stimulated cell response, including cell cell adhesion. This suggests that these rapidly synthesized arachidonic acid monohydroxides, when *intracellular*, can affect extracellular responses.

HYPOTHESIS

These and other observations have led us to postulate that the *intracellular* levels of the linoleic acid-derived 13-HODE and the arachidonic acid-derived monohydroxides police cell cell adhesion. Two more recent studies suggest a possible explanation of how these monohydroxides might effect this role. First, we reported that 13-HODE and the vitronectin receptor colocalize in unstimulated endothelial cells at a time when the endothelial cells are not adhesive. However, these two moieties dissociate following endothelial cell stimulation and the vitronectin receptor relocates on the apical surface of the cells, rendering the cell surface adhesive [11]. Second, Conforti *et al*[12] demonstrated that ability of the vitronectin receptor to bind its various ligands is influenced by its environment.

These observations suggested to us that the interaction of a specific lipid monohydroxide, i.e. those derived from linoleic and arachidonic acids, with the cell surface adhesion receptors directly influences the ability of the cell surface receptor to bind with its specific ligand(s). If so, however, it has been unclear whether 13-HODE and the HETEs act indirectly as secondary messengers or act directly by some physiochemical interaction to downregulate cell cell adhesion.

DIRECT EFFECTS OF MONOHYDROXIDES ON INTEGRIN/LIGAND BINDING

To test the possibility that monohydroxides have a direct effect on integrin/ligand binding independent of any enzymatic pathway, purified monohydroxides (0.5 to 2 parts) were inserted into liposomes (33 parts) containing purified endothelial cell-derived vitronectin receptor (6 parts) or purified platelet-derived glycoprotein IIbIIIa (6 parts). [The liposomes consisted of 6 % phosphatidylinositol, 40 % phosphatidylcholine, 12 % phosphatidylserine, 32 % phosphatidylethanolamine and 10 % cholesterol to simulate the lipid milieu of the endothelial cell and platelet plasma membranes]. Then, the ability of the integrin/liposomes to bind specific ligands was measured by adding 400 ng of [monohydroxide/integrin/liposomes, (0.5-2:6:33)] into microtitre wells coated with each specfic ligand (100 to 400 ng/well). The ligands included: i) vitronectin; ii) fibronectin; and iii) fibrinogen.

When the [liposomes] containing no integrin or monohydroxide were added to the ligand-coated wells, approximately 8 to 11 ng of [liposomes] bound to each well, irrespective of the amount of ligand bound. This [liposomes]-ligand binding was considered background and was subtracted from all subsequent values within each experiment.

When [liposomes containing only the vitronectin receptor] were added to the ligand-coated wells, the [vitronectin receptor/liposomes] bound to the wells in a concentration-dependent manner, plateauing at a ligand concentration of 250 ng/well. More importantly, the [vitronectin receptor/liposomes] bound more avidly to the vitronectin than they bound to the fibronectin, and bound more to the fibronectin than they bound to the fibrinogen. The data obtained with the [monohyroxide/vitronectin receptor/liposomes, (2:6:33)] and the wells coated with 400 ng are shown in **Table 1**.

Table 1. Effect of monohydroxides on [vitronectin receptor/liposomes] binding to wells coated with 400 ng of vitronectin, fibronectin or fibrinogen.

	LIGAND		
	Vitronectin	***Fibronectin***	***Fibrinogen***
[VnR/Liposomes]	73	56	10
+ 13-HODE	45 (-38%)*	39 (-30%)*	9 (-10%)
+12(R)HETE	72 (-1%)	56 (0%)	10 (0%)
+12(S)HETE	82 (+12%)	67 (+20%)*	9 (-10%)
+15(S)HETE	103(+42%)*	66 (+18%)*	9 (-10%)

*Data are expressed as ng [.../liposomes]/well, means; the numbers in () represent the % increase or decrease in receptor/liposome-ligand binding. *p<0.001. SD>6%, n=12. VnR = Vitronectin receptor.*

When 12(S)- or 15(S)HETE was added to the [vitronectin receptor/liposomes], ligand binding increased significantly (Table 1). The addition of racemic 12(R)HETE had no effect. Thus, adding the vitronectin receptor to the liposomes resulted in specific receptor-ligand binding which was enhanced in the presence of the active arachidonic acid monohydroxides forms; receptor-ligand binding decreased in the presence of the linoleic acid monohydroxide, 13-HODE.

Similar results were obtained with the platelet glycoprotein IIb/IIIa fibrinogen receptor. Namely, adding the glycoprotein IIb/IIIa receptor to the liposomes resulted in greater binding to fibrinogen than to fibronectin, and greater binding to fibronectin than to vitronectin. Adding 12(S)HETE or 15(S)HETE increased [glycoprotein IIb/IIIa/liposomes]-ligand binding whereas the inactive 12(R)HETE had no effect. [The effect of 13-HODE was not studied as 13-HODE is only present in platelets in an *acylated* form[6]]. Data obtained using the [monohydroxide/glycoprotein IIb/IIIa/liposomes, (2:6:33)] and wells coated with 400 ng ligand/well are shown in **Table 2**.

Table 2. Effects of monohydroxides on [glycoprotein IIb/IIIa/liposomes] binding to wells coated with 400 ng vitronectin,fibronectin or fibrinogen.

	LIGAND		
	Vitronectin	***Fibronectin***	***Fibrinogen***
[GPIIb/IIIa/Liposome]	17	26	70
+12(R)HETE	16 (-6%)	28 (+7%)	69 (-1%)
+12(S)HETE	38 (+123%)*	69 (+165%)*	136 (+136%)*
+15(S)HETE	45 (+264%)*	69 (+165%)*	153 (+219%)*

*Same legend as per **Table 1.** GPIIb/IIIa=glycoprotein IIb/IIIa.*

CONCLUSIONS

The observations that vitronectin receptor-vitronectin and glycoprotein IIb/IIIa-fibrinogen binding are increased by the arachidonic acid monohydroxides and that vitronectin receptor-ligand binding is decreased by the linoleic acid monohydroxide, are consistent with the hypothesis that lipoxygenase-derived monohydroxides synthesized under basal (13-HODE) and stimulated (the HETEs) conditions play a direct physical role in modulating platelet endothelial cell adhesion[7]. Given the specificity of the binding characteristics in these studies, and the observations that the relative ratio of 13-HODE:12-HETE in a variety of tumour cells also influences their abilities to bind to the endothelial cell surface [4], suggest that regulation of cell integrin adhesivity by lipoxygenases-derived monohydroxides is a common mechanism in vascular and circulating blood cells. If so, modulating 13-HODE and HETEs synthesis under basal and stimulated conditions *in vivo* may be a useful approach to altering cell cell-mediated outcome events in the clinical situation. One possible application and the success of such an approach in patients undergoing cardiac surgery is discussed by us in a following paper.

REFERENCES

1. M.R. Buchanan, T.A. Haas, M. Lagarde and M. Guichardant, 13-Hydroxy-octadecadienoic acid is the vessel wall chemorepellant factor, LOX. *J Biol Chem* 260: 16056 (1985).
2. T.A. Haas, M.-C. Bertomeu, E. Bastida and M.R. Buchanan, Cyclic AMP regulation of endothelial cell triglyceride turnover, 13-hydroxyoctadecadienoic acid (13-HODE) synthesis and endothelial cell thrombogenicity. *Biochim Biophys Acta* 1031: 174 (1990).
3. R.J. Soberman, T.W. Harper, D. Betteridge, R.A. Lewis and K.F. Austin, Characterization and separation of the arachidonic 2-lipoxygenase and linoleic acid *w*-6 lipoxygenase (arachidonic acid 15-lipoxygenase) of human polymorphonuclear leukocytes. *J Biol Chem* 260: 4508 (1985).
4. E, Bastida, M.-C. Bertomeu, T.A. Haas, L. Almirall, D. Lauri, F.W. Orr and M.R. Buchanan, Regulation of tumor cell adhesion by intracellular 13-HODE:15-HETE ratio. *J Lipid Med* 2: 281 (1990).
5. V.A. Ziboh and R.S.Chapkin, Metabolism and function of skin lipids. *Prog Lipid Res* 27: 81 (1988).
6. P. Blin, D. Daret and J. Larrue, Inhibition of linoleic acid oxidation by arachidonic acid in human platelets: 13-HODE is not synthesized after thrombin stimulation. *Thromb Haemorrh Dis* 7: 27 (1993).

7. M.R. Buchanan and S.J. Brister,. Other essential fatty acids, their metabolites and cell/cell interactions. In: *The Handbook of Immunopharmacology*, F. Cunningham, ed, London, 1994.
8. T.l. Kaduce, P.H. Figard, R. Leifur and A.A. Spector, Formation of 9-hydroxyoctadecadienoic acid from linoleic acid in endothelial cells. *J Biol Chem.* 264: 6823 (1989).
9. S. Lopez, L.Vila, F. Breviario and C. de Castellarna,. Interleukin-1 increases 15-hydroxy-eicosatetraenoic acid in cultured human endothelial cell. *Biochim Biophys Acta* 1170: 17 (1993).
10. E. Bastida, L. Almirall, A. Ordinas, M.R. Buchanan, F.W. Orr, M.-C. Bertomeu and T.A. Haas, Effects of endothelial cell treatment on 13-HODE and prostacyclin synthesis and its correlation with tumor cell-vascular endothelial cell adhesion. *Invas Metast* 11: 273 (1991).
11. M.R. Buchanan, M.-C. Bertomeu, T.A. Haas, F.W. Orr and L. Eltringham-Smith, Localization of 13-hydroxyoctadecadienoic acid and the vitronectin receptor in human endothelial cells and endothelial cell/platelet interactions. *Blood* 81: 3303 (1993).
12. G. Confort, A. Zanetti, I. Pasquali-Ronchetti, D. Quglino Jr, P. Negroz and E. Dejana, Modulation of vitronectin receptor binding by membrane lipid composition. *J Biol Chem* 265: 4011 (1990).

EFFECT OF LINOLEIC ACID SUPPLEMENTS ON VESSEL WALL HYPERPLASIA IN RABBITS

V. Riambau,[1,2] C. García-Madrid, [2], M.Taulés, [2] M.Camacho, [2] L.Vila,[2] and C. de Castellarnau,[2]

[1]Cardio-Vascular Surgery, Hospital Clínic, University of Barcelona, and
[2]Institut of Research of Santa Creu i Sant Pau Hospital,
Barcelona, SPAIN.

INTRODUCTION

Intimal hyperplasia is a major complication of vascular surgical procedures.[1-5] In addition, intimal hyperplasia contributes significantly to the development of atherosclerosis.[6] Hyperplasia is a result of platelet/leukocyte/injured vessel wall interactions, followed by the release of vasoactive, chemotactic and mitogenic substances which promote smooth muscle cells migration and proliferation. [6,7] Fatty acid metabolites synthesized by vascular wall and blood cells are thought to modulate, in part, the onset of intimal hyperplasia and athero-sclerosis. For example, 13-hydroxyoctadecadienoic acid (13-HODE), derived from linoleic acid via the lipoxygenase pathway, has been shown to inhibit platelet adhesion to the subendothelial layer. [8,9] Since platelet adhesion to the vessel wall is a prerequisite to the secretion of platelet-derived smooth muscle cell, and linoleic acid is metabolized into 13-HODE the vessel wall, we determined whether a linoleic acid-rich diet, given before injury, was associated with decreased intimal hyperplasia following vessel wall balloon catheter injury in normo- and hypercholesterolemic rabbits.

EXPERIMENTAL DESIGN

New Zealand white rabbits (males, 2.5-3 kg) were allocated to 4 diet groups (6 rabbits/ group): Grp 1 - control (26% linoleic acid); Grp 2 - linoleic acid-rich (40%); Grp 3 - cholesterol control (1% cholesterol); and Grp 4 - cholesterol + 40 % linoleic acid. The animals were fed each diet *ad libitum* for 4 weeks. Then, each rabbit was anesthetized, and the abdominal aorta were denuded of its endothelium, using a 3F balloon catheter. Six weeks later, the rabbits were killed and their abdominal and thoracic aortas were removed for biochemical and histological evaluations.

Recent Advances in Prostaglandin, Thromboxane, and Leukotriene Research
Edited by Sinzinger *et al.*, Plenum Press, New York, 1998

Histology

The vessel wall segments processed for histology, were incubated immediately in 10% formaldehyde. Subsequently, each segment was stained with hematoxylin-eosin and elastin stain, and >3 transverse (5 microns) cross-sections were analyzed, using a computer-assisted image analyser (Kontron®, USA; Microm®, Spain). Intimal hyperplasia was expressed as the *intimal area* ÷ *media area* for each section.

Fatty Acid Metabolite Assays

Free vessel wall 13-HODE was measured in extracts obtained from additonal vessel wall segments of injured and uninjured aorta, prepared immediated after obtaining the segments, using straight phase-HPLC as previously described.[10] Vessel wall 15-HETE and PGI_2 were measured, using commercially-available specific radioimmunoassays (Advanced Magnetics™, Atom®, Spain).

Table 1. Increased 13-HODE, $PG1_2$ and 15-HETE synthesis in vessel wall segments 6 weeks after injury.

Diet	13-HODE	PGI_2	15-HETE
Group 1	383 ± 4	21 ± 7	390 ± 80
Group 2	*160 ± 25***	*67 ± 30**	*62 ± 40***
Group 3	175 ± 29	1 ± 4	8 ± 20
Group 4	*1714 ± 92***	*92± 30***	*- 28 ± 18***

*#Data are expressed as the % increase in metabolite synthesis in the injured vessel wall as compared to the unijured vessel wall, means ± sem. Grp 2 vs Grp 1 and Grp 3 vs Grp 4 - * p <0.05; **p < 0.001.*

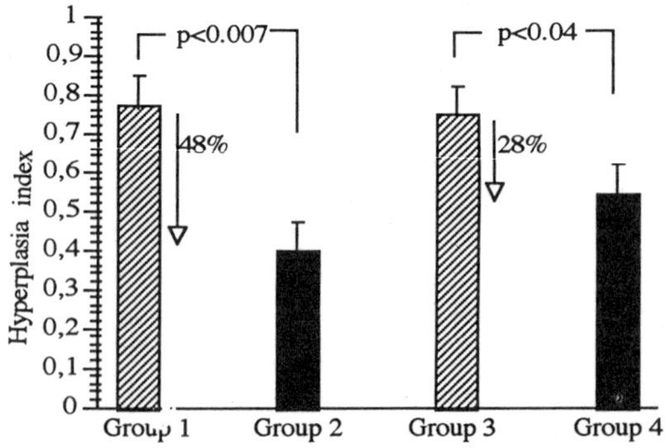

Figure 1. Hyperplasia index in normo- (Grp 1 & 2) and hypercholesterolemic rabbits (Grp 3 & 4 fed control (Grp 1 & 3) or linoleic acid-rich (Grp 2 & 4) diets.

RESULTS

There was significant and comparable intimal hyperplasia in normo- and hypercholesterolemic rabbits (Grps 1 and 3) 6 weeks after injury **(Figure 1)**. When rabbits were fed the linoleic acid-rich supplement, intimal hyperplasia was markedly decreased; i.e. decreased 48 % in normocholesterolemic rabbits (Grp 2 vs Grp 1, $p < 0.007$); and decreased 28 % in hypercholesterolemic rabbits (Grp 4 vs Grp 3, $p < 0.04$), respectively **(Figure 1)**

Fatty acid metabolism in the injured vessel walls was increased in all treatment groups. Thus, there was increased 13-HODE, $PG1_2$ and 15-HETE synthesis in all injured vessel wall segments as compared to uninjured vessel walls, regardless of treatment group **(Table 1)**. Feeding normo- and hypercholesterolemic rabbits the linoleic acid-rich supplement significantly altered fatty acid metabolism, albeit not always in the same direction.

CONCLUSIONS

We have demonstrated that a linoleic acid-rich diet, given for 1 month before vessel wall injury, results in decreased intimal hyperplasia in both normo- and hypercholesterolemic rabbits. The observations that the reduction in vessel wall hyperplasia in the linoleic acid-rich diet fed rabbits, was associated with significant changes in fatty acid metabolism, are consistent with results by others.[8,9] However, we found that the level of vessel wall free 13-HODE 6 weeks after injury did not correlate with the 6 week intimal hyperplasia. It is possible that the linoleic acid, given before injury, increased 13-HODE synthesis in the vessel wall *at the time of injury;* but by the time that we harvested the vessel wall segments; i.e. six weeks later, the vascular wall 13-HODE levels already had returned towards baseline levels. If so, these results support earlier observations that altering vessel wall 13-HODE, thereby decreasing platelet vessel wall interactions *at the time of injury* [9] can ameliorate long-term events such as hyperplasia as seen in our study.

Our data also suggest that linoleic acid can attenuate arachidonic acid-derived 15-HETE synthesis. This effect may also contribute to the decreased hyperplasia since 15-HETE has been shown to exacerbate smooth muscle cell prolferation. [11]

Finally, the observations that $PG1_2$ synthesis in the injured vessel walls, was increased in the linoleic acid-rich diet fed animals, are consistent with the observations that elevating vessel wall cyclic AMP (which $PG1_2$ does), increases vessel wall 13-HODE.

We conclude, therefore, that altering vessel wall linoleic acid metabolism *at the time of injury* can decrease *chronic* vessel wall hyperplasia. However, further studies are required to better elucidate a clearer explanation of the exact mechanism through by which linoleic acid or its 13-HODE metabolite attemuates intimal proliferation.

REFERENCES

1. V. Riambau, E. Latorre, Nuevas Terapéuticas en Cirugía Vascular, In: *Temas Actuales en Angiología y Cirugía Vascular.* M.A. Cairols, ed, Espax, Barcelona (1993).
2. A.W. Clowes, Pathologic intimal hyperplasia as a response to vascular injury and reconstruction. In: *Vascular Surgery, R.B.* Rutherford, ed. WB Saunders Comp, Philadelphia, (1989).

3. J.M. Cook, B.W. Thompson and R.W. Barnes, Is Routine duplex examination after carotid endarterectomy justified? *J Vasc Surg* 12:334 (1990).
4. R. Adar, G.C. Critchfield, D.M. Eddy, A confidence profile analysis of the results of femoropopliteal percutaneous transluminal angioplasty in the treatment of lower-extremity ischemia, *J Vasc Surg* 10:57 (1989).
5. K. Detre, R. Holubkov, S. Kelsey, M. Cowley, K. Kent, D. Williams *et al,* Percutaneous transluminal coronary angioplasty in 1985-86. The National Heart, Lung and Blood Institute Registry. *N Engl J Med* 318:265 (1988).
6. R. Ross,. The pathogenesis of atherosclerosis: a perspective for the 1990s, *Nature* 362: 801 (1993).
7. R. Ross, J.A. Glomset, B. Kariya, L.A. Harker,. A platelet-dependent serum factor that stimulates the proliferation of arterial smooth muscle cells *in vitro*. *Proc Natl Acad Sci USA:*71:1207 (1974).
8. M.R. Buchanan, T. Haas, M. Lagarde and M. Guichardant, 13-hydroxy-octadecadienoic acid is the vessel wall chemorepellent, LOX, *J Biol Chem* 260:16056 (1985).
9. M.-C. Bertomeu, G.L. Crozier, T. Haas, M. Fleith and M.R. Buchanan, Selective effects of dietary fats on vascular 13-HODE synthesis and platelet/vessel wall interactions. *Thromb Res* 59:819 (1990).
10. M. Camacho, N. Godessart, R. Antón García, L. Vila, Interleukin-1 enhances the ability of cultured human umbilical vein endothelial cells to oxidize linoleic acid. *J Biol Chem* 270:17279 (1995).
11. A.W. Ford-Hutchinson, Arachidonate 15-lipoxygenase; characteristics and potential biological significance. *Eicosanoids* 4:65 (1991).

EFFECTS OF LINOLEIC ACID AND/OR MARINE FISH OIL SUPPLEMENTS ON VESSEL WALL THROMBORESISTANCE IN PATIENTS UNDERGOING CARDIAC SURGERY

S.J. Brister,[1] and M.R. Buchanan,[2]
Departments of [1] Surgery and [2] Pathology,
McMaster University,
Hamilton, Ontario, Canada,
L8L 8E7

INTRODUCTION

Platelet/vessel wall interactions following injury play an important role not only in hemostasis, but also in acute thrombosis and subsequent atherogenesis [1]. In cardiac surgery, particularly coronary artery bypass grafting (CABG), these platelet/vessel wall interactions are thought to contribute to acute thrombus formation and chronic vessel wall hyperplasia and subsequent vessel wall restenosis. To date, the most common approach to attenuate these responses is the short-term administration of high dose heparin during cardiac surgery (to inhibit thrombin activity) and the long-term administration of low dose aspirin to inhibit platelet thromboxane A_2 (TxA_2) generation (which is thought to promote platelet aggregation and vasoconstriction [2]. Despite these approaches, acute thrombosis and chronic restenosis remain significant problems in 5 - 40 % of these patients within 6 - 12 months post surgery.

Recent studies in our laboratories suggest an alternate approach to attenuate these events; namely, to increase vessel wall thromboresistance *before* surgery in an attempt to attenuate the platelet/vessel wall interactions *at the time of injury; i.e. at the time of surgery,* thereby ameliorating the consequences of these interactions following surgery, such as decreasing the release of hyperplastic growth factors and vasoconstrictors.

Background

Buchanan *et al* [3] reported that 13-hydroxyoctadecadienoic acid (13-HODE) is synthesized from linoleic acid via the lipoxygenase pathway in vascular wall cells. 13-HODE is synthesized under basal conditions, i.e. by the healthy vessel wall which is nonthrombogenic. These and other observations led us to hypothesize that 13-HODE regulates blood cell/vessel wall interactions [4]. Consistent with that possibility, we demonstrated in rabbits that platelet/vessel wall adhesion following vessel wall injury can be increased or decreased by manipulating vessel wall 13-HODE synthesis *in vivo*. These effects were achieved by either pretreating the rabbits

Recent Advances in Prostaglandin, Thromboxane, and Leukotriene Research
Edited by Sinzinger *et al.*, Plenum Press, New York, 1998

with an "anti-platelet drug", persantine, or with dietary fat supplements rich in linoleic acid (the substrate for 13-HODE) [5-6] . The latter dietary effects were associated with the selective incorporation of linoleic acid into the vessel wall. Riambau *et al* (*see elsewhere in this volume*) report that a similar linoleic acid-rich dietary supplement given to rabbits also attenuates chronic vessel wall hyperplasia and restenosis following vessel wall injury. More importantly, we have reported that human vessel wall tissues obtained from patients undergoing coronary artery bypass grafting (CABG) also synthesize 13-HODE, and the level of vascular wall 13-HODE correlates with vessel wall thromboresistance[7]. Our most recent data suggest that vascular wall cell 13-HODE contributes to vessel wall thromboresistance by downregulating the adhesivity of the endothelial cell integrins such as the vitronectin receptor, thereby reducing platelet/vessel wall interactions (*see Buchanan and Brister elsewhere in this volume*).

HYPOTHESIS

These studies prompted us to test the hypothesis that having patients ingest a dietary supplement rich in linoleic acid *before* CABG will increase vessel wall 13-HODE synthesis, thereby decreasing vessel wall adhesivity to platelets *at the time of* surgery[4].

EXPERIMENTAL DESIGN

Patient population and treatment: Sixty patients undergoing elective CABG were randomly assigned to receive a total daily dose of 3.2 gm of one of four dietary fatty acid supplements for 4 weeks before CABG: Group 1 - Evening Primrose, [(EVP, rich in both linoleic (15 %) and gamma linolenic (7-10%) acids]; Group 2 - Marine Fish Oil, [MFO, containing 18% eicosapentanoic acid (EPA) and 12% docosahexanoic acid (DHA)], total dose, EPA 1.9 gm, DHA 1.26 gm. MFO was included since other studies have suggested that MFO inhibits platelet function, thereby attenuating platelet/vessel wall interactions [8]; Group 3 - [EVP + MFO] in a ratio of 3:1; and a placebo control, Group 4 - Gelatin capsules.

All dietary supplements were discontinued following CABG. All patients received the standard treatment, including 325 mg aspirin daily up until 7 - 10 days before surgery, and restarted 24 -48 hours post surgery, and any other recommended therapy. This study design was approved by an Institutional Ethics Review committee and all patients signed an approved Informed Consent form.

Experimental interventions: Blood samples were collected before starting the dietary supplements and at one week intervals until the day of operation. These samples were used to analyze their platelet TxA_2 and 12-hydroxyeicosatetraenoic acid (12-HETE) levels, platelet aggregation and platelet adhesivity. On the day of surgery, discarded segments of internal mammary artery (IMA) and saphenous vein (SV) not used for CABG, were assessed for their thromboresistance (described below) and their 13-HODE levels. The levels of 13-HODE and 12-HETE were measured in segments of IMA and SV and platelets, using the automated HPLC method previously described by Haas *et al* [9].

To assess vessel wall thromboresistance, segments of the IMAs and SVs were split longitudinally and pinned on a dissecting mat, endothelial cell side up. A standard amount (400 uL ^{3}H-labelled autologous platelet-rich plasma (PRP, ~ 2 x 10^5 platelets/μL) was pipetted onto the endothelial cell surface and incubated at 37°C . Thirty 30 minutes later, the vessel wall was rinsed of the PRP and the number of ^{3}H-labelled platelets remaining adherent to the vessel wall surface was assessed by measuring its radioactivity and the PRP specific activity.

To assess platelet adhesivity *in vitro,* the patient blood was collected into 3.8 % sodium citrate and centrifuged at 200 g for 12 minutes to prepare PRP. The PRP was transferred into a separate tube and the remaining blood was centrifuged at 2000 g for 30 minutes to obtain platelet-poor plasma (PPP) which was used to dilute the PRP to a final concentration of 2.5 x 10^8/mL. The PRP was incubated with 4 μCi ^{3} H-adenine/mL for 30 minutes at 22 ° C. [Unlabelled adenine (0.12 mM final concentration: i.e. 100 fold excess of ^{3}H-adenine) was added to block the uptake of any free ^{3}H-adenine onto the fibronectin-coated discs during the adhesion assay.] Then, 500 μL ^{3}H-PRP were decanted into Costar 24 plastic wells containing Thermanox plastic discs pre-coated with 400 ug fibronectin, and incubated for 30 minutes at 37°C. The discs were removed from the ^{3}H-PRP, rinsed 3 times to remove any non-adherent platelets, and then the discs were placed into a liquid scintillation vial containing the toluene base scintillation fluid, and their radioactivities were measured in a β counter. The number of platelets adherent to each fibronectin-coated disc was calculated from the ^{3}H-PRP specific activity and disc radioactivity.

To assess platelet aggregation, the whole blood was collected into 3.2% trisodium citrate (9:1, v/v), and PRP prepared by centrifugation at 180 g for 12 minutes at 37°C. The PRP was transferred to another tube and the remaining blood was recentrifuged at 1700 g for 12 minutes. The resulting PPP was transferred to another tube and stored at 37°C. The PPP was used to dilute the PRP to a final concentration of 2.5 x 10^6 platetets/μl. Platelet aggregation in response to known concentrations of adenosine diphosphate and type III collagen were performed, using the standard optical density/turbidity method[10].

RESULTS

Platelet aggregation and adhesivity were unchanged in any of the dietary supplement groups and did not differ among groups, relative to these tests performed at the onset of study. There were also no differences in platelet TxA_2 and 12-HETE synthesis among groups. In contrast, there were significant differences in 13-HODE synthesis by the IMAs and SVs obtained from the patients in the different dietary supplement-treated groups. The 13-HODE levels in IMAs obtained from the placebo-treated group were 640 $\pm$ 321 ng/gm/cm^2, mean $\pm$ SD, (**Table 1**). IMA 13-HODE levels were significantly increased in the patients who received the EPO dietary supplement but remained unchanged in the IMAs of the [EPO+MFO]- or MFO- treated patients. Similar results were obtained for the SVs, but the absolute values of 13-HODE were > 3 fold less than for the IMAs (data not shown).

Table 1. 13-HODE levels in IMAs obtained from CABG patients receiving the different dietary fatty acid supplements for 28 days.

	Placebo	EPO	[EPO+MFO]	MFO
13HODE (ng/gm/cm^2)	640 $\pm$321	1210 $\pm$ 343*	799 $\pm$ 149	487 $\pm$ 139

*Data are expressed as means $\pm$ SD, n = 15. * = p < 0.01 vs Placebo.*

The increased levels of 13-HODE in the IMAs of the EPO-treated patients were associated with increased vessel wall thromboresistance; i.e. the higher the 13-HODE level, the lower the number of platelets adherent to the IMA surface, $r = -0.4942$, $*p < 0.05$.

CONCLUSIONS

First, these results are consistent with our previous findings that IMAs and SVs obtained from CABG patients synthesize 13-HODE and that the IMAs synthesize more 13-HODE than SVs[7]. It is interesting to note that IMAs have a >90% patency rate over 10 years , while SVs have only a 60% patency rate over 10 years post CABG[2]. Second, the current results indicate that human vessel wall 13-HODE synthesis can be increased by dietary fatty acid supplementation, thereby increasing vessel wall thromboresistance. If such effects result in comparable long-term outcomes as those which have been reported in animals (*Riambau et al*), such a dietary intervention before (cardio)vascular surgery may be an inexpensive and warranted therapy to attenuate chronic hyperplasia and subsequent vessel wall restenosis.

REFERENCE

1. A. Chervu and WS Moore, An overview of intimal hyperplasia.*Surg* 171: 433 (1990).
2. RITA Trial Participants, Coronary angioplasty versus coronary artery bypass surgery: The randomized intervention treatment of angina (RITA) trial. *Lancet* 341: 573 (1993).
3. M.R. Buchanan, T. Haas, M. Lagarde and M. Guichardant, 13-hydroxyoctadecadienoic acid is the vessel wall chemorepellant, LOX. *J Biol Chem* 25:16056 (1985).
4. M.R. Buchanan and S.J. Brister, Other essential fatty acids, their metabolites and cell/cell interactions. In: *The Handbook of Immunopharmacology*, F Cunningham, ed, London, (1994).
5. E. Weber, T.A. Haas, T.H. Mueller, W.G. Eisert, J. Hirsh and M.R. Buchanan, Relationship between vessel wall 13-HODE synthesis and vessel wall thrombogenecity following injury. Influence of salicylate and dipyridamole treatment. *Thromb Res*, 57: 383 (1990).
6. M.-C. Bertomeu, G.L. Crozier, T.A. Haas, M. Fleith and M.R. Buchanan, Dietary fats, vessel wall 13-HODE synthesis and vessel wall thrombogenecity. *Thromb Res* 59: 819 (1990).
7. S.J. Brister, T.A. Haas, M.-C. Bertomeu, J. Austin and M.R. Buchanan, 13-HODE synthesis in internal mammary arteries and saphenous veins: Implications in cardiovascular surgery. *Adv Prost Thromb Leuk* 21: 667 (1990).
8. B.J. Holub, Dietary fish oils containing eicosapentaenoic acid and the prevention of atherosclerosis and thrombosis. *Can Med Assoc J* 139: 377 (1989).
9. T.A. Haas and M.R. Buchanan, An automated extraction and quantification procedure for lipoxygenase metabolites. *J Chromat* 430: 1 (1988).
10. G.V.R. Born and H.J. Cross, Aggregation of blood platelets. *J Physiol* 168: 178 (1963).

13-HYDROXYOCTADECADIENOIC ACID REVERSES EPIDERMAL HYPERPROLIFERATION VIA SELECTIVE INHIBITION OF MEMBRANE PROTEIN KINASE C-β ACTIVITY

Vincent A. Ziboh

Department of Dermatology
University of California, Davis
Davis, California 95616

The most abundant polyunsaturated fatty acids (PUFAs) in human skin are linoleic acid (LA, 18:2n-6) and arachidonic acid (AA, 20:4n-6) (1,2). LA is metabolized by epidermal 15-lipoxygenase into predominantly 13-hydroxyoctadecadienoic acid (13-HODE). An analysis of lipoxygenase products in the model of docosahexaenoic acid (22:6n-3. DHA-induced hyperproliferative skin epidermis revealed a selective and markedly depressed level of 13-hydroxyoctadecadienoic acid (13-HODE), (3). The topical application of 13-HODE on the DHA-induced hyperproliferative guinea pig (GP) epidermis caused a reversal to a normal state. The reversal paralleled the in vivo replenishment of tissue with 13-HODE (3). These findings suggested that 13-HODE, may be involved in the regulation of epidermal hyperproliferation.

Since a number of reports indicate that lipoxygenase products of arachidonic acid (HETEs) and linoleic acid (13-HODE) are incorporated into tissue and cellular phospholipids (4,5,6) and can be released by receptor-mediated stimulation of the cells as putative substituted diacylglycerol (DAG), we tested the hypothesis that the mechanism of 13-HODE reversal of epidermal hyperproliferation may be due, at least in part to its incorporation into epidermal phosphatidyl 4, 5-bisphosphate (PtdIns4, 5- P_2) followed by its hydrolytic cleavage by epidermal phospholipase C to yield putative 13-HODE-substituted DAG in epidermis which functions to modulate selective epidermal protein kinase C (PKC) and epidermal hyperproliferation. Protein kinase C is a calcium/dialcylglycerol/phospholipid-dependent kinase which is recognized to play an important role in the regulation of cellular proliferation and differentiation (7,8,9,10). The PKC isotypes: PKC-α and PKC-β were identified as the main PKC isozymes in normal epidermis (11).

Methods/Results

Employing a docosahexaenoic acid (22:6n-3) induced model of hyperproliferative guinea pig epidermis (5), we demonstrated the reversal of hyperproliferation by topical 13-HODE as shown in Fig. 1.

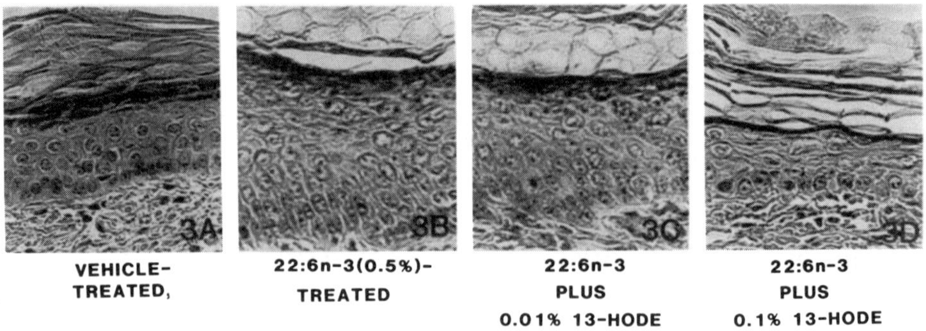

Figure 1. Skin in 3A was treated with vehicle (ethanol/propylene glycol [30:70] containing as antioxidant 1% vitamin E); Skin in 3B was treated with 0.5% of DHA in vehicle containing 1% vitamin E. Note acanthosis and hyperkeratosis in. 3B; Skin in 3C is with low conc. of 13-HODE. Data in. 3D represents skin in. 1B treated with 0.1% 13-HODE in vehicle. Note the reversal of acanthosis in 3D.

Histologic Evaluation and Epidermal Hyperproliferation of DHA-treated and 13-HODE-treated Epidermis: Histologically, the DHA-treated skin (3B) revealed marked epidermal hyperplasia characterized by acanthosis (thickening of the spinous layer) when compared with control ethanol/propylene glycol (vehicle) treated skin (3A). Dose-dependent topical application of 13HODE (0.01 and 0.1%) to DHA-treated skin reversed the hyperproliferative state (3C, and 3D). Uptake of [^{3}H]thymidine ([^{3}H]Td) revealed that DHA-treated epidermis markedly incorporated [^{3}H]Td (233%) when compared to vehicle treated skin.

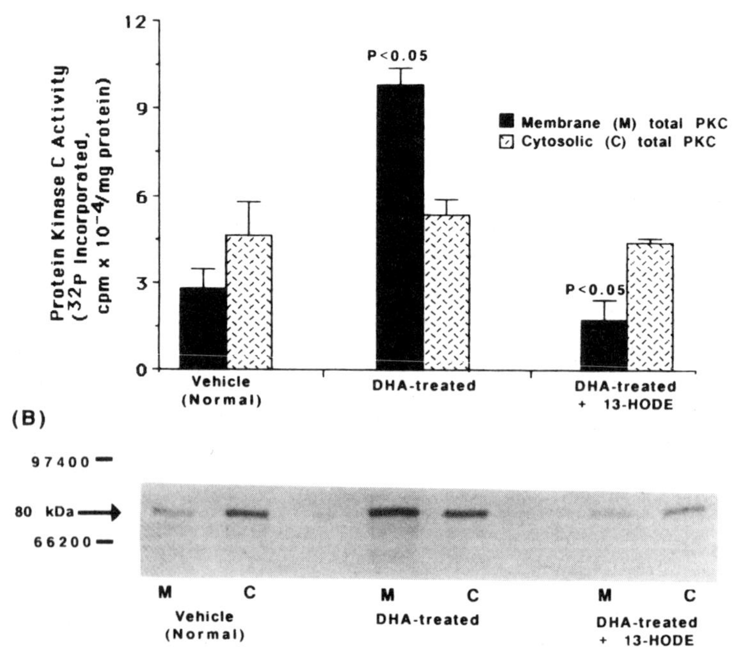

Figure 2. Subcellular distribution of total PKC activities in vehicle treated (normal), DHA-treated (hyperproliferation) and DHA-treated plus 13-HODE (reversal) Guinea Pig Epidermis.

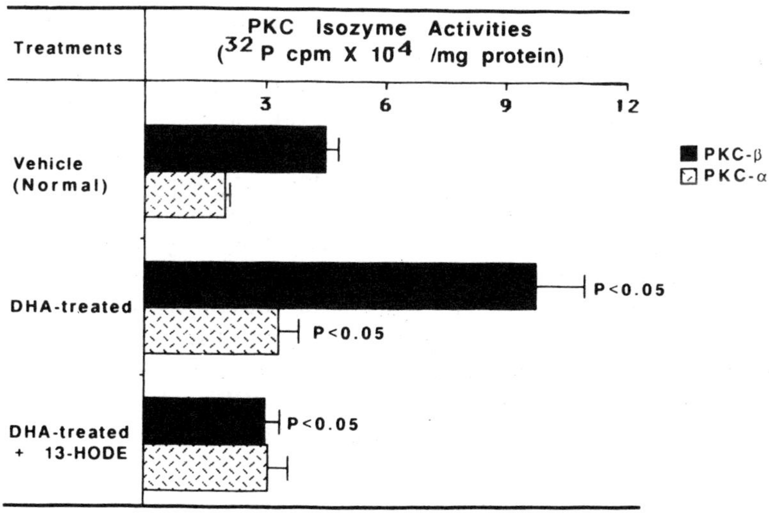

Figure 3. Activities of PKC isozymes (β and α) in vehicle treated (normal), DHA-treated (hyperproliferative) and DHA-treated plus 13-HODE (reversal) guinea pig epidermis.

Altered Total PKC Activity in DHA-Treated and 13-HODE-treated Epidermis

To delineate a possible mechanism for 13-HODE effect, we removed keratomed slices of treated epidermis and demonstrated that topical 13-HODE was incorporated into PtdIns 4, 5-P_2 and released as 13-HODE-containing diacylglycerol (13-HODE-DAG). To discern whether or not total tissue PKC was altered during the various skin treatments, μg protein aliquots from each of the epidermal homogenates of the treated animals were subjected to SDS-PAGE/Western-blot analysis. The data in Fig 2B revealed a strong expression of an 880 KD membrane-associated (M) total PKC in DHA-treated skin epidermis when compared to normal control epidermis. Topical application of 13HODE to the DHA-treated skin epidermis revealed diminished expression of the 80 KD membrane-associated PKC as shown in [2B]. Interestingly, the expression of 80 KD protein in [2B] corresponded with the elevated PKC activity shown above.

Altered PKC Isozyme activities in DHA-treated and 13-HODE-treated Epidermis:

To determine whether the altered effects of total PKC activity was on selective PKC isozymes, protein aliquots (30 μg) from each epidermal preparation was subjected to gel electrophoresis (SDS-PAGE) followed by electrophoretic transfer of the proteins onto nitrocellulose membrane. Incubation of the fractions with specific monoclonal antibodies revealed that the activity of PKC-β was markedly elevated in DHA-treated epidermis (approximately 257% over control) hen compared to vehicle-treated epidermis as shown in Fig. 3. The activity of PKC-α was also moderately elevated (approximately 136%). Interestingly, topical application of 13-HODE to the DHA-treated epidermis resulted in tissue replenishment of 13-HODE-DAG which was associated with the selective suppression of the elevated PKC-β activity while exerting negligible effect on PKC-α

Taken together, these studies revealed that membrane PKC-β isozyme activity is markedly elevated in the model of DHA-induced hyperproliferative epidermis (Fig. 3) and this was significantly suppressed after treatment of the skin with free 13-HODE. In contrast, PKC-α activity although was elevated in DHA-treated epidermis was only moderately modulated after 13-HODE treatment. These studies, therefore, underscore the selective response of membrane PKC-β to natural putative 13HODE-DAG in modulating epidermal hyperproliferation, further suggesting that the activities of other PKC-isoforms may be regulated by other selective inhibitors.

The importance of the 18-carbon 13-monohydroxy fatty acid (13-HODE) in the biology of normal epidermis underscores the dietary essentiality of linoleic acid in this issue. Data from these studies underscore the possibility that a monohydroxy fatty acid (13-HODE) derived from 18-carbon linoleic acid singly, or in association with other molecules, can modulate epidermal hyperproliferation in vivo. In a separate study, we also demonstrated that the in vivo regeneration of 13-HODE-DAG after refeeding the essential fatty acid deficient animals with safflower oil containing LA was coupled with the selective down regulation of membrane associated PKC-β, (12) thus supporting our view that 13-HODE-DAG is biologically generated in normal epidermis and functions in vivo to modulate epidermal growth.

ACKNOWLEDGMENT

Supported in part by Research Grant AM-30679 from the National Institutes of Health of the US Public Health Service.

REFERENCES

1. H.J. Yardley and R. Summer. Lipid Composition and Metabolism in Normal and Diseased Epidermis. *Pharmac. Ther.* 13: 357-383. (1981).
2. R.S. Chapkin, V.A. Ziboh, C.L. Marcelo and J.J. Voorhees. Metabolism of Essential Fatty Acids by Human Epidermal Enzyme Preparations: Evidence of Chain Elongation. *J. Lipid Res*. 27: 945-954. (1986).
3. C.C. Miller and V.A. Ziboh. Induction of Epidermal Hyperproliferation by Topical n-3 Polyunsaturated Fatty Acids on Guinea Pig Skin Linked to Decreased Levels of 13-hydroxyoctadecadienoic acid (13HODE). *J. Invest. Dermatol.* 94: 353-358. (1990).
4. C.C. Miller, W. Tang, Y. Cho and V.A. Ziboh. Selective Incorporation of HETES into Epidermal Phospholipids. In: Prostaglandins, Leukotrienes, Lipoxins, and PAF. Mech. of Action. Mol. Biol. and Clin Applications, Plenum Publ. Corp. (J.M. Bailey, ed.), New York, pp. 409-418. (1991).
5. A.B. Legrand, J.A. Lawson, B.O. Meyrick, I.A. Blair and J.A. Oates. Substitution of 15-Hydroxyeicosaenoic Acid in the Phosphoinositide Signaling Pathway. *J. Biol. Chem.* 266: 7570-7577. (1991).
6. M.E. Brezinski and C.N. Serhan. Selective Incorporation of (15S)-hydroxyeicosatraenoic Acid in Phosphatidyl Inositol of Human Neutrophils: Agonist-induced Deacylation of Transformation of Stored Hydroxyeicosanoids. *Proc. Natl. Acad. Sci.* 87: 6258-6262. (1990).
7. Y. Nishizuka. Studies and Perspectives of Protein Kinase C. *Science.* 233: 305-312. (1986)
8. L.P. Molleyres and R.R. Rando. Structural Studies on the Diacylglycerides-mediated Activation of Protein Kinase C. *J. Biol. Chem.* 263: 14831-14838. (1988).
9. N.R. Murray, G.P. Baumgardner, D.J. Burns and A.P. Field. Protein Kinase C Isotypes in Human Erythroleukemia (K562) Cell Proliferation and Differentiation. *J. Biol. Chem.* 268: 15847 -15853.

10. M.S. Matsui, S.L. Chew and V.A. DeLeo. Protein Kinases C in Normal Human Epidemal Keratinocytes During Proliferation and Calcium Induced Differentiation. *J. Invest Dermatol.* 99: 565-572. (1992)
11. S. Inohara, Y. Tatsumi, H. Tanaka, H. Tateishi, S. Sagami. Immunohistological Identification of Protein Kinase C Isozymes in Normal and Psoriatic Epidermis. *Arch. Dermatol. Res.* 280: 454-455. (1988).
12. Y. Cho and V.A. Ziboh. Nutritional Modulation of Skin Hyperproliferation in Essential Fatty Acid-Deficiency is Associated with Selective Down-Regulation of Protein Kinase C-β. *J. Nutrition.* 125: 2741-2750.

EFFECTS OF LINOLEIC ACID SUPPLEMENTS ON ATOPIC DERMATITIS

Ana Giménez-Arnau,[1] Carlos Barranco,[2] Margarita Alberola,[2] Cathy Wale,[3] Sergio Serrano,[2] Michael R. Buchanan,[3] José G. Camarasa,[1]

[1] Department of Dermatology. Hospital del Mar, IMIM , Universitat Autònoma de Barcelona, and
[2] Department of Pathology. Hospital del Mar, IMIM., Universitat Autònoma de Barcelona, Spain.
[3] Department of Pathology. MacMaster University, Hamilton, Canada.

INTRODUCTION

Atopy is primarily a condition of hypersensitivity. Atopic disease include atopic eczema, allergic bronchial asthma or hay fever. Atopic dermatitis signs and symptoms are dry itching skin, and eczematous inflammation. Pathophysiology involve: genetics, trigger factors as aeroallergens, superantigens or stress, Th2 lymphocyte cytokine profile (Interleukin (IL)-4,-5,-10) dominance, hyper IgE, monocyte phosphodiesterase increased activity, impaired epidermal barrier function and abnormal linoleic acid metabolism.

Fish oil [eicosapentaenoic (EPA) + docosahexaenoic (DHA) acids] and vegetable oil [linoleic acid (LA)] supplements are thought to be beneficial to treat atopic dermatitis. N-3 unsaturated fatty acid is transformed by lipoxygenase to leukotriene (LT)A5 and LTB5 or by cycloxygenase to prostaglandin (PG)E5, PGI3 and thromboxane (TX) TXA3. These compounds have been shown to have lower inflammatory activity than PGs of the 2 or LTs of the 4 series.[1] N-6 fatty acids, have been implicated in biological functions as immunomodulation, epidermal proliferation, epidermal barrier function, carcinogenesis or atherosclerosis. These effects have been atributed, in part, to the metabolites, 13-hydroxyoctadecadienoic acid (13-HODE), and prostaglandin E1, (see Horrobin and Ziboh, this isssue). Some clinical trials support LA+GLA efficacy to treat atopic dermatitis.[2] However, other studies have not. Essential fatty acids relevance is demonstrate by the "Essential Fatty Acid deficiency syndrome", characterized in mice by dry skin and dermatitis, weigh loss, cardiomegalia, loss of fertility and cholesterol deposit in lungs.[3]

The aim of this study is to demonstrate that atopic dermatitis in initially, also an essential fatty acid deficiency related disease. N-6 fatty acids (LA) should be more effective than n-3 (EPA+DHA) fatty acids to treat severe atopic dermatitis. If so,whether any beneficial effect should be associate with increased levels of 13-HODE.[4]

STUDY DESIGN & METHODS

Study design

Forty eight adult patients with chronic and severe atopic dermatitis (Rajka score ≥7) were randomly included in a pilot, parallel group, double blind, placebo controlled clinical trial during 1992-93. All patients signed the informed consent. One month wash out period preceded patient inclusion. Any other treatment was not permitted. Patients ingested either LA (3 gr/day), EPA +DHA (2.0+1.3 gr/day) or placebo (3.0 gr/day, oleic acid). Oil content in hard gelatine capsules has a 35ºC -40ºC fusion point. Our source of LA was sunflower oil (LA 63%) provided by Serra Pamies S.A (Reus, Spain). Each capsule of sunflower oil

Recent Advances in Prostaglandin, Thromboxane, and Leukotriene Research
Edited by Sinzinger *et al.*, Plenum Press, New York, 1998

contained 250 mg of LA in tryglyceride form.Twelve capsules per day were ingested during three months. We chose sunflower oil rich in LA and poor in GLA to measure 13-HODE, a direct metabolite from LA by the 15-lypoxygenase pathway. 13-HODE has a demonstrated antiproliferative capacity.[4]

Clinic evaluation

Clinic evaluation was performed in forty eight patients by two physicians who were blinded to patient treatment before and after 6 and 12 weeks. Disease severity was assessed by "Rajka score" which takes into consideration disease extension, course of eczema and itch intensity. Score ranking is between 3 and 9. We include ≥ 7 Rajka score patients. "Rule of 9" also was used to assess eczema extension (top 73).

Pathologic evaluation

Two 4 mm punch biopsies each patient (n=48) was performed. First biopsy on eczematous skin before and secondly after 6 weeks of treatment. Two morphometric parameters, epidermal area (mm^2) and epidermal thickness (mm) were useful to evaluate pathological improvement. Special method semiautomatic with the image analysis Videoplan 2 (Zeiss) and the microscope Polyvar (Reichert Jury) was used. Each result was the media of ten measurements. Inflammatory infiltrate characteristics (type, distribution and intensity) was evaluated. mRNA IL-10 and ICAM-1 expression, by *in situ* hybridization in paraffin blocked samples, before and after the LA supplementation was also studied.[5,6] Proteinase K (5μg/ml) and biotin IL-10 (800 ng/slice) and ICAM-1 (200 ng/slice) oligonucleotide from R&D Systems was employed. Each biopsy was tested with and without probe and with polydT as control of mRNA preservation.

Biochemistry assessment

Four 6 mm punch biopsies on eczematous skin and healthy skin, before and after 6 weeks treatment were obtained in 30 patients. Tissue was collected in HPLC tubes and extracted in 2 ml methanol, always manipulated at 4°C with nitrogen. Epidermis/dermis 15- hydroxyeicosatetraenoic acid (HETE) and 13-HODE levels in skin biopsies were measured as indices of [EPA+DHA] and LA skin uptake respectively. We used reverse phase HPLC.[7] Results were expressed as ng/cm^2/gr.

RESULTS

Forty eight adult atopic patients with chronic and severe eczema (Rajka score ≥7) were included. Female/male distribution was 3:1. Mean age was 24.2±6.8 (S.D.). Most patients showed family background of atopic disease. Thirty three patients only had atopic dermatitis and 15 patients presented a combination of eczema and other associated atopic disease. "Besnier prurigo" was the main manifestation of atopic dermatitis in 3 cases. IgE levels were increased in 30 patients ranging between 106 UI/ml and more than 5000 UI/ml. There were no differences in atopic dermatitis severity among the three groups before treatment.

Clinic

Clinic evaluations results showed a significant percentage reduction of both clinical scores at 6 weeks (Rajka score 73±10.6 SD, Rule of nine 81±9.7 SD) and at 12 weeks (Rajka score 80.6±12 SD, Rule of nine 90±6.7 SD) in patients treated with dietary supplements rich in LA, (Table I). Significant differences in the percentage of Rajka score and Rule of 9 reduction between patients treated with LA and [EPA+DHA] ($p< 0.0001$) and also in patients treated with LA and placebo ($p< 0.0001$) were observed. All patients treated with EPA+DHA and seven patients treated with placebo refused to continue the study at six weeks due to severe itch and eczema worsening. Nauseas, vomiting and bad flavour was the most frequent complain in this group of patients. Any adverse event was referred in patients treated with LA.

Pathology

Epidermal area (35.8±8.8 mm2 SE) and epidermal thickness reduction (26.1±13.0 mm SE) in LA treated group was observed (Table I). (Fig.1) We found significant differences in epidermal area reduction, between LA (Fig.1) and EPA+DHA treated patients ($p< 0.005$) and between LA and placebo ($p< 0.011$). Significant differences in epidermal thikness reduction between LA and EPA+DHA ($p< 0.05$) were also

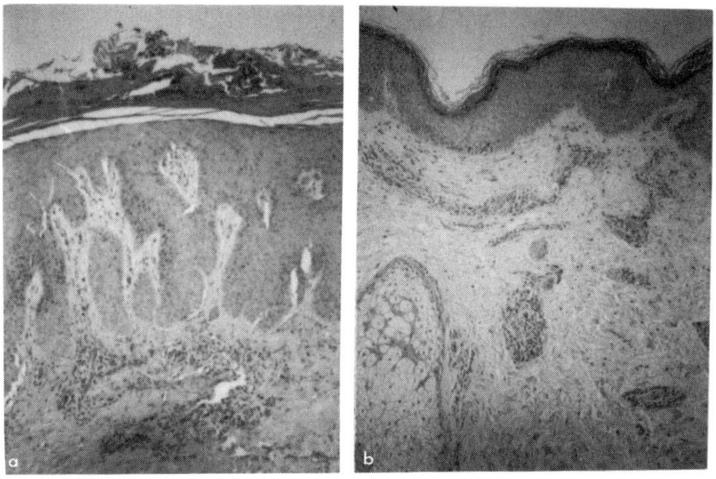

Fig.1 a) Chronic atopic eczema pathology before to be treat and b) six weeks after LA supplies.

observed. Chronic AD strongly expressed mRNA ICAM-1 in the cytoplasm of epidermal keratinocytes located close to the dermal and intraepidermal inflammatory infiltrate. Chronic atopic dermatitis expressed mRNA IL-10 in the cytoplasm of epidermal cells in all epidermal layers and in the inflamatory infiltrate. After LA intake both mRNA ICAM-1 and IL-10 expression dissapeared.

Biochemistry

Significant differences between basal levels of 13-HODE and 15-HETE in healthy skin ($p< 0.016$) and sick skin ($p< 0.015$) were obtained. 13-HODE levels were higher. The decreases in atopic dermatitis in the LA patients, as demonstrated both histologically and clinically, were associated with an increase in cutaneous 13-HODE levels. The increase of 13-HODE was observed in healthy skin of atopic patients, (Table I), specially after LA treatment.

TABLE I - Effects of LA,(EPA+DHA) and placebo on clinical scores,morphology and cutaneous 13-HODE levels,as measured after 6 weeks of treatment.

	Linoleic ac.	EPA+DHA	Oleic ac.
% Reduction of clinical scores			
Rajka Score	74.6±3.2*	5.3±5.1	8.8±8.5
Rule of Nine	81.3±3.1*	-8.6±5.3	-2.7±15.0
% Morphometric reduction			
Epidermal area	35.8±8.8*	-5.4±8.9	2.6±6.4
Epidermal thikness	26.1±13.0*	-2.1±6.9	12.9±14.8
% Increase 13-HODE			
Healthy atopic skin	186.0±92.3*	13.9±60	8.4±35

Data expressed media ± S.E., as compared to before treatment values * $p< 0.01$.

DISCUSSION

Hansen was the first to suggest that atopic dermatitis is an essential fatty acid deficiency disorder. He observed a reduced blood levels of unsaturated fatty acids. Addition of maize oil to the diet improved the eczema.[8] Uncontrolled trials of dietary fat supplements indicate linoleic acid to be more effective than lard.[9] Controlled trials using small doses of linolenic acid (130 mg daily) and linoleic acid (270 mg/day) was not benefit.[3] Only high doses of LA (2880-4320 mg/day) and GLA (360-540 mg/day) are effective to treat atopic dermatitis.

Our clinic and pathological results demonstrate sunflower oil efficacy, rich in LA and poor in GLA to treat severe and chronic atopic dermatitis. At least, 3 gr LA per day during 6 weeks is necessary. Animal fish oils, rich in EPA+DHA and olive oil (placebo) were ineffective to treat severe and chronic atopic dermatitis. We speculate if epidermal area and thickness reduction after LA rich oil supplies is due to a direct or indirect action on keratinocyte proliferation. Also we observed less inflammatory infiltrate and negative mRNA IL-10 and ICAM-1 expression after treatment probably due to a direct or indirect Th2 response immunomodulation.

Lack of Δ 6 and 5 desaturase in normal epidermis has been described. Skin arachidonic acid must be extraepidermal.[10] Increase LA and decreased GLA levels in serum, erythrocytes, fat or T lymphocytes from umbilical cord in atopic patients, suggest a Δ6 desaturase deficiency.[11] However, Zevenverger et al [12] and Pfeiffer A et al [13] demonstrate that Δ6 desaturase is not the only enzyme implicated in polyunsaturated fatty acids tissue levels and they did not found its deficiency in atopic patients. Δ6 desaturase deficiency is still controversial. How sunflower oil rich in LA and poor in GLA acts is difficult to explain. We can speculate with some hypothesis.

Atopic dermatitis is characterized by dry skin and an impaired barrier function. LA ingestion or LA topical application contribute to restore transepidermal water loss.[14] In normal skin LA sterifies skin stratum corneum ceramides. An epidermal fatty acid-binding protein with high affinity fo C18 series fatty acids (in decreasing order of affinity : stearic acid>LA>oleic acid) has been described. LA is necessary to maintain lipid barrier function. A decreased stratum corneum ceramides 1 (rich in LA) and 3 could explain itching and dry skin in atopic patients.[15-16] LA supplies could contribute to restore barrier function .

13-HODE observations with an increase in 13-HODE synthesis is compatible with Ziboh´s observations that epidermal proliferation can be attenuated by increasing epidermal 13-HODE levels.These results also are compatible with the hypothesis of Buchanan and Brister [17] who postulated that 13-HODE downregulates cell adhesion, thereby attenuating not only thrombosis and metastasis, but also inflammation.

We conclude that LA is useful to treat atopic eczema. Atopic dermatitis can be considered initially an essential fatty acid related disease. Further studies, are required to better understand the interactions of 13-HODE and LA and or GLA-derived cycloxygenase metabolites and their effects on barrier function, epidermal proliferation or immunological response in atopic dermatitis.

Acknowledgments : Fondo Investigación Sanitaria (FISS) n° exp 92/0301.

REFERENCES

1. E.Søyland,G. Rajka, A. Bjørneboe, G.E. Bjørneboe and C.A. Devron, The effect of eicosapentaenoic acid in the treatment of atopic dermatitis. A clinical study, Acta Derm Venereol (Stockh).Suppl 144: 139 (1989).
2. S.J.Taub and S.J.Zakon,Use of unsaturated fatty acids in the treatment of eczema,JAMA.105:1675 (1935)
3. G.O.Burr and M.M. Burr, A new deficiency disease produced by rigid exclusion of fat from the diet, J Biol Chem. 82: 345-367 (1929).
4. C.C.Miller and V.A. Ziboh, Induction of epidermal hyperproliferation by topical n-3 polyunsaturated fatty acids on guinea pig skin linked to decreased levels of 13-hydroxyoctadecadienoic acid (13-HODE), J Invest Dermatol. 94: 353-358 (1990).
5. A.Giménez-Arnau, C.Barranco, C.Pla, M.Arumí, E.Mato, S.Serrano and J.G.Camarasa, mRNA interleukin-10 expression by *in situ* hybridization in atopic dermatitis. Immunological significance, J Invest Dermatol 106: 467 (1996).
6. A.Giménez-Arnau, C.Barranco,C.Pla , M.Arumí, E.Mato, S. Serrano and J.G.Camarasa, mRNA intercellular adhesion molecule-1 (ICAM-1) expression by *in situ* hybridization in severe atopic dermatitis. Influence of linoleic acid, J Invest Dermatol. 105: 511 (1995).
7. T.A.Hass and M.R.Buchanan, Automated high-performance liquid chromatographic extraction and quantification procedure for lipoxygenase metabolites, J Chromatogr. 430: 1-10 (1989).
8. A.E.Hansen, Serum lipid changes and therapeutic effects of various oils in infantile eczema, Proc Soc Exp Biol Med. 31: 160-161 (1933).
9. T.Cornbleet, The use of maize oil (unsaturated fatty acids) in the treatment of eczema, Arch Dermatol Syph (Chicago). 31: 224-234 (1935).
10. V.A.Ziboh and R.S.Chapkin, Metabolism and function of skin lipids, Prog Lip Res. 27:81-105(1988).
11. S.Oliwiecki, J.L.Burton, Elles and D.F.Horrobin, Levels of essential and other fatty acids in plasma and red cells phospholipids from normal controls and patients with atopic eczema, Acta Derm Venereol (Stockh). 71: 224-228 (1990).
12. J.L.Zevenbergen and U.M.T.Houtsmuller, Effect of dietary fats on linoleic acid metabolism. A radiolabel study in rats, Biochim Biophys Acta. 1002: 312-323 (1989).

13. A.Pfeiffer, H. Demmelmair, F. Rueff, B. Przybilla, D.Reinhardt and B.Koletzko, Normal *in vivo* activity of Δ6-desaturase in patients with atopic eczema, International Conference on Highly Unsaturated Fatty Acids in Nutrition and Disease Prevention Barcelona Spain 4-6 November, 124 (1996).
14. P.M.Elias, B.E.Brown and V.A.Ziboh, The permeability barrier in essential fatty acid deficiency: Evidence for a direct role for linoleic acid in barrier function, J Invest Dermatol. 74: 230-233 (1980).
15. L.Shäfer and K.A.Kragballe, Abnormalities in epidermal lipid metabolism in patients with atopic dermatitis, J Invest Dermatol. 96:10-15 (1991).
16. G.Imokawa, A.Abe, K.Jin, Y.Higaki, M.Kawashima and A.Hidano, Decreased level of ceramides in stratum corneum of atopic dermatitis. an etiologic factor in atopic dry skin ?, J Invest Dermatol. 96: 523-526 (1991).
17. M.R.Buchanan and S.J. Brister. Endothelial cell 13-HODE and blood cell adhesion to the vessel wall. In: Adhesion molecules and cell signaling: Biology and clinical applications. P.C. Weber, R.Lorenz,eds,Raven Press Inc,New York (1995)

THE IMPORTANCE OF LINOLEIC ACID METABOLITES IN CANCER METASTASIS AND IN THE SYNTHESIS AND ACTIONS OF 13-HODE

David F. Horrobin, Scotia Research Institute, Castle Business Park, Stirling, Scotland FK9 4TZ.

Vincent A. Ziboh, Department of Dermatology, University of California School of Medicine, Davis CA 95616, U. S. A.

SUMMARY

Large scale human epidemiological studies indicate that high intakes of linoleic acid protect against the development of cancer. One mechanism may be the generation of 13-HODE from linoleic acid. 13-HODE prevents cell adhesion to endothelial cells and can inhibit cancer metastasis. 13-HODE synthesis is enhanced by cyclic AMP. Gamma-linolenic acid, a desaturated metabolite of linoleic acid, causes substantial stimulation of 13-HODE synthesis. A fall in gamma-linolenic acid synthesis with age may be related to the age-related fall in 13-HODE formation.

INTRODUCTION

There is a substantial literature describing animal studies, primarily rat studies on mammary cancer, which demonstrates that linoleic acid promotes cancer growth. This literature presents a barrier to the idea that any metabolite of linoleic acid, including 13-hydroxydienoic acid (13-HODE) might have anti-metastatic or other anti-cancer actions.

Two large epidemiological studies have demonstrated that the animal work is misleading when used as a guide to what might happen in humans. In over 80,000 American nurses studied over five years, division into quintiles of linoleic acid intake demonstrated no significant effect of dietary linoleic acid intake on breast cancer[1]. There was a non-significant trend for women with the lowest linoleic acid intakes to have the most cancer. In a study of 3000 healthy male civil servants, linoleic acid intake and linoleic levels in plasma cholesteryl esters were assessed and the individuals then followed for an average of over nine years[2]. When the individuals were divided into tertiles, those with the highest tertile of intake of linoleic acid had a cancer death risk only 39% of that in those with the lowest tertile. Those with the highest tertile blood levels of linoleic acid had a cancer risk only 16% of that found in those in the lowest tertile. An increased intake of linoleic acid is protective against human cancer development. If this is so, then more interest is attached to the possible mechanism whereby linoleic acid may be protective. One of these mechanisms may be the production of 13-HODE.

Recent Advances in Prostaglandin, Thromboxane, and Leukotriene Research
Edited by Sinzinger *et al.*, Plenum Press, New York, 1998

13-HODE AND ANTI-METASTATIC ENDOTHELIAL EFFECTS

One mechanism whereby 13-HODE may interact with cancer is in the regulation of cancer cell motility, and adhesion to and migration through the endothelial barrier presented by capillaries. If cancer cells cannot adhere to the endothelium and cannot migrate through the barrier, metastatic foci cannot become established.

Two groups, those of Buchanan and Honn, have been responsible for the development of the concept that 13-HODE is likely to play an important role in the regulation of metastasis by modulation of endothelial cell behaviour. Buchanan et al proposed that the chemorepellent property of the endothelium could be explained by its production of substantial amounts of 13-HODE[3]. Agents which enhanced platelet and cellular adhesion to endothelial cells, such as f-Met-Leu-Phe, endotoxin, thrombin or C5, all reduced endothelial 13-HODE production[4]. Moreover, production of 13-HODE by the tumour cells themselves also had an effect on vascular adhesion. Those tumour cell lines which produced the least 13-HODE were more likely to adhere. The ratio of 13-HODE, the 15-lipoxygenase metabolite of linoleic acid, to 15-hydroxy-eicosatetraenoic acid (15-HETE), the arachidonate metabolite produced by the same enzyme, appeared important[4,5]. Cyclic AMP stimulated the preferential synthesis of 13-HODE as compared to 15-HETE[5,6]. Various cytokines (including interleukin 1 alpha, interleukin 1 beta and tissue necrosis factor) and thrombin, which enhanced endothelial reactivity appeared to do so by regulating cyclic AMP and the ratio between 13-HODE and 15-HETE[5,6]. Dipyridamole which inhibits endothelial reactivity raised cyclic AMP levels[6]. The vitronectin receptor may play a central role in these changes in endothelial function. 13-HODE may change its conformation, so decreasing its adhesivity[7].

Studies which lead to similar conclusions have emerged from Honn's laboratory. Using a Lewis lung carcinoma model, they found that tetradecanoylphorbolacetate (TPA) and 12-HETE could both enhance adhesion of cancer cells to endothelial cells and also to subendothelial matrix and fibronectin[8]. The effects of both agents on all three substances could be inhibited by 13-HODE, apparently working via a plasma membrane receptor similar to the platelet aggregation receptor complex[8]. 12-(S-)-HETE had similar effects in enhancing adhesion of Walker 256 carcinosarcoma cells, by increasing translocation of protein kinase C to membrane sites. 13-HODE inhibited the effects of 12-HETE on both tumour cell adhesion and PKC[9]. In vivo the development of lung metastases after intravenous injection of two different melanoma lines was studied[10]. When the cells were pre-treated with 12-HETE before injection, the numbers of lung metastases were approximately doubled, whereas when the cells were pre-treated with 13-HODE, the numbers of lung metastases were more than halved[10].

13-HODE AND CELL DIFFERENTIATION AND PROLIFERATION

Ziboh and colleagues demonstrated that 13-HODE may be important in cellular proliferation[11]. Docosahexaenoic acid (DHA) applied to guinea pig skin stimulated epidermal hyperproliferation. This was associated with a sharp fall in 13-HODE levels. Topical 13-HODE reversed the hyperproliferation[11]. The mechanism appeared to depend on incorporation of 13-HODE into a diacylglycerol with subsequent selective inhibition of the protein kinase-beta isoenzyme without effect on the alpha isoenzyme[12].

In most tissues, induction of ornithine decarboxylase (ODC) is an early step in cellular proliferation. When rat colon mucosa is placed in tissue culture there is a rapid increase in ODC activity which may precede proliferation[13]. Both dexamethasone and 13-HODE were able to suppress the ODC induction[13]. 13-HODE dehydrogenase, which may either inactivate 13-HODE or convert it to further active metabolites, is reduced during differentiation of HT-29 and Caco-2 cultured cells[14]. Calcium is another modulator of 13-HODE since 13-HODE can raise cyclic GMP and intracellular calcium levels in vascular smooth cells[14]. The cellular actions of 13-HODE are complex and may involve both 13-HODE metabolites and a variety of different cell signalling systems.

REGULATION OF 13-HODE FORMATION BY GLA

Since 13-HODE seems to be desirable, it is important to consider how its biosynthesis may be regulated. Provision of linoleic acid is one possibility although may not be important because it is so abundant and because the synthetic process for 13 - HODE may be regulated by feedback reactions. The stimulation of 13-HODE synthesis by cyclic AMP is of interest, as is the decline in 13-HODE synthesis with aging[15].

Cyclic AMP may be regulated by the formation of E prostaglandins (PGs) such as PGE_1. The conversion of linoleic acid to its next essential fatty acid metabolite, gamma-linolenic acid (GLA), can be impaired by aging[16,17]. GLA is rapidly converted to dihomogammalinolenic acid, the precursor of PGE_1. These observations led to the hypothesis that GLA and its metabolites may regulate of 13-HODE formation.

Blackcurrant seed oil containing GLA was more effective than walnut oil containing linoleic acid in reducing in vivo vascular wall thrombogenicity[18]. However, there are many differences between these two oils and there was little increase of 13-HODE in the blackcurrant as compared to the walnut group. A better test may be the direct comparison between corn oil and evening primrose oil. These oils are similar in composition and in linoleic acid content but differ in the fact that primrose oil contains GLA but corn oil does not. Each oil was fed to 14 hamsters in the laboratory of one of us (VZ). After four weeks the 13-HODE production by the lungs was measured. In the corn oil fed animals 13-HODE concentration was 610 ± 20 mg/lung whereas in the primrose oil fed animals it was 2400 ± 40 mg/lung, approximately four fold higher ($p<0.001$). This experiment strongly suggests that GLA or one of its metabolites is an important regulator of 13-HODE formation. It also may explain the fall in 13-HODE production with aging since GLA formation also falls with aging. The GLA may act by conversion to PGE_1 and the consequent stimulation of cyclic AMP formation.

Results consistent with these were obtained by Brister et al in humans awaiting coronary by-pass surgery[18]. When primrose oil was given to patients for 4 weeks prior to surgery, and internal mammary artery 13-HODE levels at surgery compared with those in patients who received placebo capsules, the GLA-containing primrose oil approximately doubled vascular 13-HODE.

CONCLUSIONS

Linoleic acid and its metabolites, 13-HODE and GLA, may be important regulators of cancer spread in humans. 13-HODE inhibits metastasis by reducing adhesion of tumour cells to the endothelium and may have direct anti-proliferative actions. GLA enhances synthesis of 13-HODE, possibly via cyclic AMP. In addition, GLA has anti-cancer actions in its own right being selectively cytotoxic to cancer cells and stimulating the synthesis of the anti-metastatic, E-cadherin, which is down-regulated in many cancer cells[19,20].

Conventional cytotoxic approaches to cancer have produced only modest benefits. A recent editorial in the Lancet noted that a 10% improvement in survival was all that could be expected from chemotherapy and that for every patient helped another nine would be exposed to drug toxicity without benefit[21]. In this context the possibility of novel and safe ways of controlling cancer by using metabolites of linoleic acid is an exciting one.

REFERENCES

1. D.J. Hunter, D. Spiegelman, H-O. Adami *et al.*, Cohort studies of fat intake and the risk of breast cancer - a pooled analysis, *N Engl J Med.* 334:356 (1996).
2. M. Zureik, P. Ducimetiere, J.M. Warnet, and G. Orssaud, Fatty acid proportions in cholesterol esters and risk of premature death from cancer in middle aged French men, *Br Med J.* 311:1251 (1995).
3. M.R. Buchanan, T.A. Haas, M. Lagarde, and M. Guichardant, 13-Hydroxyoctadecadienoic acid is the vessel wall chemorepellant factor LOX, *J Biol Chem.* 260:16056 (1985).
4. M.R. Buchanan and E. Bastida, The role of 13-HODE and HETE's in vessel wall/circulating blood cell interactions, *Agents Actions.* 22:337 (1987).
5. M.R. Buchanan and E. Bastida, Endothelium and underlying membrane reactivity with platelets, leukocytes and tumor cells: regulation by the lipoxygenase-derived fatty acid metabolites, 13-HODE and HETES, *Med Hypotheses.* 27:317 (1988).
6. E. Bastida, M.C. Bertomeu, T.A. Haas *et al.*, Regulation of tumor cell adhesion by intracellular 13-HODE: 15-HETE ratio, *J Lipid Mediat.* 2:281 (1990).
7. M.R. Buchanan, M.C. Bertomeu, T.A. Haas, S. Gallo, and L. Eltringham-Smith, Endothelial cell 13-HODE synthesis and tumor cell endothelial cell adhesion, *Adv Prost Thromb Leuko Res.* 20:909 (1991).
8. I.M. Grossi, L.A. Fitzgerald, L.A. Umbarger *et al.*, Bidirectional control of membrane expression and/or activation of the tumor cell IRGpIIb/IIIa receptor and tumor cell adhesion by lipoxygenase products of arachidonic acid and linoleic acid, *Cancer Res.* 49:1029 (1989).
9. B. Liu, J. Timar, J. Howlett, C.A. Diglio, and K.V. Honn, Lipoxygenase metabolites of arachidonic and linoleic acids modulate the adhesion of tumor cells to endothelium via regulation of protein kinase C, *Cell Regul.* 2:1045 (1991).
10. K.V. Honn, K.K. Nelson, C. Renaud, R. Bazaz, C.A. Diglio, and J. Timar, Fatty acid modulation of tumor cell adhesion to microvessel endothelium and experimental metastasis, *Prostaglandins.* 44:413 (1992).
11. C.C. Miller and V.A. Ziboh, Induction of epidermal hyperproliferation by topical n-3 polyunsaturated fatty acids on guinea pig skin linked to decreased levels of 13-hydroxyoctadecadienoic acid (13-HODE), *J Invest Dermatol.* 94:353 (1990).
12. Y Cho and V.A. Ziboh, Expression of protein kinase C isozymes in guinea pig epidermis: selective inhibition of PKC-beta activity by 13- hydroxyoctadecadienoic acid-containing diacylglycerol, *J Lipid Res.* 35:913 (1994).
13. A.W. Bull, S.M. Earles, and M.L. Blackburn, Regulation of the induction of ornithine decarboxylase in short-term rat colon organ culture by dexamethasone and 13-hydroxyoctadecadienoic acid (13-HODE), *Life Sci.* 53:377 (1993).
14. L.L. Stoll, M.R. Morland, and A.A. Spector, 13-HODE increases intracellular calcium in vascular smooth muscle cells, *Am J Physiol.* 266:C990 (1994).
15. S.J. Brister, T.A. Haas, M.C. Bertomeu, J. Austin, and M.R. Buchanan, 13-HODE synthesis in internal mammary arteries and saphenous veins: Implications in cardiovascular surgery, *Adv Prost Thromb Leuko Res.* 21:667 (1990).
16. P.L. Biagi, A. Bordoni, S. Hrelia, M. Celadon, and D.F. Horrobin, Gamma-linolenic acid dietary supplementation can reverse the aging influence on rat liver microsome delta 6-desaturase activity, *Biochim Biophys Acta.* 1083:187 (1991).
17. R. Takahashi, H. Ito, and D.F. Horrobin, Fatty acid composition of serum phospholipids in an elderly institutionalized Japanese population, *J Nutr Sci Vitaminol Tokyo.* 37:401 (1991).
18. S.J. Brister and M.R. Buchanan, Effects of linoleic acid and/or marine fish oil supplements on vessel wall thromboresistance in patients undergoing cardiac surgery, *Prostaglandins Leukot EFAs.* 55 (Suppl):48 (1996).
19. M.E. Begin, G. Ells, U.N. Das, and D.F. Horrobin, Differential killing of human carcinoma cells supplemented with n-3 and n-6 polyunsaturated fatty acids, *J Natl Cancer Inst.* 77:1053 (1986).
20. W.G. Jiang, S. Hiscox, M.B. Hallett, D.F. Horrobin, R.E. Mansel, and M.C.A. Puntis, Regulation of the expression of E cadherin on human cancer cells by gamma linolenic acid (GLA), *Cancer Res.* 55:5043 (1995).
21. Editorial, And now all this, *Lancet.* 349:1 (1997).

THE ROLE OF PROSTAGLANDINS AND NITRIC OXIDE IN THE RESPONSE OF BONE TO MECHANICAL STIMULATION

Timothy J. Chambers, Jade W. Chow, Simon W. Fox,
Christopher J. Jagger, Jenny M. Lean.

Histopathology Department
St George's Hospital Medical School
Cranmer Terrace
London SW17 ORE, UK.

INTRODUCTION

The primary function for which bone has evolved is to act as mechanical support. Thus, the shape of bones is determined by the genetic programme, upon which is superimposed adaptation to the mechanical environment. The distinct contribution of these two influences is most clearly seen in limb bones, in which mechanical usage causes a substantial change in shape, and an increase in the quantity of bone, compared to that determined genetically [1]. Such mechanically adapted bones show a remarkably consistent strain response to mechanical usage: peak strains of 2,000-3,000 microstrain (με) are observed over the cortical surface of bones in a wide variety of species during physiological activity[see 2].

Understanding the mechanisms by which mechanical forces regulate the structure and quantity of bone could provide opportunities to mimic or amplify the responses of bone to mechanical stimuli, as a strategy to prevent fractures in diseases such as osteoporosis. Approaches used to analyse this responsiveness include overload of the radius by ulnar osteotomy; external force application through implanted pins; external force application using four-point bending; and cyclic ulnar compression. Using these approaches, it has been found that strains of physiological magnitude are sufficient to induce new bone formation in both avian and mammalian species[3-6].

MECHANICAL RESPONSES IN RAT TAIL VERTEBRAE

The experimental model we have developed uses pins, inserted through the 7th and 9th caudal vertebrae of rats, to exert a compressive load on the 8th vertebra. A single, 10 min application of external loading, sufficient to cause strains within the physiological range, is followed by a substantial osteogenic response over the ensuing 7-10 days[5]. Osteogenesis occurs through the sensation of strains imposed in an unaccustomed direction or distribution. This is believed to represent the mechanism by which bones adapt their structure to match changes in the mechanical environment. Thus, while the forces to which the rat tail vertebra is normally exposed are unknown, when the bone is exposed to strains known to be experienced by other bones during mechanical usage, osteogenesis occurs. This is presumably analagous to the process whereby mechanical forces modify the shape of bones from that which would develop in their absence[eg 1].

We have used this model to analyse the sequence of events after a temporally-defined stimulus, that leads to the initiation of bone formation. We assessed expression of an early response gene (*c-fos)*[7], and genes for a growth factor (IGF-I) and matrix macromolecules (collagen and osteocalcin), by *in situ* hybridisation[8]. We found early expression of *c-fos* in osteocytes and on bone surfaces, such that, while we have never detected *c-fos* in the cortex of non-loaded bones, a substantial proportion of osteocytes in mechanically stimulated vertebrae strongly express *c-fos* within 30 mins of loading. Osteocytes also express IGF-I shortly after mechanical stimulation, but induction of expression is more gradual, reaches a peak after 6 hrs, and persists for 1-2 days. IGF-I mRNA is subsequently detected on bone surfaces, followed by bone surface expression of matrix proteins, osteocalcin and collagen 1. Matrix protein synthesis is maximal 72 hrs after a mechanical stimulus, and returns towards basal levels 5 days after loading. These increases in gene expression are accompanied by the induction of increased numbers of osteoblasts on bone surfaces, and increases in mineralising surfaces, as measured through the use of double fluorochrome labels.

Increased expression of *c-fos* and IGF-I has also been detected after mechanical stimulation by Northern analysis of extracts of bone in the four-point bending model of mechanical stimulation[9]. Early expression of RNA in osteocytes, before an osteogenic response is detected on bone surfaces, strongly implicates osteocytes in the early transduction of mechanical strain signals.

The available evidence suggests that distinct signalling systems are responsible for the expression of *c-fos* and IGF-I. Thus, expression of both genes is independent of protein synthesis, suggesting that expression of one protein does not account for expression of the other[7]. Moreover, the spatial distribution of osteocytes expressing *c-fos* differs from that of IGF-I: IGF-I is expressed primarily in osteocytes in the central parts of the mid-diaphysis, while osteocytes expressing *c-fos* are evenly distributed from periosteum to endosteum in the mid-diaphysis; and bone lining cells and cortical vascular channels also show very early expression of *c-fos* but not IGF-I. The genes also differ in their sensitivity to inhibitors of prostaglandin (PG) synthesis (see below).

IGF-I is one of the most abundant growth factors secreted by bone cells. It induces proliferation and differentiation in osteoblastic cells, and it has anabolic actions *in vivo*[see 8]. The early expression of this growth factor in osteocytes is thus consistent with a model whereby IGF-I, generated by osteocytes in response to mechanical stimulation, diffuses along osteocytic canaliculi to bone surfaces where it might participate in the induction and/or regulation of bone formation.

PGs have been shown to stimulate IGF-I production in osteoblastic cells[10]. This raises the possibility that they could be responsible for the observed IGF-I expression, and subsequent bone formation. There is a substantial body of evidence showing a role for PGs in the mechanical responsiveness of bone. For example, an early response of osteoblastic cells to mechanical strain is PG production[11]. Moreover, exogenous PGs stimulate bone formation *in vivo*[see 12] and bone nodule formation *in vitro*[13]; and indomethacin inhibits mechanically-induced osteogenesis *in vivo*[14,15]. We found[15] that the induction of bone formation by mechanical stimulation in the 8th caudal vertebra was inhibited by the administration of a single dose of indomethacin 3 hrs before loading; the same dose given 6 hrs after loading had no effect. This is consistent with a role for PG early in the osteogenic response. We also found that osteocytic IGF-I and *c-fos* expression were suppressed by indomethacin[7]. IGF-I seemed to be more strongly suppressed than *c-fos*.

Our results suggest that cycloxygenase (COX) is involved in the transduction of osteocytic IGF-I expression and other early osteogenic signals. However, we also noted a distinct, later phase of indomethacin sensitivity for the osteogenic process. It seems likely that events in the brief loading period entrain complex, sequential interactions between bone cells, one or more of which may involve PG-dependent steps. For example, growth factors can stimulate PG production, and PGs induce growth factor production, by osteoblastic cells. *In vivo,* heterotopic bone formation by growth factors present in demineralised bone implants is inhibited by indomethacin. Production of COX-2 is increased by PGs, and has been shown to be induced by mechanical stimulation of bone *in vivo*[16]. We have found the same, in unpublished observations. Thus, PGs seem to be important not only during the immediate-early response to loading, whereby osteocytic IGF-I is induced, but also in subsequent interactions between cells and growth factors.

The differential effects of indomethacin on osteocytic IGF-I and *c-fos* expression implies the existence of a second signal mediating mechanical responsiveness in osteocytes. Mechanical stimulation has been found to induce glucose-6-phosphate dehydrogenase (G6PD) activity in osteocytes[17]. This enzyme generates NADPH, which is a co-enzyme for, among other enzymes, nitric oxide synthase (NOS). We therefore tested the possibility that NO production plays a role in the osteogenic response. We found that the competitive NOS inhibitor L-NMMA, given to rats 15 min before mechanical stimulation, completely inhibited the increase in bone formation induced by mechanical stimulation in rat vertebrae[18]. The same dose administered 6 hr after loading had no effect. This suppression was prevented by the co-administration of L-arginine, the natural substrate of NOS, but not by the co-administration of the inactive stereo-isomer, D-arginine. Bone formation in non-loaded vertebrae or tibiae was unaffected by any of these protocols. Independently, Turner *et al*[19] also found inhibition of bone formation using L-NAME to inhibit NOS in the 4-point bending model of mechanical osteogenesis.

The dosage of L-NMMA we used inhibits NO production for 30-60 min. Thus, in animals given L-NMMA 15 min before loading, NO synthesis would be inhibited for up to about 45 min after loading. This suggests that cNOS mediates NO production in the response of bone cells to mechanical stimulation. This is consistent with recent detection of cNOS in bone cells[20,21]. iNOS has also been identified in osteoblastic cells *in vitro* after stimulation with cytokines. However, suppression of the osteogenic response by brief inhibition of NOS, around the time of loading itself, is not consistent with a role for iNOS, for which the lag time for transcription is around 2 hr; and whereas high concentrations of NO, as typically generated by iNOS, inhibit proliferation, the relatively small quantities likely to be produced in bone by cNOS stimulate osteoblast proliferation *in vitro*.

Production of NO, which has been shown to activate *c-fos* gene expression[22], might be causally linked to *c-fos* expression in bone. In unpublished experiments, we have found partial suppression of *c-fos* by L-NMMA, by Northern analysis of RNA extracted from the cortex of loaded vertebrae. Indomethacin causes a similar suppression.

Because both the induction of bone formation and the increase in G6PD activity after mechanical stimulation are suppressed by indomethacin[15,23], there is the possibility that mechanically-induced bone formation is dependent on NO production, which in turn is dependent on PG production. Alternatively, PG and NO might both be required for mechanically-induced osteogenesis, but they might be generated independently of each other. Whatever the explanation, both agents are crucial, and further analysis of their role should provide insights into the mechanisms by which mechanical stimuli induce bone formation.

REFERENCES

1. L.E. Lanyon, The influence of function on the development of bone curvature. An experimental study on the rat tibia, *J. Zoology* 192:457 (1980).
2. C.T. Rubin and L.E. Lanyon, Dynamic strain similarity in vertebrates: an alternative to allometric limb bone scaling, *J. Theoret. Biol.* 107:321 (1984).
3. C.T. Rubin and L.E. Lanyon, Regulation of bone formation by applied dynamic loads, *J. Bone Joint Surg.* 66-A:397 (1984).
4. C.H. Turner, M.P. Akhter, D.M. Raab, D.B. Kimmel and R.R. Recker, A noninvasive, *in vivo* model for studying strain adaptive bone modeling, *Bone* 12:73 (1991).
5. J.W.M. Chow, C.J. Jagger and T.J. Chambers, Characterization of osteogenic response to mechanical stimulation in cancellous bone of rat caudal vertebrae, *Am. J. Physiol.* 265:E340 (1993).
6. A.G. Torrance, J.R. Mosley, R.F. Suswillo and L.E. Lanyon, Noninvasive loading of the rat ulna *in vivo* induces a strain-related modeling response uncomplicated by trauma or periosteal pressure, *Calcif. Tissue Int.* 54:241 (1994).
7. J.M. Lean, A.G. Mackay, J.W.M. Chow and T.J. Chambers, Osteocytic expression of mRNA for *c-fos* and IGF-I: an immediate early gene response to an osteogenic stimulus, *Am. J. Physiol.* 270:E937 (1996).
8. J.M. Lean, C.J. Jagger, T.J. Chambers and J.W.M. Chow, Increased insulin-like growth factor I mRNA expression in rat osteocytes in response to mechanical stimulation, *Am. J. Physiol.* 268:E318 (1995).
9. D.M. Raab-Cullen, M.A. Thiede, D.N. Petersen, D.B. Kimmel and R.R. Recker, Mechanical loading stimulates rapid changes in periosteal gene expression, *Calcif. Tissue Int.* 55:473 (1994).
10. T.L. McCarthy, M. Centrella, L.G. Raisz and E. Canalis, Prostaglandin E_2 stimulates insulin-like growth factor I synthesis in osteoblast-enriched cultures from fetal rat bone, *Endocrinology* 128:2895 (1991).
11. C.K. Yeh and G.A. Rodan, Tensile forces enhance prostaglandin E synthesis in osteoblastic cells grown on collagen ribbons, *Calcif. Tissue Int.* 36:S67 (1984).
12. R.W. Norrdin, W.S.S. Jee and W.B. High, Review: The role of prostaglandins in bone *in vivo*, *Prostagl. Leuk. Ess. Fatty Acids* 41:139 (1990).
13. A.M. Flanagan and T.J. Chambers, Stimulation of bone nodule formation *in vitro* by prostaglandins E_1 and E_2, *Endocrinology* 130:443 (1992).
14. M.J. Pead and L.E. Lanyon, Indomethacin modulation of load-related stimulation of new bone formation *in vivo*, *Calcif. Tissue Int.* 45:34 (1989).
15. J.W.M. Chow and T.J. Chambers, Indomethacin has distinct early and late actions on bone formation induced by mechanical stimulation, *Am. J. Physiol.* 267:E287 (1994).
16. M.R. Forwood, Inducible cyclo-oxygenase (COX-2) mediates the induction of bone formation by mechanical loading in vivo, *J. Bone Miner. Res.* 11:1688 (1996).
17. T.M. Skerry, L. Bitensky, J. Chayen and L.E. Lanyon, Early strain-related changes in enzyme activity in osteocytes following bone loading *in vivo*, *J. Bone Miner. Res.* 4:783 (1989).
18. S.W. Fox, T.J. Chambers and J.W.M. Chow, Nitric oxide is an early mediator of the increase in bone formation by mechanical stimulation, *Am. J. Physiol.* 270:E955 (1996).
19. C.H. Turner, Y. Takano, I. Owan and G.A. Murrell, Nitric oxide inhibitor L-NAME suppresses mechanically induced bone formation in rats, *Am. J. Physiol.* 270:E634 (1996).
20. A.A. Pitsillides, S.C.F. Rawlinson, R.F.L. Suswillo, S. Bourrin, G. Zaman and L.E. Lanyon, Mechanical strain induced NO production by bone cells: a possible role in adaptative bone (re)modelling?, *FASEB J.* 9:1614 (1995).
21. J.A. Riancho, E. Salas, M.T. Zarrabeitia, J.M. Olmos, J.A. Amado, J.L. Fernandez-Luna and J. Gonzalez-Macias, Expression and functional role of nitric oxide synthase in osteoblast-like cells, *J. Bone Miner. Res.* 10:439 (1995).
22. C. Abate, L. Patel, F.J. Rauscher III and T. Curran, Redox regulation of fos and jun DNA-binding activity in vitro, *Science* 249:1157 (1990).
23. A.J. El Haj, S.L. Minter, S.C.F. Rawlinson, R. Suswillo and L.E. Lanyon, Cellular responses to mechanical loading *in vitro*, *J. Bone Miner. Res.* 5:923 (1990).

MICE LACKING 5-LIPOXYGENASE HAVE INCREASED CORTICAL BONE THICKNESS

Lynda F. Bonewald, Monica Flynn, Mei Qiao, Mark R. Dallas, Gregory R. Mundy, Brendan F. Boyce

University of Texas Health Science Center at San Antonio, Department of Medicine, San Antonio, TX 78284-7877

INTRODUCTION

5-lipoxygenase (5-LO) metabolites have complex effects on bone cells. We have found that the leukotrine LTB_4 and the peptido leukotrienes stimulate osteoclast formation in vitro and in vivo, and are likely involved in the generation of active osteoclast-like cells in giant cell tumors of bone. Recent preliminary observation in mice lacking the gene for 5-LO have not as yet demonstrated any effect on osteoclast function, but these mice have increased cortical bone, suggesting that 5-LO is a negative regulator of bone formation. 5-LO metabolites thus have unique effects on bone cell function, and clarifying their role may be important for the understanding of physiological as well as pathological bone remodeling.

EFFECTS OF 5-LO METABOLITES ON OSTEOCLASTS

Our studies first began with a search for factors that could be direct activators of mature osteoclasts. We previously found that a cell line called C433 established from a giant cell tumor of bone was a potent stimulator of bone resorption [1]. The conditioned media from the cell line stimulated the release of ^{45}Ca from isolated neonatal mouse calvaria, but not from fetal rat long bones. This conditioned media also had the ability to stimulate the formation of resorption lacunae by isolated rat osteoclasts and to stimulate isolated avian osteoclast-like cells to form resorption lacunae on sperm whale dentin with a concomittant increase in the tartrate resistant acid phosphatase activity in the cells. This activity was different from other known stimulators of bone resorption such as parathyroid hormone. The activity in this conditioned media was compared to seven other known bone resorbing cytokines and found to behave differently in four different assays of bone resorption.

To identify the activity in the C433 conditioned media that was responsible for inducing isolated osteoclasts to resorb bone, a series of purification and characterization steps were initiated [2]. It was found that the activity in the C433 conditioned media had a molecular weight between 500 to 1000 daltons, it was not a peptide, and it eluted on HPLC at the same

Recent Advances in Prostaglandin, Thromboxane, and Leukotriene Research
Edited by Sinzinger *et al.*, Plenum Press, New York, 1998

position as known metabolites of 5-LO. This activity was identified and verified to be metabolites of the 5-LO pathway by gas chromatography/mass spectrometry.

After it was determined that the bone resorbing activity in the C433 conditioned media was due to the 5-LO metabolites, 5HETE, and the peptido-leukotrienes LTC_4, D_4, and E_4, the effects of synthetic and commercially available peptido-leukotrienes on bone were examined [3]. LTD_4 was shown to stimulate giant cells from human giant cell tumors to form resorption lacunae on sperm whale dentin and to increase tartrate resistant acid phosphatase activity and pit formation by avian osteoclast-like cells. LTD_4 increased calcium release from murine calvarial bone organ cultures as had been reported previously by Meghji and co-workers [4] but did not induce resorption in fetal rat long bone cultures. This observation correlates with the activity observed in the C433 conditioned media. To determine whether LTD_4 could be having a direct effect on mature osteoclasts, receptor assays were performed on isolated avian osteoclast-like cells. The peptido-leukotriene chosen was LTD_4 as this metabolite has been found to have specific receptors. (LTC_4, in addition to binding to the LTD_4 receptor, also binds to glutahione-S-transferase.) A single affinity binding site was found with a dissociation constant of 0.53 nM and a receptor density of 5,200 receptors per cell. Competition binding studies showed receptor specificity using a specific LTD_4 receptor antagonist.

The more stable 5-LO metabolite, leukotriene B_4 (LTB_4), has been implicated in a number of inflammatory diseases such as osteoarthritis and periodontal disease. We examined the effects of LTB_4 on osteoclastic bone resorption and found that it stimulates resorption both in vitro and in vivo. LTB_4 when injected over the calvaria of mice, resulted in increased bone resorption, osteoclast numbers, and eroded surface [5]. LTB_4 significantly increased ^{45}Ca release in cultures of neonatal mouse calvaria and stimulated isolated rat osteoclasts to resorb bone. LTB_4 was also found to be equivalent to $1{,}25(OH)_2D_3$ in supporting murine and human osteoclast formation. LTB_4 was 10 to 100 times more potent than the peptido-leukotrienes.

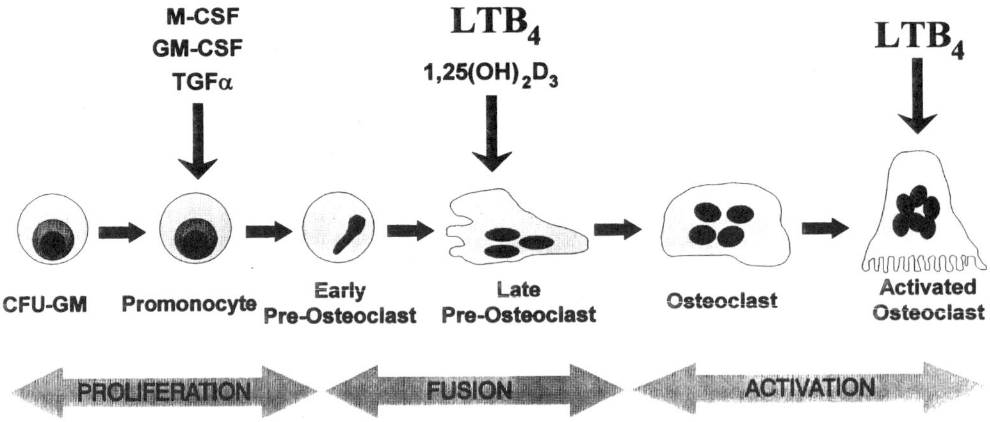

Figure 1. Diagram showing the effects of LTB_4 at various stages of osteoclast differentiation. Osteoclast precursors are derived from haemopoetic stem cells which are responsive to colony stimulating factors. Granulocyte-macrophage colony forming units (CFU-GM) respond to proliferation factors such as macrophage-CSF (M-CSF), granulocyte-macrophage CSF (GM-CSF) and transforming growth factor alpha (TGFα). and fuse to form multinucleated cells. Attachment of the multinucleated cell to the bone surface causes activation of the mature cell which forms resorption lacunae in the mineralized matrix. LTB_4 is not a proliferation factor for osteoclasts, but promotes the fusion or formation of osteoclasts in a manner similar to 1,25 dihydroxyvitamin D_3 and also activates the mature cell

EFFECTS OF 5-LO METABOLITES ON OSTEOBLASTS

Others have shown leukotrienes can also have effects on osteoblasts. In 1991, Ren and Dziak [6] showed that LTB_4 inhibited proliferation of normal osteoblastic rat calvarial cells and the osteoblast-like cell lines $SAOS_2$ and G292. Also, Sandy and coworkers [7] found that LTB_4 increased intracellular calcium in osteoblasts released from neonatal mice calvaria. No effect of LTB_4 on *c-myc* expression was found in any of the osteoblastic cell lines tested [8].

Leukotrienes can be produced by bone cells. Interleukin-1 was found to increase the production of 5-HETE in G292 cells but not $SAOS_2$ cells [9]. A series of abstracts were presented by Collins and coworkers in 1987 showing that leukotrienes were produced by bone tissue and that inhibition of leukotriene synthesis combined with mechanical stress enhanced bone formation and blocked bone resorption [10,11]. The 5-LO inhibitor piriprost was used in force-induced bone development where mechanical stress is imposed across the midsagittal sections of osteotomized weanling rats. They concluded that 5-LO inhibition strongly promotes calvarial bone formation under mechanical stress. Enhanced osteoblast formation was observed as was a increase in bone formation of osteoblast line matrix invaginations and calvarial medullary spaces. Dziak and coworkers [12] found significant inhibition of tooth movement using the leukotriene inhibitor AA861 even though enhanced levels of prostaglandins were detected with this treatment. They postulated that inhibition of leukotriene synthesis might influence tooth movement via inhibition of bone cell calcium movement.

MICE LACKING 5-LO HAVE INCREASED BONE

Recently, mice have been generated which lack the functional 5-lipoxygenase gene [13]. These mice were examined to determine the importance of 5-lipoxygenase in inflammation and anaphylaxis. According to these investigators, these 5-LO "knockout" mice show no apparent abnormalities except for slightly smaller spleens. The animals lacking 5-LO resist the lethal effects of shock induced by platelet activating factor whereas there is no difference in their reaction to endotoxin shock compared to normals. Reaction to phorbal ester is normal whereas inflammation induced by arachidonic acid is reduced compared to normal mice.

We have begun to examine properties of bones from these animals. These 5-LO "knock-out" mice were generated by disruption of exon 6 of the 5-LO gene and established on a C57B1/6Jx129sv background [13]. These same investigators have produced knock-out mice on a single genetic background, the 129sv. We have found using quantitative radiology and bone histomorphometry that both male and female mice from both the mixed and single background with the deletion of the 5-LO gene have increased cortical bone compared with their wild-type controls.

Many knock-out mice have been generated which are osteopetrotic [14, 15] but few have been described to have an increase in cortical bone. This change in cortical bone thickness has been observed in another "knockout" mouse lacking the osteocalcin gene [16]. However, mice lacking osteocalcin do not have an increase in cancellous bone formation or cortical bone formation until 6-9 months of age compared to changes observed in the 5-LO knockout mice at 6-7 weeks. The osteocalcin "knock-out" animals were not osteopetrotic and no change was observed in osteoblast surface. It was concluded that osteocalcin is an inhibitor of osteoblast function. Recently, mice with deletions of either the prostaglandin synthase 1 or 2, (also known as cyclooxygenase 1 and 2) genes have been generated [17, 18]. Mice homozygous for deletions of COX-1 produce few live offspring suggesting COX-1 is necessary for normal parturition, and mice lacking COX-2 have severe renal nephropathy. Nothing has been reported at this time concerning the bones in these mice.

In summary, investigation into the role of 5-lipoxygenase in normal bone remodeling and the role of 5-LO metabolites in bone resorption under pathological conditions is in its infancy. The total number of published manuscripts addressing the effects of 5-LO metabolites on bone cells is less than ten, indicating the limited amount of information available in this area. Evidence of bone changes in mice lacking the 5-LO gene indicates that further investigation into this area is necessary and may provide important information for therapy aimed at increasing bone mass.

REFERENCES

1. Oreffo ROC, Marshall GJ, Kirchen M, Garcia G, Gallwitz WE, Chavex J, Mundy GR, Bonewald LF. Characterization of a cell line derived from a human giant cell tumor that stimulates osteoclastic bone resorption. Clin Orthop Related Res 296:229-241, (1993).
2. Gallwitz WE, Mundy GR, Lee CH, Qiao M, Roodman GD, Raftery M, Gaskell SJ, Bonewald LF. 5-lipoxygenase metabolites of arachidonic acid stimulate isolated osteoclasts to resorb calcified matrices. J Biol Chem 268:10097-10094, (1993).
3. Garcia C, Qiao M, Chen D, Kirchen M, Gallwitz W Mundy GR, Bonewald LF. Effects of synthetic peptido-leukotrienes on bone resorption in vitro. J Bone Miner Res, 11:521-529, (1996).
4. Meghji S, Sandy JR, Scutt AM, Harvey W, Harris M. Stimulation of bone resorption by lipoxygenase metabolites of arachidonic acid. Prostaglandins 36:139-147, (1988).
5. Garcia C, Boyce BF, Gilles J, Dallas M, Qiao M, Mundy GR, Bonewald LF, Leukotriene B_4 stimulates osteoclastic bone resorption both in vitro and in vivo. J Bone Miner Res, 11:1619-1629, (1996).
6. Ren W, Kinniburgh AJ, Dziak R. Regulation of c-myc protooncogene expression in osteoblastic cells by arachidonic acid metabolites: relationship to proliferation. Calcif Tissue Int 50:372-377, (1992).
7. Sandy JR, Meikle MC, Martin BR, Farndale RW. Leukotriene B4 increases intracellular calcium concentration and phosphoinositide metabolism in mouse osteoblasts via cyclic adenosine 3', 5'-monophosphate-independent pathways. Endocrin 129:582-590, (1991).
8. Ren W and Dziak R. Effect of leukotrienes on osteoblastic cell proliferation. Calcif Tissue Int 49:197-201, (1991).
9. Zhang W and Dziak R. Effects of interleukin-1 alpha on arachidonic acid metabolism in human osteosarcoma osteoblastic cells. Prostaglandins Leukot Essent Fatty Acids 49:521-526, (1993).
10. Collins JL, Daniel JW, Cederquist R, Simmelink JW, Enlow DH. Stimulation of bone development by mechanical stress and inhibition of leukotriene biosynthesis. J Dent Res 66:1771, (1987).
11. Collins JL, Daniel JW, Cerdeuist R, Enlow DH. The role of leukotrienes in force-induced bone development J Dent Res 66: 1722, (1987).
12. Mohammed AH, Tatakis DN, Dziak R. Leukotrienes in orthodontic tooth movement. Am J Orthod Dentofac Orthop 95:231-237, (1989).
13. Chen XS, Sheller JR, Johnson EN, Funk CD. Role of leukotrienes revealed by targeted disruption of the 5-lipoxygenase gene. Nature 372:179-182, (1994).
14. Soriano P, Montgomery C, Geske R, Bradley A. Targeted disruption of the c-src proto-oncogene leads to osteopetrosis in mice. Cell 64:693-702, (1991).
15. Grigoriadis AE, Wang ZQ, Cecchini MG, Hofstetter W, Felix R, Fleisch HA, Wagner EF. C-Fos: A key regulator of osteoclast-macrophage lineage determination and bone remodeling. Science 266:443-448, (1994).
16. Ducy P, Desbois C, Boyce B, Pinero G, Story B, Dunstan C, Smith E, Bonadio J, Goldstein S, Gundberg C, Bradley A, Karsenty G. Increased bone formation in osteocalcin-deficient mice. Nature 382:448-452, (1996).
17. Langenbach R, Morham SG, Tiano HF, Loftin CD, Ghanayem BI, Chulada PC, Mahler JF, Lee CL, Goulding EH, Kluckman KD, Kim HS, Smithies O. Prostaglandin synthase 1 gene disruption in mice reduces arachidonic acid-induced inflammation and indomethacin-induced gastric ulceration. Cell 83:483-492, (1995).
18. Morham SG, Langenbach R, Loftin CD, Tiano HF, Vouloumanos N, Jennette JC, Mahler JF, Kluckman KD, Ledford A, Lee CA, Smithies O. Prostaglandin synthase 2 gene disruption causes severe renal pathology in the mouse. Cell 83:474-482, (1995).

PARATHYROID HORMONE INDUCTION OF CYCLOOXYGENASE-2 EXPRESSION IN HUMAN OSTEOBLASTS DEPENDS ON BOTH CYCLIC AMP AND CALCIUM-DEPENDENT PATHWAYS

Artur J.B. Fernandes, Maximilien Lora, Caroline Patry, Sophie Morisset, Patrice Sarrazin, Flávio Maciel

Rheumatic Diseases Unit
Faculty of Medicine, University of Sherbrooke
Sherbrooke, Québec, Canada, J1H 5N4

INTRODUCTION

Parathyroid hormone (PTH) increases serum Ca^{2+} concentration, and one of the mechanisms leading to this result is the induction of osteoclast activity and consequent bone resorption. Although the presence of PTH in osteoclasts is polemic, functional studies have shown that induction of bone resorption by this hormone depends on the presence of osteoblasts[1], on soluble factors released by these cells[2], and probably on direct contact between osteoblasts and osteoclasts[3]. Prostaglandins are probably implicated in the cross-talk between osteoblasts and osteoclasts: their production by osteoblasts is induced by mechanical stress[4], IL-1 and TNF[5], and PTH[6], and PGE_2 can stimulate both bone formation and resorption[7]. The mechanism implicated in the control of prostaglandin production by PTH, however, has not been elucidated. Since availability of cyclooxygenases is a rate-limiting step in the synthesis of prostaglandins, our objective was to verify if cyclooxygenase-2 (COX-2) expression in human osteoblasts in culture could be regulated by PTH, and to study the intracellular mechanisms implicated in this action.

MATERIAL AND METHODS

Human osteoblasts were cultured from explants of trabecular bone obtained from the femoral heads of patients undergoing hip arthroplasty as previously described[5]. An informed consent was obtained from each patient before surgery. Confluent osteoblasts in 10 cm Petri dishes were incubated 24 hours in MEM with 0.2% FBS. The medium was changed, the different compounds applied, and the cells further incubated for 3 hours. Total RNA was extracted using Trizol, according to the manufacturer's protocol. Total RNA was fractionated on agarose-formaldehyde gels and transferred to Hybond-N+ membranes overnight by capillarity. Human COX-2 cDNA probes were prepared by random primer [^{32}P]dCTP with an oligolabeling kit. The membranes were hybridized 400 mM NaH_2PO_4, 1 mM EDTA, 1 mg/ml BSA, 5 % SDS, 50 % formamide, pH 7.2 overnight at 42°C with the radioactive probes, washed twice at 55°C for 20 minutes in 1 mM EDTA, 0.1% SDS, 1 X SSC pH 7.0, and then washed once at 55°C for 5 minutes in 1 mM EDTA, 0.1% SDS, and 0.2 X SSC. The membranes were autoradiographed on Kodak XAR-5 film at -80°C with intensifying screens. PGE_2 concentration in the supernatants was determined using a commercially available ELISA assay (Cayman Chemical). All reagents, unless stated otherwise, were purchased from Sigma Chemical Co.

RESULTS AND DISCUSSION

As we have shown in a previous publication[5], human cells obtained with this technique keep their osteoblastic phenotype in primary culture. This is shown by their production of alkaline phosphatase and of osteocalcin[5], and by their response to PTH with an increase in intracellular cyclic AMP. To further characterize the cells in our experimental system we used an *in situ* hybridization technique with DIG-11-dUTP-marked human osteocalcin probes and fluorescein-coupled anti-digoxigenin antibodies. Controls included incubation with the antibodies without the probe, RNAase-treated cells, and osteocalcin-negative cells (COS-7). More than 99% of the cells examined by confocal microscopy systematically showed significant fluorescence, showing that these cells express osteocalcin mRNA, an osteoblast marker (data not shown).

In the present study human osteoblasts in culture stimulated with PTH fragment 1-34 increased their levels of COX-2 mRNA as well as the production of PGE_2. This increase in the specific mRNA level was time-dependent and reached its maximum after stimulation for 2 to 4 h, declining thereafter to reach control, undetectable levels 24 h after stimulation (data not shown). As shown in Figure 1, this response was also concentration-dependent and in the range corresponding to the described affinity of the PTH fragment 1-34 for its receptor[8], known to be around 50 nM.

It is important to point out that both in the time-response and in the concentration-response curves the increase in PGE_2 levels in the supernatants paralleled the changes in COX-2 mRNA levels. More importantly, the increases in PGE_2 were completely inhibited by compound NS-398, a specific COX-2 inhibitor. This inhibition was comparable to that of dexamethasone, and no additive effect was seen when both dexamethasone and compound NS-398 were added to the preparation (data not shown). Although an effect of PTH on other points of the arachidonic acid can not be excluded based on the present results, they

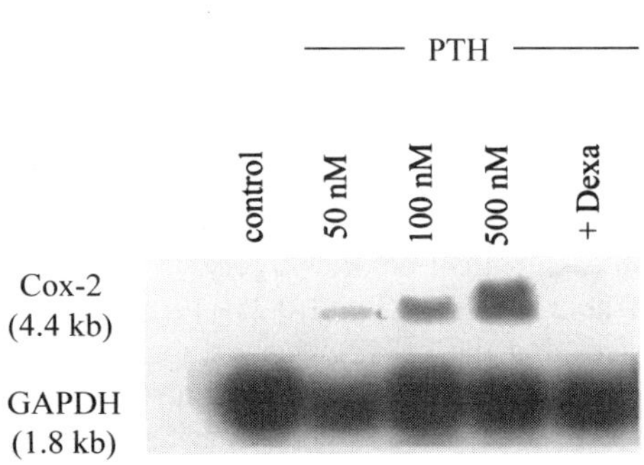

Figure 1. Induction of COX-2 mRNA by PTH. Confluent human osteoblast cultures were incubated during 4 h with different concentrations of 1-34 PTH. GAPDH was added as a control for the quantity of RNA. Dexa = dexamethasone 10^{-6}M.

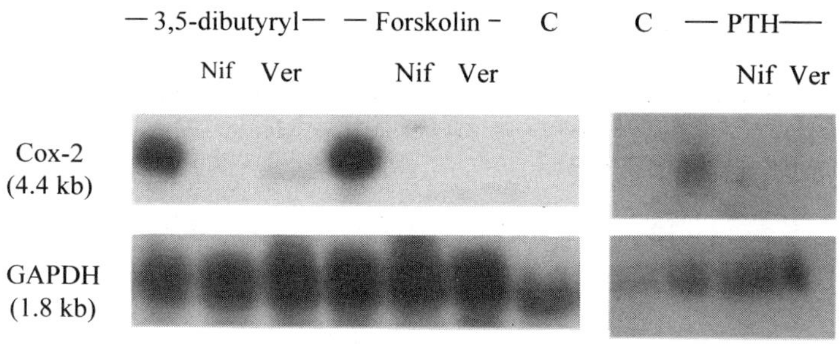

Figure 2. Effect of calcium channel blockers nifedipine 10^{-6}M and verapamil 10^{-6}M on the induction of COX-2 by 1-34 PTH 100 nM, 3'-5'-dibutyryl cAMP 10^{-5} M, and forskolin 10^{-5} M. Nif = nifedipine, Ver = verapamil.

suggest that the increase in PGE_2 production by human osteoblasts is due mainly to the increase in COX-2 activity in the present experimental conditions.

Another objective of the present study was to determine the intracellular second messengers implicated in the induction of COX-2 by PTH. As shown in Figure 2, both PTH, forskolin (a direct activator of adenylate cyclase), and 3'-5'-dibutyryl-cyclic AMP (a cAMP stable analog) were able to induce an increase in COX-2 mRNA expression. Interestingly, however, nifedipine and verapamil, two distinct L-type calcium channel blockers, completely inhibited the effect of PTH, forskolin and 3'-5'-dibutyryl cAMP. These results suggest that activation of L-type calcium channels is necessary to induce the observed response to PTH. Whether this activation is direct by the PTH receptor or indirect through activation of PKA by cAMP can not be deduced from these results. The actions of forskolin and 3'-5'-dibutyryl cAMP, however, suggest that a direct activation of the calcium channels by the PTH receptor may not be necessary to induced the observed phenomenon.

To determine if increases in intracellular calcium could induce COX-2 expression in the absence of increased cAMP levels we treated the cells with tapsigargin 10^{-6} M, which increases the release of intracellular stored Ca^{2+}, or with compound Bay K-8644 10^{-6} M, a direct activator of calcium channels. Neither tapsigargin nor Bay K-8644, alone or in combination, were able to induce COX-2 expression in human osteoblasts (data not shown), showing that increases in intracellular calcium alone are not able to stimulate the expression of this gene.

One of the conditions for prostaglandins to be considered as messengers implicated in bone response to PTH is that this hormone can somehow control the synthesis of these autacoids. The present study confirms that PTH induces prostaglandin production by human osteoblasts in culture and shows, for the first time, that the hormone induces COX-2 mRNA expression in these cells. Although action on other points of the arachidonic acid cascade can not be excluded, our results strongly suggest that the main mechanism implicated in this action of PTH is the induction of COX-2 expression, since the increase in prostaglandin production was abolished by the specific COX-2 inhibitor compound NS-398. This study also shows that the effect of PTH can be inhibited by L-type calcium channel blockers, but that substances known to increase intracellular Ca^{2+} levels are not able to induce COX-2 expression, suggesting that an increase in intracellular calcium is a necessary but not sufficient condition for the induction of COX-2 expression by PTH. Further studies using specific protein kinase inhibitors are needed to elucidate the relative roles of PKA and calcium channels pathways on this effect of PTH. These studies are presently being done in our laboratory.

ACKNOWLEDGMENTS

The authors would like to thank Dr. Timothy Hla for kindly supplying the COX-2 cDNA. This study was supported by grants from the Medical Research Council of Canada and The Arthritis Society of Canada.

REFERENCES

1. P.M. Barling, and N.J. Bibby, Study of the localization of [^{3}H]bovine parathyroid hormone in bone by light microscope autoradiography. *Calcif. Tissue Int.* 37:442 (1985).
2. G.L. Wong, Paracrine interaction in bone-secreted products of osteoblasts permit osteoclasts to respond to parathyroid hormone, *J. Biol. Chem.* 259:4019 (1984).
3. E. Jimi, I. Nakamura, H. Amano, Y. Taguchi, T. Tsurukai, N. Takahashi, and T. Suda, Osteoclast function in activated by osteoblastic cells through a mechanism involving cell–to–cell contact, *Endocrinol.* 137:3446 (1996).
4. D. Somjen, I. Binderman, E. Berger E *et al*, Bone remodeling induced by physical stress is prostaglandin mediated, *Biochim. Biophys. Acta* 629:91 (1984).
5. A.J.B. Fernandes, S. Laporte, M. Heroux, *et al*, Expression of PGH_2 synthase–1 and PGH_2 synthase–2 in human osteoblasts in culture, *Biochem. Biophys. Res. Commun.* 198:955 (1994).
6. M. Mitnick, C. Isales, I. Paliwal, and K. Insogna, Parathyroid hormone–related protein stimulates prostaglandin E_2 release from human osteoblast–like cells: modulating effect of peptide length, *J. Bone Min. Res.* 7: 887 (1992).
7. L.G. Raisz, The role of prostaglandins in the local regulation of bone metabolism, *Progr. Clin. Biol. Res.* 332: 195 (1990).
8. G.V. Segre, M. Rosenblatt, B.L. Reiner, J.E. Mahaffey, and J.T. Potts Jr, Characterization of parathyroid hormone receptors in canine renal cortical plasma membranes using a radioiodinated sulfur–free hormone analogue, *J. Biol. Chem.* 254:6980 (1979).

INVOLVEMENT OF PROSTAGLANDIN ENDOPEROXIDE H SYNTHASE-2 IN OSTEOCLAST FORMATION INDUCED BY PARATHYROID HORMONE

Takako Tokushima, Takahiro Sato, Ikuo Morita, Sei-itsu Murota

Department of Physiological Chemistry, Graduate School
Tokyo Medical and Dental university
Yushima 1-chome 5-45, Bunkyo-ku, Tokyo 113, Japan

ABSTRACT

Prostaglandin (PG)E_2 is one of the most important endogenous bone resorbing factors. In the previous study, we demonstrated that osteoclast formation induced by IL-1β was mediated by PGE_2 produced by induced prostaglandin endoperoxide H synthase-2 (PGHS-2) in osteoclastic cells. In the same bone marrow culture system, indomethacin also suppressed the osteoclast formation induced by PTH. The inhibition was abolished by exogenously added PGE_2 at dose as low as $3x10^{-9}$M, which was too low to elevate the intracellular cAMP and calcium levels and also it was too low to cause osteoclast formation by itself. In order to estimate what kind of cell produced such small amount of PGE_2 in the PTH treatment, we carried out antibody staining of PGHS-1&PGHS-2 and PGHS activity in the intact bone marrow cells. PTH was found to induce PGHS activity in tartrate-resistant acid phosphatase (TRACP) positive mononuclear cells and the PGHS activity was inhibited by NS-398, a specific inhibitor of PGHS-2. Immunocytochemical staining supported the expression of PGHS-2 in TRACP-positive mononuclear cells. These findings suggest that PGHS-2 induced by PTH may regulate osteoclast formation by different mechanism from that induced by IL-1β.

INTRODUCTION

Osteoclast have been known to be derived from hematopoietic stem cells[1]. In the course of the development of osteoclasts from stem cells, stromal cells are known to play some important roles in it by providing various kinds of associated factors with the osteoclast formation[2]. On the other hand, a mouse bone marrow cell culture provides an ideal in vitro experimental system capable of examining the processes of the osteoclast development from stem cells. By using this mouse bone marrow cell culture, we found that IL-1β induced osteoclast formation through PGHS induction, large amount of PGE_2which activated A-kinase in stromal cells through EP-2 receptor and provide some osteoclast differentiation factors.

RESULTS AND DISCUSSION

PTH was able to induce osteoclast formation, and it was blocked by indomethacin similar to the case of IL-1β induced osteoclast formation, suggesting that the production of PGE_2 was also essential in the PTH-induced osteoclast formation, while the 1α,25 $(OH)_2D_3$ induced osteoclast formation had nothing to do with PGE_2 production. Although both PTH and IL-1β induced osteoclast formation were equally blocked by indomethacin, there was a big difference between them in the role of PGE_2 playing there. Because in the PTH-induced osteoclast formation, the indomethacin block was completely recovered by an extremely small amount of PGE_2[3], i.e., as low as 1ng/ml to 10ng/ml of PGE_2, however,the dose of this 10ng/ml was too small to abolish the indomethacin block in the case of IL-1β induced osteoclast formation, where 100 times more PGE_2 was required to get a complete recovery. During the PTH induced osteoclast formation, a very small amount of PGE_2 was produced, i.e.,the pg order. This amount of PGE_2 was 100 times less than that produced during the IL-1β treatment ,where PGE_2 was produced in the ng order. The PTH-induced osteoclast formation was also blocked by a PGHS-2 specific inhibitor NS-398. These data suggest that PTH has mouse bone marrow cells induce PGHS-2. The PTH-induce osteoclast formation was also blocked by other type of NSAID than indomethacin,i.e., DF in a dose dependent fashion and each of these inhibition was abolished by an extremely small amount of PGE_2 , i.e.,10 ng/ml. Such a small amount of PGE_2 alone was not able to induce osteoclast formation at all by itself. If we have PGE_2 alone induce osteoclast formation, a dose as high as 1000 ng/ml was required. 10ng/ml PGE_2 could not induce osteoclast formation at all. According to our previous studies, it was elucidated that the role of PGE_2 in this PGE_2-induced or IL-1b-induced osteoclast formation was in the elevation of the intracellular cAMP level in stromal cells through their PGE_2 receptor EP-2. In the PTH-induced osteoclast formation, PTH alone showed an enough activity to elevate the intracellular cAMP level by itself and the elevated cAMP level was not affected by indomethacin at all. The 10ng/ml PGE_2 which was required to recover the indomethacin block in the PTH induced osteoclast formation had nothing to do with this cAMP production, suggesting that the role of this small amount of PGE_2 must be independent from cAMP production in stromal cells.

Next, we examined which type of cells produced PGE_2 most during the PTH induced osteoclast formation. Real time activity of PGHS in each cell was assayed according to the new method that we established recently[4] . In this assay system, once PGG_2 is produced in the cells, instantly it causes fluorescence emission on the spot. PGHS activity was examined during the PTH induced osteoclast formation. 4days after the PTH treatment, a very weak PGHS activity was observed. The activity was not inhibited by NS-398, PGHS-2 specific inhibitor, suggesting that at this stage an almost negligible amount of PG are produced by PGHS-1 alone. On the other hand, 6days after the PTH treatment, a remarkably strong PGHS activity was observed, especially, in TRACP positive mononuclear cells that were in contact with stromal cells. Moreover, this PGHS activity was inhibited by PGHS-2 specific inhibitor, suggesting that PTH can induce PGHS in TRACP positive mononuclear cells at the late stage of the treatment.

In summery, PTH can induce PGHS-2 in stem cell lineage, the small amount of PGE_2 produced by this PGHS-2 may facilitate stem cell differentiation. Moreover, PTH can also increase the cAMP level in stromal cells, followed by inducing osteoclast differentiation factors.

EXPERIMENTAL PROCEDURES

Bone marrow mononuclear cells were isolated from 7 weeks old male ddy mice (Saitama Jikken dobutsu ,Japan) and were cultured as described previously[5]. After being cultured for

8day, aliquots of conditioned medium were assayed for the levels of PGE_2 using an enzyme immunoassay system (Amersham, USA) and the cell layers were stained with tartrate-resistant acid phosphatase, a marker enzyme of osteoclast as described[6]. The total cellular RNA was isolated as described previously[7].The PGHS activity in individual intact cells were measured as described previously.

CONCLUSIONS

1) Both PTH-induced and IL-1b-induced osteoclast formation in the mouse bone marrow cell culture were equally blocked by indometacin. However, there was a big difference between them in the role of PGE_2 playing there.
2) The inhibition of osteoclast formation by indomethacin was completely recovered by PGE_2 at a dose as low as 1 to 10 ng/ml in the case of PTH-induced osteoclast formation, while in the case of IL-1β induced osteoclast formation 100 to 1000 fold more PGE_2 was required to get a complete recovery.
3) An extremely small amount of PGE_2 , 1 to 10 ng/ml, was essential in the PTH-induced osteoclast formation. This PGE2 was mainly produced by PGHS-2 which was induced by PTH in TRACP positive mononuclear cells , exclusively in those adhered to stromal cells.
4) In the PTH-induced osteoclast formation , one of the important roles of PTH is to induce PGHS-2 in stem cell lineage to help their differentiation . The other role of PTH is to increase the cAMP level in stromal cells to help their production of osteoclast differentiation factors.

REFERENCES

1. Udagawa N, Takahashi N, Akatsu T, Tanaka H, Sasaki T, Nishihara T, Koga T, Martin TJ and Suda T.: Origin of Osteoclasts: Mature Monocytes and Macrophages are capable of Differentiating into Osteoclasts under a Suitable Microenvironment Prepared by Bone Marrow-derived Stromal Cells. Proc. Natl. Acad. Sci USA . 87: 7260-7264 (1990).
2. Suda T,Takahashi N and Martin TJ. Endocrine Rev.:Modulation of Osteoclast Differentiation. 13: 66-80 (1992).
3. Sato T., Morita I., Sakaguchi K., Nakahama K., Smith W.L. , DeWitt D.L. and Murota S.: Involvement of Prostaglandin Endoperoxide H Synthase-2 in Osteoclast-like Cell Formation Induced by Interleukin-1β. J. Bone Miner. Res.; 11: 392-400 (1996).
4. Morita I, Smith W.L.,DeWitt D.L.,and Schindler M.: Expression-Activity Profiles of Cells Transfected with Prostaglandin Endoperoxide H Synthase Measured by Quantitative Fluorescence Microscopy. Biochemistry. 34: 7194-7199 (1995).
5. Takahashi N, Yamana H, Yoshiki S, Roodman G.D., Mundy G.R., Jones S.J., Boyde A and Suda T.: Osteoclast-Like Cell Formation and its Regulation by Osteotropic Hormones in Mouse bone Marrow Cultures. Endocrinology .122: 1373-1382 (1988).
6. Minkin C.: Bone Acid Phosphatase: Tartrate-Resistant Acid Phosphatase as a Marker of Osteoclast Function. Calcif Tissue Int. 34: 285-290 (1982).
7. Chomczynski P, Sacchi N.: Single-step Method of RNA Isolation by Acid Guanidinium Thiocyanate-Phenol-Chloroform Extraction. Anal Biochem.162: 156-159 (1987).

MECHANOTRANSDUCTION IN BONE CELLS: INDUCTION OF NITRIC OXIDE AND PROSTAGLANDIN SYNTHESIS BY FLUID SHEAR STRESS, BUT NOT BY MECHANICAL STRAIN

Rolf Smalt[1], Fred T. Mitchell[2], Ron L. Howard[2], and Timothy J. Chambers[1]

[1] Histopathology Department
St. George's Hospital Medical School
London SW17 ORE, UK
[2] Medical Physics Department
Atkinson Morley's Hospital
London SW20 ONE, UK

INTRODUCTION

The skeleton acts as a support for the rest of the body, and it has long been known that bone tissue adapts to its local mechanical environment: mechanical loading leads to an increase in the bone mass and density, whereas disuse results in a net loss of bone. It has been found that bones respond primarily to dynamic rather that static loads[1]. Mechanical loading of bone cells is associated with increased prostaglandin E_2 (PGE_2) production[2], and inhibition of prostaglandin synthesis suppresses the osteogenic response[3]. Inhibition of nitric oxide synthase (NOS) also abolishes the loading response, indicating that nitric oxide (NO) production has an essential early role in bone remodeling, too[4].

Although the response of bone tissue to mechanical loading has been extensively studied, little is known about the stimulus that is involved at the cellular level, and it is not even clear which cell type is responsible for the loading response. The surfaces of bone are covered in cells, mainly of the osteoblastic lineage. When dynamic mechanical loads are applied to the bone, these cells could perceive the deformation of the substrate (mechanical strain). However, during the process of bone formation, a proportion of the bone forming cells become embedded as osteocytes in the newly formed bone. These osteocytes remain connected via a network of cellular processes that run in narrow canaliculi throughout the bone. Upon compression of the bone, these cells could perceive the aforementioned mechanical strain, but in addition, they are exposed to compression-generated flow of the interstitial fluid along their cell membranes. Increased fluid flow has been shown to induce NO and PGE_2 production in vascular endothelial cells as part of the mechanism involved in the regulation of vascular tone[5,6], and it is intriguing that the response of bone to mechanical loading also requires NO and PGE_2 production.

Here we present a study of the responses of osteoblastic cells to mechanical strain and fluid flow *in vitro,* and compare their effects on NO and PGE_2 production.

Recent Advances in Prostaglandin, Thromboxane, and Leukotriene Research
Edited by Sinzinger *et al.*, Plenum Press, New York, 1998

METHODS AND RESULTS

Primary culture of osteoblasts, isolated from newborn rat calvaria or long bones, and the osteoblastic cell lines MC3T3-E1, ROS 17/2.8 and UMR 106.01 were used in both strain and flow experiments. For the strain experiments, cells were grown on a tissue-culture treated strip of polystyrene film in flexibly-mounted wells. The film was linearly stretched with a trapezoidal cycle of 1 Hz by the use of an electromagnet, thus allowing for precise electronic control. Strains and strain rates in the substrate were optically calibrated. PGE_2 was measured by RIA[7] and NO levels were determined electrochemically[8]. Subjection of the osteoblastic cells to 20 minutes of cyclic strains between 500 and 5,000 με did not induce significant changes in either their PGE_2 or NO production within 2 hours after stimulation.

The same cell types were exposed to a quanitifed laminar fluid flow: cells were grown on a glass slide, and mounted into a parallel plate flow chamber, as previously described[9]. The medium flow was controlled with a syringe pump. Osteoblastic cells, exposed to a 1 minute pulse of wall-shear stresses between 1.0 and 21.5 dyne/cm^2, showed a flow-dependent increase in the NO production immediately after the stimulus, returning to baseline levels within 30 minutes. This response was seen in all osteoblastic cells used, but not in primary rat skin fibroblasts. Due to its short half-life time, it was not practical to measure NO levels directly. Instead, the amount of NO_2^-, which is the stable oxidation product of NO, formed in the medium was assessed. To confirm that the increases in the NO_2^- concentration we found were actually the result of increased NO production, cells were stimulated with shear stress in the absence and presence of the NOS inhibitor L-NAME. The inhibitor completely blocked the NO_2^- response, demonstrating that the changes in the NO_2^- concentration were indeed the direct result of changes in the NO production rate.

Fluid flow can transfer its effect on cells in at least two distinct ways. The first is mechanical in nature; the wall shear stress, expressed in dyne/cm^2 (or Pa; 1 Pa = 10 dyne/cm^2). The second possible effect of flow on cells is electrical, and is caused by the movement of charged molecules in the fluid flow along the cell surface. This can generate potential differences across the cell membrane, which could either directly affect the cellular ion balance, or activate voltage-gated ion channels. To differentiate between these two fluid-flow generated stimuli, the viscosity of the medium was increased 80-fold with 1.2% ($^w/_v$) methylcellulose. This compound is routinely used at similar concentrations in semi-solid tissue culture, without toxic effects on the cells. Methylcellulose does not contain charged groups, and should therefore not affect the electrical properties of the fluid. When the flow was kept constant, and normal medium was replaced with the viscous medium, there was a distinct increase in NO production by osteoblastic cells, indicating that the flow effects are due to mechanical rather than electrical stimuli.

Besides this immediate increase in the NO concentration, we also found a flow-induced increase in the PGE_2 secretion, first observed at 5 minutes after the start of stimulation. The increase leveled off at 30 minutes, and showed a second, more pronounced increase, starting after 60-90 minutes.

DISCUSSION

The nature of the stimulus involved in mechanotransduction by bone cells has recently been the subject of extensive debate. Many experiments have shown that bone cells are sensitive to mechanical strain *in vitro.* Most such studies, though, exposed the cells to very large strains. For example, cells in cultures on flexible membranes experience strains between 3 and 24%, in contrast to the strains up to 3,000 με (0.3%) measured under physiological loading conditions. Many cell types are sensitive to strain at these very high levels, so it is uncertain whether the responses induced by such high strains in bone cells reflect the mechanisms by which bone normally senses mechanical stimuli, or whether they reflect responses also observed in non-osseous cells, unrelated to bone physiology.

As an alternative to strain, bone cells might detect mechanical signals through fluid flow. Mechanical loading of bone causes flow of interstitial fluid through the canalicular network. Experimental evidence has been obtained that interstitial fluid flow affects the rate of bone formation in rats[10]. Weinbaum, Cowin & Zeng[11] have proposed that fluid flow through canaliculi provides the mechanism by which bone cells experience the very small strains measured in bone during mechanical stimulation. Such a mechanism implies that bone cells must be sensitive to fluid-flow-induced wall-shear stresses of about 8-30 dynes/cm^2, that have been predicted to result from normal mechanical usage, which are similar to those acting on vascular endothelial cells. Reich & Frangos[6] have reported induction of osteoblastic PG production by fluid flow in this range. Klein-Nulend et al[12] noted PG production by osteocytes subjected to wall-shear stress; and Frangos' group[13] noted that a wall-shear stress of 6 dynes/cm^2 caused NO production continuously for 12 hours.

In this communication, we attempted to compare the ability of mechanical strain and fluid flow to induce NO and PG production in bone cells. To do this we developed an experimental model in which bone cells are exposed to measurable and physiological strains, in the absence of significant levels of fluid flow, by incubating the cells on strips of polystyrene film. For fluid flow experiments, we adapted a model commonly used in the assessment of endothelial cell responses. We could detect no induction of either PG or NO production by cyclic strains. In contrast, low levels of fluid flow induced rapid production of both agents. Since NO and PG production appear to be essential for mechanical responsiveness *in vivo,* this observation that even relatively large mechanical strains do not cause detectable production *in vitro,* while fluid flow induces both, in the same cell populations, suggests that fluid flow is more likely to be the stimulus acting directly on bone cells *in vivo.*

The mechanisms by which detection of fluid flow by bone cells could be translated into information that allows bone to adapt is structure in a way appropriate for the mechanical environment is unknown. The remodeling of capillary beds by fluid flow during development suggests that information from fluid flow can be translated into morphogenetic decisions. While PG and NO production might both represent signals that directly regulate bone formation and resorption, it is equally possible that they, and other signals generated by fluid flow, are involved in the processing of mechanically-generated information by bone cells. Whatever their role, our experiments suggest that fluid flow is the primary stimulus whereby bone cells respond to mechanical stimulation. If this is so, analysis of the effects of fluid flow on the behavior of bone cells should provide insights into the processes whereby the skeleton adapts its structure to meet the challenges of the mechanical environment.

REFERENCES

1. L.E. Lanyon and C.T. Rubin, Static vs dynamic loads as an influence on bone remodelling, *J. Biomechs.* 17:897 (1984).
2. I. Binderman, U. Zor, A.M. Kaye, Z. Shimshoni, A. Harell and D. Somjen, The transduction of mechanical force into biochemical events in bone cells may involve activation of phospholipase A_2, *Calcif. Tissue Int.* 42:261 (1988).
3. M.J. Pead and L.E. Lanyon, Indomethacin modulation of load-related stimulation of new bone formation *in vivo, Calcif. Tissue Int.* 45:34 (1989).
4. S.W. Fox, T.J. Chambers and J.W.M. Chow, Nitric oxide is an early mediator of the increase in bone formation by mechanical stimulation, *Am. J. Physiol.* 270:E955 (1996).

5. M.J. Kuchan and J.A. Frangos, Role of calcium and calmodulin in flow-induced nitric oxide production in endothelial cells, *Am. J. Physiol.* 266:C628 (1994).
6. K.M. Reich and J.A. Frangos, Effect of flow on prostaglandin E_2 and inositol triphosphate levels in osteoblasts, *Am. J. Physiol.* 261:C428 (1991).
7. B.M. Jaffe and H.R. Behrman, Prostaglandins and prostaglandin metabolites. in: *Methods of Hormone Radioimmunoassay.* B.M. Jaffe and H.R. Behrman eds., Academic Press, New York (1974).
8. K. Shibuki, An electrochemical microprobe for detecting nitric oxide release in brain tissue, *Neurosci. Res.* 9:69 (1990).
9. L.V. McIntire and S.G. Eskin, Mechanical and biochemical aspects of leukocyte interactions with model vessel walls, *KROC Foundation Series* 16:209 (1984).
10. C.H. Turner, M.R. Forwood and M.W. Otter, Mechanotransduction in bone: do bone cells act as sensors of fluid flow?, *FASEB J.* 8:875 (1994).
11. S. Weinbaum, S.C. Cowin and Y. Zeng, A model for the excitation of osteocytes by mechanical loading-induced bone fluid shear stresses, *J. Biomechs.* 27:339 (1994).
12. J. Klein-Nulend, A. Van der Plas, C.M. Semeins, N.E. Ajubi, J.A. Frangos, P.J. Nijweide and E.H. Burger, Sensitivity of osteocytes to biomechanical stress *in vitro*, *FASEB J.* 9:441 (1995).
13. D.L. Johnson, T.N. McAllister and J.A. Frangos, Fluid-flow stimulates rapid and continuous release of nitric oxide in osteoblasts, *Am. J. Physiol.* 34:E205 (1996).

HISTAMINE POTENTIATES LEUKOTRIENE B_4-INDUCED LEUKOCYTE ADHESION: A POTENTIAL ROLE OF HISTAMINE$_1$-RECEPTOR ANTAGONISM IN ALLERGIC LEUKOCYTE RECRUITMENT

Henrik Thorlacius, Johan Raud, Lennart Lindbom, and Per Hedqvist
Department of Physiology & Pharmacology
Karolinska Institutet
S-17177, Stockholm, Sweden

INTRODUCTION

The inflammatory process is vital for the survival of all complex organisms, and it plays an important role in health and disease. Acute inflammation is the result of a series of overlapping changes in the microcirculation, ultimately leading to the accumulation of plasma fluid and leukocytes in the extravascular tissue. These changes are evoked by chemical mediators released from various tissue cells, such as mast cells, and from leukocytes that are recruited from the circulating blood during the inflammatory reaction. Allergic reactions are characterized by IgE-mediated mast cell activation leading to the release of a large number of substances, including histamine and leukotrienes. Beside their direct effects, these mediators may functionally interact and potentiate the effects of each other.

This chapter summarizes recent work on interactions between histamine and leukotriene B_4 (LTB_4) with respect to the leukocyte response in acute inflammation. The results are based on intravital microscopy of the microcirculation in the hamster cheek pouch.

HISTAMINE POTENTIATES LTB_4-INDUCED LEUKOCYTE ADHESION

Histamine can induce some of the microvascular changes characteristic of inflammation, *i.e.* increased local blood flow and vascular permeability, but is considered not to cause overt recruitment of circulating leukocytes. Leukotriene B_4, on the other hand, rapidly activates leukocytes to promote firm adhesion and emigration[1]. Interestingly, our recent observations show that histamine can act synergistically with chemoattractants such as LTB_4 to potentiate leukocyte adhesion[2,3]. Thus, it seems reasonable to suggest that histamine may contribute to leukocyte recruitment in allergic inflammation.

Recent Advances in Prostaglandin, Thromboxane, and Leukotriene Research
Edited by Sinzinger *et al.*, Plenum Press, New York, 1998

COMPLETE HISTAMINE$_1$-RECEPTOR ANTAGONISM REDUCES ALLERGIC LEUKOCYTE RECRUITMENT

Knowing that histamine can potentiate chemoattractant-induced leukocyte adhesion, it was of interest to characterize this synergism further and evaluate the therapeutic potential of histamine receptor antagonists in reducing allergic leukocyte recruitment. To address these topics, intravital microscopy of the hamster cheek pouch was used. Topical challenge with histamine dose-dependently increased the leakage of plasma from postcapillary venules in the hamster cheek pouch (Figure 1). Furthermore, challenge with a threshold dose of LTB_4 (1 nM) induced only a minor increase in plasma extravasation. However, coadministration of this low dose of LTB_4 (1 nM) with the different concentrations of histamine (0.3, 1 and 10 μM) resulted in a striking synergism with respect to increased vascular permeability between these mediators (Figure 1).

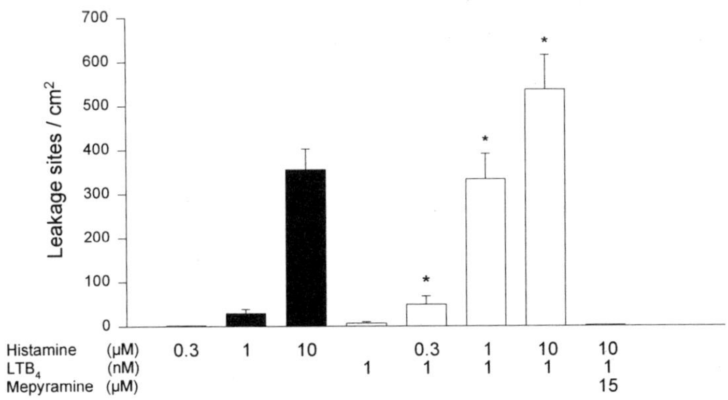

Figure 1. Peak (0-10 min) number of plasma protein leakage sites in hamster cheek pouch evoked by topical challenge with 0.3 μM, 1 μM, 10 μM histamine and/or 1 nM LTB_4 for 5 min. Mepyramine (15 μM) was applied topically 15 min before and throughout agonist challenge. Mean values ± SEM (n = 3-9). *($P < 0.05$) denotes differences between histamine + LTB_4 and the corresponding histamine doses alone. (Used with permission from Clin. Exp. Allergy 1996, vol 26).

Topical administration of 1 nM LTB_4 caused no significant increase in adherent leukocytes, whereas 3 nM LTB_4 increased leukocyte adhesion by 325 %. Histamine alone did not cause leukocyte adhesion, whereas simultaneous challenge with histamine and LTB_4 (1 nM) synergistically increased the adhesion of leukocytes (Figure 2). These results demonstrate that a low dose of histamine (*i.e.* 1 μM, with only minor effect on venular permeability) was sufficient to markedly potentiate the LTB_4-induced leukocyte response, again suggesting that histamine greatly improves the conditions for chemoattractant-induced adhesion of leukocytes to the venular endothelium.

These data indicate that even very low concentrations of histamine may contribute significantly to the recruitment of leukocytes in allergic inflammation, and they imply a therapeutic potential for antihistamines in this respect.

This possibility was examined by the use of a model for allergic inflammation in the cheek pouch of hamsters sensitized to ovalbumin. The time course and magnitude of the initial (0-5 min) increase in plasma extravasation after ovalbumin challenge corresponded well to that of topical application of 10 μM histamine (Figure 3).

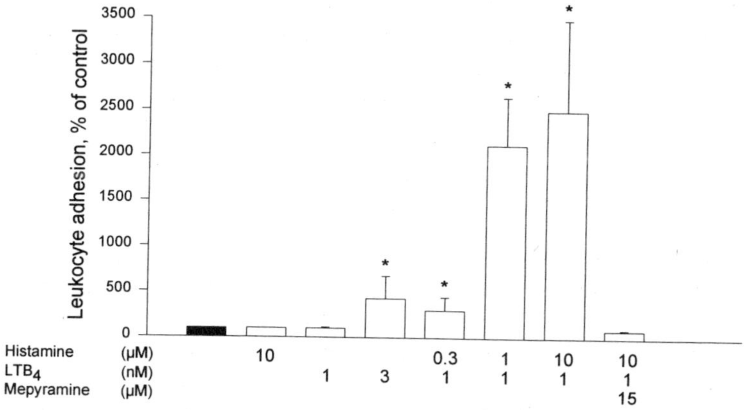

Figure 2. Venular leukocyte adhesion in hamster cheek pouch 20 min after challenge with different combinations of 0.3 μM, 1 μM, 10 μM histamine and 1 nM or 3 nM LTB_4. Mepyramine (15 μM) was applied topically 15 min before and throughout agonist challenge. Effects of agonists are expressed as % of control values (filled bar) before challenge. Mean values ± SEM (n = 3-8). *($P < 0.05$) denotes differences between the three combinations of histamine + LTB_4 and 1 nM LTB_4 alone. (Used with permission from Clin. Exp. Allergy 1996, vol 26).

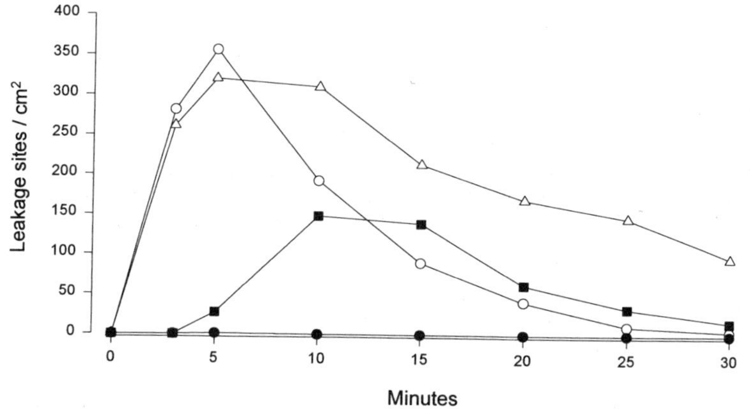

Figure 3. Time course of plasma protein leakage sites in hamster cheek pouch after 5 min topical challenge with 10 μg/ml ovalbumin alone (Δ), ovalbuminn + 15 μM mepyramine (■), 10 μM histamine alone (O), or 10 μM histamine + 15 μM mepyramine (●). Mean values, error bars omitted for clarity (n = 5-7). (Used with permission from Clin. Exp. Allergy 1996, vol 26).

Moreover, ovalbumin increased leukocyte emigration from 6.4 ± 1.4 to 50.0 ± 7.0 extravasated cells/unit tissue area ($P < 0.05$). Local pretreatment with the H_1-receptor antagonist mepyramine (15 μM) totally blocked the plasma leakage induced by 10 μM histamine, with or without coapplication of LTB_4. Mepyramine also abolished the initial (0-5 min) leakage of plasma provoked by ovalbumin challenge. The remaining, and somewhat delayed, plasma leakage (5-30 min) after ovalbumin challenge has previously been shown to be mainly leukotriene dependent[4]. Furthermore, the treatment with mepyramine completely inhibited the histamine-induced potentiation of leukocyte adhesion evoked by LTB_4, and also significantly reduced the marked ovalbumin-induced leukocyte emigration. Separate control experiments showed that mepyramine (15 μM) did not interfere with the leukocyte adhesion induced by 10 nM LTB_4 *per se*.

CONCLUSION

Taken together, these data demonstrate that threshold histamine activity may greatly potentiate chemoattractant-induced leukocyte recruitment, indicating an important role for histamine in the allergic cellular response. In order to abolish this component in allergic inflammation and to reduce leukocyte recruitment, it may be necessary to achieve tissue concentrations of histamine antagonists sufficient to completely inhibit histamine-induced microvascular actions. Accordingly, these findings may also help explain the poor clinical effects of antihistamines on allergic leukocyte accumulation.

Acknowledgments

We thank Ulla Palmertz for excellent technical assistance. This study was supported by the Swedish Medical Research Council (14X-4342, 04P-10738), the Swedish Foundation for Health Care Sciences and Allergy Research (A95093), the Swedish Society for Medical Research, and Karolinska Institutet.

REFERENCES

1. S.E. Dahlen, J. Björk, P. Hedqvist, K.E. Arfors, S. Hammarström, J.Å.Lindgren, and B. Samuelsson, Leukotrienes promote plasma leakage and leukocyte adhesion in postcapillary venules: in vivo effects with relevance to the acute inflammatory response, *Proc. Natl. Acad. Sci.* 78:3887 (1981).

2. H. Thorlacius, J. Raud, P. Hedqvist, and L. Lindbom, Microvascular mechanisms of histamine-induced potentiation of leukocyte adhesion evoked by chemoattractants, *Br. Journal Pharmacol.* 116:3175 (1995).

3. H. Thorlacius, L. Lindbom, P. Hedqvist, and J. Raud, Microvascular actions of histamine: synergism with leukotriene B4 and role in allergic leukocyte recruitment, *Clin. Exp. Allergy* in press (1996).

4. P. Hedqvist, H. Thorlacius, L. Lindbom, and J. Raud, Microvascular actions and interactions of eicosanoids and histamine in inflammation, in: *Eicosanoids: From Biotechnology to Therapeutic Application,* G. Folco, B. Samuelsson, J. Maclouf, G.P. Velo, eds., Plenum Press, New York (1996).

GASTRIC PROSTACYCLIN (PGI_2) PREVENTS STRESS-INDUCED GASTRIC MUCOSAL INJURY IN RATS BY INHIBITING LEUKOCYTE ACTIVATION

Naoaki Harada,[1] Kenji Okajima,[2] Kazunori Murakami,[2] Wenge Liu,[2] Hirotaka Isobe,[2] Keiichi Tanaka,[1] and Hiroaki Okabe[2]

[1]Department of Emergency and Critical Care Medicine, School of Medicine, Fukuoka University, Fukuoka, [2]Department of Laboratory Medicine, Kumamoto University School of Medicine, Kumamoto 860, Japan

INTRODUCTION

Prostaglandins play an important role in preventing gastric mucosal injury, referred to as "gastric cytoprotection". Although PGE_1 and PGE_2 inhibit gastric secretion, cytoprotection is considered independent of the antisecretory mechanism. PGE_2 and PGI_2 in gastric mucosa, have prevented gastric mucosal injury from various noxious stimuli. This mucosal cell protection is thought to involve increased gastric mucus secretion, bicarbonate production, and mucosal blood flow, though its precise mechanism has not fully been elucidated. Activated leukocytes have been implicated in gastric mucosal injuries induced by non-steroidal anti-inflammatory drugs, water-immersion restraint stress,[1] hemorrhagic shock and ethanol. Activated leukocytes release a variety of inflammatory mediators, including granulocyte elastase and reactive oxygen species that damage the endothelial cells and other tissues. PGI_2, the most abundant PG in gastric mucosa, is synthesized in endothelium and has various physiologic actions at the interface between blood and tissue. Since PGI_2 inhibits leukocyte activation *in vitro* ,[2] it is possible that the cytoprotective action of gastric PGI_2 may involve leukocyte activation inhibition, though direct evidence is lacking. Therefore we examined whether PGI_2 prevented the stress-induced gastric mucosal injury through leukocyte activation inhibition in the rat.

METHODS

Water immersion-restraint stress-induced gastric mucosal lesion formation in rats

Adult male Wistar rats (Nihon SLC, Hamamatsu, Japan), weighing 280-320 g, were placed in a restraint cage and immersed up to the xiphoid process in water at 22°C. At the indicated times during water immersion-restraint stress (WIR), gastric 6-keto-$PGF_1\alpha$ levels were determined by specific enzyme immunoassay. Gastric leukocyte accumulation evaluated by microscopy and gastric myeloperoxidase activity determination. Mucosal blood flow was determined by laser-Doppler flowmetry. Gastric lesion index was calculated by measuring the total length of gastric mucosal linear lesion.

Statistical analysis

All data are expressed as the mean ± SD. Comparisons among different groups of data were performed using analysis of variance with the Scheffé's (post hoc) test. A p value of ≤ 5% was considered significant.

Recent Advances in Prostaglandin, Thromboxane, and Leukotriene Research
Edited by Sinzinger *et al.*, Plenum Press, New York, 1998

RESULTS

Changes in gastric 6-keto-PGF$_1\alpha$ level, gastric myeloperoxidase activity, gastric mucosal blood flow and gastric lesion index in rats subjected to water immersion-restraint stress

Gastric levels of 6-keto-PGF$_1\alpha$ were significantly increased after 30 min of WIR compared to pre-WIR levels (Fig. 1A). These levels began to decrease after 1 h of WIR, and declined to below pre-WIR levels after 6 and 8 h of WIR (Fig.1A). Gastric accumulation of leukocytes as evaluated by gastric myeloperoxidase (MPO) activity significantly increased after 8 h of WIR compared to pre-WIR levels (Fig. 1B). The gastric mucosal blood flow began to decrease after 2 h of WIR, and decreased to 45% of pre-WIR level after 8 h of WIR (Fig. 1C). The gastric lesion index began to increase after 4 h of WIR and reached a maximum after 8 h of WIR (Fig. 1D).

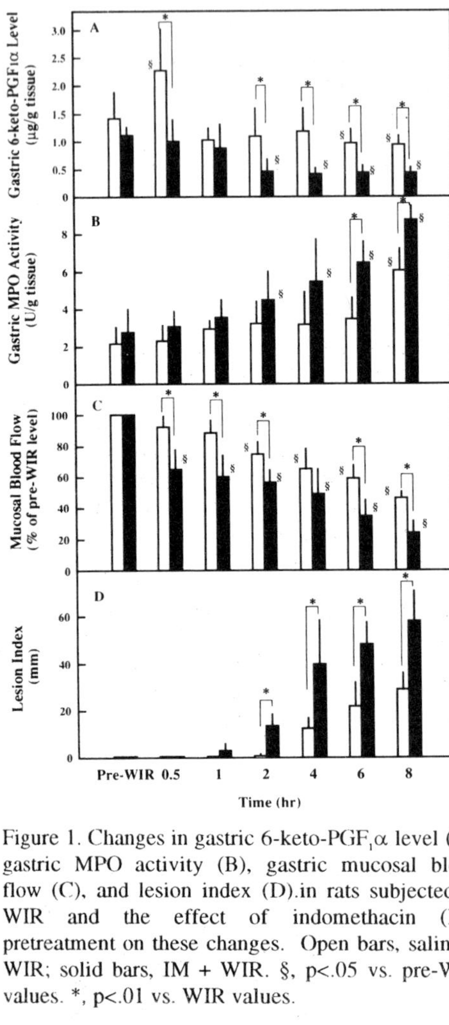

Figure 1. Changes in gastric 6-keto-PGF$_1\alpha$ level (A), gastric MPO activity (B), gastric mucosal blood flow (C), and lesion index (D).in rats subjected to WIR and the effect of indomethacin (IM) pretreatment on these changes. Open bars, saline + WIR; solid bars, IM + WIR. §, p<.05 vs. pre-WIR values. *, p<.01 vs. WIR values.

Effects of indomethacin pretreatment on changes in the gastric 6-keto-PGF$_1\alpha$ level, gastric accumulation of leukocytes, gastric mucosal blood flow and gastric lesion index in rats subjected to WIR

Subcutaneous IM, 5 mg/kg administered 30 min prior to WIR completely inhibited the WIR-induced increase in gastric 6-keto-PGF$_1\alpha$ level (Fig. 1A). Furthermore, IM pretreatment decreased gastric 6-keto-PGF$_1\alpha$ after 2 to 8 h of WIR to a level significantly lower than pre-WIR (Fig. 1A). In animals pretreated with IM gastric 6-keto-PGF$_1\alpha$ levels at 30 min, 2, 4,6, and 8 h of WIR were significantly low compared to those of animals subjected to WIR without IM pretreatment (Fig. 1A). Gastric MPO activity began to increase 2 h after WIR, reaching maximum after 8 h of WIR in the animals pretreated with IM (Fig. 1B). Gastric MPO activity at 2 to 8 h of WIR in animals pretreated with IM were significantly higher than those of control animals (Fig. 1B). Gastric mucosal blood flow was significantly low in animals subjected to WIR with IM pretreatment compared to that of animals subjected to WIR without IM pretreatment (Fig. 1C). WIR-induced gastric mucosal lesion formation was significantly exacerbated in animals pretreated with IM compared to that observed in control animals (Fig. 1D).

Effects of iloprost and nitrogen mustard-induced leukocytopenia on IM-induced changes in gastric accumulation of leukocytes, gastric mucosal blood flow and gastric mucosal lesion formation in rats subjected to WIR

The continuous 8-h intravenous infusion of iloprost (100 ng/kg/min), a stable analog of PGI$_2$, inhibited IM-induced increases in both gastric MPO activity (Fig. 2A) and gastric lesion index (Fig. 2B) at 8 h of WIR. Iloprost infusion inhibited the decrease in gastric

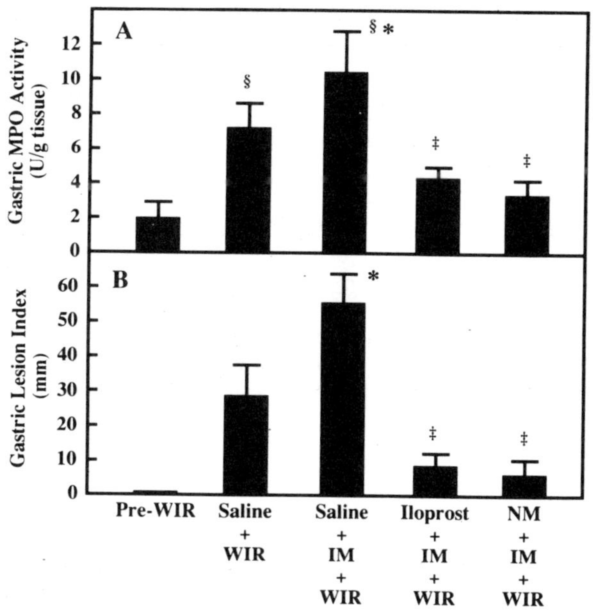

Figure 2. Effects of iloprost and leukocyte depletion on gastric MPO activity (A) and the legion index (B) in rats subjected to WIR pretreated with indomethacin (IM) pretreatment. §, p<.01 vs . pre-WIR values, *, p<.01 vs. WIR values. ‡, p<.01 vs. IM + WIR values.

mucosal blood flow induced by IM pretreatment in animals subjected to WIR (Fig. 3). Both the IM-induced increase in gastric lesion index and the IM-induced increase in gastric MPO activity at 8 h of WIR were significantly inhibited in animals with nitrogen mustard (NM)-induced leukocytopenia (Fig. 2, A and B). However, NM-induced leukocytopenia did not inhibit the IM-induced decrease in the gastric mucosal blood flow in animals subjected to WIR (Fig. 3).

Effects of iloprost and leukocytopenia on the IM-induced gastric leukocyte infiltration in animals subjected to WIR

Gastric mucosa in rats 8 h after WIR was examined under light microscopy (Table 1). The number of polymorphonuclear leukocytes (PMNs) infiltrating gastric tissue in stressed animals was significantly higher than in nonstressed animals (Table 1). IM pretreatment

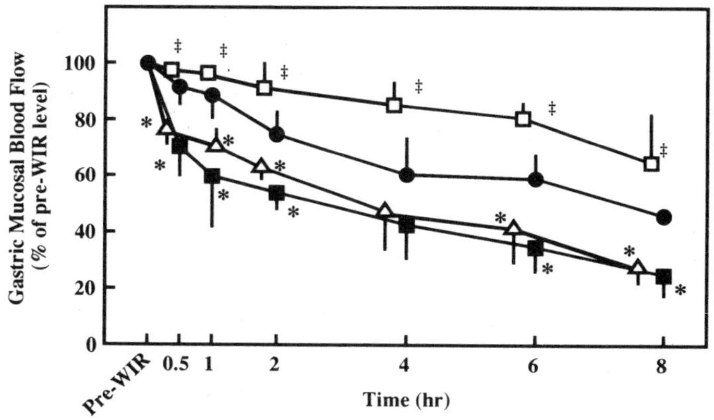

Figure 3. Effect of IM on WIR-induced changes of gastric mucosal blood flow in rats. (●) saline + WIR, (■) saline + IM + WIR, (□) iloprost + IM + WIR (△) NM + IM + WIR. *, p<.01 vs. WIR values. ‡, p<.01 vs. IM + WIR values.

Table 1. Numbers of polymorphonuclear leukocytes (PMN) in rat stomach wall

Group	Number of PMN (per high power field)
Non-stress	7.4 ± 2.3
Water-immersion restraint stress (WIR)	13.1 ± 3.2 §
Indomethacin (IM) + WIR	21.5 ± 4.1 § *
Iloprost + IM + WIR	5.1 ± 2.1 ‡
NM-induced leukocytopenia + IM + WIR	3.3 ± 2.3 ‡

§, p<.01 vs. non-stress values. *, p<.01 vs. WIR values. ‡, p<.01 vs. IM + WIR values.

significantly increased the numbers of infiltrating PMN compared to of animals subjected to WIR alone (Table 1). Iloprost or NM-induced leukocytopenia significantly reduced the IM-induced increase in PMN infiltration of gastric mucosa in rats subjected to WIR (Table 1).

DISCUSSION

In the present study, inhibition of the stress-induced increase in gastric synthesis of PGI_2 by indomethacin (IM) exacerbated the gastric mucosal lesion formation and increased the gastric accumulation of leukocytes. Administration of iloprost, a stable derivative of PGI_2, reversed the IM-induced pathologic events in rats subjected to WIR. NM-induced leukocytopenia significantly inhibited both the gastric accumulation of leukocytes and the exacerbation of the gastric mucosal lesion formation induced by IM in rats subjected to WIR, suggesting that inhibition of gastric accumulation of leukocytes by iloprost can be an effector rather than an effect of lesion prevention by iloprost. These observations suggest that increased gastric PGI_2 synthesis may act to prevent stress-induced gastric mucosal lesion formation by inhibiting the gastric leukocyte accumulation. Gastric mucosal blood flow decreased with WIR in rats. Pretreatment of animals with IM further decreased gastric mucosal blood flow in animals subjected to WIR in the present study. Since iloprost administration significantly inhibited the IM-induced decrease in the gastric mucosal blood flow as well as exacerbation of gastric mucosal lesion formation, it is possible that gastric PGI_2 might prevent gastric mucosal injury by maintaining the gastric mucosal blood flow. However, IM-induced mucosal lesion exacerbation was markedly reduced in the stressed animals with nitrogen mustard-induced leukocytopenia without a change in the IM-induced decrease in the gastric mucosal blood flow. This observation suggests that maintenance of gastric mucosal blood flow may not be necessary for reduction of stress-induced gastric mucosal injury. This notion is consistent with the observation that PGE_1 and PGE_2, gastric cytoprotective prostaglandins, decrease gastric mucosal blood flow secondarily due to their inhibitory effect on parietal cells. Furthermore, the fact that PGF_2a, a cytoprotective PG, is a vasoconstrictor also supports the notion that the mechanisms of PG cytoprotection need not depend on maintenance of mucosal blood flow. Thus, it is likely that gastric PGI_2 may prevent the stress-induced gastric mucosal injury mainly by inhibiting the leukocyte activation. This notion is also consistent with the report by Wallace et al.[3] demonstrating that gastric mucosal injury with nonsteroidal anti-inflammatory drugs is neutrophil-dependent.

REFERENCES

1. N. Harada, K. Okajima, K. Murakami, et al., Leukocyte depletion and ONO-5046, a specific inhibitor of granulocyte elastase, prevent a stress-induced decrease in gastric prostaglandin I_2 in rats. *Biochem. Biophys. Res. Commun.* 231:52 (1997).
2. M. Kainoh, R. Imai, and T. Umetsu et al., Prostacyclin and beraprost sodium as suppressors of activated rat polymorphonuclear leukocytes. *Biochem. Pharmacol.* 39:477 (1990).
3. J.L. Wallace, C.M. Keenan, and D.N. Granger, Gastric ulceration induced by nonsteroidal antiinflammatory drugs is a neutrophil-dependent process. *Am. J. Physiol.* 259:G462 (1990).

ANTI-P-SELECTIN ANTIBODY EXACERBATED INFLAMMATORY RESPONSES IN ACETIC ACID-INDUCED COLITIS

Wong, P.Y-K, Yue, G., Yin, K., Liu, P., Lin, M., Rodriguez, A., Nomen, M., Spur, B.W., Geng, J.G.*, Spokas, E.G. and Sun, F.F.

Department of Cell Biology, University of Medicine and Dentistry-School of Osteopathic Medicine, Stratford, NJ 08084 and *Department of Cell Biology and Inflammation Research, Pharmacia Upjohn, Inc., Kalamazoo, MI 49001

INTRODUCTION

The emigration of leukocytes from the vascular compartment to the site of injury is a key event in the development of inflammation. This transendothelial emigration of leukocytes is a complex series of events involving the expression and function of several adhesion molecules. The series of steps is started by tethering of leukocytes to the endothelium which initiates rolling, this in turn activates the leukocytes and leads to firm adherence and transendothelial migration[1]. The use of specific antibodies to particular adhesion molecules has been reported to reduce the inflammatory response in ischemia/reperfusion injury[2,3] and acute lung injury[4]. Increased expression of intercellular adhesion molecule-1 (ICAM-1), and its counter receptors the integrins LFA-1 and Mac-1 has been reported in mucosal biopsies of patients with IBD[5]. We have demonstrated, that monoclonal antibodies to ICAM-1 (1A29) greatly ameliorated the inflammatory response in rats, seven days after acetic acid (6%) administration[6]. In this communication, we report the role of P-selectin on early inflammatory response in acetic acid-induced colitis in rats. Expression of P-selectin is believed to be critical in the tethering of leukocytes to vascular endothelium and subsequent rolling[1]. In addition, P-selectin expression is increased in activated platelets where it mediates leukocyte-platelet adhesion[7]. In-vivo, selectin has been reported to be involved in lactoferrin induced intestinal mucosal injury[8], splanchnic ischemia-reperfusion[3] and acute lung injury[4]. P-selectin is stored in intracellular granules in endothelial cells (and platelets). It can be mobilized to the plasma membrane within minutes of activation of endothelial cells by thrombin, histamine and oxygen radicals or peroxides[9,10]. On the other hand, P-selectin has been shown to inhibit superoxide ion release from human neutrophils stimulated by TNFα[11]. Thus, the exact role of P-selectin in inflammation has not been firmly established. In this study, we have examined the effect of anti-P-selectin antibody in an acetic acid-induced model of acute colitis.

MATERIALS AND METHODS

Male Sprague-Dawley rats were fasted for 24h before use. On the day of the experiment, rats were anesthetized with metofane and given acetic acid (HOAc; 6%) or saline (0.9%) intracolonically. The rats given HOAc were then further subdivided and given either saline

Recent Advances in Prostaglandin, Thromboxane, and Leukotriene Research
Edited by Sinzinger *et al.*, Plenum Press, New York, 1998

(0.9%, i.v.), control antibody (rabbit IgG; 1mg/kg, i.v.) or polyclonal antibody to P-selectin (1mg/kg, i.v., P-sel Ab), administered through the lateral tail vein. All intravenous injections were administered immediately after HOAc was introduced. The rats were then placed back in their cages for 0.5h, 2h and 8h. We did not observe any significant difference between control rats sacrificed at 4h compared to 8h and thus the results were combined.

At time of sacrifice, rats were anesthetized by metofane. Blood was withdrawn into a syringe containing sodium citrate (0.38%, final concentration). The blood was centrifuged at 100g to obtain platelet rich plasma (PRP) for platelet aggregation.

After blood withdrawal, a midline incision was made to expose the distal colon. Approximately 8cm of colon was isolated and removed. Strips of colon, approximately 1cm in length (approximately 100mg) were cut from the colon for determination of superoxide generation, myeloperoxidase (MPO) activity and tissue NOS activity.

RESULTS AND DISCUSSION

Platelet aggregation: There was no change in platelet aggregation to 5μM ADP in platelets taken from HOAc-treated rats compared to saline controls.

Superoxide generation: At 0.5h and 2h tissue superoxide levels of HOAc treated positive control rats were similar to sham controls. However, at 0.5h, superoxide generation of colon tissue from P-sel Ab treated rats significantly increased compared to saline controls as well as HOAc treated positive controls (Table 1). At 8h, superoxide generation was increased by 140% in tissues from HOAc positive control rats compared to saline controls. Superoxide generation in tissue from rats given HOAc + P-sel Ab was increased by approximately 6-fold compared to saline controls and increased by 2.6-fold compared to HOAc positive controls. When the tissues were stimulated with fMLP (11.4μM), the trends remained, i.e. there was an increase in superoxide generation in tissues from HOAc treated rats as compared to saline controls at 8h.This increase was further augmented in tissues from rats given P-sel Ab Table 1.

MPO activity : Tissues taken from given HOAc were not different from saline control rats at 0.5h and 2h. AT 8h, there was a 2.8-fold increase in MPO activity in tissues taken from HOAc positive controls and a 3.3-fold increase in HOAc + P-sel tissues compared to saline controls ($P<0.05$). However, MPO activities of tissues taken from rats given HOAc + P-sel Ab were not different from MPO values of rats given HOAc positive controls.

NOS activity : At 0.5h and 2h, total NOS an iNOS activities were not altered in tissues of HOAc positive controls of HOAc + P-sel Ab rats as compared to saline controls. At 8h however, there was a 2-fold increase in total NOS activity of tissues from HOAc positive controls and HOAc + P-sel as compared to saline controls. Total NOS activity of P-sel Ab

Table 1: Tissue Superoxide generation (cpm/mg wet weight; n = 6 - 11 for all groups)

				+ fMLP (11.4μM)		
	0.5h	**2h**	**8h**	**0.5h**	**2h**	**8h**
HOAc	706 ± 154	1306 ± 188	1978 ± 234*	1088 ± 171	2214 ± 223	3456 ± 391*
HOAc + Psel AB	1398 ± 181*	1566 ± 405	5066 ± 1071*#	1374 ± 178	2208 ± 223	8523 ± 1645*#
Saline controls		822 ± 156			972 ± 153	

*** Denotes significantly different from saline control, $P < 0.05$**
 # Denotes significantly different HOAc positive control, $P < 0.05$, Gr. = groups,
NB: Saline controls were sacrificed at 4h and 8h. As results were not significantly different between these time points, they have been combined. Similarly, HOAc positive controls are the combined results of HOAc + control antibody and HOAc + saline vehicle.

treated rats at 8h showed a tendency to be lower (approximately 28%) compared to HOAc positive controls, however, this difference did not reach significance. Activity of iNOS in positive control rats was increased by approximately 4-fold. P-sel Ab treatment reduced the iNOS activity 30% compared to HOAc positive controls.

These results suggest that use of anti-P-selectin antibody augmented the tissue generation of superoxide in HOAc treated rats. This observation is consistent with the hypothesis that P-selectin expression is correlated with a reduction in superoxide ion generation in activated neutrophils[11]. Interestingly, MPO activities of P-sel Ab rats were similar to that of positive controls suggesting that the increased superoxide generation in colon tissues was not due to an increased number of emigrated neutrophils. Thus, our results indicate that the administration of P-sel Ab resulted in increased superoxide generation by individual neutrohils.

The exact mechanism for the failure of P-sel Ab administration to decrease neutrophil emigration in the acetic acid induced colitis model is unclear. There are several possible explanations. The blockade of P-selectin-ligand interaction may reduce certain compensatory antiinflammatory molecules generated by the interacting cells. For example, P-selectin dependent platelet neutrophil interaction generates a putative antiinflammatory lipid mediator, lipoxin A_4 (LXA_4). LXA_4 is formed from arachidonic acid by the transcellular biosynthesis[12]. It has been reported to be a potential inhibitor of neutrophil-endothelial cell adhesion[13,14]. Our preliminary data (results not shown) indicated that LXA_4 was reduced in colon tissues of rat given P-sel Ab as compared with positive controls. Furthermore, in an acute basement membrane glomerulonephritis model of the mouse, aggravated neutrophil influx was observed in glomeruli of P-selectin deficient mice with a concomitant decrease in LXA_4 levels[15]. These results suggest that the roles of P-selectin may have an antiinflammatory component and its blockade can cause exacerbation of the proinflammatory response.

From our studies, it is unclear if the modest decrease in iNOS activity in P-sel antibody treated rats could account for the substantial increase in superoxide generation. It has been shown that NO donors can inhibit chemotaxis and reduce the respiratory burst in human neutrophils[16]. Hence, the possibility that the decrease in iNOS activity seen with P-sel Ab treatment may lead to the enhanced superoxide generation cannot be ruled out.

In conclusion, the results of our studies with the P-sel Ab were in stark contrast to the results obtained using 1A29 (anti-ICAM-1 monoclonal antibody)[7]. Our results suggested that in the acetic acid induced colitis model, inhibition of P-selectin function did not affect transendothelial migration of neutrophils. In contrast to the inhibition of ICAM-1, the blockade of P-selectin caused enhanced superoxide generation. The cause of the increase in superoxide generation after anti-P-selectin antibody administration is unknown.

REFERENCES

1. J.C.Paulson, Selectin/carbohydrate-mediated adhesion of leukocytes, in: Adhesion:Its Role in Inflammatory Disease, J.M.Harlan and D.Y.Liu, eds., W.H.Freeman and Co., New York Pgs. 19 - 42 (1992).
2. R.Scalia, T.Murohara, J.A.Delyani, T.O.Nossuli, and A.M.Lefer, Myocardial protection by N,N, N-trimethylsphinosine in ischemia-reperfusion injury is mediated by inhibition of P-selectin, J.Leukoc.Biol. 59:317-324(1996).
3. T.W.Gauthier, K.L.Davenpeck, and A.M.Lefer, Nitric oxide attenuates leukocyte-endothelial interaction via P-selection in splanchnic ischemia-reperfusion, Am. J. Physiol. 267:G562-568 (1994).
4. M.S.Mulligan, M.J.Polley, R.J.Bayer, M.F.Nunn, J.C.Paulson, and P.A.Ward, Neutrophil-dependent acute lung injury: Requirement for P-selectin, J. Clin. Invest. 90:1600-1607(1992).
5. S. Nakamura, H.Ohtani, Y.Watanabe, K.Fukushima, T.Matsumoto, A.Kitano, K.Kobayashi and H.Nagura, In-situ expression of the cell adhesion molecules in inflammatory bowel disease, Lab.Invest. 69:77-85(1993).
6. P.Y-K.Wong, G.Yue, K.Yin, M.Miyasaka, C.L.Lane, A.M.Manning, D.C.Anderson, and F.F.Sun, Antibodies to intercellular adhesion molecule-1 ameliorate the inflammatory response in acetic acid-induced inflammatory bowel disease, J.Pharmacol. Exp.Ther. 274:475-480(1995).

7. E.Larsen, A.Celi, G.E.Gilbert, B.C.Furie, J.K.Erban, R.Bonfati, D.D.Wagner, and B.Furie, PADGEM protein: A receptor that mediates the interaction of activated platelets with neutrophils and monocytes, Cell. 59:305-312(1989).
8. I. Kurose, T.Yamada, R.Wolf and D.N.Granger, P-selectin-dependent leukocyte recruitment and intestinal mucosal injury induced by lactoferrin, J.Leukoc. Biol. 55:771-777(1994).
9. D.E. Lorant, K.D.Patel, T.M.McIntyre, R.P.McEver, S.M.Prescott and G.A.Zimmermann, Coexpression of GMP-140 and PAF by endothelium stimulated by histamine or thrombin: A juxtacrine system for adhesion and activation of neutrophils, J.Cell. Biol. 115:223-234(1991).
10. K.D. Patel, G.A.Zimmermann, S.M.Prescott, R.P.McEver and T.M.McIntyre, Oxygen radicals induce human endothelial cells to express GMP-140 and bind neutrophils, J.Cell. Biol. 112:749-759(1991).
11. C.S. Wong, J.R.Gamble, M.P.Skinner, C.M.Lucas, M.C.Berndt and M.A.Vadas, Adhesion protein GMP 140 inhibits superoxide anion release by human neutrophils, Proc. Natl. Acad. Sci. USA 88:2397-2401(1991).
12. C.N. Serhan, Lipoxin biosynthesis and its impact in inflammatory and vascular events, Biochim.Biophys.Acta. 1212:1-25(1994).
13. A.Papayianni, H.Rennke, C.N.Serhan and H.R.Brady, Platelet-leukocyte adhesion through P-selectin facilitates transcellular lipoxin biosynthesis in acute glomerulonephritis, Kidney Int.47: 1295-1302(1995).
14. P.Hedqvist, J.Raud, J.Palmertz, J.Haeggstrom, K.C.Nicolaou and S.E.Dahlen, Lipoxin A_4 inhibits leukotriene B_4-induced inflammation in the hamster cheek pouch, Acta Physiol.Scand. 137:571-581(1989).
15. T.N. Mayadas, D.L.Mendrick, H.R.Brady, T.Tang, A.Papayianni, K.J.M.Assmann, D.D.Wagner, R.O.Hynes and R.S.Cotran, Acute passive anti-glomerular basement membrane nephritis in P-selectin-deficient mice, Kidney Int. 49:1342-1349(1996).
16. E.Moilanen, P.Vuorinen, H.Kankaanranta, T.Metsa-Ketela and H.Vapatalo, Inhibition of nitric oxide-donors of human leukocyte functions, Br.J.Pharmacol. 109:852-858(1993).

HETEROGENEOUS DISTRIBUTION OF COX-2 OVER-EXPRESSION IN HUMAN COLON CARCINOMA CELLS

Jörg Berg*, Thomas Christoph, Angelika Bodenteich and Robertson Towart

Department of Pharmacology
Nycomed Austria GmbH
A-4021 Linz, Austria

INTRODUCTION

The two isoforms of cyclooxygenase (COX-1, COX-2) are considered the rate-limiting enzymes in the pathway of the conversion of arachidonic acid to eicosanoids (1). COX-1 is constitutively expressed whereas COX-2 is highly up-regulated by various stimuli in inflammatory controlling cells leading to increased eicosanoid formation. Increased eicosanoid formation has been observed in breast cancer cells, skin cancer cells as well as in colorectal carcinoma cells (2). Furthermore, COX-2 over-expression has been demonstrated in skin and colorectal cancer cells (3, 4) suggesting this to be the cause for the increased eicosanoid formation. Thus, non steroidal anti-inflammatory drugs have been tried and found to reduce the risk of developing colorectal cancer in animals as well as in humans (5). However, the molecular mechanisms for these effects remain unknown as yet.

In this study we analyse 14 colon carcinoma cell lines for COX-activity and COX-2 protein expression. We note a heterogeneous distribution of COX-2 protein in the cell populations of those cell lines that express COX-2 and discuss possible implications of this finding for the carcinogenesis of colon carcinomas and other malignancies.

MATERIALS AND METHODS

Cell Culture

The human colon carcinoma cell lines were either obtained from the American Type Culture Collection (Rockville, MI, USA) or the German Collection of Microorganisms (Braunschweig, FRG). The cells were cultured either in Dulbecco's modified Eagle's

* Corresponding address: Zentral-Labor, Allgemeines Krankenhaus der Stadt Linz, Krankenhausstr.9 A-4020 Linz, Austria. Tel.: +43-70-7806-1841; FAX.: +43-70-7806-1815.

Recent Advances in Prostaglandin, Thromboxane, and Leukotriene Research
Edited by Sinzinger *et al.*, Plenum Press, New York, 1998

medium (DMEM) or RPMI 1640 supplemented with 10% FCS, sodium pyruvate (1 mM), glutamine (2 mM), penicillin (100 U/ml), and streptomycin (100 μg/ml). Standard cell culture conditions were applied.

Cytoplasmic immunofluorescence

Cytoplasmic immunofluorescence was carried out as described previously (6). About 10^5 cells were centrifuged on a slide, fixed in ethanol, rehydrated in PBS with BSA and incubated with mouse IgG_1 to human COX-2 (7) or with mouse IgG_1 of irrelevant specificity (MOPC 21; Sigma, Munich, FRG) each followed by FITC-labelled goat anti-mouse IgG (Southern Biotechnology Inc., Birmingham, AL, USA). The slides were analysed by fluorescence microscopy. At least 250 cells were analysed per slide.

Western blot analysis

Western blot analysis for COX-1 and COX-2 protein was performed as previously described (8). Briefly, cells were lysed with NET (150 mM NaCl, 5 mM EDTA, 0.1% NaN_3, 0.5% Triton-X 100, 0.05 M Tris pH 7.4 and 1 mM phenylmethylsulfonyl fluoride), subjected to Tricine-SDS-polyacrylamide gel electrophoresis (T-SDS-PAGE) (9) followed by immunoblotting with mouse IgG_1 to human COX-2 (7) or with mouse IgG_1 of irrelevant specificity (MOPC 21) both followed by alkaline phosphatase-labeled goat anti-mouse IgG (Accurate, Westbury, NY, USA). For COX-1 detection mouse IgG_{2b} to human COX-1 (10), or mouse IgG_{2b} of irrelevant specificity (MOPC 141) were used. Development was with NBT/BCIP solution (Sigma, Munich, FRG).

Assessment of COX activity in intact cells

For the assessment of COX activities in intact cells (11) culture medium was renewed and arachidonic acid (30 μM; Sigma, Munich, FRG) was added for 15 min. at 37°C. Thereafter, PGE_2 was assessed in cell free culture supernatants by ELISA (Cayman, Ann Arbor, MI, USA) according to the manufacturer's instructions.

RESULTS AND DISCUSSION

When we analysed a panel of 14 colon carcinoma cell lines for COX protein by Western blot using monoclonal antibodies (mAb) directed to human COX-1 and COX-2, the COX-2 specific double band of 72/74 kD was detectable in 10 out of 14 cell lines. Among the 10 positive cell lines the intensity of the COX-2 specific double bands varied from cell line to cell line. COX-1 was detectable by Western blot analysis in only one cell line (table 1). In order to find out whether the COX-2 protein found in these cell lines is also enzymatically active COX activities were assessed by intact cell assay using PGE_2 as a measure for COX activity. In 9 out of these 10 cell lines PGE_2 formation was detectable, again to varying degrees. The COX activities found correlated to the results obtained by Western blot analysis of COX-2 protein. However, the correlation was not perfect. Besides PGE_2 other eicosanoids could have been synthesised in these cells, too. Cells of the COX-2 negative cell lines did not show any detectable PGE_2 formation.

To study the distribution of COX-2 protein in the cell population of these cell lines at the level of single cells, cytoplasmic immunofluorescence with the mAb to human COX-2 was performed. The cell lines found negative by Western blot analysis did not stain with the anti-COX mAb (table 1). The cell lines showing low levels of COX-activities showed only

Table 1. Cytoplasmic Immunofluorescence, COX-activity and Western blot analysis of colon carcinoma cell lines.

Cell Lines	Cytoplasmic IFL of COX-2 [% stained cells]	COX Activity [pg $PGE_2/10^6$ cells/min]	Western Blot Analysis of COX-2	Western Blot Analysis of COX-1
COLO 206 F	0	0	neg.	neg.
COLO 205	0	0	neg.	neg.
COLO 320	0	0	neg.	neg.
T 84	0	1.4 ± 0.7	neg.	+
COLO 678	0.5	1.2 ± 0.5	+	neg.
CX-1	1	5.3 ± 1.9	+	neg.
DLD-1	50	6.5 ± 1.8	+	neg.
LS174T	5	31.3 ± 3.6	+	neg.
SK-CO-1	50	90.7 ± 3.9	+	neg.
SW 1417	2	33.2 ± 4.7	+	neg.
CACO-2	10	554.5 ± 31	++	neg.
HT-29	98	52.4 ± 2.7	++	neg.
LoVo	20	149.5 ± 22	++	neg.
WiDr	5	28.2 ± 2.6	++	neg.

Abbr.: neg. is no COX protein staining; + is weak COX protein staining; ++ is prominent COX protein staining; IFL is Immunofluorescence;

few cells expressing COX-2 (0.5 to 1%) exhibiting bright fluorescence. In the cell lines with strong COX activities more than 10 to 20% of the cells were stained bright positive with anti-COX-2 (table 1). Thus a heterogeneous distribution pattern of COX-2-expressing cells within the cell lines was observed.

Carcinogenesis (including colon carcinogenesis) is thought to comprise several steps of genetic alterations that accumulate over time and lead to increasing degrees of malignancy. These genetic alterations relate to tumor suppressor genes and include losses of their allelic wild types. Furthermore, mutations in oncogenes occur. In colon carcinomas this happens most frequently in the ras gene leading to de-controlled cellular activation (12). Since Ras has been suggested to be involved in the signal transduction pathway that controls transcription of the COX-2 gene (13), mutated Ras could be the reason for the over-expression of COX-2 gene in colon carcinomas. Since the peroxidase activity of COX catalyses the oxidation of a wide range of xenobiotics including several carcinogens, the resultant products can react with DNA which then could lead to further genetic alterations (14). Thus, COX-2 over-expressing tumor cells would eventually arrive at a further increased stage of malignancy.

In case COX-2 over-expression represents just a side effect of a mutated Ras activity, then COX-2 over-expression would play a symptomatic rather than a mechanistic role in colon carcinogenesis. Under these conditions the observed cancer preventive effects of NSAIDs would rely on other properties of the compounds rather than on COX inhibition. However, several structurally quite different NSAIDs (Aspirin, Indomethacin, Sulindac and Piroxicam) have been observed to be effective in preventing colorectal cancer (5), making this explanation less likely. In addition, over-expression of COX-2 in intestinal cells has been demonstrated to lead to phenotypic changes involving increased adhesion to extracellular matrix proteins, resistance to butyrate-induced apoptosis, and reduced transforming growth factor $beta_2$ receptors levels (15). We, therefore, hypothesise that the COX-2 enzyme does play a role in pathogenesis of colon cancer (and perhaps in other types of malignancy) and suggest that over-expression of COX-2 occurs in some single tumor cell at some time. COX-2 over-expression provides this cell with some selective advantage that eventually leads to overgrowth of the tumor by the descendants of this cell, shifting the tumor cell population from COX-2 negative to COX-2 positive. Thus, the tumor in total arrives at an increased stage of malignancy.

In conclusion, we find over-expression of COX-2 in colon carcinoma cells to be accompanied by increased PGE_2 formation. The distribution pattern of COX-2 protein within the tumor cell population was found to be quite heterogeneous for most of the cell lines suggesting a heterogeneous distribution pattern of COX-2 gene expression. It will be interesting to see if the cells of those tumor cell lines that over-express COX-2 also show mutated ras genes. In addition, it will be interesting to investigate whether in one or the other colon carcinoma cell line the COX-2 over-expressing cells will eventually overgrow the whole tumor cell population when cultured for an extended period of time.

Acknowledgments

We are grateful to M. Widerna for expert technical help and to Dr. H. Blaschke for helpful discussions. We would like thank Drs. S. Engelsen and D. Stimmeder for support and encouragement.

REFERENCES

1. Hershman, H. R., Xie, W. & Reddy, S. (1995) BioEssays 17: 1031-1037.
2. Hubbard, W. C., Alley, M. C., McLemore, T. L. & Boyd, M. R. (1988) Cancer Res. 48: 4770-4775.
3. Mueller-Decker, K., Scholz, K., Marks, F. & Fuerstenberger, G. (1995) Molecular Carcinogenesis 12: 31-41.
4. Gustafson-Svärd, C., Lilja, I., Hallböök, O. & Sjödahl, R. (1996) Gut 38: 79-84.
5. Giardiello, F. M., Offerhaus, G. J. A. & DuBois, R. N. (1995) Eur. J. Cancer 31: 1071-1076.
6. Berg, J., Doe, B., Steimer, K. S. & Wabl, M. (1991) J. Virol. Meth. 34: 173-180.
7. Creminon, C., Habib, A., Maclouf, J., Pradelles, P., Grassi, J. & Frobert, Y. (1995) Biochim. Biophys. Acta 1254: 341-348.
8. Christoph, T., Widerna, M., Bodenteich, A. & Berg, J. (1996) Inflammopharmacology 4: 209-221.
9. Schägger, H. & von Jagow, G. (1987) Analytical Biochemistry 166: 368-379.
10. Creminon, C., Frobert, Y., Habib, A., Maclouf, J., Pradelles, P. & Grassi, J. (1995) Biochim. Biophys. Acta 1254: 333-340.
11. Mitchell, J. A., Akarasereenont, P., Thiermermann, C., Flower, R. & Vane, J. R. (1994) Proc. Natl. Acad. Sci. USA 90: 11693-11697.
12. Fearon, E. R. & Vogelstein, B. (1990) Cell 61: 759-767.
13. Xie, W. & Herschman, H. R. (1995) J. Biol.Chem. 270, 27622-27628.
14. Smith, B.J., Curtis, J.F. & Eling, T.E. (1991) Chem. Biol. Interact. 79: 245-264.
15. Tsujii, M. & DuBois, R.N. (1995) Cell 83: 493-501.

HSP70: A SUBUNIT OF THE CYTOSOLIC 12(S)-HETE BINDING COMPLEX

Tobias Kühme, Helena Herbertsson and Sven Hammarström

Division of Cell Biology, Linköping University
S-58185 Linköping, Sweden

Introduction

Eicosanoids are formed from polyunsaturated fatty acid that are present in phospholipids of cell membranes. These acids can be released by the action of phospholipase A_2. The free acids may be metabolized e.g. by lipoxygenase or cyclooxygenase. The cyclooxygenase pathway gives rise to prostaglandins and thromboxanes while the lipoxygenase pathway gives hydroperoxyeicosatetraenoic acid (HPETE), hydroxyeicosatetraenoic acid (HETE) and leukotrienes. Three distinct lipoxygenase enzymes produce 5-, 12- and 15-HPETE respectively. These three hydroperoxy acids are rapidly reduced to the corresponding hydroxy acids. 5-HPETE may be further converted to leukotrienes (1).

12(S)-HETE is produced by several kinds of cells e.g. platelets, macrophages, smooth muscle cells, tumor cells, keratinocytes as well as by brain tissue (2). Thromboxane A_2, another arachidonic acid metabolite formed by platelets mediates thrombin- and collagen-induced platelet

Recent Advances in Prostaglandin, Thromboxane, and Leukotriene Research
Edited by Sinzinger *et al.*, Plenum Press, New York, 1998

aggregation. Arachidonic acid metabolism is also important for tumor cell induced platelet function (3). It has been shown that tumor cell adhesion to endothelial cells, subendothelial matrix and fibronectin is stimulated by 12(S)-HETE (4). This effect appears to be mediated by the cell surface glycoprotein receptor GpIIb/IIIa which is also abaundantly present in platelets. GpIIb/IIIa is a member of a family of adhesive protein receptors called integrins. These receptors interact with several adhesion proteins such as fibronectin, fibrinogen, vonWillebrand factor and vitronectin (5). Metastasis includes a variety of cellular interactions; cell-cell, tumor cell-platelet, tumor cell-endothelial cell and cell-intercellular matrix. These interactions are mediated by adhesion molecules or adhesion molecule receptors on tumor cells, on normal cells and in the subendothelial matrix. The glycoprotein receptor GpIIb/IIIa has recently been found on the surface of human, rat and murine tumor cells. Adhesion mediated by this receptor is a critical step in metastasis. 12(S)-HETE has been shown to increase surface expression of GpIIb/IIIa and 12(S)-HETE may thus be involved in the spreading of cancer.

The detailed mechanism by which 12(S)-HETE affects the glycoprotein complex is unknown but it demonstrates both stereospecificity and geometric selectivity : 12-(S)-HETE does affect the glycoprotein while 12-(R)-HETE, 5-(S)-HETE and 15-(S)-HETE have no effect (4, 6).

BINDING OF 12(S)-HETE TO LEWIS LUNG CARCINOMA CELLS

Characteristics of the binding reaction

We have earlier published data showing that 12(S)-HETE binds to Lewis lung carcinoma cells (7). The binding is saturable, reversible and specific for 12(S)-HETE. Equilibrium was reached within 90 min at 4°C after the addition of ligand. Addition of a thousandfold excess of unlabeled 12(S)-HETE resulted in an 80% reduction of cellbound radioactivity, suggesting that 12(S)-HETE binding is reversible. Scatchard analysis of the data indicated a single population of binding sites with apparent K_d and B_{max} values of 0.3-0.4 nM and 66,000 sites per cell, respectively. The metabolic stability of 12(S)-HETE in the binding assay was analylyzed by HPLC and about 90% of the radioactivity recovered from the incubation buffer had the same retention time as 12(S)-HETE.

Table 1. Competition by various lipoxygenase products on specific binding of 12(S)-HETE to LLC cells (7).

Compound		IC_{50}
Abbreviated name	Systematic name	
5(S)-HETE	(6E,8Z,11Z,14Z)-(5(S))-5-hydroxyeicosa-6,8,11,14-tetraenoic acid	*ca* 5×10^{-7} M
12(S)-HETE	(5Z,8Z,10E,14Z)-(12(S))-12-hydroxyeicosa-5,8,10,14-tetraenoic acid	*ca* 5×10^{-8} M
12(R)-HETE	(5Z,8Z,10E,14Z)-(12(R))-12-hydroxyeicosa-5,8,10,14-tetraenoic acid	$>>10^{-6}$ M
15(S)-HETE	(5Z,8Z,11Z,13E)-(15(S))-15-hydroxyeicosa-5,8,11,13-tetraenoic acid	*ca* 5×10^{-7} M
5(S),12(S)-diHETE	(6E,8Z,10E,14Z)-(5(S),12(S))-5,12-dihydroxyeicosa-6,8,10,14-tetraenoic acid	$>>10^{-6}$ M
6-*trans*-LTB_4	(6E,8E,10E,14Z)-(5(S),12(R))-5,12-dihydroxyeicosa-6,8,10,14-tetraenoic acid	$>>10^{-6}$ M
6-*trans*-12-epi-LTB_4	(6E,8E,10E,14Z)-(5(S),12(S))-5,12-dihydroxyeicosa-6,8,10,14-tetraenoic acid	$>>10^{-6}$ M
LTB_4	(6Z,8E,10E,14Z)-(5(S),12(R))-5,12-dihydroxyeicosa-6,8,10,14-tetraenoic acid	$>10^{-6}$ M
13(S)-HODE	(9Z,11E)-(13(S))-13-hydroxyoctadeca-9,11-dienoic acid	*ca* 5×10^{-7} M

Specificity of the Binding Reaction

To examine the specificity of binding, competition experiments were performed. LLC cells were incubated with 12(S)-[^{3}H]HETE (0.1 nM) in the presence of different concentrations (0-10^{-6} M) of structurally related, unlabeled compounds. Table 1 shows the IC_{50} values obtained. Two positional isomers of 12(S)-HETE, 5(S)-HETE and 15(S)-HETE, formed from arachidonic acid by distinct lipoxygenases, and 13(S)-HODE, derived from linoleic acid by one type of 12- or by 15-lipoxygenase [1], gave IC_{50} values of around 5×10^{-7} M. Unlabeled 12(S)-HETE, analyzed under the same conditions, had an IC_{50} value of ca 5×10^{-8} M. 12(R)-HETE, the enantiomer of 12(S)-HETE, did not compete for binding even at a concentration of 10^{-6} M. Thus, the stereospecificity of binding was greater than the positional specificity. Two dihydroxy metabolites of arachidonic acid both having a 12(S)-hydroxyl group and in addition a 5(S)-hydroxyl group, 5(S),12(S)-diHETE and 6-*trans*12-epi-LTB_4, also did not compete for binding at concentrations up to 10^{-6} M. The effects of LTB_4 and 6-*trans*-LTB_4 were also determined. These two compounds have 5(S)- and 12(R)-hydroxylgroups. No competition was observed for 6-*trans*-LTB_4 (IC_{50}>>10^{-6} M) whereas LTB_4 at 10^{-6} M reduced the binding to 64% of the initial value(7).

It was reported (8) that Lewis lung carcinoma cells contain specific binding sites for 12(S)-HETE which are present in cytosol (52%), mitochondria (30%) and in nuclei (18%). No 12(S)-HETE binding sites were found in plasma membranes where a number of eicosanoid receptors are localized. This for example has been shown for prostaglandin E (9), prostaglandin $F_{2\alpha}$ (10) and thromboxane A_2 receptors (11) as well as for leukotriene B_4 (12) and leukotriene D_4 receptors (13).

The binding in cytosol has been further studied via gel permeation chromatographed on Superdex™200 (8). The apparent molecular weight of the binding complex was found to be about 650 000 Da (8). To confirm the high value found by gel permeation chromatography, a sedimentation analysis was performed by sucrose density gradient centrifugation (8) and the sedimentation coefficient was found to be 20.5 S. The corresponding values for thyroglobulin are 669 000 Da and 19.2 S. Degradation studies

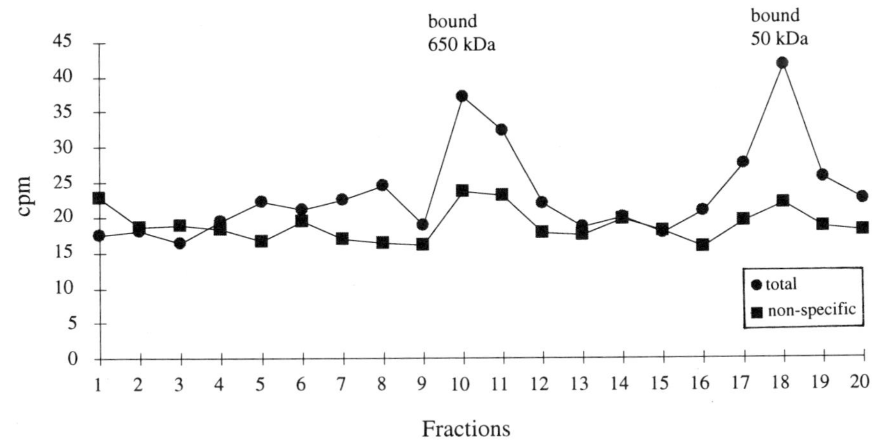

Fig. 1. Cell-free binding of 12(S)-HETE. Cytosol from $2 \cdot 10^7$ cells was fractionated on Superdex™200 and the fraction containing the 650 kDa complex was incubated for 1 h at 4°C with 1 nM 12(S)-[^{3}H]HETE in the absence or presence of 1 μM unlabeled 12(S)-HETE . The samples were then rechromatographed on Superdex™200. The results are shown as specific binding and non-specific binding in corresponding chromatography fractions.

	650 kDa Fraction 11	170 kDa Fraction 14	50 kDa Fraction 18
- ATP	23,8	-7,9	-44,9
+ ATP	-23,2	-6,9	14

% change relative to control

Fig. 2. ATP depletion enhances formation of the 650 kDa heterocomplex. LLC cells were incubated with 0.1 nM 12(S)-[^{3}H]HETE for 2 hrs at 4°C. Cytosol was prepared and divided into three parts; a) control, b) sample in which ATP was depleted and c) sample in which ATP was first depleted and then regenerated. The three samples were analyzed on Superdex™200. Data are presented as mean values ± S.E. from n=5.

showed that the complex could be cleaved with trypsin and chymotrypsin suggesting that at least part of the complex consists of a polypeptide.

SUBUNITS OF THE 12(S)-HETE BINDING COMPLEX

Binding of 12(S)-HETE to fractionated cytosol

Cytosol from non-incubated LLC-cells was fractionated by gel permeation chromatography (8). The fraction containing the 650 kDa complex was incubated with tritium-labeled 12(S)-HETE in the presence or absence of unlabeled 12(S)-HETE. As shown in Fig. 1 specific binding was observed. Non-specific binding was considerably higher than when whole cells were incubated with 12(S)-HETE.

Effects of ATP depletion and regeneration on the apparent size of the binding complex

To deplete ATP (14) in cytosol from cells prelabeled with 12(S)-[^{3}H]HETE, the cytosol fraction was incubated with apyrase. An increase (24%, mean value; n=5) of the 650 kDa complex was observed compared to a non-treated control (Fig. 2). When ATP was regenerated in apyrase-treated cytosol (12) there was a decrease (23%, mean value; n=5) in the amount of the 650 kDa complex. Two additional radiolabeled components appeared (apart from the 650 kDa complex) of molecular weight 120 kDa and 50 kDa. The relative amount of the 120 kDa binding component was not significantly influenced by the ATP concentration but the 50 kDa component increased (14%, mean value; n=5) when ATP was regenerated. A 50 kDa subunit was also observed in the control cytosol that had been incubated for 30 minutes at room temperature without apyrase or ATP regeneration. ATP depletion resulted in a 45% reduction (mean value; n=5) of the 50 kDa component. These results suggest that ATP destabilizes the association of subunits in the 650 kDa 12(S)-HETE binding complex.

Identification of hsp70 as a 12(S)-HETE binding complex subunit

Cytosol was prepared from LLC cells which had been preincubated with 12(S)-HETE or from non-incubated control cells. After fractionation on

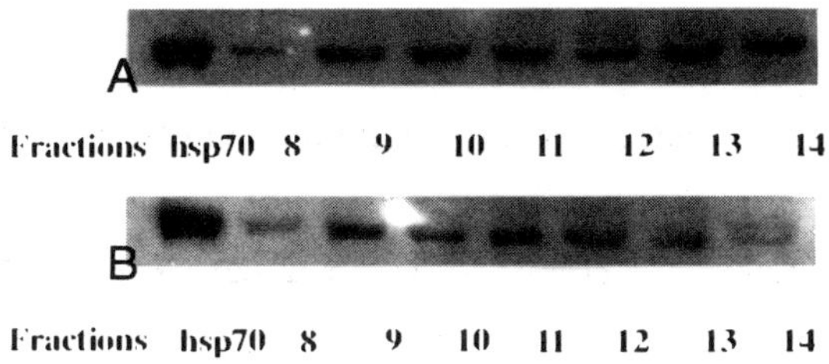

Fig. 3. Detection of hsp70 by Western blot analyses. Cytosol was prepared and fractionated on Superdex™200. Fractions 8-14 were analyzed by Western blot using a monoclonal hsp70 antibody. Lane 1: pure hsp70 control. Panel **A:** Cytosol fractions from cells that were preincubated with 12(S)-HETE (0.1 nM, 2 hrs, 4°C). Panel **B:** Cytosol fractions from nonincubated control cells.

Superdex™200, the high molecular weight fractions were analyzed on 7.5% SDS-PAGE, electrotransferred to nitrocellulose and incubated with a monoclonal hsp70 antibody followed by HRP-conjugated anti-mouse IgG for ECL-detection (Fig.3). Densitometry showed that the intensity of the hsp70 band was highest in fractions exhibiting specific 12(S)-HETE binding (hsp70 immunoreactivity and binding both had maxima in fraction 9) (15). In control cells a lower concentration of hsp70 was detected compared to cells incubated with 12(S)-HETE.

Acknowledgments

This work was supported by grants from the Swedish Medical Research Council (03X-5914).

References

1. A.W. Ford-Hutchinson; Federation Proceedings 44(1):25-29, 1985
2. K.V. Honn, I.M. Grossi, C.A. Diglio, M. Wojtukiewicz, J.D. Taylor; FASEB Journal 3:2285-2293, 1989
3. K.V. Honn, B.W. Steinert, K. Moin, J.M. Onoda, J.D. Taylor, B.F. Sloane; Biochemical and Biophysical Research Communications 145(1):384-389, 1987
4. K.V. Honn, I.M. Grossi, L.A. Fitzgerald, L.A. Umbarger, C.A. Diglio, J.D. Taylor; Proceedings of the Society for Experimental Biology and Medicine189:130-135, 1988

5. K.V. Honn, I.M. Grossi, B.W. Steinert, H. Chopra, J. Onoda, K.K. Nelson, J.D. Taylor; Advances in Prostaglandin, Thromboxane and Leukotriene Research19:439-443, 1989

6. S. Silletti, J. Timar, K.V. Honn, A. Raz; Cancer Research 54:5752-5756, 1994

7. Herbertsson, H. , and S. Hammarström. 1992. *FEBS Letters.* **298:** 249-252.

8. Herbertsson, H. , and S. Hammarström. 1995. *Biochim. Biophys. Acta.* **1244:** 191-197.

9. Sugimoto, Y., T. Namba, A. Honda, Y. Hayashi, M. Negishi, A. Ichikawa , and S. Narumiya. 1992. *J. Biol. Chem.* **267:** 6463-6466.

10. Powell, W. S., S. Hammarström , and B. Samuelsson. 1976. *Eur. J. Biochem.* **61:** 605-611.

11. Hirata, M., Y. Hayashi, F. Ushikubi, Y. Yokota, R. Kageyama, S. Nakanishi , and S. Narumiya. 1991.*Nature.* **349:** 617-620.

12. Goldman, D. W., L. A. Gifford, T. Marotti, C. H. Koo , and E. J. Goetzl. 1987. *Federation Proc.* **46:** 200-203.

13. Watanabe, T., T. Shimizu, I. Miki, C. Sakanaka, Z.-i. Honda, Y. Seyama, T. Teramoto, T. Matsushima, M. Ui , and K. Kurokawa. 1990. *J. Biol. Chem.* **265:** 21237-21241.

14. Hansen, W. J., V. R. Lingappa, and W. J. Welch. 1994. *J. Biol. Chem.* **269:** 26610-26613.

15. Herbertsson, H., T. Kühme, U. Evertsson , and S. Hammarström. 1997. *Manuscript in preparation.*

16. Heyman, R. A., Mangelsdorf, D. J., Dyck, J. A., Stein, R. B., Eichele, G., Evans, R. M., and Thaller, C. (1992) *Cell* **68,** 397-406

17. Mangelsdorf, D. J., and Evans, R. M. (1995) *Cell* **83,** 841-850

APOPTOSIS INDUCTION AND INHIBITION OF COLON-26 TUMOUR GROWTH AND ANGIOGENESIS: FINDINGS ON COX-1 and COX-2 INHIBITORS *IN VITRO* & *IN VIVO* and TOPICAL DICLOFENAC IN HYALURONAN

Michael P. Seed, Claire N. Freemantle, Chandan A.S. Alam, Paul R. Colville-Nash, Joanne R. Brown, Jonathan L. Papworth, Kevin W. Somerville, Derek A Willoughby.

Experimental Pathology
Environmental & Preventive Medicine, Saint Bartholomew's and the Royal London School of Medicine and Dentistry, Charterhouse Square, London EC1M 6BQ, United Kingdom

INTRODUCTION

The chronic administration of non-steroidal anti-inflammatory drugs (NSAIDs) results in reduced colonic adenocarcinoma incidence[1]. When administered to existing tumours, NSAIDs induce tumour regression[2]. Other states susceptible to NSAIDs include basal cell carcinoma (BCC) and the precancerous lesion actinic keratosis (AK)[3] using a topical delivery system, HYAL CT1101 (3% diclofenac in 2.5% hyaluronan). The advent of inducible cyclooxygenase (COX-2) raises the question as to its role in tumour growth and development. The role of COX of any isoform remains speculative. Tumour derived prostaglandin-E_2 (PGE_2) may result in inhibition of macrophage and NK cell function or there may be direct effects on tumour cells. COX-2 has been reported in many human colonic adenocarcinomas but not all[4,5]. Studies with the dual COX-1/2 inhibitor sulindac have shown that familial intestinal polyposis is reduced in the clinic[6], whilst the murine equivalent is modified by selective COX-2 inhibition or gene deletion[7].

We aimed to study the role of COX-2 in tumour biology by using murine colon-26 tumour cells in vitro, and to compare these effects *in vivo* taking advantage of the topical drug delivery system utilising HA induced targeting (HIT™). This allows the delivery of large cumulative doses of diclofenac to the underlying tumour without compromising safety[8], and PG synthesis is half that from orally administered diclofenac[9].

METHODS

Colon-26 cells, were maintained in antibiotic supplemented DMEM (10% FCS). Proliferation was assessed after 24 hrs. in 0.5% serum, the drugs added with 5% FCS for 24hrs and the ^{3}H-thymidine for the final 2 hours. Cells were washed with ice cold 10% TCA, disrupted, and scintillation counted. Cell viability was measured as dehydrogenase

Recent Advances in Prostaglandin, Thromboxane, and Leukotriene Research
Edited by Sinzinger *et al.*, Plenum Press, New York, 1998

activity measured as the dehydrogenation of (3-[4,5-dimethylthiazol-2-yl]-2,5-diphenyltetrazolium bromide (MTT) to form the formazan product and read at 570nm.

The identification of apoptotic and necrotic cells *in vitro* utilised acridine orange and ethidium bromide staining. Colon-26 cells were grown to 80% confluence on coverslips. NSAIDs were added for 24 hours, and the coverslips washed in HBSS, incubated with 100μg/ml acridine orange and 100μg/ml ethidium bromide in HBSS and washed again. The coverslips were mounted and the cells viewed by fluorescence microscopy. Apoptotic cells had dense fluorescent green nuclei stained with acridine orange, often lacking cytoplasm. Necrotic cells were stained as diffuse orange staining with ethidium bromide.

In vivo, polyvinyl polyurethane sponge discs were implanted subcutaneously into the dorsum of anaesthetised female Balb/c mice, and seeded with $1x10^6$ colon-26 cells. Drugs were administered orally one hour prior to seeding, or applied topically in 1mL 2.5% HA (Mw. 500,000, Hyal Pharmaceutical Co., Canada). Nimesulide and NS-398 were split into two daily doses. Tumour vascularity was assessed by vascular casts incorporating carmine, as previously described[9].

Proliferating cells within colon-26 tumours were visualised as mitotic figures, whilst apoptosis and zones of necrosis were assessed using the Apoptag DNA fragmentation detection kit (Apoptag, Oncor USA). Nuclei were stained with haematoxylin. Cells (>6,000) were counted from 15 fields from each of 2 sections from 4-5 animals. The indices were calculated as mean $\pm$ s.e.m. percentage ratio of the number of positive cells to total. Areas of necrosis were observed as diffuse staining with Apoptag, and excluded from analysis. COX-2 expression *in vitro* and *in vivo* was determined by Western Blot analysis utilising COX-2 -specific rabbit anti-mouse antibody (Cayman). The same antibody was used for the immuno-histological visualisation of COX-2 in 5μm cryostat sections from colon-26 tumours *in vivo* using Vectorstain ABC peroxidase development. The absolute data of treatment groups were compared to controls by ANOVA and post hoc Bonferroni t-test (Graphpad Instat™).

RESULTS

Western blotting of colon-26 cells *in vitro* and *in vivo* showed that COX-2 protein is expressed. Immunohistology showed that staining of the tumour cells *in vivo* was far from ubiquitous. Infiltrating macrophages and a small proportion of vascular endothelium (when compared to CD31 expression) showed clear cytoplasmic expression, but a minority of tumour cells revealed perinuclear staining for COX-2.

The COX-1 inhibitors aspirin and piroxicam, and the COX-2 inhibitors nimesulide and NS398, had no effect on proliferation *in vitro* (Table 1). Piroxicam and the COX-2 inhibitors showed some dose related inhibition of viability up to 300μM. The dual inhibitor diclofenac inhibited proliferation and viability to a greater extent at 100 and 300μM. HA inhibited proliferation and viability at 1mg/ml at which substantial intermolecular association occurs. Diclofenac also induced colon-26 apoptosis at 300μM *in vitro*. On the other hand, the COX-1 inhibitors piroxicam and aspirin appeared to induce changes in morphology at 300μM suggesting that apoptosis could occur at higher concentrations. Nimesulide and NS-398 appeared to have no effect when compared to untreated controls. Necrosis was not seen with any drug at any concentration.

In vivo, the apoptotic index within tumours was reduced from 0.32 ± 0.05 at day 7 to 0.04 ± 0.02% ($p<0.001$) at day 14, whilst the proliferation indices were unchanged (day 7: 5.59 ± 0.4; day 14: 3.49 ± 0.40%). The drop in apoptotic index correlates with the induction of angiogenesis in this model. Topical diclofenac in HA increased apoptosis to 0.15 ± 0.01 ($P<0.05$) and 0.32 ± 0.04% ($p<0.001$) on days 14 and 20 respectively compared to HA applied alone with indices of 0.05 ± 0.01 and 0.05 ± 0.02%

Table 1. The effect of NSAIDs on colon-26 proliferation and viability cultured *in vitro*.

DRUG	[μM]	% Control MTT Formazan	% Control [^{3}H]thymidine
Piroxicam	[30]	5.04±7.28	- 7.93±4.44
	[100]	- 1.91±6.21	-15.82±4.81***
	[300]	0.28±5.68	-19.94±4.14***
Aspirin	[30]	3.72±2.53	- 6.67±3.90
	[100]	3.30±3.25	- 3.81±5.91
	[300]	0.95±4.90	- 6.82±6.81
Nimesulide	[30]	2.46±3.98	-12.22±3.60
	[100]	- 4.06±4.36	-14.99±2.91**
	[300]	-10.51±3.44*	-38.23±3.14**
NS398	[30]	7.19±5.99	- 6.39±4.04
	[100]	- 5.80±2.83	-12.91±4.00
	[300]	- 9.28±4.55	-24.78±4.11**
Diclofenac	[30]	- 2.54±4.84	- 7.42±4.36
	[100]	- 7.59±5.65	-46.76±5.29***
	[300]	-30.69±4.20***	-80.37±6.23***
HA (mg/mL)	0.01	- 2.81±3.93	-12.34±6.25
	0.1	- 2.92±1.72	- 9.70±4.31
	1.0	-22.72±5.16**	-33.18±7.97***

Table 2. NSAIDs and tumour growth (dry mass) and angiogenesis (carmine) *in vivo*.

	mg dry mass		μg carmine/mg dry mass)	
Drug (mg/kg)	day 10	day 14	day 10	day 14
Control	*53± 3*	*102±10*	*10.3±0.9*	*10.9±1.1*
Nimesulide (@10)	52±6	70±5*	7.8±0.8	12.0±1.2
Piroxicam (@2)	48±3	52±5***	10.9±1.0	10.5±0.9
NS398 (@6)	50±3	87±8	9.0±0.7	12.0±1.4
Control	*62±12*	*162±18*	*11.5±0.9*	*16.5±1.9*
Diclofenac (@6)	25± 1***	85±12	5.9±0.7***	16.2±1.9
Control	*43± 4*	*122±12*	*7.8±0.7*	*5.9±1.4*
Aqueous cream (0.1mL)	38± 2	117± 7	6.1±0.4	5.9±1.4
2.5% HA (0.1mL)	44± 5	74± 9**	7.6±0.5	4.2±0.6
Diclofenac/HA (@6/0.1mL)	31± 3**	36± 7***	5.0±0.4**	6.5±1.3

Table 2 illustrates Colon-26 growth during the oral administration of COX-1 and COX-2 inhibitors. NS398 appeared ineffective, whilst nimesulide afforded some inhibition of tumour growth. Piroxicam inhibited tumour growth potently. However, none of the selective NSAIDs tested, whether antitumour or not, inhibited tumour angiogenesis. This is unlike the dual inhibitor diclofenac which delayed angiogenesis when administered both topically and p.o.. Topical diclofenac in HA totally prevented tumour growth. HA vehicle also retarded tumour growth. NS398 (6mg/kg) administered to mice bearing subcutaneous sponge granulomas for 7 days did not significantly alter granuloma PGE_2 synthesis, (control 791±70, treated 913±100 pg/mg protein, n=9 and 10 respectively).

DISCUSSION

Murine colon-26 cells express COX-2 *in vivo* and *in vitro*, but to a much lesser extent *in vivo* as detected by through immunohistology (data not shown). This is similar to expression in murine familial intestinal polyposis where *in situ* methods had to be used to detect COX-2 protein *in vivo*[7]. However, expression in macrophages and endothelial cells is strong. *In vivo*, nimesulide retarded tumour growth, whilst NS-398 was inactive. This could be due to bioavailability, since both drugs have short $t_{1/2}$s, and NS-398 at 6mg/kg did not inhibit granuloma PGE_2 synthesis *in vivo*. The COX-1 inhibitor piroxicam and topical diclofenac were the most effective, but *in vitro* diclofenac was the most potent at reducing proliferation and viability, and the only NSAID to induce apoptosis. However, the levels of 100 and 300μM are unlikely to be achieved p.o. *in vivo*. Increasing tissue diclofenac bioavailability through local delivery in hyaluronan[8] mimicked this apoptotic effect *in vivo*. This was coupled with angiogenesis inhibition as it was with p.o. administration. In addition, the anti-tumour effect was greater and more prolonged when compared with oral administration. Under these circumstances, PGE_2 synthesis inhibition is doubled[9]. These results may explain the efficacy of this formulation against BCC and AK in the clinic[3]. COX-2 inhibition by nimesulide and NS-398 at the doses used was not as effective as topical diclofenac or systemic piroxicam and may warrant more selective COX-2 inhibitors with improved pharmacokinetics or the utilisation of HIT™ technology.

REFERENCES

1. C.W. Heat, M.J. Thun, E.R. Greenberg, B. Levin, L.J. Marnett. Nonsteroidal antiinflammatory drugs and human cancer - Report of an interdisciplinary research workshop. *Cancer* 74:2885 (1994).
2. J.C. Valdez, G. Perdigon. Piroxicam, indomethacin and aspirin on a murine fibrosarcoma - effects of tumour associated and peritoneal macrophages. *Clin. Exp. Immunology* 86:315 (1991).
3. J. Rivers, D.I. McLean, S. Maddin S. A novel topical treatment for actinic keratosis. *Roy. Soc. Med. Round Table Ser.* 40:122 (1995).
4. N. Yoshimi, N. Ino, M. Suzui, T. Tanaka, S. Nakashima, M. Nakamura, Y. Nozawa, H. Mori. The messenger rna over-expression of inflammatory enzymes, phospholipase A2 and cyclo-oxygenase, in the large bowel mucosa and neoplasms of F344 rats treated with naturally occurring carcinogen, 1-hydroxyanthraquinone. *Cancer Letters* 97:75 (1995).
5. H. Sano, Y. Kawahito, R.L. Wilder, A. Hashiramoto, S. Mukai, K. Asai, S. Kimura, H. Kato, M. Kondo, T. Hla. Expression of cyclooxygenase-1 and cyclooxygenase-2 in human colorectal cancer. *Cancer Res* 55:3785 (1995).
6. R.N. DuBois, A. Radhika, J. Shao, M. Tsujii, H. Sheng, O. Kobayashi, R.D. Beuchamp, C.S. Williams. Cyclooxygenase 2 in intestinal cancer, in: *New Targets in Inflammation* N. Bazan, J. Botting & J.R. Vane eds., William Harvey Press, Kluwer Academic Publishers. pp125-130 (1996).
7. M. Oshima, J.E. Dinchuk, S.L. Kargman, H. Oshima, B. Hancock, E. Kwong, J.M. Trzaskos, J.F. Evans, M.M. Taketo. Suppression of intestinal polyposis in $Apc^{delta716}$ knockout mice by inhibition of cyclooxygenase 2 (COX-2). *Cell* 87:803 (1996).
8. M.P. Seed, C.A.S. Alam, J.R. Brown, C.N. Freemantle, D.A. Willoughby. The delivery of diclofenac to pathological tissues by hyaluronan (HYAL EX-0001). *R. Soc. Med. Round Table Ser.* 40:74 (1995).
9. J. Papworth, M.P. Seed, D.A. Willoughby. Resident granulomatous tissue and tumour prostaglandin synthesis inhibition by topical diclofenac in hyaluronan (HYAL EX-0001). *R. Soc. Med. Round Table Ser.* 45:54 (1996).
10. P.R. Colville-Nash, C.A.S. Alam, I. Appleton, J.R. Brown, M.P. Seed, D.A. Willoughby. The Pharmacological modulation of angiogenesis in chronic granulomatous inflammation. *J. Pharmacol. Exp. Ther.* 274:1463 (1995).

We thank Hyal Research Foundation (Canada), and ONO Pharmaceutical Co. (Japan) for financial support.

PHOSPHOLIPASE A_2 INHIBITORY ACTIVITY IN LUNG CANCER CELL LINES

Masayoshi Abe,[1] Tsugutomo Sakurada,[1] Hiroshi Kuwata,[2] Yoshihito Nakatani,[2] Ichiro Kudo,[2] and Tatsuo Furukawa[1]

[1]Dapartment of Pharmacology, Fukuoka University School of Medicine, Fukuoka 814-80, Japan and [2]Department of Health Chemistry, Showa University School of Pharmaceutical Sciences, Tokyo 142, Japan.

It has been suggested that cancer patients exhibit pathophysiologically different responses from normal subjects (1). One of such the differences may be susceptible to severe infections in patients with advanced cancers (2). In order to understand such the mechanisms, we studied cellular interactions between polymorphonuclear leukocytes (PMN) and human lung cancer cell lines, especially for LTB_4 and superoxide anion production.

To examine cellular interactions between PMN and cancer cells, we first evaluated LTA_4 hydrolase and LTC_4 synthase activity in the cancer cell lysates. Nine lung cancer cell lines established from human lung cancers were suspended in Hepes buffer containing 1 mM EDTA at the density of 10^8 cells/ml and sonicated. Each lysate was incubated with LTA_4 in the presence of 3 mM GSH and the products were extracted and then analyzed by RP-HPLC as previously reported (3). As the result, none of the lysates converted LTA_4 to LTC_4, but several lysates did LTA_4 to LTB_4. H23 showed the highest LTA_4 hydrolase activity.

Then, we used H23 cell line (adenocarcinoma) to examine cellular interaction between PMN and cancer cells for LTB_4 production by stimulation with A23187. When 10^6 PMN were coincubated in Tyrode's buffer with various numbers of RBC (red blood corpuscles) and stimulated by A23187 (2 mM) more LTB_4 were produced in comparison with PMN alone. On the contrary, nonenzymatic products of LTA_4, 6t-LTB_4 diastereoisomeres were reduced, as consistent with the previous report (4). On the other hand, the coincubation of 2 x10^6 PMN and various numbers of H23 (10^6 - 10^7) reduced LTB_4 production in cell-number-dependent manner. The coincubation of PMN and H23 also decreased 5-HETE and 12-HETE production when compared with PMN alone. Other two cell lines, Q9 (squamous cell carcinoma) and N417 (small cell carcinoma) inhibited LTB_4 production in the similar manner with H23. The

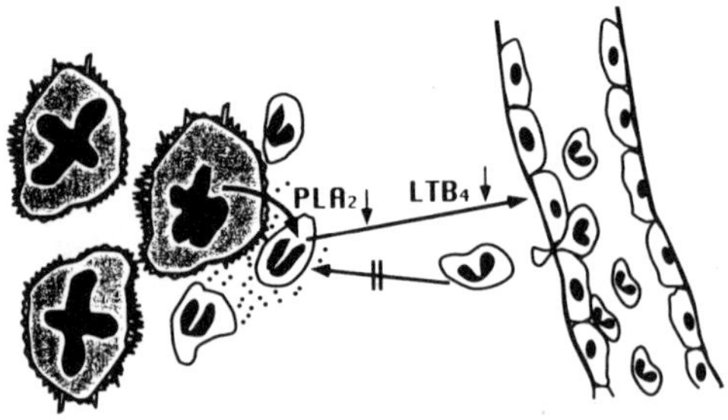

Fig. 1. Cancer cells release factors which inhibit phospholipase A_2 activity. Consequently, LTB_4 production by PMN is inhibited. The inhibition of LTB_4 production of PMN by the cancer cells may result in the suppression of the infiltration of circulating cells to the inflammatory focus such as bacterial infection.

addition of H23 lysate also inhibited the production of LTB_4 from PMN stimulated by A23187.

Since it is suggested that some factors released by the cancer cells inhibit the production of LTB_4, we conducted the following experiment. After an H23 cell suspension in the Tyrode's buffer was incubated with A23187 (2 mM) at 37 °C for 15 min, each tube was centrifuged at 1700 x g for 15 min. The addition of the resulting supernatant into the incubation mixture inhibited the LTB_4 production by PMN to 54 - 61 % of the control values. This study suggests the release of inhibitory factors from the cancer cells. On the contrary, the inhibition of LTB_4 production by the coincubation with H23 was not restored by the treatment with a cyclooxygenase inhibitor, indomethacin up to 1 mg/ml. This result suggests PGE_2 to be an unlikely inhibitory factor in this reaction.

When 2 x10^6 PMN and 7.5 x10^6 H23 were costimulated by A23187 (2 mM), LTB_4, 5-HETE, and 12-HETE productions were almost completely inhibited. Under these conditions, when various amounts of arachidonic acid were added to the incubation mixtures, LTB_4, 5-HETE, and 12-HETE productions were recovered in a dose-dependent manner of arachidonic acid. These results suggest that the inhibition of LTB_4 production from PMN by the cancer cells is at the level of phospholipase A_2. Then we examined whether or not the coincubation of the lung cancer cell line Q9 inhibited ^{14}C-release from ^{14}C-arachidonic acid-prelabelled PMN by stimulation with A23187. Under either the absence or the presence of 0.2 % BSA, Q9 inhibited ^{14}C-release from the prelabelled PMN in the cell -number dependent manner. Finally, we directly examined whether or not Q9 lysates inhibited phospholipase A_2 activity that was estimated by the modified Dole's method using partially purified cytosolic PLA_2 as the enzyme (5, 6). Negative control that contained only the substrate, ^{14}C-AA-phosphatidylethanolamine but without the enzyme showed 745 $\pm$ 141 DPM of ^{14}C in the heptane layer of the incubation mixtures . The addition of the enzyme, $cPLA_2$, cleaved 14902 $\pm$ 442 DPM of ^{14}C from 74000 DPM of ^{14}C-AA-phosphatidylethanolamine. Then, the addition of the 10^4 x g supernatant of Q9 lysates dose-dependently inhibited ^{14}C-release from the PMN. Consequently, it has been confirmed that the lung cancer cells inhibit LTB_4 production by the inhibition of phospholipase A_2 activity.

On the other hand, the cancer cell line, H23 did not inhibit superoxide anion production from PMN stimulated by phorbormyristate acetate over the same ranges of both the cells ratios (PMN : H23 = 5.0) as those for LTB_4 production. This suggests that phospholipase A_2 activity of PMN is inhibited by the cancer cells with specificity.

In conclusion, the following hypothesis may be speculated (see Fig. 1). The lung cancer cells release some factors that inhibits phospholipase A_2 activity which results in reduced production of LTB_4 by PMN, an important chemotactic factor for PMN and monocytes in the lung, under the stress such as bacterial infection (7,8). The reduced production of LTB_4 by PMN results in impared accumulation of PMN to the infectious locus. This leads to susceptibility to severe infection in patients with advanced cancers.

REFERENCES

1. O.J. Plescia, A.H. Smith, and K. Grinwich, Subversion of immunesystem by tumor cells and role of prostaglandins, Proc. Natl. Acad. Sci. USA. 72: 1848 (1975).
2. J.J. Pagani, and H.I. Libshitz, Opportunistic fungal pneumonias in cancer patients, Am. J. Roentgenol. 137: 1033 (1981).
3. M. Abe, N. Hara, H. Muranishi, T. Ikeda, N. Nagata, and N. Shigematsu, Enhanced leukotriene C_4 synthase activity in thioglycollate-elicited peritoneal macrophages, Biochem. Biophys. Res. Commun. 171: 1344 (1990).
4. J.E. McGee, and F.A. Fitzpatrick, Erythrocyte-neutrophil interactions: formation of leukotriene B_4 by transcellular biosynthesis, Proc. Natl. Acad. Sci. USA. 83: 1349 (1986).
5. K. Horigome, M. Hayakawa, K. Inoue, and S. Nojima, Selective release of phospholipase A_2 and lysophosphatidylserine-specific lysophospholipase from rat platelets, J. Biochem. 101: 53 (1987).
6. Y. Fujimori, M. Murakami, D.K. Kim, S. Hara, K. Takayama, I. Kudo, and K. Inoue, Immunochemical detection of arachidonoyl-preferential phospholipase A_2, J. Biochem. 111: 54 (1992).
7. A.W. Ford-Hutchinson, M.A. Bray, M.V. Doig, M.E. Shipley, and M.J.H. Smith, Leukotriene B: a potent chemokinetic and aggregating substance released from polymorphonuclear leukocytes, Nature. 286: 264 (1980).
8. M. Abe, H. Matsuki, M. Domae, H. Kuwata, I. Kudo, Y. Nakanishi, N. Hara, T. Mitsuyama, and T. Furukawa, Lung cancer cell lines inhibit leukotriene B_4 production by human polymorphonuclear leukocytes at the level of phospholipase A_2, Am. J. Respir. Cell Mol. Biol. 15: 565 (1996).

PROSTAGLANDIN D SYNTHASE, β-TRACE AND SLEEP

Osamu Hayaishi

Osaka Bioscience Institute
Osaka 565, Japan

Prostaglandin synthase catalyzes isomerization of PGH_2 to PGD_2 . Two types of enzymes are known. One is the so-called brain-type PGD synthase, which is a member of the lipocalin superfamily and was recently identified as β-trace. The other is the so-called spleen-type PGD synthase, which is a kind of glutathione s-transferase and therefore requires glutathione for activity. The product of the reaction, PGD_2 had long been considered as a minor and biologically inactive prostanoid. However, in the late 1970s, we found that PGD_2 is the major PG in the CNS of rats and other mammals including humans indicating the possibility that it may play some important role in the brain. We soon found out that PGD_2 is an endogenous sleep promoting agent and began to study its somnogenic activity in more detail.

Before I proceed further, I would first like to say a few words about sleep and how sleep is determined for the benefit of the nonspecialists in the audience. As all of your are well aware, sleep is a very complex phenomenon. Wakefulness and the two major types of sleep, namely, slow-wave sleep or SWS and rapid eye movement or REM sleep, can be judged by the animal's behavior. However, more precise qualitative and quantitative assessment of the sleep-wake pattern requires the use of EEG, or electroencephalography, namely the recording of brain waves, EOG, or electro-oculography, the recording of eye movement, and EMG, or electromyography, that is, the recording of muscle tension. In the sleep bioassay system currently in use in my laboratory, samples are infused continuously and slowly into the third ventricle of a freely moving rat through a chronically implanted cannula by means of a microinjection pump. The sleep stages are determined on the basis of polygraphic recordings of EEG and EMG. Brain temperature as well as food and water intake is also monitored, and behavior of the rat is recorded by a video-recording system.

Rats are nocturnal animals and normally sleep during the day, but are awake most of the time during the night. When PGD_2 was infused at a rate of 0.6 pmol per minute during the night, both SWS and REM sleep increased significantly. The effect was specific for PGD_2 and dose-dependent. As little as one femtomole (10^{-15} mol) of PGD_2 per second was effective in inducing excess sleep. Most importantly, sleep induced by PGD_2 was indistinguishable from physiological sleep as judged by EEG,

Recent Advances in Prostaglandin, Thromboxane, and Leukotriene Research
Edited by Sinzinger *et al.*, Plenum Press, New York, 1998

EMG, locomotor activities, brain temperature, heart rate, and general behavior of the rat, indicating that PGD_2 is the endogenous sleep inducing agent (Hayaishi, 1991).

In order to further prove that the effect of PGD_2 was not pharmacological, but physiological, we determined the specific activity of the enzyme that produces PGD_2 in the rat brain, namely PGD synthase, and PGD_2 in the CSF during the sleep-wake cycle. The PGD synthase activity in the rat brain as well as the amount of PGD_2 in the CSF exhibited a circadian fluctuation in parallel with the sleep/wake cycle. In contrast, the specific activities of other related enzymes, for example, PGD_2 dehydrogenase, remained almost the same during the day and night. If PGD synthase is indeed involved in the regulation of physiological sleep, when the synthase activity during the day is inhibited by a specific enzyme inhibitor, sleep should also be inhibited and the amount of diurnal sleep should decrease. In order to find a specific inhibitor of PGD synthase, we then purified this enzyme and studied its properties.

We purified the enzyme from human, rat and frog brains, isolated cDNAs encoding PGD synthase, and determined the nucleotide and amino acid sequences. By a computer-assisted homology search in data bases of protein primary structure, PGD synthase was shown to be a member of the lipocalin superfamily, a group of small secretory proteins that bind and transport small lipophilic molecules. These data provided several lines of important and unexpected information, which as I shall tell you later, opened up a new insight into our understanding of the mechanism of the somnogenic activity of PGD_2 (Urade et al., 1996).

Islam and coworkers in my laboratory then found that inorganic quadrivalent (4^+) selenium compounds are potent, specific, non-competitive, and reversible inhibitors of brain prostaglandin D synthase, presumably by interacting with the free sulfhydryl group in the active center of the enzyme. Ki values are about 10 mM. This inhibition is reversible and can be reversed by the addition of excess amounts of sulfhydryl compounds such as glutathione or DTT (Islam et al., 1991).

When selenium chloride was infused into the third ventricle of a rat during the day, both SWS and REM sleep were inhibited time- and dose-dependently. After about 2 hours from the start of the infusion, sleep was almost completely inhibited. When the infusion was interrupted, sleep was restored. Furthermore, the effect was reversed by the simultaneous infusion of sulfhydryl compounds such as DTT and reduced glutathione, as in the case of the in vitro enzyme activity. These results taken together clearly show that PGD_2 is the endogenous sleep-regulating substance in the rat under physiological conditions and that PGD synthase is the key enzyme in sleep regulation (Mtsumura et al., 1991). Questions then arose as to ; where in the rat brain the PGD synthase is located and in what type of cells. In an attempt to answer these questions, Urade and coworkers have employed three independent approaches, namely, in situ hybridization, immunohistochemical staining, and the enzyme activity determination. The results were quite striking and unexpected.

A typical result of in situ hybridization of a rat brain and spinal cord indicates that mRNA for PGD synthase is expressed faintly and diffusely in the white matter of the brain. However, a much stronger signal is seen in the leptomeninges, namely, the pia mater and arachnoid membrane, and in the choroid plexus, in the ventricles. Furthermore, the leptomeninges and choroid plexus are loaded with PGD synthase mRNA and enzyme protein. Oligodendrocytes are also positively stained; but the other cells, including neurons, show little, if any, activity (Urade et al., 1993).

We then determined the enzyme activity in these tissues and also in the CSF of rats and humans. The specific activities in the arachnoid membrane and choroid plexus were much higher than that activity in the whole brain. Furthermore, both rat and human CSF contained a remarkably high PGD synthase activity. β-trace showed a very high specific activity, almost comparable to that of the most purified enzyme preparation from rat brain.

β-trace was discovered in 1961, but its function remained unknown for more than 30 years. Quite recently, Zahn's group in Göttingen and Conradt's group in Braunschweig determined N-terminal and total amino acid sequences of β-trace and reported that β-trace and PGD synthase were structurally identical. Subsequently, Watanabe showed that β-trace and PGD synthase were structurally, enzymatically and immunologically identical. These results indicate that PGD synthases is produced in the membrane system surrounding the brain and is secreted into the CSF to become β-trace (Watanabe et al., 1994). Questions then arose as to: How and where PGD_2 exerts its somnogenic activity?

Matsumura and coworkers applied a small amount of PGD_2 to more than 200 different areas in the rat brain, but the results were mostly negative, except the preoptic area. However, bilateral infusion of PGD_2 into the subarachnoid space in the surface area of the medial ventral region of the rostral basal forebrain was found to be the most effective. SWS was increased more than 120 min. above the base line values of about 100 min (Matsumura et al., 1994).

All these experimental observations taken together led us to arrive at the tentative postulate as follows; PGD synthase, the key enzyme in sleep regulation, that produces PGD_2 in the rat brain, is present mainly, if not exclusively, in the membrane system surrounding the brain rather than in the brain parenchyma. PGD synthase is a secretory protein, and is secreted into the CSF to become β-trace. β-trace, as well as the PGD_2 thus produced, circulates in the ventricular system and subarachnoidal space.

PGD_2 interacts with the chemosensor on the ventromedial surface of the rostral basal forebrain, thus turning the switch on to initiate the signal to make the whole brain sleep.

REFERENCES

Hayaishi, O., 1992, Molecular mechanisms of sleep-wake regulation: roles of prostaglandin D_2 and E_2, *FASEB J.* 5: 2575.

Islam, F., Watanabe, Ya., Morii., H., and Hayaishi, O., 1991, Inhibition of rat brain prostaglandin D synthase by inorganic selenocompounds, *Arch. Biochem. Biophys.* 289: 161.

Matsumura, H., Takahata, R., and Hayaishi, O.,1991, Inhibition of sleep in rats by inorganic selenium compounds, inhibitors of prostaglandin D synthase, *Proc. Natl. Acad. Sci. USA*. 88: 9046.

Matsumura, H., Nakajima, T., Osaka, T., Satoh, S., Kawase, K., Kbo, E., Sri Kantha, S., Kasahara, K., and Hayaishi, O., 1994, Prostaglandin D_2-sensitive, sleep-promoting zone defined in the ventral surface of the rostral basal forebrain, *Proc. Natl.Acad. Sci, USA*. 91:11998.

Urade, Y., Kitahama, K., Ohishi, H., Kaneko, T., Mizuno, N., and Hayaishi, O., 1993, Dominant expression of mRNA for prostaglandin D synthase in leptomeninges, choroid plexus, and oligodendrocytes of the adult rat brain, *Proc. Natl. Acad. Sci, USA* 90: 9070.

Urade, Y., Hayaishi, O., Matsumura, H., and Watanabe, K., 1996, Molecular mechanism of sleep regulation by prostaglandin D_2, *J. Lipid Meiators Cell Signalling*, 14: 71.

Watanabe, K., Urade, Y., Mäder, M., Murphy, C., and Hayaishi, O., 1996, Identification of β-trace as prostaglandin D synthase, *Biochem. Biophys. Res. Commun.* 203: 1110.

HUMAN BETA-TRACE IN NORMAL AND PATHOLOGICAL CNS TISSUES, GENITAL ORGANS AND BODY FLUIDS

Jan-Edvin Olsson

Department Neurology
University Hospital
S-581 85 Linköping
Sweden

INTRODUCTION

Beta-trace protein (BTP) was first described by Clausen 1961 (1) as one or two extra bands found in cerebrospinal fluid (CSF) when tested with immunoprecipitation using an antiserum against human CSF. Later on, BTP could also be discovered in blood, urine and other body fluids (2). The concentration of BTP was found to be about 10 x that in serum and in healthy adult persons between 26 and 40 mg/l (3). The serum concentration of BTP was less than 5 mg/l and the 24 hour urine excretion was found to be less than 100 mg/l (4). BTP is one of the major proteins of the CSF constituting about 6% of the total protein content, a migrates as three or more bands in the gammaregion of a regular CSF electrophoresis (5). Almost equal amounts of CSF-BTP have also been found in seminal and amniotic fluids (6, 7). In seminal fluid there is a correlation between the BTP concentration and the number of spermatozoa with lowest BTP values found in patients with azospermia (6). In samples of amniotic fluid the BTP concentration increases with gestational age, but no correlation could be found to neural tube defects (8).

Chemical Data. BTP can be isolated from pooled, concentrated human CSF by means of ion cromotography and stepwise elution (9). An even more pure fraction can be isolated by adding gelfiltration (10). This fraction is slightly yellow-green and has a molecular weight of 31 000 calculated from the aminoacid composition, which mainly is trionine, serine, alanine, leucine and glutamic acid. The sedimentation coefficient was found to be 2.3S at ultracentrifugation (5). The 3-5 bands seen on agarose gel electrophoresis could be homogenous in SDS gel electrophoresis after treatment with neuroaminidase (5). This indicates that BTP is a glycoprotein and the total carbohydrate content has been found to be about 9% with the major constituent: 4.9% glucosamine, 0.9% sialic acid, 0.6% fucose, 1.4% manose and 1.0% galactose (11).

Human Organs. By immunoprecipitation methods BTP was early found in extracts of water-soluble proteins of the human brain but also in other organs without the central nervous system (CNS) as well as extracts of the kidneys, pancreas and both male and female genital organs (13).

Recent Advances in Prostaglandin, Thromboxane, and Leukotriene Research
Edited by Sinzinger *et al.*, Plenum Press, New York, 1998

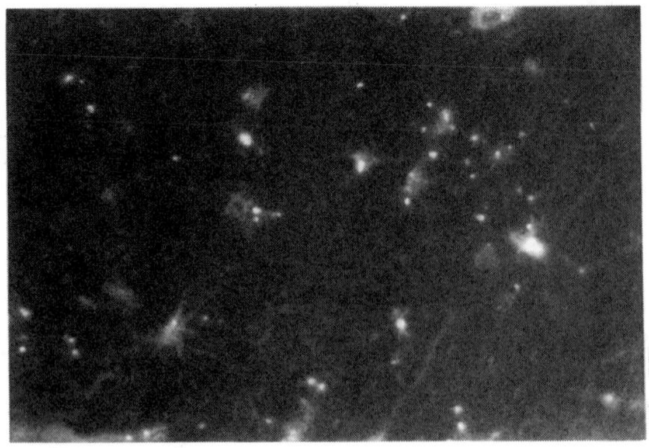

Fig. 1. Cryostat section through cerebral matter of normal squirrel monkey. The section is treated with antiserum vs. human BTP. Specific immunofluorescence s seen in glial cells (x250).

Central Nervous System. Using immunological qualitatively investigations BTP could be found in both human and monkey brains with 3 x higher concentrations in water-soluble protein extracts from white CNS matter compared with grey. (14). Even higher amounts were found in pooled material of the choroid plexus and the lowest values in peripheral (femoral) nerve. In fetal brains the BTP concentration increased with age from the 26th to the 35th weeks (14). Using immunofluorescence specific fluorescence could be found in both cerebrum and cerebellum, basal ganglia and spinal cord with brightest fluorescence in the white CNS matter (15). Washed sections of the choriod plexus and the femoral nerve did not show any specific immunofluorescence. Fraction enriched in glial cells contained about three times as much BTP as fractions enriched in neuronal cells (16). Special sections of CNS matter from squirrel monkeys showed very specific fluorescence in astrocytes and oligodendrocytes and in sections of cerebellum matter a bright fluorescence in the granular layer whereas no immunoflourescence could be seen in the Purkinje cells (16) (Fig. 1). In water-soluble protein extracts from various human brain tumors the highest amounts of BTP were found in gliomas whereas extracts from neurinomas didn't show any BTP content. In human CSF the BTP concentration increased with age and was correlated to the total protein content. There was no sure correlation to any disease (17).

Genital Organs. An immunological identity was found between CNS extracts and most human organs but especially the male genital organs testes and epididymis. Qualitative immunofluorescence studies on this tissues showed very specific and bright immunofluorescence for BTP in the stroma of epididymis (15). Quantitative determination of BTP showed as high concentrations in epididymis as in white CNS matter. In normal human seminal fluid the BTP concentration varied between 10-175 mg/l with a mean concentation of 66 mg/l decreasing with the number of spermatozoes (6).

Metabolism. The intravascular metabolism of BTP was studied in three patients after i.v. injections of ^{125}I-labelled BTP (18). One patient had a stroke and two patients benign MS with normal total CSF protein content and normal serum creatinine. The protein-bound radioactivity in serum decreased rapidly. The injected ^{125}I-labelled BTP was excreted in urine, mostly as free iodide. The excretion rate was rapid and after 24 hours about 80% of the injected dose was recovered in urine. When the study was terminated after 96 hours almost all of the injected radioacitivty hade been excreted in urine.

In two patients with stroke, the ^{125}I-labelled BTP was injected intrathecally. The protein-bound radioactivity decreased rather rapidly in the CSF, appeared in serum, and was excreted in urine. Urinary excretion rate, however, was slower than after i.v. injections and about 40% of the injected dose was recovered in urine when the study was terminated after 114 hours.

Based of these studies the turnover rate was found to be very rapid with a mean turnover time in serum of 1-2 h. After intrathecal injection the intravascular turn overtime was increased and calculated to about 3 h. With a mean concentration of BTP in serum of 4 mg/l and a serum volume of 3 litre about 240 mg BTP will be metabolized per 24 h.
The studies above point indirectly to a synthesis of BTP within the CNS. Direct probes of a CNS synthesis were obtained by culturing different tissues of squirrel monkeys in Eagle's minimum essential medium together with 14 C-labelled valine, trionine and leucine, i.e. aminoacids choosen among those most common in BTP. The synthesis of BTP in the supernatants of the cultures were precipitated with a rabbit antiserum against BTP and the incorporated radioactivity was measured (19). Synthesis of BTB was found in supernatants of white CNS matter, whereas a probably synthesis was found in supernatants of spinal cord, basal ganglia, liver, testis, epididymis, spermatic cord and fimbriae ovary. Supernatants of cultures of grey CNS matter, kidney, sceletal muscle and peripheral nerve did not show any sign of synthesis. The amount BTP synthesized within the CNS was calculated to be about 30 mg BTP/24 h, which would be about one eight of the daily amount (240 mg) beeing metabolized in the whole body per 24 h. In one culture from three patients with MS a probably synthesis was found (19).

Summary. The function of BTP is still unknown. In CNS, the high amount of BTP in the white CNS matter and the glial cells as well as in genital organs of the stroma of epididymis suggest that BTP has a supportive function. Slight evidence of a synthesis in patiens with MS and slightly increased CSF values in stroke patients may suggest that BTP is involved in repair mechanism of damaged brain tissue, or may be released from destroyed brain tissue. In samples of CSF, determination of BTP could be of value as differentiating glial cell tumors from tumors of other kinds, as well as the recently suggested diagnostic value of the Creutzfeldt-Jakob disease (20). However, also other functions of BTP has been suggested. (21, 22).

REFERENCES

1. J. Clausen. Proteins in normal cerebrospinal fluid not found in serum, Proc. Soc. Exptl. Biol. Med. (NY) 107:170 (1961).
2. G.M. Hochwald and G.J. Thorbecke. Use of an antiserum against cerebrospinal fluid in demonstration of trace proteins in biological fluids, Proc. Soc. Extpt. Biol. Med. (NY) 109:91 (1962).
3. A.J. Pepe and G.M. Hochwald. Trace proteins in biological fluids. III. Quantitation of gamma-trace and major spinal fluid proteins including beta-trace, Proc. Soc. Exptl. Biol. Med. 126:630 (1967).
4. J. Ericsson, H. Link and O. Zettervall. Urine excretion of a low molecular weight protein in cerebral damage, Neurology 19:606 (1969).
5. H. Link. Immunoglobulin G and low molecular weight protein in human cerebrospinal fluid. Chemical and immunological characterization with special reference to multiple sclerosis, Acta Neurol Scand. 43:suppl 28 (1967).
6. J.E. Olsson. Correlation between the concentration of beta-trace protein and the number spermatozoa and human semen, J. Reprod. Fert. 42:149 (1975).
7. D.J.H. Brock and J.E. Olsson. Comparison of alfafetoprotein and beta-trace protein and the antenatal diagnosis of anencephaly and spina bifida, Clin. Genetics. 9:385 (1976).
8. J. E. Olsson, M. Sherman and B. Kjessler. Beta-trace protein and neural tube defects, Lancet ii:347 (1974).
9. G. Hochwald and G.J. Thorbecke. Trace proteins in cerebrospinal fluid and other biological fluids. I. Effect of various fractionation procedures on beta-trace and gamma-trace proteins and methods for isolation of both proteins, Arch. Biochem.Biophys. 101:325 (1963).
10. H. Link. Isolation and partial characterization of "trace" proteins and immunoglobulin G from cerebrospinal fluid, J. Neurosurg. Psychiat. 28:552 (1965).
11. J.E. Olsson and I. Carlstedt. The carbohydrate composition of beta-trace protein from the human cerebrospinal fluid, Vth international meeting of the ISN, Barcelona, Abstract p. 314 (1975).

12. E.C. Laterre, J.F. Heremans ad A.Carbonara. Immunological comparizon of some proteins found in cerebrospinal fluid, urine and extracts from brain and kidney, Clin. Chim. Acta 10:197 (1964).
13. R. Penny and E.F. Osserman. Studies on beta-trace protein, Aust.J. exp. Biol. med. Sci, 49:111 (1971).
14. J.E. Olsson and H. Link. Distribution of serumproteins and beta-trace protein within the nervous system, J. Neurochem. 20:837 (1973).
15. J.E. Olsson and L. Nord. Immunochemical and immunofluorescence studies of beta-trace protein in different species and organs, with special reference to the central nervous system, J. Neurochem. 21:625 (1973).
16. J.E. Olsson, C. Blomstrand and K.G. Haglid. The cellular distribution of beta-trace protein in CNS and brain tumours, J. Neurol. Neurosurg: Psychiat. 37:302 (1974).
17. H. Link and J.E. Olsson. Beta-trace protein concentration in CSF in neurological disorders, Acta. Neurol.Scand. 48:57 (1972).
18. J.E. Olsson, H. Link and B. Nosslin. Metabolic studies on ^{125}I beta-trace protein, with special reference to synthesis within the central nervous system, J. Neurochem 21: 1153 (1973).
19. J.E Olsson and M. Sandberg. Demonstration of synthesis and beta-trace protein in different tissues of squirrel monkey, Neurobiology 5:270 (1975).
20. M.G. Harrington, C.R. Merril, D.M. Asher and D.C. Gajdusek. Abnormal proteins in the cerebrospinal fluid of patients with Creutzfeldt-Jakob disease. N. Engl. J. Med. 315:279 (1986).
21. O. Hayaishi. Prostaglandin D synthase, beta-trace and sleep. Prostaglandins 51:281 (1996).
22. A. Hoffmann, E. Grabenhorst, M. Nimtz and H.S. Conradt. Characterization of human beta-trace proteins from natural body fluids, from patients and from recombinant sources. Prostaglandins 51:284 (1996).

EFFECT OF IMMUNIZATION WITH β-TRACE PROTEIN ON TESTOSTERONE LEVELS IN ANIMALS

G.M. Hochwald[1], R.M. Schwarcz[2], C. Huang[2], B.S. Bhogal[2] and G.J. Thorbecke[2,3]

[1]Dept. of Neurology
[2]Dept. of Pathology
[3]Kaplan Cancer Center
New York University School of Medicine
New York, NY 10016

INTRODUCTION

Interest in β-trace (β-tr) has been rekindled after the realization that its N-terminal amino acid sequence[1] was identical to that of glutathione independent prostaglandin D synthase, sequenced by Urade and coworkers[2]. It has now been firmly established that β-tr from cerebrospinal fluid (CSF) and this prostaglandin D synthase from brain are identical [3,4,5]. β-tr is found in most abundance in the CSF, where it constitutes 8-10% of the total protein[6,7], in seminal fluid, where it is higher in concentration although much lower as a percentage of total protein[8], and in urine[9]. β-tr is synthesized in the choroid plexus and brain as shown by incorporation of labeled amino acids into β-tr protein, as well as in testis and fimbriae of the fallopian tubes[6]. By staining with fluorescent antibody, localization of β-tr was shown in oligodendrocytes and in epididymis[10]. Moreover, by *in situ* hybridization with antisense RNA, Urade et al[11] have shown synthesis of β-tr in leptomeninges and choroid plexus as well as in oligodendrocytes of the adult rat brain. The enzymatic activity association with β-tr also occurs in the epididymis[12].

Functions attributed to β-tr and its product prostaglandin D_2 (PGD_2) in CNS include regulation of the sleep-wake cycles and luteinizing hormone (LH) release[13,14]. In both these functions PGD_2 appears to have the opposite effect of prostaglandin E_2 (PGE_2), i.e. PGD_2 induces sleep and suppresses LH release, whereas PGE_2 has a waking effect [13] and enhances LH release[15]. Like β-tr, PGD_2 circulates in the spinal fluid and serves as a neurohormone [16]. An extraneural functional significance of β-tr for spermatogenesis has been suggested by the correlation between seminal fluid β-tr content and degrees of azospermia in men[8] and by a reduction in the β-tr content of seminal

Recent Advances in Prostaglandin, Thromboxane, and Leukotriene Research
Edited by Sinzinger *et al.*, Plenum Press, New York, 1998

plasma from low fertility bulls[17]. A local effect of PGD_2 and thus of β-tr on testosterone production by Leydig cells in the testis is possible in view of the fact that these cells exhibit receptors for PGs through which testosterone production is regulated[18]. In addition, in view of the importance of LH for testosterone production by Leydig cells[19], a centrally mediated effect of β-tr on testosterone production is also possible. Therefore, in the present study we have attempted to induce autoimmunity to β-tr in animals in order to determine whether 1) the presence of circulating autoantibodies to β-tr would affect testosterone production and 2) autoimmune reactions in gonadal tissues could be induced by such immunization.

MATERIALS AND METHODS

Animals: Adult male mice (BALB/c) and rats (Lewis and Sprague Dawley) were obtained from Taconic Farms (Germantown, NY) and Charles River Labs.(Wilmington, MA), respectively. Mice were immunized with a conjugate of the N-terminal peptide of rat β-tr (sequence FPQTPAQGHDTVQPNFQQDKFLG) to ovalbumin (OVA) or of human β-tr (sequence APEAQVSVQPNFQQDKFLG) to keyhole limpet hemocyanin (KLH) emulsified in complete Freund's adjuvant (CFA). They received 200 μg per injection. Rats were immunized with purified human β-tr protein in either CFA or incomplete Freund's adjuvant (IFA, containing 20-25 μg of β-tr per injection). Control animals received OVA or KLH and/or adjuvant alone.

Materials: β-tr was isolated from CSF by affinity chromatography using sepharose to which monoclonal anti-β-tr, 6F5 (kindly donated by Dr. Y. Urade, Osaka Biochemical Institute, Japan) had been conjugated. The N-terminal β-tr peptides were synthesized on a peptide synthesizer (430 Applied Biosystems Inc., Foster City, CA), using solid phase fluorenyl methyloxy carbonyl (Fmoc/HBTU) as described by Stewart and Young in Solid Peptide Synthesis, Pierce Chemical Co., Rockford, IL, 1984. They were conjugated to protein using glutaraldehyde. Antibodies used for Western blotting included alkaline phosphatase labeled goat anti-rat and anti-rabbit Ig, purchased from Gibco (Gaithersburg, MD) and Sigma (St. Louis, MI), respectively.

Testosterone Measurements: Total testosterone levels in sera were measured by a solid phase radioimmuno assay, using testosterone-specific antibody and ^{125}I-labeled testosterone. The assay was done according to the manufacturer's instructions (Diagnostic Products Corp., Los Angeles, CA).

RESULTS AND DISCUSSION

Effect of immunization with β-tr N-terminal peptide conjugates on testosterone levels in mice. The results in Table 1 show that immunization with antigen in CFA caused a significant decrease in testosterone levels below the pre-immunization levels, regardless of the antigen used.

Table 1. **TESTOSTERONE LEVELS IN BALB/c MICE IMMUNIZED WITH β-TRACE PEPTIDE-CONJUGATES**

Mice Immunized With*	Testosterone (ng/ml) 6 wks after 1st Injection	8 wks after 1st Injection
Rat β-tr-pept-OVA + CFA	7.3 ± 5.8 (n = 5)**	13.2 ± 8.3 (n = 5)
Hu β-tr-pept-KLH + CFA	2.6 ± 3.6 (n = 5)	3.6 ± 2.8 (n = 5)
OVA or KLH + CFA	1.1 ± 1.4 (n = 9)	7.5 ± 9.6 (n = 9)

* Preimmunization levels in male mice were 10.3 ± 4.4 ng/ml (n=5); levels in unimmunized female mice were 0.19 ± 0.18 ng/ml (n=5). The mice were immunized twice, the second injection was at 4 wks. They were bled at 6 and 8 wks.

** Significantly different from control value.

This is presumably due to the induction of IL-1 production by the adjuvant used, since IL-1 is known to inhibit testosterone production in Leydig cells [20]. It was therefore difficult to get an exact evaluation of the specific effects of immunization with β-tr peptide. However, in the mice bled 2 weeks after the second immunization, a significantly higher serum testosterone level was observed in the group immunized with the rat β-tr peptide conjugate, whereas the group immunized with human β-tr peptide conjugate showed no difference from the group immunized with control proteins. At 4 weeks after immunization this difference was still seen, but was not statistically significant. It seemed, therefore, that immunization with the rat, but not with the human peptide had caused a transient deregulation/enhancement of testosterone synthesis.

Effect of immunization with β-tr protein on testosterone levels in male rats. In these experiments, β-tr was isolated from human CSF and injected with IFA into Sprague Dawley rats or with CFA or IFA into Lewis rats (Table 2). Control rats received adjuvant emulsified with saline. In the experiment with Lewis rats, two rats were left unimmunized to examine their testosterone levels during the same time period. It is clear that the decrease in testosterone levels induced by the immunization procedure is also observed in rats. In the experiment with the Sprague Dawley rats, one or two of the immunized rats bled 2 weeks after the second immunization had higher testosterone levels than any of the other immunized or control rats, but no significant changes in testosterone levels could be detected in the immunized group as compared to the control rats. In the Lewis rats (Table 2), there was a significant increase in testosterone levels in the combined groups immunized with β-tr in IFA or CFA as early as 3 weeks after the first immunization. At longer intervals after immunization, there was no longer any difference between the rats immunized with adjuvant with or without β-tr, but the testosterone levels in the unimmunized control rats increased more than did those in the rats injected with adjuvant alone. As in the mice, these data are therefore suggestive of a transient deregulating effect of immunization with β-tr on testosterone levels, followed by a nonspecific decreasing effect due to the adjuvant used during the immunization.

Table 2. **TESTOSTERONE LEVELS IN LEWIS RATS IMMUNIZED WITH PURIFIED HUMAN β-TRACE**

Rats Immunized With*	Serum Testosterone (ng/ml)		
	3 wks After 1st Inj	5 wks After 2d Inj	2 wks After 3d Inj
Hu β-tr + CFA or IFA	14.2 ± 9.1 (n=6)**	5.4 ± 2.9 (n=6)	7.1 ± 4.0 (n=6)
CFA or IFA alone	6.1 ± 2.8 (n=6)	8.0 ± 4.1 (n=6)	10.3 ± 2.3 (n=6)
None	5.0 ± 2.2 (n=2)	7.1 ± 0.9 (n = 2)	19.4 ± 4.5 (n=2)@

* The rats were immunized 3 x, the second injection was at 3 wks and the third injection at 13 wks. They were bled at 3, 8, and 15 wks after the primary immunization. Since there was no significant difference between IFA and CFA immunized rats, the results for these groups are reported together.
** For comparison with all control groups combined, $p = 0.027$.
@ These values are higher than any of the values obtained after immunization with IFA or CFA ± β-tr.

The results of Western blotting showed that, already at the time of the first bleeding, all the rats immunized with β-tr plus CFA or IFA, but not the rats which received adjuvant without β-tr, had serum antibody to human β-tr. No attempt was made, however, to quantitate the β-tr in seminal fluid or CSF of the immunized animals. Testis, epididymis and prostate from immunized rats were examined histologically, but no evidence of infiltration with lymphoid cells or morphological destruction was obtained. There was also no significant difference in the weights of these organs determined at the end of each experiment.

If β-tr and PGD_2 have a regulating effect on LH[14] and/or local effects in the male sex organs[19,21,22], which under normal conditions tend to decrease testosterone synthesis, a temporary increase in testosterone levels as a result of circulating anti-β-tr might be caused by a neutralizing effect of the antibody. Since there was no evidence in the present studies that an autoimmune disease of the genital organs could be induced by immunization with β-tr, escape from the effect of the circulating antibody may have been caused by a compensatory increase in local production of β-tr.

REFERENCES

1. A.P. Kuruvilla, G.M. Hochwald, J. Ghiso, E.M. Castano, M. Pizzolato, and B. Frangione, Isolation and amino terminal sequence of beta-trace, a novel protein from human cerebrospinal fluid, *Brain Res*, 565: 337 (1991).
2. A. Nagata, Y. Suzuki, M. Igarashi, N. Eguchi, H. Toh, Y. Urade, and O. Hayaishi, Human brain prostaglandin D synthase has been evolutionarily differentiated from lipophilic-ligand carrier proteins, *Proc Natl Acad Sci U S A*, 88: 4020 (1991).
3. M. Zahn, M. Mader, B. Schmidt, E. Bollensen, and K. Felgenhauer, Purification and N-terminal sequence of beta-trace, a protein abundant in human cerebrospinal fluid, *Neurosci Lett*, 154: 93 (1993).
4. A. Hoffmann, H.S. Conradt, G. Gross, M. Nimtz, F. Lottspeich, and U. Wurster, Purification and chemical characterization of beta-trace protein from human cerebrospinal fluid: its identification as prostaglandin D synthase, *J Neurochem*, 61: 451 (1993).
5. K. Watanabe, Y. Urade, M. Mader, C. Murphy, and O. Hayaishi, Identification of beta-trace as prostaglandin D synthase, *Biochem Biophys Res Commun*, 203: 1110 (1994).
6. G.M. Hochwald, A.J. Pepe, and G.J. Thorbecke, Trace proteins in biological fluids. IV. Physicochemical properties and sites of formation of gamma-trace and beta-trace proteins, *Proc. Soc. Exp. Biol. Med.*, 124: 961(1967).
7. H. Link, and J.E. Olsson, Beta-trace protein concentration in CSF in neurological disorders, *Acta Neurol Scand*, 48: 57 (1972).
8. J.E. Olsson, Correlation between the concentration of beta-trace protein and the number of spermatozoa in human semen, *J Reprod Fertil*, 42: 149 (1975).
9. G.M. Hochwald, and G.J. Thorbecke, Use of an antiserum against cerebrospinal fluid in demonstration of trace proteins in biological fluids., *Proc Soc Exp Biol Med*, 109:91 (1962).
10. J.E. Olsson, and L. Nord, Immunochemical and immunofluorescence studies of beta-trace protein in different species and organs, with special reference to the central nervous system, *J Neurochem*, 21: 625 (1973).
11. Y. Urade, K. Kitahama, H. Ohishi, T. Kaneko, N. Mizuno, and O. Hayaishi, Dominant expression of mRNA for prostaglandin D synthase in leptomeninges, choroid plexus, and oligodendrocytes of the adult rat brain, *Proc Natl Acad Sci U S A*, 90:9070 (1993).
12. M. Ujihara, Y. Urade, N. Eguchi, H. Hayashi, K. Ikai, and O. Hayaishi, Prostaglandin D_2 Formation and characterization of its synthetases in various tissues of adult rats, *Arch Bioch Biophys*, 260:521 (1988).
13. O. Hayaishi, Molecular mechanism of sleep-wake regulation: roles of prostaglandins D_2 and E_2, *FASEB J*, 5:2575 (1991).
14. F. Kinoshita, Y. Nakai, H. Katakami, H. Imura, T. Shimizu, and O. Hayaishi, Suppressive effect of prostaglandin (PG)D2 on pulsatile luteinizing hormone release in conscious castrated rats, *Endocrinology*, 110:2207 (1982).
15. P.G. Harms, S.R. Ojeda, and S.M. McCann, Prostaglandin involvement in hypothalamic control of gonadotropin and prolactin release, *Science*, 181:760 (1973).
16. Y. Urade, O. Hayaishi, H. Matsumura, and K. Watanabe, Molecular mechanism of sleep regulation by prostaglandin D2, *J Lipid Mediat Cell Signal*, 14:71 (1996).
17. G. Killian, R. Gerena, D. Chapman, and Y. Cao, A fertility-associated protein in bull seminal plasma is prostaglandinD_2 synthase, *Prostaglandins*, 51:296 (1996).
18. A. Erichsen, F.O. Levy, S. Golf, and V. Hansson, A transplantable rat Leydig cell tumor--5. Cellular localization of beta-adrenergic and prostaglandin receptors and their effects on testosterone production, *J Steroid Biochem*, 31:41 (1988).
19. D.S. Keeney, S.M. Mendis-Handagama, B.R. Zirkin, and L.L. Ewing, Effect of long term deprivation of luteinizing hormone on Leydig cell volume, Leydig cell number, and steroidogenic capacity of the rat testis, *Endocrinology*, 123:2906 (1988).
20. J.H. Calkins, M.M. Sigel, H.R. Nankin, and T. Lin, Interleukin-1 inhibits Leydig cell steroidogenesis in primary culture, *Endocrinology*, 123:1605 (1988).
21. K. Yamada, N. Ohkura, and T. Satoh, Reduction of testicular testosterone levels in rats by prostaglandin E2, *Res Commun Chem Pathol Pharmacol*, 47:153 (1985).
22. K. Yamada, [The effect of prostaglandin D2 on rat testicular testosterone levels], *Nippon Naibunpi Gakkai Zasshi*, 65:1180 (1989).

EICOSANOIDS MEDIATE INSECT CELLULAR IMMUNE REACTIONS TO BACTERIAL INFECTIONS

David W. Stanley

Insect Biochemical Physiology Laboratory
University of Nebraska
Lincoln, NE 68583-0816

INTRODUCTION

The occurrence of eicosanoids in invertebrate tissues has been recognized for nearly thirty years, beginning with the discovery of certain prostaglandins (PGs) in some species of octacorals[1]. This discovery attracted more commercial than zoological attention because the large amounts of PGs in the corals represented a natural source of material for research and other pharmaceutical interests[2]. The initial finding prompted efforts to locate other natural sources of PGs and other eicosanoids, and it is now commonly thought that eicosanoids occur in the tissues of virtually every animal taken under appropriate scrutiny. This idea carries considerable significance, because it suggests eicosanoids were recruited into biological signaling systems before evolution of the Metazoa. I suppose eicosanoids occur in most, if not all, animals, and eicosanoids exert important biological actions in animals. Some eicosanoid actions are limited to a few species. For example, PGs release egg-laying behavior in some, but not all, cricket species[3]. Similarly, F-series PGs serve as sex pheromone in some, by no means all, fish species[4]. On the other hand, other eicosanoid actions may be fundamental to animal systems. Eicosanoids seem to be involved in ion transport physiology in every vertebrate and invertebrate suitably studied[5]. Regulating ion transport physiology may be a fundamental eicosanoid action. In a similar vein, we will suggest in this chapter that mediating cellular defense reactions may also be a fundamental eicosanoid action.

The presence and biological actions of eicosanoids in invertebrates have been treated, from various perspectives, in a spate of reviews. I cite these reviews here as a convenient entry to the literature[2,3,5,6-16]. My group has been considering the possibility that cellular immune defense is another area of animal biology in which eicosanoids serve as mediators. In this essay I review the main lines of research into this hypothesis. Examination of this, and similar, hypotheses is of inclusive interest because eicosanoids exert profound biological actions: their biological significance lends tremendous explanatory power to eicosanoids.

EICOSANOIDS MEDIATE INSECT CELLULAR IMMUNE REACTIONS TO BACTERIAL INFECTIONS

The hypothesis that eicosanoids mediate invertebrate, particularly insect cellular immune reactions to bacterial infections emerged from the two main points just mentioned. First, eicosanoids exert many actions in mammalian immunity. Second, some eicosanoid actions may be fundamental to animal systems. For the initial explorations, we selected a large insect, the tobacco hornworm, *Manduca sexta*. The hornworms were reared under semi-sterile conditions on standard culture medium. We tested the hypothesis using an elementary manipulation of Koch's postulate[17]. Experimental hornworms were first injected with an inhibitor of eicosanoid biosynthesis, then injected with a standard dosage of a red pigmented strain of the bacterium, *Serratia marcescens*. Control hornworms were injected ethanol (the drug vehicle), then with bacteria. After selected incubation periods, usually 1 hr, hemolymph (blood) samples were withdrawn, serially diluted and plated on standard methods agar. We interpreted recovery of red pigmented colony forming units (CFU) as impaired ability to clear bacterial infections from circulating hemolymph.

The results of these experiments showed that recovery of red pigmented CFUs increased in a dose-dependent way with increased dosages of dexamethasone. We also performed similar experiments using other eicosanoid biosynthesis inhibitors, including indomethacin and esculetin. All results supported the hypothesis that eicosanoids mediate one or more aspects of insect cellular immune reactions to bacterial infections[17].

This suggestion opened a more detailed question: which of the hemocytic defense mechanisms are mediated by eicosanoids? Three hemocytic defense reactions are recognized, nodulation, encapsulation and phagocytosis. Nodulation begins with formation of microaggregates, or clusters of hemocytes that entrap bacteria cells in a cellular matrix. The microaggregates grow, and ensnare more bacteria cells, by attracting additional hemocytes. The process ends with melanization of the mature nodules[18]. The nodules and their entrapped bacteria are not cleared from the body cavity; they usually attach themselves to a solid structure within the body. The encapsulation reaction is similar in some ways to nodulation, in that hemocytes are attracted to the foreign material. As the term suggests, larger invaders that can not be entrapped in nodules, such as eggs of parasitoids, are encapsulated by hemocytes. Phagocytosis is another defense reaction against bacterial infections.

Nodulation is the temporally and quantitatively predominant cellular defense reaction to bacterial infections in insects[19]. On the basis of its importance to clearing bacterial infections from circulating hemolymph, we hypothesized that nodulation is one of the insect cellular defense mechanisms mediated by eicosanoids. We tested this idea using tobacco hornworms in a line of experiments similar to the ones just described. Experimental hornworms were treated with an eicosanoid biosynthesis inhibited, then injected with a standard dosage of bacteria. Control hornworms were injected with the drug vehicle, ethanol, then injected with a standard dosage of bacteria. After selected incubation periods, the insects were anesthetized and nodulation was assessed by exposing the hemocoels, and counting dark, melanized nodules under a stereomicroscope. After the initial counting, the alimentary canal was excised, and nodules in the previously unexposed areas were counted. We also assessed microaggregation reactions in some experiments. In these determinations, hemolymph was collected onto a hemacytometer and the number of cellular aggregates in each sample was estimated by direct counting of every field under phase-contrast optics.

Compared to ethanol-injected, control hornworms, nodulation was severely impaired in experimental hornworms. The influence of the eicosanoid biosynthesis inhibitors, including dexamethasone, esculetin, indomethacin, naproxin and ibuprofen, was expressed in a dose-dependent manner. Moreover, the influence of dexamethasone was reversed by treated experimental hornworms with arachidonic acid. Together, these findings supported the hypothesis that eicosanoids mediate an identified cellular defense reaction to bacterial infections, particularly nodulation, in insects.

However, as presented, our information on nodulation did not allow us to distinguish between two possibilities. On one hand, the inhibitors could have impaired the final step in nodulation, that is, the melanization step, with no direct influence on nodulation, *per se*. In the absence of the melanization step, many undarkened nodules, impossible to see under a stereomicroscope, could have been present. On the other hand, the eicosanoid biosynthesis inhibitors could have impaired nodulation in earlier phases of the overall process. To gain insight into these possibilities, we also considered the influence of eicosanoid biosynthesis inhibitors on microaggregation reactions to bacterial infections. In these experiments, hornworms were as just described to assess the influence of eicosanoid biosynthesis inhibitors on nodulation. At 1 hr post infection, hemolymph samples were withdrawn, and microaggregates were counted on a hemacytometer. We recovered about 8.3 million microaggregates/ml of hemolymph from ethanol-injected control hornworms, compared to about 1.9 million from dexamethasone-injected experimental hornworms[18]. These data show that eicosanoids act in early phases of the nodulation process, rather than exclusively the last phase, that is, the melanization step. We will return to this point just ahead, in discussion of recent findings from another group. The main thrust at this juncture is recognition of a specific insect cellular defense reaction to bacterial infections, specifically nodulation, is mediated by eicosanoids[18].

The idea that insect cellular defense reactions are mediated by eicosanoids rests on experimental results from one insect species, namely tobacco hornworms, *Manduca sexta*. Insects represent the largest and most diversified group of animals, with nearly 1 million known species and at least that many more species still unknown. Insects inhabit an amazing array of habitat types, including terrestrial, aquatic, and marine habitats, as well as some extreme environments, such as hypersaline waters and oil pools. These comments are meant to emphasize the conceptual difficulties associated with forming sweeping generalizations on the basis of data from one experimental model. The point is not trivial. Discovery that PGs releases egg-laying behavior in newly mated crickets marks the first known eicosanoid action in insects[3,6]. However, experiments to test the idea that PGs uniformly release egg-laying behavior in other insect groups have yielded mostly negative results. Again, it is difficult to generalize from one or a few insect species to the class as a whole.

We considered the possibility that eicosanoids mediate cellular immune reactions in a fairly broad range of insect groups. We conducted a line of experiments similar to the ones described for tobacco hornworms. So far, we have considered the possibility in a tenebrionid beetle, *Zophobas atratus*[19], in two moths of agricultural importance, black cutworms and true armyworms[20], in silkworms[21], and in a cricket, *Gryllus assimulus* (unpublished observations). All experiments yielded similar, positive results, from which we offer this generalization: eicosanoids mediate cellular defense reactions in most, if not all, insect species.

Nodulation is a complex process, entailing many distinct cellular actions, some of which remain to be identified. Let us imagine one or another of the eicosanoids act sequentially in some of these steps, and imagine all steps are essential to the overall process. If so, then inhibition of the cyclooxygenase pathway, or separately, the lipoxygenase pathways would be expected to impair the insects' ability to form nodules in response to bacterial infections. This is consistent with results of our experiments with eicosanoid biosynthesis inhibitors in all the species we have tested so far. We return, now, to the final step in nodulation, melanization of the mature nodule. In their work with hemocytes from the waxmoth, *Galleria mellonella*, Mandato et al.[22] were able to dissect two separate steps in the nodulation process. They observed that inhibition of eicosanoid biosynthesis impaired cell spreading, an early step in the process. Moreover, they observed that prophenoloxidase activation, the first step in the melanization process, also depends on eicosanoid biosynthesis. Hence, these two steps, cell spreading and melanization, are separately mediated by eicosanoids. Beyond dissecting two steps in nodulation, these authors also demonstrated that phagocytosis of bacterial cells is mediated by eicosanoids[22].

Acknowledgments

Thanks to Dr. Ralph W. Howard for critical comments on a draft of this article. Thanks also to Dr. Andrew F. Rowley for organizing the symposium on invertebrate and plant eicosanoids. Research in my laboratory is funded by the Agricultural Research Division, University of Nebraska and by specific cooperative agreement # 58-5430-5-115.

REFERENCES

1. A.J. Weinheimer and R.L. Spraggins, The occurrence of two new prostaglandin derivatives (15-epi-PGA2 and its acetate methyl ester) in the gorgonian Plexaura homomalla. Chemistry of coelentrates XV. *Tetrahedron lett.* 59:5185 (1969).
2. G.L. Bundy, Nonmammalian sources of eicosanoids, in: Advances in Prostaglandin, Thromboxane, and Leukotriene Research, Volume 14, J.E. Pike and D.R. Morton, Jr. eds., Raven Press, New York (1985).
3. D.W. Stanley-Samuelson and W. Loher, Prostaglandins in insect reproduction. *Ann. Entomol. Soc. Amer.* 79:841 (1986).

4. P.W. Sorensen, T.J. Hara, N.E. Stacey, and F.Wm. Goetz, F prostaglandins function as potent olfactory stimulants that comprise the postovulatory female sex pheromone in goldfish. *Biol. Reprod.*39:1039 (1988).
5. D.W. Stanley-Samuelson and V.K. Pedibhotla, What can we learn from prostaglandins and related eicosanoids in insects? *Insect Biochem. Molec. Biol.* 26:223 (1996).
6. E.U. Brady, Prostaglandins in insects. *Insect Biochem.* 13:443 (1983).
7. D.W. Stanley-Samuelson, Physiological roles of prostaglandins and other eicosanoids in invertebrate animals, *Biol. Bull.* 173: 92 (1987).
8. D.W. Stanley-Samuelson, Comparative eicosanoid physiology in invertebrate animals. *Am. J. Physiol.* 260:R849 (1991).
9. D.W. Stanley-Samuelson, The biological significance of prostaglandins and related eicosanoids in insects, in: *Insect Lipids: Chemistry, Biochemistry and Biology*, D.W. Stanley-Samuelson and D.R. Nelson, eds., University of Nebraska Press, Lincoln, NE (1983).
10. D.W. Stanley-Samuelson, Assessing the physiological significance of prostaglandins and other eicosanoids in insect physiology. *J. Insect Physiol.* 40:3 (1994).
11. D.W. Stanley-Samuelson, Prostaglandins and related eicosanoids in insects. *Advances in Insect Physiol.* 24:115 (1994).
12. D.W. Stanley-Samuelson, The biological significance of prostaglandins and other eicosanoids in invertebrates. *Amer. Zool.* 34:589 (1994).
13. J.R. Sauer, A.S. Bowman, M.M. Shipley, C.L. Gengler, M.R. Surdick, J.L. McSwain, C. Luo, R.C. Essenberg, and J.W. Dillwith, Arachidonate metabolism in tick salivary glands, in: : *Insect Lipids: Chemistry, Biochemistry and Biology*, D.W. Stanley-Samuelson and D.R. Nelson, eds., University of Nebraska Press, Lincoln, NE (1983).
14. L. DePetrocellis and V. DiMarzo, Aquatic invertebrates open up new perspectives in eicosanoid research: Biosynthesis and bioactivity. *Prostaglandins Leukotrienes Essential Fatty Acids* 51:215 (1994).
15. M. Lamacka and J. Sajbidor, The occurrence of prostaglandins and related compounds in lower organisms. *Prostaglandins, Leukotrienes Essential Fatty Acids* 52:357 (1995).
16. D.W. Stanley and R.W. Howard, The biology of prostaglandins and related eicosanoids in invertebrates: Cellular, organismal, and ecological actions. *Amer. Zool.* , in press (1997).
17. D.W. Stanley-Samuelson, E. Jensen, K.W. Nickerson, K. Tiebel, C.L. Ogg, and R.W. Howard, R.W. Insect immune response to bacterial infection is mediated by eicosanoids. *Proc. Natl. Acad. Sci. USA*, 88:1064 (1991).
18. J.S. Miller, T. Nguyen, and D.W. Stanley-Samuelson, Eicosanoids mediate insect nodulation responses to bacterial infections. *Proc. Natl. Acad. Sci., USA* 91:12418 (1994).
19. D.W. Horohov and P.E. Dunn, Phagocytosis and nodule formation by hemocytes of *Manduca sexta* larvae following injection of *Pseudomonas aeruginosa*. *J. Invert. Pathol.* 41:203 (1983).
20. J.S. Miller, R.W. Howard, T. Nguyen, A. Nguyen, R.M.T. Rosario, and D.W. Stanley-Samuelson, Eicosanoids mediate nodulation responses to bacterial infections in larvae of the tenebrionid beetle *Zophobas atratus*. *J. Insect Physiol.* 42:3 (1996).
21. R.A. Jurenka, J.S. Miller, V.K. Pedibhotla, R.L. Rana, and D.W. Stanley-Samuelson, Eicosanoids mediate microaggregation and nodulation responses to bacterial infections in black cutworms, *Agrotis ipsilon*, and true armyworms, *Pseudaletia unipuncta*. *J. Insect Physiol.* 43:125 (1997).
22. D.W. Stanley-Samuelson, V.K. Pedibhotla, R.L. Rana, A.R. Nor Aliza, W.W. Hoback and J.W. Miller, Eicosanoids mediate nodulation responses to bacterial infections in larvae of the silkmoth, *Bombyx mori*. *Comp. Biochem. Physiol.*, in press (1997).
23. C.A. Mandato, W.L. Diehl-Jones, S.J. Moore and R.G.H. Downer, The effects of eicosanoid biosynthesis inhibitors on prophenoloxidase activation, phagocytosis and cell spreading in *Galleria mellonella*. *J. Insect Physiol.* 43:1 (1997).

ARACHIDONIC ACID AND EICOSANOIDS IN *HYDRA*: POSSIBLE ENDOGENOUS SIGNALS INVOLVED IN CHEMORECEPTION AND MODULATION OF THE FEEDING BEHAVIOUR

Paola Pierobon[1], Luciano De Petrocellis[1*], Rosario Minei[1], and Vincenzo Di Marzo[2]

[1]Istituto di Cibernetica and [2]Istituto per la Chimica di Molecole di Interesse Biologico C.N.R., Via Toiano 6, 80072, Arco Felice, Napoli, Italy

INTRODUCTION

Arachidonic acid (AA) exerts many biological actions in different tissues. Modulation of ion channels by AA has received considerable attention in studies of the role of this fatty acid in the nervous system. In mammalian central nervous system polyunsaturated fatty acids (PUFAs) have been found to regulate the activity of $GABA_A$ receptor-coupled Cl^- channels[1,2]. AA serves also as a precursor of a variety of eicosanoids, among which lipoxygenase metabolites with *R* stereochemistry have been identified in several species of marine and freshwater invertebrates[3]. AA and its ω10-lipoxygenase metabolites, as well as phospholipase A_2 (PLA_2), the enzyme mostly responsible for PUFA release from membrane phospholipids have been found in *Hydra*[3], one of the most primitive organisms to evolve a nervous system[4]. The freshwater polyp *Hydra* is a predator whose tentacles have stinging capsules named nematocysts; when the prey touches a tentacle a complex behavioral phenomenon, the feeding response, is initiated by the outflow of reduced glutathione (GSH) from the wounded prey[5]. This response consists of a sequence of events, including tentacle writhing, mouth opening, ingestion of prey and closing of the mouth. Duration of response, i.e. the interval between mouth opening (T_i) and mouth closure (T_f) time is dependent on GSH concentration through the interaction with an external receptor[6] and can be quantitatively measured. At present, mechanisms of termination of the response are still poorly understood, though we have recently shown that γ-aminobutyric acid (GABA) prolongs the duration of response through a receptor-mediated action[7]. Further biochemical and physiological studies have shown that neuroactive steroids, benzodiazepines, bicuculline, picrotoxin and the specific Cl^- channel ligand *t*-butylbicyclophosphorothionate (TBPS) also modulate response duration, suggesting that a population of $GABA_A$-like receptors and/or GABA-gated Cl^- channels are involved[8]. Here we present the results of an *in vivo* study on the participation of AA and some of its metabolites in the GSH-induced feeding response in *Hydra*.

MATERIALS AND METHODS

Hydra vulgaris (Cnidaria, Hydrozoa) were cultured asexually in our laboratory. The feeding response was studied by the procedure previously described[7]. The test was started by adding several GSH doses in a 1-10 μM concentration range to different samples of 6-8 polyps. The effect of several substances on the GSH-induced response was then studied.

* Author to whom all correspondence should be addressed.
Fax: +39-81-5267654. E-mail: +LUCIANO@CELL.NA.CNR.IT

Duration of the response was independently measured for each animal, and statistical significance of mean differences was evaluated by ANOVA followed by Scheffè's test. In normal, otherwise untreated animals, duration of the response to GSH varies from about 10 minutes at 1 μM GSH to 20-25 minutes at 10 μM GSH and is linearly correlated to GSH concentration in this range of concentrations.

RESULTS AND DISCUSSION

In a previous report we have shown that the specific PLA_2 inhibitor oleoyloxyethylphosphorylcholine (OOPC) prolonged the GSH-induced response by delaying the times of mouth closure in *Hydra* [9]. It is noteworthy that the effect of OOPC was further potentiated by GABA, suggesting that the PLA_2 inhibitor and the neurotransmitter independently contribute to prolonging response duration. We also examined the effect of GSH on the release of radiolabelled AA from live polyps previously labelled with [^{3}H]AA, and found that GSH stimulation always resulted in AA release in an OOPC-inhibitable fashion; the kinetics of this release were dependent upon GSH doses and, therefore, upon the time of mouth closure, with higher AA levels always being found before mouth closure[9]. The aim of the present work was to study the effects of AA, a very abundant PUFA esterified to membrane phosphoglycerids, and of metabolites derived from both oxidative and non-oxidative AA metabolism, on the GSH-induced response in *Hydra*. We found that: a) in a 10-100 μM range of concentrations, AA shortens the GSH-induced response by causing an earlier closure of the mouth; b) at ten-fold lower concentrations, AA most abundant metabolite in *H. vulgaris*, i.e. 11-*R*-hydroxyeicosa-5Z,8Z,12E,14Z-tetraenoic acid (11-*R*-HETE) (but not its 11-*S*-enantiomer) also causes an earlier closure of the mouth; this might explain why 1 μM AA inhibits the GSH-induced response only after 30 min pre-treatment of live polyps, since previous studies[10] have shown that AA, diffused through *Hydra* ectodermal cell membranes, may be enzymatically oxidized to 11-*R*-HETE in a few minutes; c) 1 μM 11-*R*-HETE and 1-100 μM AA counteract the prolongation of response induced by GABA, which acts by delaying GSH-induced mouth closure (Table 1).

Table 1. Effect of AA and its ω10-lipoxygenase metabolites on 10 μM GSH-induced feeding response (% of control). The effect of 100 μM GABA is also reported

Substance (concentration)	10 μM GSH	+100 μM GABA
None	100±2.8	127.6±4.9*
AA (1 μM)	98.4±2.1	104.2±1.6**
AA (10 μM)	85.3±3.7*	90.8±1.5**
AA (100 μM)	62.8±2.6*	68.8±4.8**
AA (1 μM) 30 min pretreatment	82.5±3.6*	87.6±3.6**
11-*R*-HETE (1 μM)	80.6±1.9*	87.6±1.2**
11-*S*-HETE (1 μM)	104.7±3.9	125.5±3.6*

*=Statistically significant from None; **=Statistically significant from None + 100 μM GABA. Data are means±S.E.M. n=6-8.

Table 2. Effect of AA in association with 1 μM TBPS on 10 μM GSH-induced response (% of control). The effect of picrotoxin is also shown.

Substance (concentration)	10 μM GSH	+1 μM TBPS
None	100±2.8	67.1±2.5*
AA (1 μM)	98.4±2.1	68.4±2.3
AA (10 μM)	85.3±3.7*	79.2±2.0
AA (100 μM)	62.8±2.6*	87.8±4.8**
picrotoxin (1 μM)	88.4±3.6*	N.D.
picrotoxin (10 μM)	76.4±5.4*	N.D.

*=Statistically significant from None; **=Statistically significant from None +1 μM TBPS; means±S.E.M..

The effect of AA on GSH response was comparable to that exerted by the $GABA_A$-coupled Cl^- channel blocker TBPS. Simultaneous TBPS and AA administration did not result in an additive response but, on the contrary, AA dose-dependently reduced the inhibitory action of TBPS, thus seemingly behaving as a partial agonist at the chloride channel sites (Table 2).

These data may lead to the hypothesis that AA and 11-*R*-HETE interfere with the putative mouth closure-inducing endogenous mechanisms at the level of $GABA_A$-coupled Cl^- channels. Indeed, inhibition of these channels by AA has been reported in mammalian CNS[1,2]. However, the data presented here do not allow to draw any definitive conclusion as to whether this effect also occurs in *Hydra.*

Another AA-derived metabolite, anandamide (arachidonoyl-ethanolamide), the first brain endogenous substance to be proposed as a physiological agonist at central cannabinoid (CB1) receptors [11], also shortened GSH-induced feeding response by causing an earlier closure of the mouth. Anandamide was more potent than AA, by eliciting its inhibitory effect in the high nanomolar concentration range (75.0$\pm$1.5 % of control at 100 nM). Another acylethanolamide, i.e. the anti-inflammatory palmitoylethanolamide[12], did not have any effect (104.6$\pm$0.8). The effect of 100 nM anandamide was suppressed by 1 μM SR 141716A, a potent and highly selective CB1 antagonist[13], indicating that ananadamide action may be receptor-mediated, and opening the way to intriguing speculations on the role of the endogenous cannabinoid system[14] in such a primitive neural network. Further studies will be required in order to: a) determine the mechanisms underlying anandamide action on the feeding response; b) assess whether this eicosanoid is biosynthesized and metabolized in *Hydra.*

In conclusion, we propose that endogenous AA, produced following GSH-induced stimulation of PLA_2, through its metabolites or on its own, might intervene in the final phase

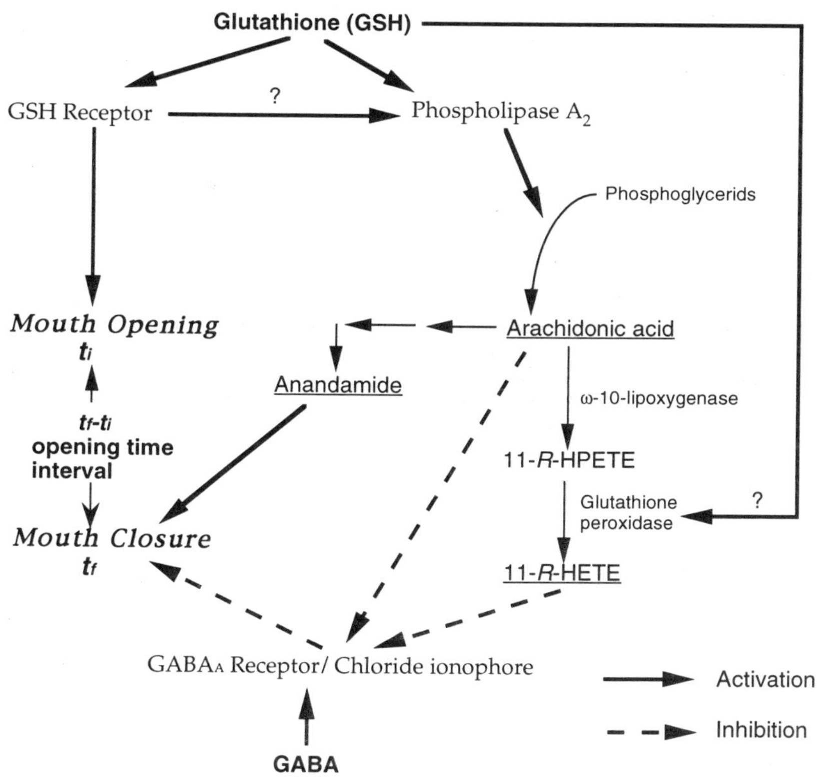

of GSH-induced feeding response, i.e. mouth closure. The duration of mouth opening may thus be determined at the molecular level by a metabolic interplay between GABA and AA and/or eicosanoids. The latter may be necessary components of feedback mechanisms started by GSH stimulation in order to achieve termination of the feeding response (Figure below). While AA/11-*R*-HETE may act on ionotrophic $GABA_A$ receptors, preliminary evidence on the biological action of anandamide rather points to the involvement of metabotrophic, G protein-coupled transduction pathways.

Acknowledgments

The authors wish to thank G. Marino for technical assistance, and the Human Frontiers in Science Organization for financial support (grant RG26/95 to VDM).

REFERENCES

1. Schwartz RD and Yu X, Inhibition of GABA-gated chloride channel function by arachidonic acid, *Brian Res*, 585:405 (1992).
2. Samochocki M and Strsznajder J, Age-related changes of GABA-activated chloride channel properties in brain cortex synaptic plasma membrane: evidence for phospholipase involvement, *Neurochem J* , 63:1522 (1994).
3. De Petrocellis L and Di Marzo V, Aquatic invertebrates open up new perspectives in eicosanoid research: biosynthesis and bioactivity, *Prostag Leuk Ess Fatty Acids* 51:239 (1994).
4. Parker GH, The Elementary Nervous System, Lippincott, Philadelphia (1919).
5. Loomis WF, Glutathione control of the specific feeding reactions of hydra, *Ann NY Acad Sci,* 62:209 (1955).
6. Bellis SL, Laux DC, Breuer A, Pertwee RG, Stevenson LA, Griffin G, Gibson D, Mandelbaum A, Etinger A and Rhoads DE, Affinity purification of *Hydra* glutathione binding proteins, *FEBS Lett* 354:320 (1994).
7. Pierobon P, Concas A, Santoro G, Marino G, Minei R, Pannaccione A, Mostallino MC and Biggio G, Biochemical and functional identification of GABA receptors in *Hydra vulgaris, Life Sci* 53:61 (1995).
8. Concas A, Pierobon P, Mostallino MC, Marino G, Minei R and Biggio G, Modulation of γ-aminobutyric acid (GABA) receptors and the feeding response by neurosteroids in *Hydra vulgaris, Neurosci* (1997) in press.
9. Pierobon P, De Petrocellis L, Minei R and Di Marzo V, Arachidonic acid as an endogenous signal for the glutathione-induced feeding response in *Hydra, Cell Mol Life Sci* 53:61 (1997).
10. Di Marzo V, De Petrocellis L, Gianfrani C and Cimino G, Biosynthesis, structure and biological activity of hydroxyeicosatetraenoic acids in *Hydra vulgaris, Biochem J* 295:23 (1994).
11. Devane WA, Hanus L, Breuer A, Pertwee RG, Stevenson LA, Griffin G, Gibson D, Mandelbaum A, Etinger A and Mechoulam R, Isolation and structure of a brain costituent that binds to the cannabinoid receptor, *Science* 258:1946 (1992).
12. Facci L, Dal Toso R, Romanello S, Buriani A, Skaper SD and Leon A, Mast cells express a peripheral cannabinoid receptor with differential sensitivity to anandamide and palmitoylethanolamide, *Proc Natl Acad Sci USA* 92:3376 (1995)
13. Rinaldi-Carmona M, Barth F, Héaulme M, Alonso R, Shire D, Congy C, Soubrie P, Breliere JC and Le Fur G, Biochemical and pharmacological characterisation of SR141716A, the first potent and selective brain cannabinoid receptor antagonist, *Life Sci* 56:1941 (1995).
14. Di Marzo V and De Petrocellis L, The endogenous cannabinoid signalling system: chemistry, biochemistry and physiology, *Internet J Sci-Biol Chem* Vol. 1 http://www.netsci-journal.com (1997)

SYNTHESIS AND ACTION OF EICOSANOIDS IN WARM AND COLD-ACCLIMATED AMPHIBIANS

Ceil A. Herman, Xinying Wang, and Jimin Sun

Department of Biology
New Mexico State University
Las Cruces, NM 88003

INTRODUCTION

Eicosanoids, including prostaglandins (PGs) and leukotrienes (LTs) are derived from 20 carbon fatty acids in membrane phospholipids and are important physiological regulators in a variety of systems. While synthesis and actions of eicosanoids are well characterized in mammals, less is known about cellular production and biological actions in poikilothermic vertebrates. As body temperatures in poikilotherms reflect those of the environment, different acclimation states may therefore dramatically affect synthesis and action of eicosanoids. In cannulated bullfrogs, prostaglandins (Herman et al, 1986) and leukotrienes (Herman et al, 1991) have less potent cardiovascular effects in cold-acclimated (5°, 30 days) animals when compared with those at room temperature.

The cellular source of leukotrienes in bullfrogs has been shown to be both erythrocytes (Gronert et al, 1995a) and thrombocytes (Gronert et al, 1995b). In these studies, the purified cells were stimulated with A23187. Interestingly, bullfrog erythrocytes and thrombocytes, unlike their mammalian homologues, contain lipoxygenase and can carry out synthesis without requiring transfer of LTA_4 from leukocytes. However, the effect of cold acclimation on synthesis was not determined. This study examines eicosanoid production during the natural stimulus of clotting in bullfrog blood as well as the cardiovascular actions of leukotrienes comparing short and long-term cold acclimation.

METHODS

American bullfrogs (*Rana catesbeiana*, n = 39, weight 396 $\pm$ 39.2 g) were acclimated to cold temperature (5°) for 1 (CA-1), 10 (CA-10), or 30 (CA-30) days and compared with warm-acclimated animals (WA) at room temperature (22°). Both clotting and cardiovascular studies utilized bullfrogs cannulated under MS-222 (3-aminobenzoic acid ethyl ester, 0.3%) anesthesia and allowed to recover for one day before testing. For clotting studies, blood was removed from a PE90 straight cannula in the right sciatic artery

Recent Advances in Prostaglandin, Thromboxane, and Leukotriene Research
Edited by Sinzinger *et al.*, Plenum Press, New York, 1998

and allowed to clot in a tube for increasing periods of time. Cells were separated from serum by centrifugation, serum was purified with Sep-Pak columns, and eicosanoids measured by radioimmunoassay as previously described (Herman and Martinez, 1988; Gronert et al, 1995a). As the antisera to LTC_4 expressed significant cross-reactivity to the LTD_4 and LTE_4, data using this antisera are expressed as peptide leukotrienes (p-LTs). For cardiovascular measurements, bullfrogs were cannulated with one T-cannula in the sciatic artery and the other in the sciatic vein as previously described (Herman et al, 1991). One group was WA, the second cannulated at room temperature and placed at 5° for 1 day (CA-1), and the third group cold acclimated for 30 days (CA-30), cannulated and tested at 5^oC. LTC_4 was infused into the venous cannula in doses from 100 to 3000 ng/kg-body weight (kg-bw) and responses measured from the arterial cannula using a P23-dB pressure transducer and a Gould Model 2000 recorder.

RESULTS AND DISCUSSION

Clotting blood from warm-acclimated (WA) and from bullfrogs exposed to 5° for 1 (CA-1), 10 (CA-10), or 30 (CA-30) days produced both lipoxygenase and cyclooxygenase products. Figure 1 compares the production of peptide leukotrienes (p-LTs), LTB_4, TXB_2 and PGE_2 in clotting blood after 15 min in the four groups. While clotting took longer in blood from cold-acclimated frogs, the clot was fully formed in all cases by 15 min. Bullfrogs placed at 5^oC for one day did not show decreased eicosanoid synthesis when compared with WA controls. Significant decreases in LTB_4, TXB_2, and PGE_2 synthesis were observed in the CA-10 group, with further decreases in the CA-30 group. The decrease in LTC_4 synthesis was only significant after 30 days. Synthesis changes were not a function of incubation temperature because blood from WA frogs incubated at 5° had equivalent synthesis when compared blood incubated at room temperature. Likewise, synthesis by CA-30 group blood was the same regardless of whether incubations were carried out at 5° or 22° (Wang and Herman, 1996).

Decreased synthesis of eicosanoids during clotting may be accounted for by the fact that blood cell numbers decreased during cold acclimation (Wang and Herman, 1996). To examine whether decreased synthesis was solely due to decreased cell number, erythrocytes

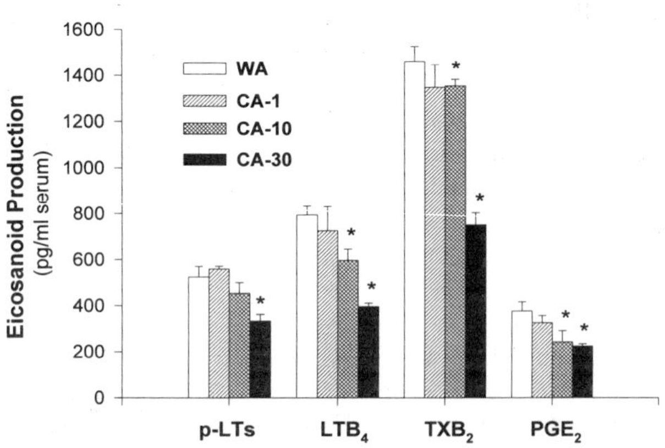

Figure 1 Eicosanoids produced by clotting bullfrog blood *in vitro*. Blood from warm acclimated (WA) or bullfrogs exposed to 5° for 1 (CA-1), 10 (CA-10), or 30 (CA-30) days was allowed to clot for 15 min and eicosanoids measured by radioimmunoassay. pLTs represent peptide leukotrienes. Results are expressed as pg/ml and represent mean $\pm$ s.e.m., n = 6 in each group. Data were redrawn from Wang, 1996.
* = significantly different ($p < 0.05$) from warm-acclimated value.

and thrombocytes were isolated from bullfrogs acclimated to warm or cold temperature for 30 days. Lipoxygenase products including leukotrienes and HETEs have been shown to be produced by isolated bullfrog erythrocytes (Gronert et al, 1995a) and thrombocytes (Gronert et al, 1995b). Thrombocytes from warm- and cold-acclimated bullfrogs were activated by A23187 and thrombin. Thrombocytes from cold-acclimated bullfrogs produced lower amounts of TXB_2 (Wang and Herman, 1997). Thus, the increase in clotting time is a function of lowered cell number and lowered thrombocyte synthesis of TXB_2.

To compare leukotriene effects due to short and long term cold exposure, bullfrogs at 22°C were compared with two groups of frogs which were placed at 5° either for 30 days or 1 day and LTC_4 dose-response curves for the two cold-exposed groups obtained while the animals were at 5° (Figure 2). WA animals had greater hypotensive responses than either of the cold-acclimated groups. However, the cold-acclimated animals (5° for one month) showed greater hypotensive responses to LTC_4 than did those animals that had been at 5° for 1 day when both groups were compared at 5°. Interestingly, the increase in heart rate in response to LTC_4 was only present in the warm frogs. There were no differences between the two cold groups.

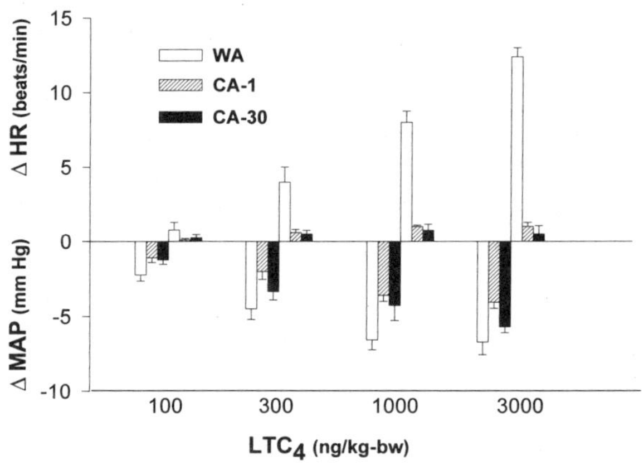

Figure 2. Effect of LTC_4 on mean arterial pressure (MAP) and heart rate (HR) in cannulated bullfrogs. Warm acclimated (WA) frogs were compared with animals at 5° for 1 day (CA-1) or 30 days (CA-30). MAP responses for the CA-1 and CA-30 were significantly lower than the WA group. However, MAP responses from the CA-30 were significantly greater than the CA-1 group. The HR responses of the CA-1 and CA-30 groups were significantly different from the WA group, but not different from each other. Results represent mean $\pm$ s.e.m., n = 5 for each group. Data were redrawn from Sun and Herman, 1995.

The increase in heart rate in response to the hypotensive efffect of leukotrienes has been shown to be an adrenergic reflex in both frogs (Herman et al, 1988) and toads (Herman et al, 1993). The lack of effect of leukotrienes on heart rate in CA animals is not due to a decreased number of beta receptors (Herman et al, 1996). The increased response to LTC_4 in animals which had been cold acclimated for 30 days compared with the 1 day group, suggest that up-regulation of LTC_4 receptors may have occurred in the cold-acclimated animals. The decreased synthesis of LTC_4 after 30 days of cold acclimation supports this hypothesis. Further studies are necessary to characterize changes in leukotriene receptor numbers or binding kinetics in response to cold acclimation.

ACKNOWLEDGMENTS

The authors would like to thank Merck Frosst Centre for Therapeutic Research, Pointe Claire, Dorval, Quebec, Canada, which provided antisera to peptide leukotrienes and to LTB_4. This work was supported by NSF grant IBN9318047 to C.A.H. and graduate student research awards to X.W. and J.S. from the Graduate School, New Mexico State University.

REFERENCES

Gronert, K., Virk, S.M. and Herman, C.A. Endogenous sulfidopeptide leukotriene synthesis and 12-lipoxygenase activity in bullfrog (*Rana catesbeiana*) erythrocytes. Biochim. Biophys. Acta. 1255:311 (1995a).

Gronert, K., Virk, S.M. and Herman, C.A. Thrombocytes are the predominant source of endogenous sulfidopeptide leukotrienes in the American bullfrog (*Rana catesbeiana*). Biochim. Biophys. Acta 1259:203 (1995b).

Herman, C.A., Robleto, D.O., Mata, P.L., and Lujan, M.D. Attenuated cardiovascular effects of prostaglandin I_2 and prostaglandin $F_{2\alpha}$ in cold-acclimated American bullfrogs, (*Rana catesbeiana*). J. Exp. Zool. 238:167 (1986).

Herman, C.A., and Martinez, J.M. Epinephrine stimulates prostaglandin synthesis by bullfrog lung from warm-acclimated, but not cold-acclimated animals. J. Exp. Zool. 248:101 (1988).

Herman, C.A., Robleto, D.O., Reitmeier, S.T., Martinez, L.A., and Herman, R.P. Cardiovascular effects of leukotrienes in the bullfrog, *Rana catesbeiana*. Biomed. Biochem. Acta 47:S178 (1988).

Herman, C.A. Charlton, G.A., and Cranfill, R.L. Metabolism and cardiovascular effects of leukotrienes in warm- and cold-acclimated American bullfrogs (*Rana catesbeiana*). Am. J. Physiol. 260:R824 (1991).

Herman, C.A., Goss, D.D., Charlton, G.A., and Pfeifer, C.A. Metabolism and cardiovascular effects of leukotrienes in the marine toad *Bufo marinus*. Gen. Comp. Endocrinol. 92:113 (1993).

Herman, C.A., Luczy, G., Wikberg, J.E.S. and Uhlén, S. Characterization of adrenoceptor types and subtypes in American bullfrogs acclimated to warm or cold temperature. Gen. Comp. Endocrinol. 104:168 (1996).

Sun, J. and Herman, C.A. The American bullfrog (*Rana catesbeiana*) exhibits differential cardiovascular responses to LTC_4 after short and long term cold exposure. Gen. Comp. Endocrinol. 97:199 (1995).

Wang, X. Eicosanoid production during clotting in the American bullfrog (*Rana catesbeiana*). M.S. Thesis, New Mexico State University (1996).

Wang, X. and Herman, C.A. Changes in blood cell number and formation of eicosanoids during clotting in warm- and cold-acclimated bullfrogs (*Rana catesbeiana*). Gen. Comp. Endocrinol. 101:211 (1996).

Wang, X., and Herman, C.A. Eicosanoid synthesis by purified thrombocytes and erythrocytes from warm- and cold-acclimated American bullfrogs (*Rana catesbeiana*). Gen. Comp. Endocrinol. 106:138 (1997).

MOLECULAR AND CATALYTIC PROPERTIES OF MAMMALIAN LIPOXYGENASES COMPARED WITH SOYBEAN LIPOXYGENASE-1

Hiroshi Suzuki and Shozo Yamamoto

Department of Biochemistry, Tokushima University School of Medicine, Tokushima 770, Japan

Soybean lipoxygenases and mammalian lipoxygenases exhibit essentially the same cztalytic properties, and conserve some essential primary structures. However, their physiological functions remain unclarified except for 5-lipoxygenase which is responsible for the biosynthesis of bioactive leukotrienes[1].

CATALYTICALLY ESSENTIAL IRON

Non-heme iron is believed to play a critical role in the lipoxygenase reaction, and almost an equimolar amount of iron has been found in each enzyme protein[1]. Earlier imidazole of histidine and caboxylate of glutamate or aspartate were proposed as ligands for the iron atom in soybean lipoxygenase-I[2], and, as the primary structures were deduced by cDNA cloning of various lipoxygenases, the conservation of several histidine residues was noted among the soybean lipoxygenases[3]. Site-directed mutation of three of the histidines reduced significantly both the enzyme activity and the iron content of porcine 12-lipoxygenase[4] and human 5-lipoxygenase[5]. Furthermore, X-ray crystallography of soybean lipoxygenase-I confirmed the three histidines as the ligands[6]. In addition, the carboxyl terminus isoleucine was proposed to be the fourth ligand, and this finding was confirmed by mutagenesis analysis of mammalian enzymes[7,8].

SUBSTRATE SPECIFICITY

Substrate specificities of various lipoxygenases were studied in detail by our group and others[1]. We noticed that lipoxygenases can be distinguished into two groups. One group has a narrow substrate specificity reacting almost selectively with carbon-20 fatty acids but only poorly with carbon-18 fatty acids. The other group has a broad substrate specificity reacting with both C-18 and C-20 fatty acids. We noted that 12-lipoxygenases of platelets and leukocytes were distinguished by the reactivity with linoleic and linolenic acids[9,10]. We found further differences between platelet and leukocyte 12-lipoxygenases, and concluded that there are two isozymes of 12-lipoxygenase referred to as leukocyte-type and platelet type 12-lipoxygenases[1,9,10].

MULTIFUNCTIONAL NATURE

As shown in Fig. 1 the major product of 12-lipoxygenases from arachidonic acid is 12-hydroperoxy acid (12-HPETE), and they also produce a small amount of 15-hydroperoxy acid (15-HPETE). We found further conversion of the 15-HPETE to dihydroperoxy acids and 14, 15-leukotriene A4, but 12-HPETE remained unchanged[11]. It should be noted that the production of the epoxide did not require a molecular oxygen and the epoxide was the major product from 15-HPETE in the absence of air. Soybean lipoxygenase-I is also a multifunctional enzyme; namely, the enzyme produces mainly 15-HPETE from arachidonic acid, and the same enzyme further oxygenates 15-HPETE into dihydroperoxy acids[12].

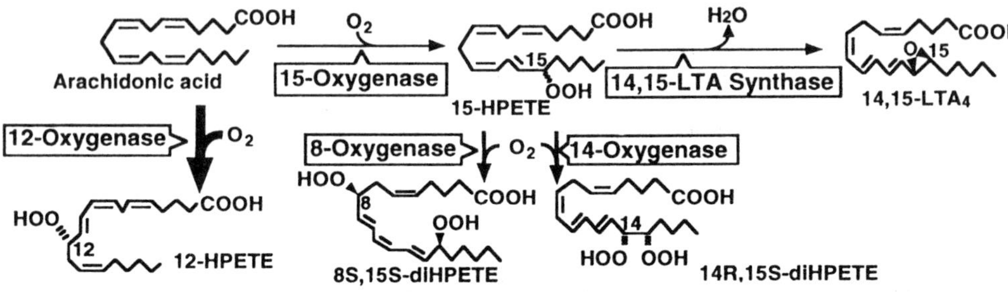

Fig.1 Arachidonate Metabolites by 12-Lipxoygenases

Qualitatively, both leukocyte and platelet 12-lipoxygenase are multifunctional enzymes with arachidonic acid, but the rates of the side reactions were much lower in plaetelet enzyme than the leukocyte enzyme[13]. First, the leukocyte 12-lipoxygenase produced 10 times more 15-HPETE than the platelet enzyme. Secondly, the rate of reaction with 15-HPETE was 10 times higher in the leukocyte enzyme than in the platelet enzyme. Thus, the production of dihydroperoxy acids and 14, 15-leukotriene A4 from arachidonic acid is about 100 times higher in the leukocyte enzyme than the platelet enzyme.

SUICIDE INACTIVATION

As shown in Fig. 2, upon reaction with arachidonic acid, the reactions of leukocyte 12-lipoxygenase and soybean lipoxygenase slowed down rapidly, and stopped within a few minutes. The enzymes were inactivated since the reactions did not start again by the addition of arachidonic acid. This phenomenon is well known as the suicide inactivation of

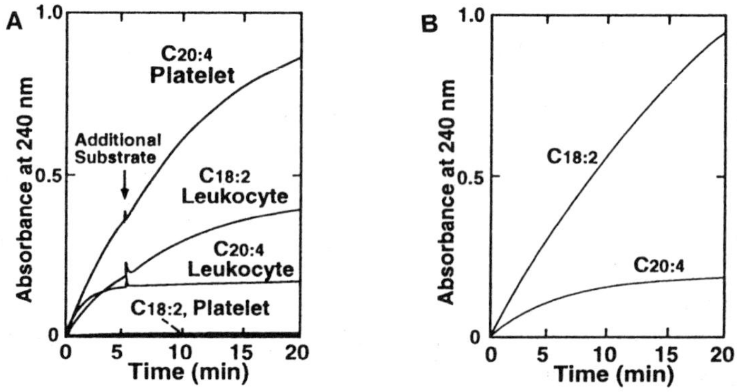

Fig.2 Reaction time courses with arachidonic acid and linoleic acid. (A) 12-lipoxygenases of human platelet and porcine leukocyte. (B) soybean lipoxygenase-I.

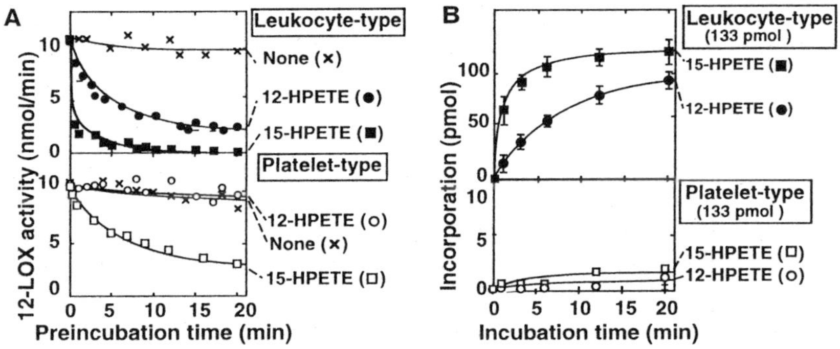

Fig.3. The two types of 12-Lipoxygenase were time-dependently inactivated by the hydroperoxides (A), and the leukocyte-enzyme incorporated the hydroperoxides (B).

lipoxygenases. In sharp contrast, the human platelet 12-lipoxygenase continued to react with arachidonic acid in an almost linear time course for up to 20 min. In contrast to arachidonic acid, such a marked suicide inactivation was not observed in the soybean[14] and leukocyte enzymes[13] reacting with linoleic acid. It should be noted that leukocyte 12-lipoxygenase produces a mixture of hydroperoxides, dihydroperoxides and epoxide from arachidonic acid whereas only 13-hydroperoxide is produced from linoleic acid[11]. In addition, we found a rapid inactivation of the leukocyte 12-lipoxygenase reacting with 15-HPETE. The inactivation of the platelet enzyme was also observed but at a slower rate[13].

In view of such a suicide reaction with 15-HPETE, 12-lipoxygenases were preincubated with various hydroperoxides in the presence or absence of air[13]. After the indicated time intervals, aliquots of the preincubation mixtures were allowed to react with arachidonic acid. As shown in Fig. 3A, 15-HPETE inactivated the leukocyte enzyme very rapidly in about 5 minutes, but 12-HPETE inactivated the enzyme very slowly. In contrast, 15-HPETE inactivated the platelet enzyme slowly, and 12-HPETE did not affect the platelet enzyme activity. Preincubation in the absence of air did not affect the inactivation time course by 15-HPETE, suggesting that the inactivation did not require molecular oxygen. Then, we were interested if the enzyme inactivation was caused by the binding of 15-HPETE to the enzyme protein. 12-Lipoxygenases in an amount of 133 pmol were incubated with isotope labeled hydroperoxides, and radioactivity of the enzyme proteins was determined (Fig. 3B). 15-HPETE was rapidly incorporated into the leukocyte 12-lipoxygenase, and the incorporation reached its plateau in about 5 minutes. Quantitatively, almost an equimolar amount of 15-HPETE was incorporated to the enzyme. Incorporation of 12-HPETE was very slow. In contrast, these hydroperoxides were incorporated into platelet 12-lipoxygenase only in a small amount. The incubation in the absence of air did not affect the 15-HPETE incorporation into the leukocyte enzyme. Moreover, other experiments suggested an almost stoichiometric and covalent binding of 15-HPETE or its derivatives to the leukocyte 12-lipoxygenase protein.

In summary, the leukocyte enzyme produces a considerable amount of 15-HPETE in addition to 12-HPETE from arachidonic acid, and the 15-HPETE or its derivative inactivates the enzyme by covalent binding to the enzyme protein, possibly near the active center. Presumably, 14, 15-leukotriene A4 produced from 15-HPETE is involved in the enzyme inactivation because both the enzyme inactivation and the 15-HPETE incorporation were observed under the anaerobic condition where the 14, 15-leukotriene A4 was the major product from 15-HPETE. In contrast, platelet-type 12-lipoxygenase produces only a small amount of the epoxide and may not be exposed to much epoxide. In agreement with this interpretation, Lepley and Fitzpatrick reported an irreversible inactivation of 5-lipoxygenase by 5, 6 leukotriene A4[15]. In soybean lipoxygenase-I, production of such a leukotriene-type epoxide from arachidonic acid has not been detected[16]. Kim and Sok postulated an unstable reactive derivative of 15-HPETE which has not been identified[17]. We are now attempting experiments using synthetic 14, 15-leukotriene A4 to further evaluate this model.

REFERENCES

1. S. Yamamoto, Mammalian lipoxygenases: molecular structures and functions. *Biochim. Biophys. Acta* 1128: 117-131, (1992).
2. S. Navaratnam, M.C. Feiters, M. Al-Hakim, J.C. Allen, G.A. Veldink, and J.F.G. Vliegenthart, Iron environment in soybean lipoxygenase-1. *Biochim. Biophys. Acta* 956: 70-76, (1988).
3. D. Shibata, J. Steczko, J.E. Dixon, P.C. Andrews, M. Hermodson, and B. Axelrod, Primary structure of soybean lipoxygenase-2. *J. Biol. Chem.* 263: 6816-6821, (1987)
4. H. Suzuki, K. Kishimoto, T. Yoshimoto, S. Yamamoto, F. Kanai, Y. Ebina, A. Miyatake, and T. Tanabe, Site-directed mutagenesis studies on the iron-binding domain and the determinant for the substrate oxygenation site of porcine leukocyte arachidonate 12-lipoxygenase. *Biochim. Biophys. Acta* 1210: 308-316, (1994).
5. Y-Y. Zhang, B. Lind, O. Radmark, and B. Samuelsson, Iron content of human 5-lipoxygenase, effects of mutations regarding conserved histidine residues. *J. Biol. Chem.* 268: 2535-2541, (1993).
6. J.C. Boyington, B.J. Gaffney, and L.M. Amzel, The three-dimensional structure of an arachidonic acid 15-lipoxygenase. *Science* 260: 1482-1486, (1993).
7. X-C. Chen and C.D. Funk, Structure-function properties of human platelet 12-lipoxygenase: chimeric enzyme and in vitro mutagenesis studies. *FASEB J.* 7: 694-701 (1993)
8. H. Suzuki, unpublished observation
9. Y. Takahashi, N. Ueda,, and S. Yamamoto, Two immunologically and catalytically distinct arachidonate 12-lipoxygenases of bovine platelets and leukocytes. *Arch. Biochem. Biophys.* 266: 613-621, (1988).
10. T. Hada, N. Ueda, Y. Takahashi, and S. Yamamoto, Catalytic properties of human platelet 12-lipoxygenase as compared with the enzymes of other origins. *Biochim. Biophys. Acta* 1083: 89-93, (1991).
11. C. Yokoyama, F. Shinjo, T. Yoshimoto, S. Yamamoto, J.A. Oates, and A.R. Brash, Arachidonate 12-lipoxygenase purified from porcine leukocytes by immunoaffinity chromatography and its reactivity with hydroperoxy-eicosatetraenoic acids. *J. Biol. Chem.* 261: 16714-16721, (1986).
12. G.S.Bild, C.S.Ramadoss, S.Lim amd B. Axelrod, Double dioxygenation of arachidonic acid by soybean lipoxygenase-1 *Biochem. Biophys. Res. Commun.* 74: 949-54 (1977)
13. K. Kishimoto, M. Nakamura, H. Suzuki, T. Yoshimoto, S. Yamamoto, T. Takao, Y. Shimonishi, and T. Tanabe, Suicide inactivation of porcine leukocyte 12-lipoxygenase associated with its incorporation of 15-hydroperoxy-5,8,11,13-eicosatetraenoic acid derivative. *Biochim. Biophys. Acta* 1300: 56-62, (1996).
14. W.L. Smith and W. Lands, Oxygenation of unsaturated fatty acids by soybean lipoxygenase. *J. Biol. Chem.* 247: 1038-47 (1972)
15. R.A. Lepley and F.A. Fitzpatrick, Irreversible inactivation of 5-lipoxygenase by leukotriene A4. Characterization of product inactivation with purified enzyme and intact leukocytes. *J. Biol. Chem.* 269: 2627-31 (1994)
16. R.W. Bryant, T. Schewe, S.M. Rapoport and J.M. Bailey, Leukotriene formation by a purified reticulocyte lipoxygenase enzyme. Conversion of arachidonic acid and15-hydroperoxyeicosatetraenoic acid to 14,15-leukotriene. *J. Biol. Chem.* 260: 3548-3555, (1985).
17. M.R. Kim and D.-E. Sok, Irreversible inhibition of soybean lipoxygenase-1 by hydroperoxy acids as substrates. *Arch. Biochem. Biophys* 288: 270-5 (1991)

LEUKOTRIENE A$_4$ HYDROLASE ACTIVITY IN *XENOPUS LAEVIS*

Filippa Strömberg[1], Urban Rosenqvist[2], Sven-Erik Dahlén[2] and Jesper Z. Haeggström[1]

[1]*Department of Medical Biochemistry and Biophysics*
[2]*Department of Physiology and Pharmacology,*
Karolinska Institutet, S-171 77 Stockholm, SWEDEN.

INTRODUCTION

Leukotriene (LT) A_4 hydrolase catalyses the final step in the biosynthesis of the lipid mediator LTB_4 (5*S*,12*R*-dihydroxy-6,14-*cis*-8,10-*trans*-eicosatetraenoic acid), which is a potent leukocyte chemotactic and aggregating agent. In the biosynthesis of leukotrienes arachidonic acid is converted by 5-lipoxygenase into the unstable epoxide intermediate LTA_4 (5*S*-*trans*-5,6-oxido-7,9-*trans*-11,14-*cis*-eicosatetraenoic acid), the common precursor for all leukotrienes [1].

Leukotriene A_4 hydrolase is a soluble zinc metalloenzyme ($M_r \approx 69\ 000$) which possesses an anion dependent peptidase activity towards several synthetic substrates [2]. The enzyme is ubiquitous in mammalian tissues [3-6], even in those apparently devoid of 5-lipoxygenase activity and thus lacking the ability to synthesize the substrate LTA_4.

Previous work has shown that lower vertebrates, e.g., frogs and fish, have the ability to synthesize leukotrienes [7, 8]. Here we report the LTA_4 hydrolase expression in organs of the African claw toad *Xenopus laevis*. An unknown enzymatic metabolite of LTA_4 was found and structurally characterized. This metabolite was biologically active and contracted guinea-pig lung parenchyma (GPLP).

MATERIALS AND METHODS

Female and male *Xenopus laevis* were decapitated. Organs (blood, heart, liver, lung, spleen, kidney, intestine, and the reproductive organs, oocytes and testis) were removed, homogenized in 50 mM Tris-HCl buffer, pH 8, and 2 mM EDTA, and centrifuged at 10 000 x g for 15 min. The resulting supernatants are referred to as organ extracts.

For further meaterials and methods see reference [9].

Recent Advances in Prostaglandin, Thromboxane, and Leukotriene Research
Edited by Sinzinger *et al.*, Plenum Press, New York, 1998

RESULTS AND DISCUSSION

Organ distribution of LTA_4 hydrolase in *Xenopus laevis*

Tissue extracts from nine different organs of *Xenopus laevis* were incubated with LTA_4 and formation of LTB_4 was determined by RP-HPLC. The highest LTA_4 hydrolase activity was observed in the intestine and the reproductive organs (oocytes and testes) followed by lung, spleen, and kidney (Fig. 1). These results agree relatively well with previous reports on organ distribution of LTA_4 hydrolase in the guinea pig [3, 6]. High levels of LTA_4 hydrolase activity has also been observed in rat intestine and lung [4], whereas the enzyme distribution in human organs follows a somewhat different pattern with high levels in lung and liver [5]. Very low enzyme activity was observed in *Xenopus laevis* whole blood.

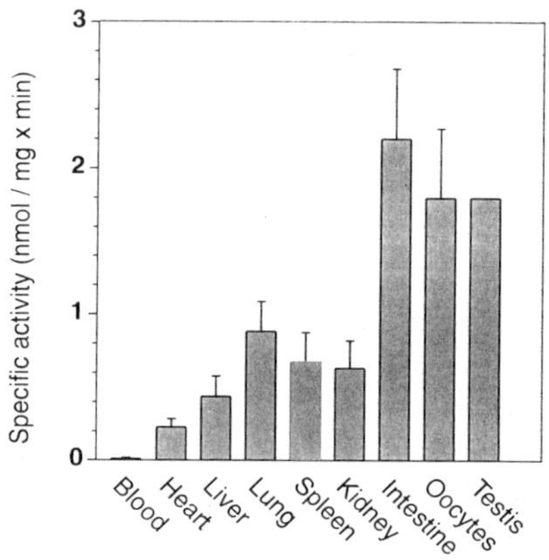

Figure 1. The distribution of LTA_4 hydrolase activity (mean $\pm$ SE, n = 5-7; for testis mean, n = 2)

Incubations of tissue extracts with 4 mM Ca^{2+} and 100 μM arachidonic acid did not reveal any significant formation of 5-lipoxygenase products in any of the organs, as judged by RP-HPLC analysis. Considering the very low activity in whole blood most, if not all, of the LTA_4 hydrolase activity was presumably associated with tissue bound cells. This conclusion was further supported by the insignificant 5-lipoxygenase activity.

Attempts to analyze levels of LTA_4 hydrolase protein and mRNA was made. However, no signals were observed in Western and Northern blots, using antibody and cDNA probes for the human enzyme. Apparently, considerable structural differences exist between the human and the frog LTA_4 hydrolase. Nevertheless, we think that our activity determinations well reflect the amounts of enzyme since earlier studies have shown a good correlation between enzyme activity and presence of LTA_4 hydrolase protein [6].

Enzymatic conversion of LTA_4 into 5S,12R-DiHETE in organs of *Xenopus laevis*

RP-HPLC analysis of organ extracts incubated with LTA_4 revealed a previously unidentified peak eluting after LTB_4, denoted compound X. Compound X was produced in all organs tested and always with the ratio 1:10 compared with LTB_4. Several lines of evidence, *e.g.*, protein-dependence and heat-sensitivity, indicated that this compound is formed via an enzymatic route. From analytical data obtained with RP-HPLC, GC/MS, ultraviolet-induced double-bond isomerization, and comparison with a synthetic standard, the structure of compound X was determined to 5*S*,12*R*-dihydroxy-6,10-*trans*-8,14-*cis*-eicosatetraenoic acid, *i.e.* 5*S*,12*R*-DiHETE.

Compound X contracts the guinea-pig lung parenchyma via a BLT receptor.

Compound X was found to be biologically active and contracted the GPLP by the same mechanism as previously documented for LTB_4 [10]. After the response to compound X had faded, the preparation responded to LTB_4 or other agonists such as histamine (not shown) or the thromboxane receptor agonist U-46,619. In contrast, the preparation was desensitized to a second exposure to compound X (Fig 2a). Pretreatment with the leukotriene B receptor (BLT-receptor) antagonist SC-41930 (10 μM) blocked the contractile effect of both compound X and LTB_4. Pretreatment with indomethacin (10 μM) completely abolished the contractile effect of compound X. This supports an indirect mode of action mediated by the release of a cyclooxygenase product, presumably thromboxane A_2 (TXA_2), as established for LTB_4.

The concentration-response relation for compound X was not perfectly parallel with that of LTB_4 (1 - 1000 nM), but it indicated that compound X was between 10 and 30 times less potent on a molar basis than LTB_4 (Fig. 2b). Although compound X and LTB_4 both were sensitive to an antagonist of the BLT-receptor, the cross-desensitization experiments raise the possibility that distinct recognition sites partly may be involved in initiation of agonist effects by compound X and LTB_4.

The *Xenopus laevis* LTA_4 hydrolase was purifed from oocytes. Preliminary results show that it has the same molecular mass as the mammalian enzyme and hydrolyzes LTA_4

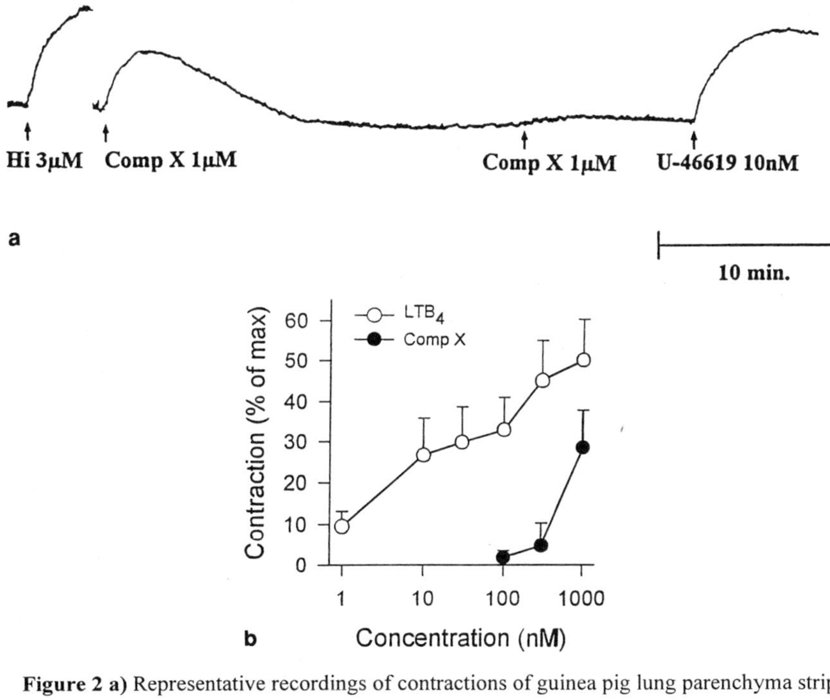

Figure 2 a) Representative recordings of contractions of guinea pig lung parenchyma strips. **b)** Concentration/response relations for compound X (•) and LTB_4 (○).

into both LTB_4 and Compound X (5*S*,12*R*-DiHETE) with the ratio of 1:10. Futher studies of the purified enzyme will reveal whether it possesses a peptidase activity, and if the kinetic properties are different from the mammalian enzyme.

ACKNOWLEDGMENTS

This work was supported by the European Union (BMH4-CT96-0229), the Swedish Medical Research Council (O3X-10350, 14X-9071,14P-11503), *Petrus & Augusta Hedlunds stiftelse, Konung Gustav V:s 80-årsfond* and the Swedish Heart Lung Foundation.

REFERENCES

1 B. Samuelsson, Leukotrienes: mediators of immediate hypersensitivity reactions and inflammation, *Science*, vol. 220, pp. 568-575, (1983).

2. J. Z. Haeggström, A. Wetterholm, J. F. Medina, and B. Samuelsson, Leukotriene A_4 hydrolase: structural and functional properties of the active center, *J. Lipid Mediators*, vol. 6, pp. 1-13, (1993).

3. T. Izumi, T. Shimizu, Y. Seyama, N. Ohishi, and F. Takaku, Tissue distribution of leukotriene A_4 hydrolase activity in guinea pig, *Biochem. Biophys. Res. Commun.*, vol. 135, pp. 139-145, (1986).

4. J. F. Medina, J. Haeggström, M. Kumlin, and O. Rådmark, Leukotriene A_4: metabolism in different rat tissues, *Biochim. Biophys. Acta*, vol. 961, pp. 203-212, (1988).

5. J. Y. Fu, J. Haeggström, P. Collins, J. Meijer, and O. Rådmark, Leukotriene A_4 hydrolase: analysis of some human tissues by radioimmunoassay, *Biochim. Biophys. Acta*, vol. 1006, pp. 121-126, (1989).

6. N. Ohishi, M. Minami, J. Kobayashi, Y. Seyama, J. Hata, H. Yotsumoto, F. Takaku, and T. Shimizu, Immunological quantitation and immunohistochemical localization of leukotriene A_4 hydrolase in guinea pig tissues, *J. Biol. Chem.*, vol. 265, pp. 7520-7525, (1990).

7. F. A. Green, C. A. Herman, R. P. Herman, H. E. Claesson, and M. Hamberg, Monohydroxyeicosatetraenoic acid and leukotriene production by the inflammatory cells of *Xenopus laevis*, *J. Exp. Zool.*, vol. 243, pp. 211-215, (1987).

8. T. R. Pettitt, A. F. Rowley, S. E. Barrow, A. I. Mallet, and C. J. Secombes, Synthesis of lipoxins and other lipoxygenase products by macrophages from the rainbow trout, *Oncorhynchus mykiss*, *J. Biol. Chem.*, vol. 266, pp. 8720-8726, (1991).

9. F. Strömberg, M. Hamberg, U. Rosenqvist, S. E. Dahlén, and J. Z. Haeggström, Formation of a novel enzymatic metabolite of leukotriene A_4 in tissues of Xenopus laevis, *Eur. J. Biochem.*, vol. 238, pp. 599-605, (1996).

10. S. E. Dahlén, P. Hedqvist, P. Westlund, E. Granström, S. Hammarström, J. Å. Lindgren, and O. Rådmark, Mechanisms of leukotriene-induced contractions of guinea pig airways: leukotriene C_4 has a potent direct action whereas leukotriene B_4 acts indirectly, *Acta Physiol. Scand.*, vol. 118, pp. 393-403, (1983).

MONOCYTE BINDING TO ENDOTHELIAL CELLS INDUCED BY OXIDIZED PHOSPHOLIPIDS PRESENT IN MINIMALLY OXIDIZED LOW DENSITY LIPOPROTEIN IS INHIBITED BY A PLATELET ACTIVATING FACTOR RECEPTOR ANTAGONIST

Norbert Leitinger, Andrew D. Watson, Kym F. Faull*, Alan M. Fogelman, Judith A. Berliner

Departments of Medicine and Pathology, Psychiatry & Biobehavioral Sciences and the Neuropsychiatric Institute*, University of California, Los Angeles, CA 90095-1679, USA.

ABSTRACT

Our group has previously demonstrated that oxidized phospholipids derived from mildly oxidized low density lipoprotein (MM-LDL) or oxidized 1-palmitoyl-2-arachidonoyl-*sn*-glycero-3-phosphorylcholine (oxPAPC) increase monocyte-endothelial interactions. There are indications that the effects of these phospholipids are receptor mediated. Oxidized phospholipids have previously been shown to activate cells via the platelet activating factor (PAF) receptor, therefore, we pretreated human aortic endothelial cells (HAECs) with a PAF-receptor antagonist (WEB 2086) prior to treatment with oxPAPC. WEB 2086 inhibited monocyte binding to endothelial cells induced by oxPAPC (200 μg/ml) at concentrations from 1 nM to 10 μM, but had no effect on the induction of monocyte binding induced by lipopolysaccharide (LPS). We were able to isolate and identify several active oxidized phospholipids by combined normal phase high performance liquid chromatography (HPLC) and electrospray mass spectrometry (LC/MS). The induction of monocyte binding to HAECs by two of these partially purified phospholipids was totally abolished by the pretreatment of HAECs with WEB2086 (100 nM). PAF itself, however, when tested at concentrations from 10 nM to 10 μM had no effect on monocyte binding. These results suggest that several of the oxidized phospholipids present in MM-LDL and oxPAPC induce monocyte binding through a receptor which is perhaps distinct from the PAF-receptor, but can be blocked by the PAF-receptor antagonist WEB 2086.

INTRODUCTION

The interaction of monocytes and endothelial cells of the vessel wall is thought to be an initiating event in atherogenesis. It has been shown that mildly oxidized low density lipoproteins (MM-LDL), but not native LDL or highly oxidized LDL stimulate endothelial cells to bind monocytes and induce their subsequent maturation into macrophages.[1,2,3] Oxidized phospholipids in MM-LDL were demonstrated to be the biologically active components and their action could be mimicked by synthetic oxidized 1-palmitoyl-2-arachidonoyl-*sn*-glycero-3-phosphocholine (oxPAPC).[4] Watson et al. were able to determine the chemical structure of two oxidized phospholipids present in MM-LDL and

Recent Advances in Prostaglandin, Thromboxane, and Leukotriene Research
Edited by Sinzinger *et al.*, Plenum Press, New York, 1998

oxPAPC which induced monocyte-endothelial interactions. Another molecule present in MM-LDL and oxPAPC with a mass to charge ratio *(m/z)* 828.6 was also shown to be biologically active.[5]

We were able to isolate these active molecules from MM-LDL and oxPAPC using normal phase HPLC combined with electrospray mass spectrometry (LC/MS) and in this report we show that the stimulation of human aortic endothelial cells to bind human monocytes by two of the three active compounds can be inhibited by the platelet activating factor receptor antagonist WEB 2086.

MATERIAL AND METHODS

Materials

L-α-1-palmitoyl-2-arachidonoyl-*sn*-glycero-3-phosphocholine was obtained from Avanti Polar Lipids Inc., Alabaster, AL or Sigma Chemical Co. (St. Lois, MO). WEB 2086 was kindly provided by Boehringer Mannheim, Germany.

HAECs at passages 4-7 were cultured in Medium 199 supplemented with fetal bovine serum (FBS) as described.[1,3]

Lipid oxidation

PAPC was oxidized by transferring 1 mg in 100 μl of chloroform to a clean glass test tube and evaporating the solvent under a stream of nitrogen. The dried lipid was allowed to oxidize exposed to air for 24 to 48 hours. The extent of oxidation was monitored by positive ion electrospray mass spectrometry.

High performance liquid chromatography

Normal phase HPLC was performed using a preparative column (Adsorbosphere, 250 mm × 10 mm, 5 μm; Alltech Associates, Inc.) at a flow rate of 5.0 ml / min. Phospholipids were applied in chloroform and eluted isocratically with acetonitrile / methanol / water (77:8:15, v/v/v). UV absorbance was detected with a diode array detector (L-3000, Hitachi, Ltd., Tokyo, Japan). Fractions were collected from the column in 1 minute intervals, dried under a stream of nitrogen, and resuspended in tissue culture medium for monocyte adhesion assays.

Electrospray mass spectrometry

An API III triple-quadrupole biomolecular mass analyzer (Perkin-Elmer Sciex Instruments, Norwalk, CT) was used for mass analysis of oxidized phosphpolipids.[5] For flow injection analysis phospholipids were dissolved in acetonitrile / water / formic acid (50:50:0.1, v/v/v) at 1 mg / ml and 20 μl were injected into the ion source of the mass spectrometer at a flow rate of 20 μl / min. The mass spectrometer was set to scan from *m/z* 450 to 950 with an orifice voltage of +65, step size 0.3, dwell time of 3 msec, and a scan speed of ca. 4 sec. Reconstructed selected ion monitoring (RSIM) was performed by software supplied by PE Sciex.

Monocyte adhesion assay

HAECs were incubated with WEB 2086 dissolved in medium 199 + 10% FBS for 1 min prior to incubation with oxidized phospholipids. After 4 hrs of incubation at 37°C the cells were rinsed with medium, a suspension of human monocytes was added for 12 minutes and unbound monocytes were removed. The number of bound monocytes was determined microscopically.

RESULTS

OxPAPC was analyzed by LC/MS. Fractions collected from this analysis were tested on their ability to stimulate HAECs to bind human monocytes. Three fractions significantly

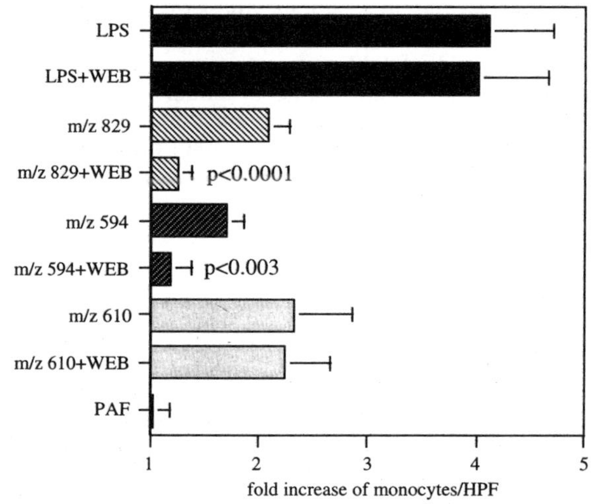

Figure 1. Induction of monocyte binding to HAECs by oxidized phospholipids and inhibition by WEB 2086. Fractions highly enriched in oxidized phospholipids with *m/z* 594.3, *m/z* 610.2, and *m/z* 828.6 were collected after normal phase LC/MS, dried under nitrogen, resuspended in tissue culture medium, and added to HAECs which have been pretreated with WEB 2086 as described in methods. After incubation for 4 hrs at 37°C HAECs were rinsed with medium and human monocytes were added for 12 min. Non-adherent monocytes were removed and bound monocytes were counted by light microscopy. Data are expressed as fold increase ± SD of monocytes / high power field (HPF) above the control. Data were analyzed using model 1 ANOVA.

increased the amount of monocytes bound to HAECs, whereas other fractions were inactive. RSIM of the LC/MS runs showed that molecules with *m/z* of 594.3, *m/z* 610.2, *m/z* 828.6 co-migrated with biological activity.[5] When the HAECs were pretreated with 100 nM of the PAF receptor antagonist WEB 2086 the biological activity associated with two of the three fractions was reduced to control levels (data not shown). The third fraction associated with biological activity, however, seemed not to be affected by pretreatment with WEB 2086.

In subsequent experiments the active compounds (*m/z* 594.3, *m/z* 610.2, *m/z* 828.6) were isolated from oxPAPC and the ability of different concentrations of WEB 2086 to inhibit their action on inducing monocyte adhesion to HAECs was tested. The action of the compounds with *m/z* 594.3 and 828.6 isolated from 3 mg oxPAPC could be inhibited by pretreatment of HAECs with 10 µM, 100 nM and 1 nM WEB 2086, while the activity of the compound with *m/z* 610.2 was not affected by the addition of WEB 2086 at these concentrations.

The ability of WEB 2086 (100 nM) to inhibit monocyte binding induced by *m/z* 594.3, *m/z* 610.2, and *m/z* 828.6 is illustrated in figure 1. WEB 2086 itself had no influence on monocyte binding at concentrations from 1 nM to 10 µM (data not shown). LPS at a concentration of 1 ng/ml, used as a positive control in these experiments, was not affected by WEB 2086 at the same concentrations. PAF itself at 100 nM (or as high as 10 µM), however, had no effect on inducing monocyte binding to HAECs.

DISCUSSION

There is evidence that the induction of monocyte adhesion to endothelial cells by MM-LDL is a receptor mediated event. It has been shown that MM-LDL stimulates G-protein linked responses on endothelial cells suggesting its interaction with a G-protein coupled receptor. Furthermore, it has been demonstrated that cyclic AMP is increased in endothelial cells after treatment with MM-LDL.[6] In addition, the intracellular signalling pathway induced by MM-LDL involves activation of the nuclear transcription factor NFκB. However, the plasma membrane receptor for MM-LDL or its active compounds on endothelial cells remains unknown.

In this study we demonstrate that biologically active phospholipids present in MM-LDL induce monocyte binding to endothelial cells and that the activity of two out of three active compounds can be blocked by a PAF receptor antagonist (WEB 2086). However, the presence of a PAF receptor on HAECs has not been demonstrated and is a matter of debate. Furthermore, PAF itself was not able to induce monocyte binding to endothelial cells in our system. These results suggest, that oxidized phospholipids present in MM-LDL and oxPAPC act via a receptor distinct from the PAF receptor .

REFERENCES

1. J.A. Berliner, M.C. Territo, A. Sevanian, S. Ramin, J.A. Kim, B. Bamshad, M. Esterson, and A.M. Fogelman, Minimally modified low density lipoprotein stimulates monocyte endothelial interactions, J. Clin. Invest. 85:1260 (1990).
2. S.D. Cushing, J.A. Berliner, A.J. Valente, M.C. Territo, M. Navab, F. Parhami, R. Gerrity, C.J.Schwartz, and A.M. Fogelman AM, Minimally modified low density lipoprotein induces monocyte chemotactic protein 1 in human endothelial cells and smooth muscle cells, Proc. Natl. Acad. Sci. USA 87:5134 (1990).
3. T.B. Rajavashisth, A. Andalibi, M.C. Territo, J.A. Berliner, M. Navab, A.M. Fogelman, and A.J. Lusis, Induction of endothelial cell expression of granulocyte and macrophage colony-stimulating factors by modified low density lipoproteins, Nature 344:254 (1990).
4. A.D. Watson, J.A. Berliner, S.Y. Hama, B.N. La Du, K.F. Faull, A.M. Fogelman, and M. Navab M, Protective effect of high density lipoprotein associated paraoxonase. Inhibition of the biological activity of minimally oxidized low density lipoprotein, J. Clin. Invest. 96:2882 (1995).
5. A.D. Watson, N. Leitinger, M. Navab, K.F. Faull, S. Hörkkö, J.L. Witztum, W. Palinski, D. Schwenke, R.G. Salomon, S. Wei, A.M. Fogelman, and J.A. Berliner, Structural identification by mass spectrometry of oxidized phospholipids in minimally oxidized low density lipoprotein that induce monocyte/endothelial interactions and evidence for their presence in vivo, J. Biol. Chem. in press.
6. F. Parhami, Z.T. Fang, A.M. Fogelman, A. Andalibi, M.C. Territo, and J.A. Berliner, Minimally modified low density lipoprotein-induced inflammatory responses in endothelial cells are mediated by cyclic adenosine monophosphate, J. Clin. Invest. 92:471 (1993).

REGULATION OF 5-LIPOXYGENASE ACTIVITY BY SELENIUM-DEPENDENT PEROXIDASES - EFFECTS OF TRANSFORMING GROWTH FACTOR-BETA AND 1,25-DIHYDROXYVITAMIN D_3

Oliver Werz and Dieter Steinhilber

Institute of Pharmaceutical Chemistry, Johann Wolfgang Goethe-University Frankfurt, Marie-Curie-Str. 9, D-60439 Frankfurt, Germany

INTRODUCTION

The 5-lipoxygenase (5-LO) pathway of myeloid cells is upregulated during cell maturation. Using the in-vitro differentiation of human leukemic cell lines as a model for myeloid cell maturation, it was previously found that differentiation of HL-60 cells by dimethylsulfoxide (ME_2SO) or Mono Mac 6 cells with either 1,25-dihydroxyvitamin D_3 (VD3) or transforming growth factor-ß (TGFß) induces 5-lipoxygenase protein expression but only low cellular 5-lipoxygenase activity. Similarly, primary B-lymphocytes or EBV-transformed B-lymphocytic cell lines express 5-lipoxygenase protein and show enzymatic activity in cell homogenates but not in intact cells. We now report that selenium-dependent peroxidases are responsible for suppression of cellular 5-LO activity in B-lymphocytes and immature myeloid cells and that VD3 and TGFß upregulate cellular leukotriene synthesis by induction of peroxidase-insensitive 5-LO catalytic activity.

METHODS

Cell culture, isolation of human granulocytes, determination of 5-lipoxygenase activity, reverse transcription and PCR analysis were performed as described (Werz and Steinhilber, 1996).

RESULTS AND DISCUSSION

ME_2SO-differentiated HL-60 cells, cultured in FCS-medium, show low cellular 5-LO activity compared to the corresponding homogenates (fig. 1). Addition of VD3 and TGFß during differentiation leads to a strong upregulation of 5-LO activity in intact cells without concomitant changes in the corresponding homogenates. After differentiation of HL-60

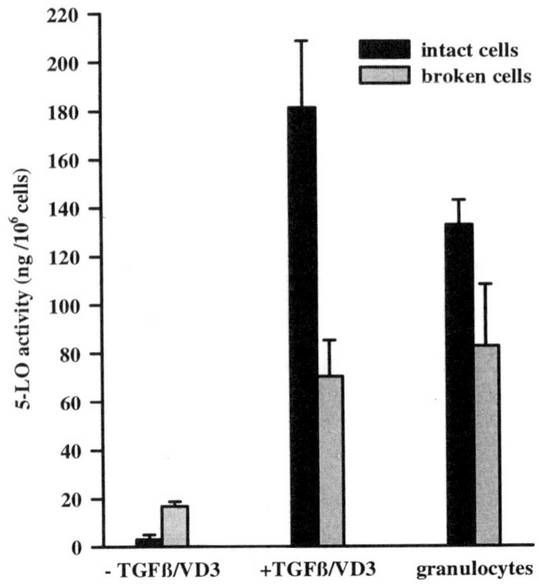

Figure 1. 5-LO activity of differentiated HL-60 cells and human granulocytes. HL-60 cells were differentiated with 1.5% ME_2SO with or without TGFß (5 ng/ml) and vitamin D3 (300 pM) in FCS-medium. After four days, 5-LO activity of intact cells and homogenates of HL-60 cells and freshly isolated granulocytes was determined. Results are expressed as mean + S.E. of three independent experiments.

cells in the presence of TGFß/VD3, 5-lipoxygenase activity of intact cells and cell homogenates is comparable to human granulocytes (fig. 1). The suppression of cellular 5-lipoxygenase activity in HL-60 cells differentiated with ME_2SO only and in the B-lymphocytic cell line BL41-E95A disappeared when the cells were transferred into serumfree medium. Then, cellular 5-LO activity was upregulated and both cell lines showed comparable 5-LO activity in intact cells and broken cell preparations, respectively (Werz and Steinhilber, 1996). Selenium mimics the inhibitory effects of serum. When selenium (at 0.1 or 1 ng/ml) was added to the culture medium for one passage, it strongly reduced cellular 5-LO activity in BL41-E95-A and in HL-60 cells, differentiated in the absence of VD3/TGFß. This inhibitory effect of selenium on cellular 5-LO activity was comparable to the reduction of 5-LO activity by FCS. However, no inhibition of 5-lipoxygenase activity or only marginal effects by selenium were observed on 5-LO activity in the corresponding cell homogenates under our standard assay conditions, i.e. in the absence of thiols like glutathione (GSH) or dithiothreitol (DTT). Addition of DTT (1mM) or GSH (5mM) slightly stimulated 5-LO activity of HL-60 cell homogenates when the cells were grown in the absence of selenium (fig. 2). However, different results were obtained when the cells were grown in selenium-containing medium one passage before the experiments were started. Then, 5-LO activity of cell homogenates derived from cells with low cellular 5-LO activity (grown in the absence of TGFß/VD3) was inhibited at millimolar concentrations of thiols while homogenates derived from cells with high cellular 5-LO activity (grown in the presence of TGFß/VD3) were not

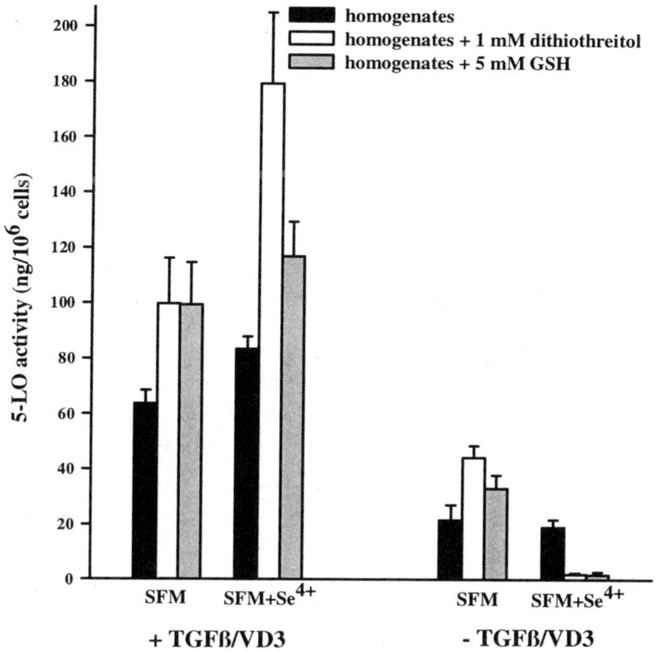

Figure 2. Influence of glutathione and dithiothreitol on 5-LO activity in cell homogenates of HL-60 cells. HL-60 cells were induced to differentiate with 1.5% ME_2SO with or without TGFß (5 ng/ml) and vitamin D3 (300 pM) in serum-free medium in the presence or absence of selenium (1 ng/ml). After four days, cells were harvested and 5-LO activity of the cell homogenates was determined in the presence of the indicated concentrations of thiols. Results are expressed as mean + S.E. of three independent experiments.

affected (fig. 2). The selenium-dependent effects of thiols suggested that these compounds inhibit 5-LO as cofactor to selenium-dependent peroxidases which represent well known inhibitors of 5-lipoxygenase activity. Selenium dependence, molecular weight, sensitivity to iodacetate, 5-LO inhibition in the presence of mercaptosuccinate and the utilization of DTT as substrate point to a selenoenzyme with properties of the phospholipid hydroperoxide GSH-peroxidase (GPx-4) as inhibitor of 5-LO activity (see (Werz and Steinhilber, 1996)).

Inhibition of cellular 5-LO activity by selenium was not observed when the cells were differentiated in the presence of VD3/TGFß (data not shown). Also, 5-lipoxygenase activity in homogenates or 100.000 × g supernatants of VD3/TGFß-differentiated HL-60 cells was not inhibited by DTT or GSH (fig. 2). From these experiments, it can be concluded that differentiation of HL-60 cells in the presence of TGFß/VD3 leads to protection of 5-LO activity against peroxidases or decreases the expression of selenium-dependent peroxidases. In order to test both hypotheses, 100,000 × g supernatants of undifferentiated Mono Mac 6 cells which are a rich source for GSH-dependent peroxidase activity but which do not express 5-lipoxygenase protein (Claesson *et al.*, 1993) were added to 100,000 × g supernatants of BL41-E95-A cells and neutrophils and 5-LO activity was determined in the presence of 0.5 mM ß-mercaptoethanol.

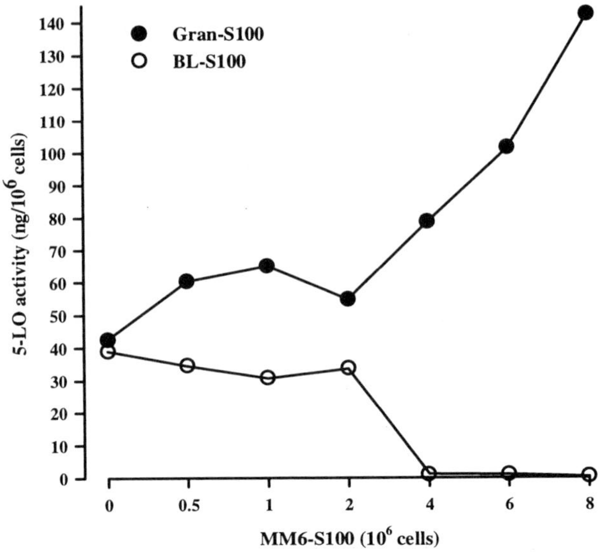

Figure 3. Effects of Mono Mac 6 cell-derived selenium-dependent peroxidases on 5-LO activity of granulocytes and BL41-E95-A supernatants. BL41-E95-A and Mono Mac 6 cells were grown in FCS-medium for 4 days. The indicated aliquots of 100,000 × g supernatants from undifferentiated Mono Mac 6 cells were added to 100.000 × g supernatants of granulocytes and BL41-E95-A cells. Then, 5-LO activity of the granulocyte and BL41-E95-A supernatants was determined in the presence of ß-mercaptoethanol (0.5 mM), respectively. The figure shows a representative experiment out of four titrations.

As can be seen from fig. 3, 5-LO activity of neutrophil supernatants was stimulated by Mono Mac 6 supernatants, whereas the enzyme activity of BL41-E95A supernatants was inhibited. The data suggest that the high 5-LO activity in granulocytes is due to stability of 5-LO catalytic activity against peroxidases but not to low peroxidase activity. Similar results were obtained when 100,000g supernatants of BL41-E95A cells and granulocytes were mixed (Werz and Steinhilber, 1996). The peroxidase-insensitive 5-LO catalytic activity could be explained by the existence of an activation mechanism that facilitates ferric enzyme generation in the presence of low peroxide concentrations. It is possible that VD3/TGFß upregulate cellular 5-LO activity during myeloid cell maturation by the induction of one or more components that are involved in the conversion of the 5-LO enzyme into the ferric state during cell activation and that enable 5-LO activation within the reducing environment inside the cell.

REFERENCES

Claesson, H.-E., Jakobsson, P.-J., Steinhilber, D., Odlander, B. and Samuelsson, B., 1993, Expression of 5-lipoxygenase and biosynthesis of leukotriene B4 in human monomorphonuclear leukocytes. *J Lipid Mediators* 6:15.

Werz, O. and Steinhilber, D., 1996, Selenium-dependent peroxidases suppress 5-lipoxygenase activity in B-lymphocytes and immature myeloid cells - the presence of peroxidase-insensitive 5-lipoxygenase activity in differentiated myeloid cells. *Eur. J. Biochem.* 242:90.

THROMBOXANE A_2 POTENTIATES THROMBIN-INDUCED PROLIFERATION OF CORONARY ARTERY SMOOTH MUSCLE CELLS

Tom-Philipp Zucker[1], Detlef Bönisch[2], Stephanie Muck[2], Artur-Aron Weber[2], Ellen Bretschneider[3], Erika Glusa[3], and Karsten Schrör[2]
Institute für Klinische Anaesthesiologie[1] and Pharmakologie[2] der Heinrich-Heine-Universität Düsseldorf, Germany, [3] Forschungszentrum für Vaskuläre Biologie, Friedrich-Schiller-Universität Jena, Erfurt, Germany

SUMMARY

The activation of thrombin is the key event in clot formation after vascular injury. Thrombin itself, but also other clot-derived factors, such as thromboxane A_2 (TXA_2), are mitogenic for vascular smooth muscle cells. We have studied the possible interactions between thrombin and TXA_2 in stimulation of coronary artery smooth muscle cell (SMC) proliferation. Thrombin (1 U/ml) caused a significant proliferatory response in SMC. U 46619, a stable TXA_2 mimetic, had only a minor stimulating effect by its own but markedly potentiated the thrombin-induced mitogenesis. A possible mechanism for these potentiating effects is provided by the demonstration of a marked (6fold) but transient (maximum after 20 min) increase in the expression of TXA_2 receptor (TP receptor) mRNA in SMC by thrombin. Since a significant clot-related TXA_2 generation was detected for at least 2 hours, the up-regulation of TP receptors by thrombin may represent a mechanism that is relevant for the in vivo situation of SMC proliferation after vessel injury.

METHODS

Cell culture: Bovine coronary artery smooth muscle cells (SMC) were isolated enzymatically from the left anterior descending coronary artery as previously described (Grosser et al., 1995). The cells were grown in Ham's F12 medium supplemented with Dulbecco's modified Eagle medium (20 %), fetal calf serum (15 %), penicillin (100 U/ml), and streptomycin (0.1 mg/ml) in a humidified atmosphere of 5 % CO_2 and 95 % air at 37°C. Passages 4-10 were used for the experiments.

DNA synthesis and cell proliferation: DNA synthesis was assessed by the measurement of [^{3}H]thymidine incorporation as previously described (Grosser et al., 1997). In addition, the incorporation of 5-bromo-2'-deoxyuridine (BrdU) was determined using a cell proliferation kit (BrdU labeling and detection kit III, Boehringer, Mannheim, Germany). For the measurement of cell proliferation, smooth muscle cells were grown to about 50 % confluence, then serum-deprived for 24 h, and stimulated. After 48 h, cells were trypsinized and total cell number was assessed by counting the cells per well using a hemocytometer.

Clot-dependent TXB_2 formation: Plasma clots were prepared from citrated platelet-rich plasma as previously described (Hauptmann & Glusa, 1995). Clot-derived TXB_2 was measured by radioimmunoassay (Schrör & Seidel, 1988).

TXA₂ receptor mRNA: TXA_2 receptor mRNA was amplified by means of RT-PCR with primers corresponding to the putative fourth and sixth transmembrane domain of the bovine TXA_2 receptor (EMBL genebank U 53484/U 52485). The PCR products were quantified by automated laser fluorescence using a 238-bp-GAPDH-fragment as an internal standard.

Statistics: Data are mean ± SEM of *n* independent experiments. Group differences were calculated by two-tailed Student's t-test for unpaired or paired samples as applicable. P values < 0.05 were considered significant.

RESULTS AND DISCUSSION

Incubation of coronary artery smooth muscle cells (SMC) with the stable TXA_2 mimetic U 46619 for 24 h did not significantly increase DNA synthesis ($p > 0.05$). In contrast, thrombin (1 U/ml) induced a significant increase in [^{3}H]thymidine incorporation. This thrombin-induced DNA synthesis was markedly potentiated by U 46619 (Fig. 1 A). The measurement of [^{3}H]thymidine incorporation was validated by the determination of 5-bromo-2'-deoxyuridine (BrdU)-incorporation into cellular DNA. Similarly, U 46619 did not significantly increase the number of BrdU-positive nuclei as compared to unstimulated controls. However, thrombin-induced BrdU-incorporation was significantly potentiated by the thromboxane mimetic (not shown). In order to determine if the observed effects on DNA synthesis also result in a further progression of the cell cycle through the M-phase, the cell number was quantified. 24 h after stimulation, no increase in cell number was observed by any of the agonists studied. However, in accordance with the data on DNA synthesis, after 48 h, no cell proliferation was observed with U 46619 but there was a significant increase in cell number by thrombin that was further potentiated by U 46619 (Fig. 1 B).

To address the question whether the potentiation of thrombin-induced mitogenesis by U 46619 is mediated by TP receptors, the effects of the selective thromboxane antagonist SQ 29,548 (1 μM) were studied. The thromboxane receptor antagonist had no effect on basal or thrombin-induced DNA synthesis. However, SQ 29,548 completely blocked the potentiation of thrombin-induced mitogenesis by U 46619.

Since the potentiation of thrombin-induced mitogenesis by U 46619 appears to be TP receptor-mediated, one possible mechanism for the potentiation effects would be an up-regulation of vascular TP receptors by thrombin, as it has recently been shown to occur in platelets (Modesti et al., 1995). We have, therefore, studied the possible effects of thrombin on TP receptor mRNA expression by RT-PCR. Stimulation of SMC with thrombin (1 U/ml) resulted in a significant enhancement of TP receptor mRNA. A marked (6fold) but transient (maximum after 20 min) increase in the expression of TP receptor mRNA was seen.

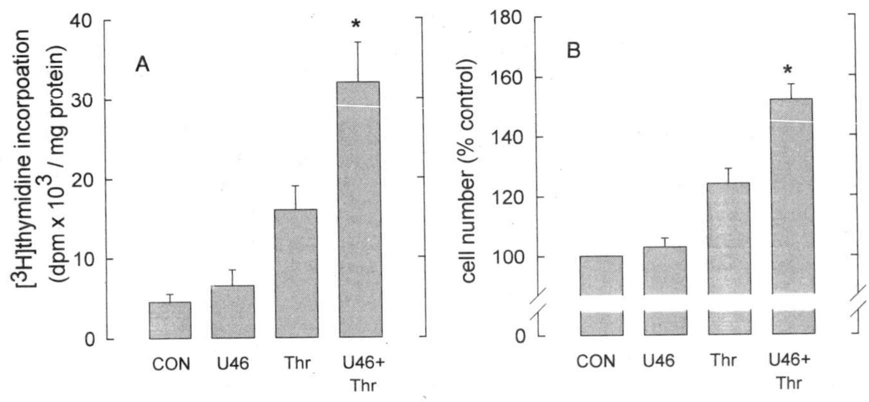

Figure 1. Potentiation of thrombin (Thr, 1 U/ml)-induced [^{3}H]thymidine incorporation by U 46619 (U46, 3 μM). Data are mean ± SEM of n = 5 experiments. * $p < 0.05$ vs. thrombin alone.

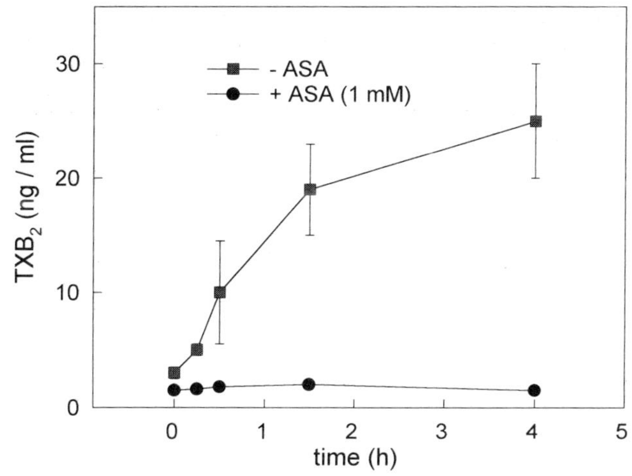

Figure 2. Time course of thromboxane A_2 generation (measured as TXB_2) in platelet-rich clots in the absence (open circles) or presence (closed circles) of aspirin (1 mM). The data are mean ± SEM of n=3 independent experiments.

A significant, aspirin-sensitive, thromboxane A_2 generation from platelet-rich clots in vitro was detected for at least 2 h after stimulation with thrombin (Fig. 2). Thus, the thrombin-induced up-regulation of TP receptors may represent a mechanism that is relevant for the in vivo situation of SMC proliferation after vessel injury.

The relationship between thrombin-induced clot formation and mitogenesis is a key issue in the pathophysiology of thrombin actions. A thrombus contains numerous peptide and non-peptide growth factors. We have shown that a sustained thromboxane A_2 generation occurs in a platelet-rich clot. In addition, enzymatically active thrombin is found within a clot that escapes inactivation by circulating antithrombins (Nadir et al., 1995). This thrombin further stimulates platelets or the clotting cascade and remains active over hours (Sarembock et al., 1991).

The major finding of our study was the demonstration of synergistic effects of thrombin and thromboxane A_2 on SMC mitogenesis that probably involves an up-regulation of vascular TP receptors by thrombin. Thus, thrombin-induced SMC proliferation appears to result from a combined, direct and indirect, i.e. thromboxane A_2-dependent, effect. These findings give further support for the use of TP receptor antagonists in combination with thrombin inhibitors for prevention of SMC proliferation after vessel injury, e.g. after balloon angioplasty.

REFERENCES

T. Grosser, D. Bönisch, T.-P. Zucker, and K. Schrör, Iloprost-induced inhibition of proliferation of coronary artery smooth muscle cells is abolished by homologous desensitization, *Agents Actions Suppl.* 45:85 (1995).

T. Grosser, T.-P. Zucker, A.-A. Weber, K. Schulte, A. Sachinidis, H. Vetter, and K. Schrör, Thromboxane A_2 induces cell signaling but requires platelet-derived growth factor to act as a mitogen, *Eur J Pharmacol.* 319:352 (1997).

J. Hauptmann, and E. Glusa, Differential effects of staphylokinase, streptokinase and tissue-type plasminogen activator on the lysis of retracted human plasma clots and fibrinolytic plasma parameters in vitro, *Blood Coagul Fibrinolysis.* 6:579 (1995).

P.A. Modesti, A. Colella, I. Cecioni, A. Costoli, D. Biagini, A. Migliorini, and G.G. Neri Serneri, Increased number of thromboxane A_2-prostaglandin H_2 platelet receptors in active unstable angina and causative role of enhanced thrombin formation, *Am Heart J.* 129:873 (1995).

A.M. Nadir, G. Villarreal-Levy, and A.I. Schafer, The role of thrombin and thrombin inhibitors in coronary angioplasty, *Chest.* 108:1409 (1995).

I.J. Sarembock, S.D. Gertz, L.W. Gimple, R.M. Owen, E.R. Powers, and W.C. Roberts, Effectiveness of recombinant desulphatohirudin in reducing restenosis after balloon angioplasty of atherosclerotic femoral arteries in rabbits, *Circulation.* 84:232 (1991).

K. Schrör, and H. Seidel, Blood vessel arachidonate metabolism and its pharmacological modification in a new in vitro assay system, *Naunyn-Schmiedeberg's Arch Pharmacol.* 337:177 (1988).

ISOFORMS OF TYPE-I 15-HYDROXYPROSTAGLANDIN DEHYDROGENASE mRNA

Régis Delage-Mourroux, Mélinée Frenkian, Franck Pichaud, Jean Louis Frendo, Elisabeth Pidoux, and Annick Jullienne

INSERM U. 349, Hôpital Lariboisière
centre Viggo Petersen, 75475 Paris
cedex 10, France

INTRODUCTION

The prostaglandins are members of the eicosanoid family and derived from arachidonic acid. These local hormones coordinate responses to circulating factors which regulate their biosynthesis. They are involved in many physiological and pathological processes including hemostasis, thrombosis, parturition, gastrointestinal secretion, smooth muscle tone and inflammation.

The prostaglandins lose in part their biological activity when they undergo a reversible oxidation of the 15-hydroxyl group which leads to a 15-keto metabolite. The enzyme responsible of this initial step of prostaglandin degradation is the NAD^+-dependent 15-hydroxyprostaglandin dehydrogenase (type-I 15-PGDH) (Änggard and Samuelson, 1964). The type-I 15-PGDH is an ubiquitous enzyme in mammalian tissues and cells and this enzyme is particularly found in the human placenta, lung and kidney (Ensor et al., 1995).
In 1990, Krook (1990) established the amino acid (aa) sequence of this enzyme. Simultaneously, the cDNA nucleic sequence was established from a human placental library (Ensor et al., 1990). The 266 aa open reading frame (ORF) confirmed the established amino acid sequence (Krook et al., 1990). Northern blot analysis of placenta RNA with a specific type-I 15-PGDH probe indicated the presence of two RNA species of 2.0 kb and 3.4 kb (Ensor et al., 1990). These authors suggested the possible existence of alternative splicing.

We reported the expression in several cell lines and tissues (placenta, liver) of a mRNA coding for a predicted isomer of type-I 15-PGDH (Pichaud et al., 1995) here by referred to as PGDHrI mRNA. Of particular interest was the high expression of this enzyme and its mRNA in the TT cell line which is derived from medullary thyroid carcinoma (MTC) (Leong et al., 1981), a human tumour existing in familial and sporadic forms. Further investigations on the expression of this isomer and type-I 15-PGDH in the TT cell line and in the MTC demonstrated the presence of a third mRNA (PGDHrII mRNA) (Delage-Mourroux et al., 1997) which also predicts another type-I 15-PGDH related protein.

PGDHrI mRNA

The nucleic acid sequence of the 15-PGDH rI mRNA was reported in HL-60 cells (Pichaud et al., 1995). The full length sequence is identical to type-I 15-PGDH, except for

the lack of 164 bp, leading to the appearance of a TGA stop codon at the position 724, due to a phase shift. This sequence presents an open reading frame coding for a predicted protein of one hundred and seventy eight amino acids. The predicted isomer PGDHrI contains all the strictly conserved residues implicated in NAD^+ binding site and active site. The predicted isomer contains the amino acids strictly conserved in all short chain dehydrogenases (Persson et al., 1991). They correspond to three glycines at positions 12, 18, and 131, Asp64, Tyr151 and Lys155 in type I 15-PGDH (Figure 1).
In order to test the stability and the presumed activity of the 15-PGDHrI protein, we have cloned the coding sequence (from ATG to TGA) in frame with a nucleic sequence coding for a 13 kD peptide biotinylated in E. coli. The recombinant protein was purified on an avidin affinity column and we then tested the ability of this new protein to convert PGE_2 into 15-Keto PGE_2 using thin layer chromatography experiments. The recombinant protein was detected as both a monomer (21kD) and a dimer (40kD) after purification from E. coli. Subsequently we tested the ability of this new protein to convert PGE_2 into 15-Keto PGE_2 with NAD^+ as cofactor since this protein retained all the strictly conserved amino acids found in the short chain alcohol dehydrogenases. This C-terminal truncated recombinant 15-PGDHrI did not convert PGE_2 into 15-Keto metabolite (Pichaud et al., 1997b).

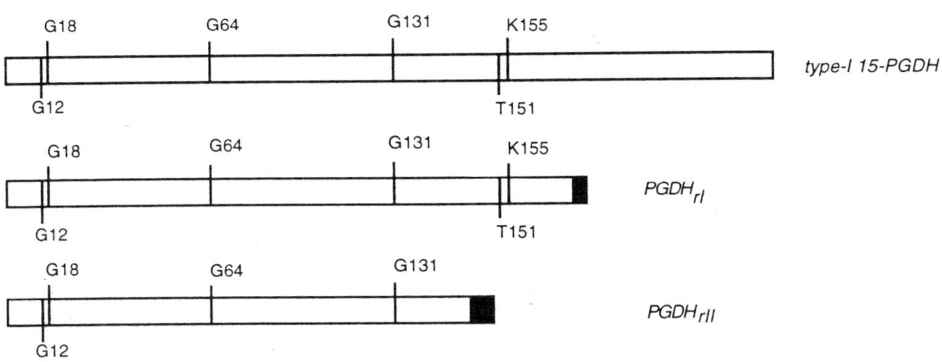

PGDHrII mRNA

In the course of our investigations on the expression of type-I 15-PGDH by the TT cell line and by MTC, we detected a third mRNA species (PGDHrII) which also codes for another 15-PGDH isomer (Delage-Mourroux et al., 1997). The 583 nt mRNA presented an ORF coding a predicted protein of 143 amino acids. We therefore used ribonuclease protection assay to detect and demonstrate the expression of this third mRNA in TT-cells. We used a full length RNA probe specific for the isomer (PGDHrII). This procedure clearly established that this isomer was expressed in quantities similar to the other two forms (Delage-Mourroux et al., 1997).
The predicted protein is identical to the first 140 aa of placental type-I 15-PGDH. The last 3 C-terminal aa differ from the known sequence due to a phase shift. The predicted isomer II contains all the highly conserved residues with the exception of tyrosine 151 and lysine 155. The tyrosine (151) and lysine (155) are the most likely candidates for active site residues (Ensor et al., 1991; Ensor et al., 1994). These authors used site-directed mutagenesis, to demonstrate the role the conserved tyrosine and lysine played in the activity of the enzyme. The N-terminal part presents a chain of three glycine residues (Gly12, Gly16, Gly 18) which could form the adenine binding fold of the NAD^+ binding site (Wierenga et al., 1986). In truncated PGDHrII, amino acids necessary for the nicotinamide binding site are present (Figure 1). This isomer, if effectively expressed in cells, could compete with active type-I 15-PGDH. Western blot of TT cells extracts using a monoclonal antibody specific of type-I

15-PGDH, revealed a band at 29 kD indicating the presence of type-I 15-PGDH. The two other forms (PGDHrII and PGDHrI) were not detected by the antibody used. We also used thin layer chromatography to test the activity of type-I 15-PGDH in TT cells. Our results showed that a PGDH activity exists in TT cells.
Our results indicated that PGDHrII mRNA was present in medullary thyroid carcinoma, normal liver, TT cells but not in breast carcinoma cell lines (T47D and MCF-7), placenta samples, and normal thyroid (Delage-Mourroux et al., 1997).

SPLICING OF THE PGDH GENE

To determine the origin of these different isomers, we characterized the genomic sequence of 15-PGDH gene. We generated a 1.1 kb genomic DNA probe by using PCR and type-I 15-PGDH-specific primers on genomic DNA. This fragment contains exons situated in the 3' sequence of type-I 15-PGDH and its highly related isomers. This work on the 3' region of human type-I 15-gene indicated the existence of at least two different splicing pathways for the type-I 15-PGDH gene (Pichaud et al., 1997a). We used this 1.1 kb genomic DNA probe to localize the 15-PGDH gene. The results established that the human PGDH gene is located on the chromosome bands 4q34-4q35 (Pichaud et al., 1997a).

ROLE OF PGDH ISOMERS

Western blot established that type-I 15-PGDH protein is expressed in TT cells and in HL-60 cells while the two other predicted protein species were not detected. This could be due to the absence of recognition of the two predicted proteins by the monoclonal antibody used. Alternatively it could be due to the lack of protein expression. These isomers, if effectively expressed in cells, could compete with active type-I 15-PGDH. The recombinant protein PGDHrI has the possibility to form a dimer. Of particular interest could be the potentiality to form inactive heterodimers of 15-PGDH, leading to a post-transcriptional regulation of the 15-PGDH activity, or a modulation of the substract specificity of this enzyme. The apparent high expression of this 15-PGDH related mRNA in medullary thyroid carcinoma could be of pathological significance as preliminary results indicate that in these tumours cyclooxygenase is strongly expressed. In situ hybridization with ^{33}P labelled probe specific for 15-PGDH demonstrated that 15-PGDH mRNA is expressed by the tumoral cells. The lack of optimal 15-PGDH activity in this tumour, could lead to high PGs levels. Prostaglandins play a role in the growth and metastasis of tumours (Honn et al., 1981).

CONCLUSION

In conclusion our results indicate that at least two different mRNA isomers of type-I 15-PGDH, the key enzyme in prostaglandins degradation are expressed in human tissues and that the pattern of expression of these mRNA isomers could differ in normal and pathological tissues. Elucidation of the sequence of these isomers opens the way to the study of the specific enzymatic activities of these forms and to their tissue distribution.

Development of specific antibodies against these predicted proteins will be necessary to establish if they are expressed and if they have enzymatic activities.

Acknowledgments

R.D.M is a recipient of an individual grant from the "Ligue Nationale contre le Cancer". F.P. and J.L.F. are the recipients of an individual grant from the French Ministry of Research and Education. This work was supported by the « Association pour la Recherche sur le Cancer », grant N°1359.

REFERENCES

Änggard, E., 1966, The biological activities of three metabolites of prostaglandin E1. *Acta. Physiol. Scand.* 66:509.

Änggard, E., and Samuelsson, B., 1964, Metabolism of prostaglandin E1 in guinea pig lung: the structure of two metabolites, *J. Biol. Chem.* 239:4097.

Chavan, A.J., Ensor, C.M., Wu, P., Haley, B.E., and Tai, H.H., 1993, Photoaffinity labeling of human placental NAD(+)-linked 15-hydroxyprostaglandin dehydrogenase with [alpha-32P]$_2$N$_3$NAD$^+$. Identification of a peptide in the adenine ring binding domain, *J. Biol. Chem.* 268:16437.

Delage-Mourroux, R., Pichaud, F., Frendo, J.L., Pidoux, E., Guliana, J.M., Moukhtar, M.S., and Jullienne, A., 1997, Cloning and sequencing of a new 15-hydroxyprostaglandin dehydrogenase related mRNA, *Gene* in press.

Ensor, C.M., Yang, Y.Y., Okita, R.T., and Tai, H.H., 1990, Cloning and sequence analysis of the cDNA for human placental NAD$^+$-dependent 15-hydroxyprostaglandin dehydrogenase, *J. Biol. Chem.* 265:14888.

Ensor, C.M., and Tai, H.H., 1991, Site-directed mutagenesis of the conserved tyrosine 151 of human placental NAD$^+$-dependent 15-hydroxyprostaglandin dehydrogenase yields a catalytically inactive enzyme, *Biochem. & Biophys. Res. Com.* 176(2):840.

Ensor, C.M., and Tai, H.H., 1994, Bacterial expression and site-directed mutagenesis of two critical residues (tyrosine-151 and lysine-155) of human placental NAD$^+$-dependent 15-hydroxyprostaglandin dehydrogenase, *Biochim. et Biophys. Acta.* 1208:151.

Ensor, C.M., and Tai, H.H., 1995, 15-Hydroxyprostaglandin dehydrogenase, *J. Lipid. Mediat. Cell Signal* 12:313.

Honn, K.V., Bockman, R.S., and Marnett, L.G., 1981, Prostaglandins and cancer: a review of tumor initiation through tumor metastasis, *Prostaglandins* 21:833.

Krook, M., Marekov, L., and Jörnvall, H., 1990, Purification and structural characterization of placental NAD$^+$-linked 15-hydroxyprostaglandin dehydrogenase. The primary structure reveals the enzyme belongs to the short-chain alcohol dehydrogenase family, *Biochemistry*, 29:738.

Leong, S.S., Horoszewicz, J.S., Shimoaka, K., Friedmann, M., Kawinski, E., Song, M.J., Ziegel, R., Chu, T.M., Baylin, S., and Mirand, E.A., 1981, *Advances in Thyroid Neoplasia.*, In Andreoli M., Monaco F. and Robbins J. (Eds.), Field Educational Italia, Rome.

Persson, B., Krook, M., and Jörnvall, H., 1991, Characteristics of short-chain dehydrogenases and related enzymes, *Eur. J. Biochem.* 200:537.

Pichaud, F., Delage-Mourroux, R., Pidoux, E., Jullienne, A., and Rousseau-Merck, M.F., 1997a, Chromosomal localization of the type-I 15-PGDH gene to 4q34-q35, *Hum Genet.* 99:279.

Pichaud, F., Frendo, J.L., Delage-Mourroux, R., de Vernejoul, M.C., Moukhtar, M.S., and Jullienne, A., 1995, Sequence of a novel mRNA coding for a C-terminal-truncated form of a human NAD$^+$-dependent 15-hydroxyprostaglandin dehydrogenase, *Gene* 162:319.

Pichaud, F., Roux, S., Frendo, J.L., Delage-Mourroux, R., Maclouf, J., de Vernejoul, M.C., Moukhtar M.S., and Jullienne, A., 1997b, 1,25-dihydroxyvitamin D3 induces NAD$^+$-dependent 15-hydroxyprostaglandin dehydrogenase in human neonatal monocytes, *Blood* 89:2105.

Wierenga, R., Terpstra, P., and Hol, W., 1986, Prediction of the occurrence of the ADP-binding beta alpha beta-fold in proteins, using an amino acid sequence fingerprint, *J. Mol. Biol.* 187:101.

KERATINOCYTE GROWTH FACTOR INCREASES LTC_4 SYNTHESIS AND EXPRESSION OF FLAP IN CULTURED RAT ALVEOLAR TYPE II CELLS

Peter H. S. Sporn,[1] Steven O. Lestrud,[2] Francine P. Fries,[1] and Ralph J. Panos[1]

Departments of [1]Medicine and [2]Pediatrics, Northwestern University Medical School and [1]Research Service, Lakeside Division, Veterans Affairs Chicago Health Care System, Chicago, Illinois 60611, USA

INTRODUCTION

Alveolar type II epithelial cells (ATII) play a key role in maintaining the integrity of the distal airspaces of the lung through the production of surfactant, transport of ions and formation of a permeability barrier for high molecular weight solutes. ATII also participate in inflammatory responses in the lung through synthesis, expression and release of adhesion molecules, cytokines, growth factors and eicosanoids. Further, ATII are essential for repair after lung injury by virtue of their ability to proliferate, re-establish an intact epithelial surface and differentiate into alveolar type I cells, which comprise the vast majority of the surface area of the normal alveolar epithelium.

ATII are highly active in metabolism of arachidonic acid (AA). Freshly isolated rat ATII stimulated with calcium ionophore A23187 produce substantial quantities of both cyclooxygenase(COX)-derived (PGE_2>PGI_2≈$PGF_{2\alpha}$≈TxA_2) and 5-lipoxygenase(5-LO)-derived products (LTC_4>LTB_4).[1,2] However, within two days of culture on a plastic surface, rat ATII lose the ability to make leukotrienes,[2] whereas their capacity to release AA and synthesize prostaglandins progressively increases for at least 7 days.[3] Transcellular metabolism of AA by ATII cultured with alveolar macrophages or with neutrophils has also been demonstrated.[4,5]

Keratinocyte growth factor (KGF) is a fibroblast-derived, heparin-binding, epithelial cell growth factor with mitogenic activity for ATII *in vitro*[6] and *in vivo*.[7] KGF protects rats from hyperoxic and other forms of oxidant lung injury.[8,9] The mechanism(s) of KGF's protective effect are not known.

We hypothesized that modulation of anti- and/or pro-inflammatory eicosanoid synthesis by ATII might be one mechanism by which KGF limits injury and enhances repair of the lung. This seemed plausible since hepatocyte growth factor (HGF), another ATII mitogen,[6] increased synthesis of PGE_2 and COX enzyme activity in gastric carcinoma cells.[10] We therefore examined the effect of KGF on eicosanoid synthesis and expression of proteins involved in AA metabolism in cultured rat ATII.

METHODS

Cells were obtained from viral antigen-free male Sprague-Dawley rats. ATII were isolated by the method of Dobbs et al[11] and cultured in Dulbecco's minimal essential

Recent Advances in Prostaglandin, Thromboxane, and Leukotriene Research
Edited by Sinzinger *et al.*, Plenum Press, New York, 1998

medium containing 10% fetal calf serum (FCS). ATII preparations were 90-92% pure at the time of plating, as indicated by the presence of lamellar bodies on Papinicolou stain. Alveolar macrophages (AM) were isolated by lung lavage and cultured in medium 199 containing 10% FCS.

ATII and AM were cultured with or without recombinant human KGF (1-100 ng/ml; a gift of W.S. Simonet, Amgen) for 24-72 h. Cells were then washed and either stimulated for 30 min in serum-free media with 5 μM A23187 or scraped into ice-cold homogenization buffer (100 mM NaCl, 50 mM K-PO_4, pH 7.1, 100 mM EDTA, 10% glycerol).

PGE_2 and LTC_4 in culture supernatants were quantitated by enzyme immunoassay (Cayman Chemical).

Proteins in ATII and AM lysates were separated by SDS-PAGE, transferred to nitrocellulose and immunoblotted using rabbit polyclonal antisera raised against purified human leukocyte 5-LO or a peptide corresponding to amino acid residues 41-52 of human 5-LO-activating protein (FLAP) (gifts of J. Evans, Merck Frosst Canada); purified human lung LTC_4 synthase (a gift of J.F. Penrose, Harvard University); or a peptide corresponding to amino acids 584-598 of murine COX-II (a gift of D.L. DeWitt, Michigan State University). Immunoreactive proteins were detected by enhanced chemiluminescence (ECL; Amersham). Bands on immunoblots were quantitated using a Bio-Rad GS-670 imaging densitometer and Molecular Analyst software.

RESULTS AND DISCUSSION

Following isolation, ATII were cultured in the absence or presence of KGF for 48 h, washed and stimulated with A23187. Supernatants from these cultures were then analyzed for PGE_2 and LTC_4 by EIA. In the absence of KGF, ATII produced appreciable quantities of PGE_2, and synthesis of PGE_2 was not affected by KGF at any of the concentrations tested (Fig. 1A). On the other hand, without growth factor, ATII produced lesser amounts of LTC_4 than PGE_2, but in contrast to its lack of effect on PGE_2 formation, KGF caused dose-dependent enhancement of LTC_4 synthesis (Fig. 1B). Maximum enhancement of LTC_4 formation was seen following exposure to KGF at 100 ng/ml for 48 h. Under the culture conditions utilized, exposure to KGF (1-100 ng/ml, 48 h) had no effect on the protein content of ATII monolayers (data not shown), indicating that KGF's effect on leukotriene synthesis was not the result of an increase in ATII numbers or size.

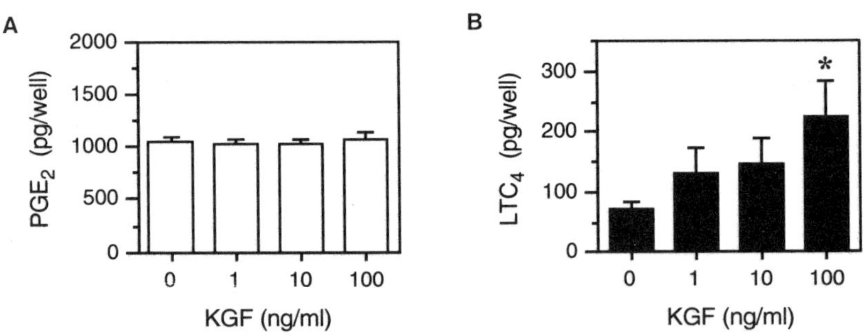

Figure 1. Effect of KGF on A23187-stimulated synthesis of eicosanoids by ATII. Following isolation, ATII were cultured in the absence or presence of KGF for 48 h, washed, and stimulated with 5 μM A23187 for 30 min. PGE_2 and LTC_4 in culture supernatants were quantitated by EIA. Data are expressed as means±SEM. A. PGE_2 (n=4). B. LTC_4 (n=8). *P<0.05 vs. control (no KGF).

To clarify the mechanism by which KGF increased LTC_4 synthesis in ATII, we analyzed expression of the three key proteins required for synthesis of this leukotriene, 5-LO, FLAP and LTC_4 synthase, by immunoblotting. Immunoreactive 5-LO and FLAP protein levels were found to decline markedly in ATII cultured for 48 h, as compared to freshly isolated cells, whereas LTC_4 synthase expression did not change over time in

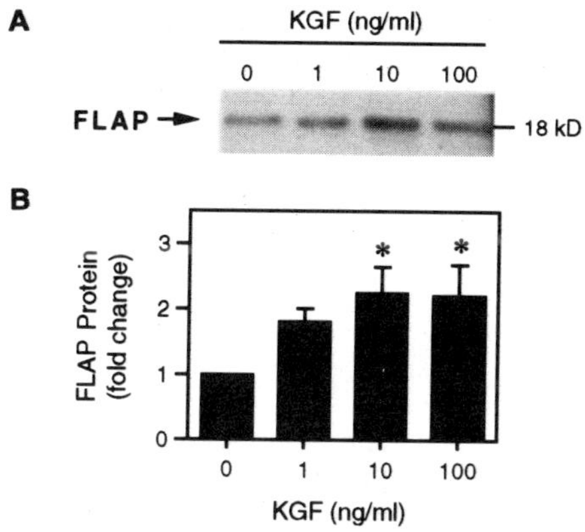

Figure 2. Effect of KGF on FLAP protein in ATII. . Following isolation, ATII were cultured in the absence or presence of KGF for 48 h, then scraped into homogenization buffer. For each condition, equal aliquots of protein were subjected to SDS-PAGE, transferred to nitrocellulolse and immunoblotted for FLAP. A. Representative immunoblot. B. Summary (mean±SEM) of densitometric quantitation of FLAP for n=7 experiments. *P<0.05 vs. control (no KGF).

culture (data not shown). The declines in 5-LO and FLAP protein are consistent with the decrease in leukotriene synthesis by ATII over time in culture, as reported previously[2] and confirmed in this investigation. Notably, exposure to KGF (1-100 ng/ml) for 48 h did not affect ATII expression of either 5-LO or LTC_4 synthase (data not shown). In contrast, KGF increased the level of FLAP in ATII in dose-dependent fashion (Fig. 2). KGF increased FLAP expression to an equivalent maximum degree at both 10 and 100 ng/ml, whereas maximum enhancement of LTC_4 synthesis was seen only at 100 ng/ml (Fig. 1B). The maximum increments in LTC_4 synthesis (3.1±0.5-fold) and FLAP protein (2.3±0.3-fold) resulting from KGF exposure were of similar magnitude.

Although the ATII preparations used in this study were over 90% pure, we wished to exclude the possibility that small numbers of contaminating cells might have been the source of LTC_4 measured in these experiments. Among the contaminating cells in ATII cultures, AM represent the major cell type with the capacity for 5-LO metabolism. Therefore, in experiments analogous to those with ATII, we isolated and cultured rat AM (99% pure) in the absence and presence of 100 ng/ml KGF for 48 h, then assessed A23187-stimulated LTC_4 synthesis and FLAP expression. In these experiments, KGF failed to increase either LTC_4 synthesis or FLAP protein in AM (data not shown). Thus, contamination by AM cannot account for the KGF-inducible increases in LTC_4 synthesis and expression of FLAP seen in ATII cultures in this study.

The major finding of our investigation is that KGF enhances synthesis of LTC_4 in ATII by increasing expression of FLAP, the arachidonate-binding protein essential for cellular leukotriene synthesis. To our knowledge, this is the first demonstration in epithelial cells of growth factor-regulated expression of a protein in the 5-LO pathway. The discordant effect of KGF on expression in ATII of FLAP and of 5-LO, which did not change, is reminiscent of the effect of 1,25-dihydroxyvitamin D_3 (1,25-$(OH)_2D_3$) on cultured human blood monocytes.[12] In the latter study, 1,25-$(OH)_2D_3$ augmented monocyte expression of FLAP, but not 5-LO; the increase in FLAP was associated with increased A23187-stimulated synthesis of 5-LO metabolites.

The failure of KGF to increase prostaglandin synthesis in ATII contrasts with the observation of Hori et al[10] that HGF, another fibroblast-derived growth factor with activity for ATII, enhanced PGE_2 synthesis and COX activity in a gastric carcinoma cell line. Based on those findings, we thought that KGF might similarly increase PGE_2 synthesis in ATII and speculated that this might occur through induction of COX-II by KGF. However, in accord with its inability to enhance PGE_2 synthesis, KGF had no effect on expression of COX-II in ATII, as assessed by immunoblotting (data not shown).

The finding that KGF enhances ATII synthesis of LTC_4 suggests that epithelial cells, in addition to inflammatory cells, may be an important source of pro-inflammatory leukotrienes in the setting of lung injury. On the other hand, since KGF protects animals against lung injury caused by various agents, it is also possible that enhancement by KGF of ATII LTC_4 synthesis is important in alveolar epithelial repair. It is noteworthy in this regard that 5-LO metabolites appear to play a role in proliferation of airway epithelial cells[13] as well as carcinoma cells of lung[14] and gastric[15] origin. This suggests that the proliferative effects of KGF on ATII may in fact be mediated by LTC_4 or other 5-LO metabolites whose synthesis KGF increases. This possibility is currently under investigation.

ACKNOWLEDGMENTS

We thank Dr. Simonet for generously supplying recombinant human KGF, and Drs. Evans, Penrose and DeWitt for generously providing antisera to 5-lipoxygenase, FLAP, LTC_4 synthase and cyclooxygenase-II. This work was supported by HL-02407 and grants from the Department of Veterans Affairs and the American Lung Association.

REFERENCES

1. J.B. Chauncey, M. Peters-Golden and R.H. Simon, Arachidonic acid metabolism by rat alveolar epithelial cells, *Lab. Invest.* 58:133 (1988).
2. G.R. Cott, J.Y. Westcott and N.F. Voelkel, Prostaglandin and leukotriene production by alveolar type II cells in vitro, *Am. J. Physiol. (Lung Cell. Mol. Physiol.)* 2:L179 (1990).
3. R.J. Lipchik, J.B. Chauncey, R. Paine, R.H. Simon and M. Peters-Golden, Arachidonate metabolism increases as rat alveolar type II cells differentiate in vitro, *Am. J. Physiol. (Lung Cell. Mol. Physiol.)* 3:L73 (1990).
4. J.B. Chauncey, R.H. Simon and M. Peters-Golden, Rat alveolar macrophages synthesize leukotriene B4 and 12-hydroxyeicosatetraenoic acid from alveolar epithelial cell-derived arachidonic acid, *Am. Rev. Respir. Dis.* 138:928 (1988).
5. F. Grimminger, I. von Kürten, D. Walmrath and W. Seeger, Type II alveolar epithelial eicosanoid metabolism: predominance of cyclooxygenase pathways and transcellular lipoxygenase metabolism in co-culture with neutrophils, *Am. J. Respir. Cell Mol. Biol.* 6:9 (1992).
6. R.J. Panos, J.S. Rubin, S.A. Aaronson and R.J. Mason, Keratinocyte growth factor and hepatocyte growth factor/scatter factor are heparin-binding growth factors for alveolar type II cells in fibroblast-conditioned medium, *J. Clin. Invest.* 92:969 (1993).
7. T.R. Ulich, E.S. Yi, K. Longmuir, S. Yin, R. Blitz, C.F. Morris, R.M. Housley and G.F. Pierce, Keratinocyte growth factor is a growth factor for type II pneumocytes in vivo, *J. Clin. Invest.* 93:1298 (1994).
8. R.P. Panos, W.S. Simonet, J.S. Rubin and L.J. Smith, Intratracheal instillation of keratinocyte growth factor decreases hyperoxia-induced mortality in rats, *J. Clin. Invest.* 96:2026 (1995).
9. E.S. Yi, S.T. Williams, H. Lee, D.M. Malicki, E.M. Chin, S. Yin, J. Tarpley and T.R. Ulich, Keratinocyte growth factor ameliorates radiation- and bleomycin-induced lung injury and mortality, *Am. J. Pathol.* 149:1963 (1996).
10. T. Hori, S. Shibamoto, M. Hayakawa, K. Takeuchi, N. Oku, K. Miyazawa, N. Kitamura and F. Ito, Stimulation of prostaglandin production by hepatocyte growth factor in human gastric carcinoma cells, *FEBS Lett.* 334:331 (1993).
11. L.G. Dobbs, R. Gonzalez and M.C. Williams, An improved method for isolating type II cells in high yield and purity, *Am. Rev. Respir. Dis.* 134:141 (1986).
12. M.J. Coffey, M. Gyetko and M. Peters-Golden, 1,25-Dihydroxyvitamin D3 upregulates 5-lipoxygenase metabolism and 5-lipoxygenase activating protein in peripheral blood monocytes as they differentiate into mature macrophages, *J. Lipid Mediators* 6:43 (1993).
13. G.D. Leikauf, H.-E. Claesson, C.A. Doupnik, S. Hybbinette and R.C. Grafström, Cysteinyl leukotrienes enhance growth of human airway epithelial cells, *Am. J. Physiol. (Lung Cell. Mol. Physiol.)* 3:L255 (1990).
14. I.M. Avis, M. Jett, T. Boyle, M.D. Vos, T. Moody, A.M. Treston, A. Martinez and J.L. Mulshine, Growth control of lung cancer by interruption of 5-lipoxygenase-mediated growth factor signaling, *J. Clin. Invest.* 97:806 (1996).
15. S. Shimakura and C.R. Boland, Eicosanoid production by the human gastric cancer cell line AGS and its relation to cell growth, *Cancer Res.* 52:1744 (1992).

TYPE-I 15-HYDROXYPROSTAGLANDIN DEHYDROGENASE: ROLE IN MACROPHAGE/OSTEOCLAST DIFFERENTIATION

Franck Pichaud, Régis Delage-Mourroux, Mélinée Frenkian, Jean Louis Frendo, Sophie Roux, Marie-Christine de Vernejoul, and Annick Jullienne

INSERM U. 349, Hôpital Lariboisière
centre Viggo Petersen, 75475 Paris
cedex 10, France

INTRODUCTION

Vitamin D3 acts on several cellular processes, including cellular growth and differentiation. Osteoclasts (OCs), are multinucleated bone-resorbing cells formed by fusion of mononuclear cells of hematopoietic origin. OCs share committed hematopoietic progenitors with monocytes and are suspected to be bone-adaptated macrophages (Udagawa et al., 1990). Studying the genes, which could be transcriptionaly regulated by vitamin D3, during the differentiation of hematopoietic precursors of OCs, we showed that type-I 15-hydroxyprostaglandin dehydrogenase (type-I 15-PGDH [EC.1.1.1.141]) (Ensor et al., 1990) gene activation was one of the major regulatory effect of the steroid on human neonatal monocytes (NMs). This member of the short chain alcohol dehydrogenase catalyses the reversible inactivation of PGs into their respective 15-keto metabolites (Ånggard, 1966). We show here that this enzyme could play an important role in the vitamin D3 *in vitro* induced differentiation of NMs through the macrophage-osteoclast lineage.

VITAMIN D3 AND PROSTAGLANDIN BIOSYNTHESIS

In order to study the molecular events which occur during vitamin D3 induced differentiation of NMs, we performed differential display reverse transcription experiments (Liang et al., 1992) on total RNA extracted from treated and untreated cells. Profiles revealed that type-I 15-PGDH gene transcription was induced by vitamin D3 (10^{-8} M), while this transcript remained undectectable in untreated NMs (Pichaud et al., 1997). The transcriptional induction of type-I 15-PGDH mRNA was accompanied by the induction of the 15-PGDH related I (15-PGDHrI) mRNA, an isomer previously reported in HL-60 cells (Pichaud et al., 1995). Interestingly, adult monocytes, constitutively expressed the 15-PGDHrI mRNA, while vitamin D3 failed to induce type-I 15-PGDH mRNA transcription in these cells.

We confirmed the 15-PGDH mRNA(s) induction in treated NMs using semi-quantitative *RT-PCR* experiments, and we showed that this induction occurs after 1 week of 10^{-8} M vitamin D3 treatment. 15-PGDH transcripts were not detected after 0, 24 or 48 hours of vitamin D3 treatment of MNs culture, suggesting that vitamin D3 could act indirectly on this gene transcription, probably via a second molecule.

The transcriptional induction of 15-PGDH transcripts preceeded the expression of specific markers of mature OCs such as vitronectin (VNR) and calcitonin (CTR) receptors which are expressed by NMs after 2 weeks of vitamin D3 treatment. This observation suggested that type-I 15-PGDH could play a role during the first step of the macrophage-osteoclast differentiation.

We extended our study to the expression of the cyclooxygenase 2 (Cox 2) gene in these cells. We showed that the level of Cox 2 mRNA was significantly decreased in the vitamin D3 treated NMs after 1 week of culture. While Cox 2 mRNA levels appeared to be decreased in both group of cells (i.e. treated or untreated) the decrease was more markedly in the vitamin D3 treated NMs.

An indirect evidence of the coordinated regulation of type-I 15-PGDH and Cox 2 by vitamin D3 was the decrease of PGE_2 [50%, ($p<0.0042$)] in the culture medium of treated NMs.

Using thin layer chromatography experiments and tritiated PGE_2, we showed that treated NMs expressed fully active type-I 15-PGDH. Lysate of NMs treated by vitamin D3 for 1, 2 and 3 weeks showed 15-PGDH activity since these cells were able to convert PGE_2 into its inactivate metabolite 15-keto-PGE_2 [mean of 15-keto-PGE_2/PGE_2/5.10^6 cells= 4.2 +/-0.4, ($p<0.001$)]. This reaction catalysed by type-I 15-PGDH was not detected in the untreated cells [mean of 15-keto-PGE_2/PGE_2/5.10^6 cells= 0.25 +/-0.1, ($p<0.001$)].

In addition, we showed that when expressed in E. coli., the predicted truncated 15-PGDH, corresponding to the 15-PGDH related I transcript was not able to catalyse the conversion of PGE_2 into 15-keto-PGE_2. Since this mRNA isomer appeared to be widely expressed (i.e. in lung, liver, placenta and adult monocytes) further studies will be necessary to elucidate the role of the predicted protein, for example in the capacity of regulating the activity of the wild type 15-PGDH, originally described as a homodimer (for review see Ensor and Tai, 1995).

INDUCTION OF TYPE-I 15-PGDH AND OSTEOCLASTOGENESIS

To test the potential implication of the type-I 15-PGDH induction in our model of osteoclastic differentiation, we evaluated the effects of exogenous PGE_2 on the appearance of several osteoclastic and/or macrophagic differentiating antigens. We tested the expression of VNR and CTR considered as specifically expressed by mature OCs. We also tested the expression of macrophage antigens such are CD-14 and CD-11b (Roux et al., 1997). Our results showed that PGE_2 inhibited the appearance of VNR positive cells in NMs treated by Vitamin D3 as assayed by immunohistochemistry [12.27 +/-2.27%, compared to the untreated NMs: 33.58 +/-2.78%, ($p<0.001$)].

We also observed morphological changes and adhesiveness in the culture treated by PGE_2. The appearance of multinucleated cells , a characteristic of the OCs *in vitro* (and *in vivo*) was dose dependently decreased by PGE_2. The presence of PGE_2 (10^{-7} or 10^{-6} M) significantly decreased the proportion of multinucleated cells (3 to 5 nucleus/ cell) [9.60 +/-1.33% and 8.24 +/-1.44% respectively ($p<0.05$); compared to the untreated NMs: 18.14 +/-2.78% ($p<0.05$)].

The proliferation of NMs, evaluated on day 7 was also significantly decreased in the presence of PGE_2 at concentrations of 10^{-8}, 10^{-7} and 10^{-6} M [1 +/-0.6%, 2.33 +/-0.6% and 2.33 +/-0.9%, respectively, of the proliferating cells, ($p<0.05$)] compared to that in the control NMs cultured without PGE_2 (i.e. 11.66 +/-3.17%)
In addition, CTR gene transcription appeared to be decreased by the PGE_2 (10^{-7} M) treatment of vitamin D3 treated NMs.

These results suggest that PGE_2 acts as an inhibitory regulator of the OCs differentiation and thus confirm our hypothesis regarding the potential importance of the vitamin D3 coordinated regulation of type-I 15-PGDH and Cox 2 expression in our culture system.

Studies performed on the expression of macrophagic markers, revealed that both the number of CD-11B and CD-14 positive cells were increased by PGE_2. The proportion of CD-14 positive NMs in control was 32.36 +/-1.1% while this propotion was 41.1 +/-3.33% ($p<0.05$), 52.95 +/-3.36% ($p<0.001$) and 55.37 +/-2.27% ($p<0.001$); respectively in the

NMs incubated with PGE_2 (10^{-8}, 10^{-7}, 10^{-6} M). The proportion of CD-11b positive NMs in control [14.22 +/-2.35%] was also increased in a dose dependent way by PGE_2 (10^{-6} M) [38.11 +/-3.93%, ($p<0.001$)] compared to that in controls.

In addition, using *RT-PCR* experiments, we showed that c-fos gene transcription was markedly reduced when NMs were cultured in the presence of PGE_2, compared to that in the control cultures. After 3 weeks of culture, the calculated ratio of c-fos/GAPDH *RT-PCR* product revealed a decrease of approximatively 8-fold in the PGE_2-treated NMs compared to that in the untreated cells [7.89 +/-2.3]. Our results are consistent with the important role that c-fos plays in OCs activity and differentiation (Grigoriadis et al., 1994) and point out the fact that PGE_2 could regulate the transcription of this gene during the macrophage-osteoclast differentiation.

CONCLUSION

Our investigation revealed for the first time that transcription of the type-I 15-PGDH gene, was triggered during the vitamin D3 induced differentiation of neonatal monocytes into osteoclast. This transcriptional induction was accompanied by neo-synthesis of the fully active enzyme. Moreso, the cyclooxygenase 2 mRNA levels were decreased in the vitamin D3 treated NMs. This coordinated regulation of the two main enzymes of PGs biosynthesis leds to a significant decrease of PGE_2 levels (50%) in the corresponding culture medium. Adult monocytes were shown to express an isomer of type-I 15-PGDH (15-PGDHrI) transcript, first characterized in the promyelocytic Hl-60 cells (Pichaud et al., 1995). In this mature cells, the type-I 15-PGDH mRNA was not inducible by vitamin D3 (Pichaud et al., 1997).

To investigate the potential role of the type-I 15-PGDH induction during osteoclastogenesis, we tested the effect of PGE_2 addition in our culture system.

Addition of exogenous PGE_2 in culture medium of NMs showed a significant inhibition of the proliferation of the osteoclastic precursors with a dose response inhibition of multinucleation and a decrease of CTR and VNR positive cells, two markers of mature osteoclasts. On the contrary, PGE_2 acted as a positive regulatory factor on the monocyte-macrophage differentiation, since this molecule enhanced the proportion of CD11-b and CD-14 positive cells in our culture (figure 1).

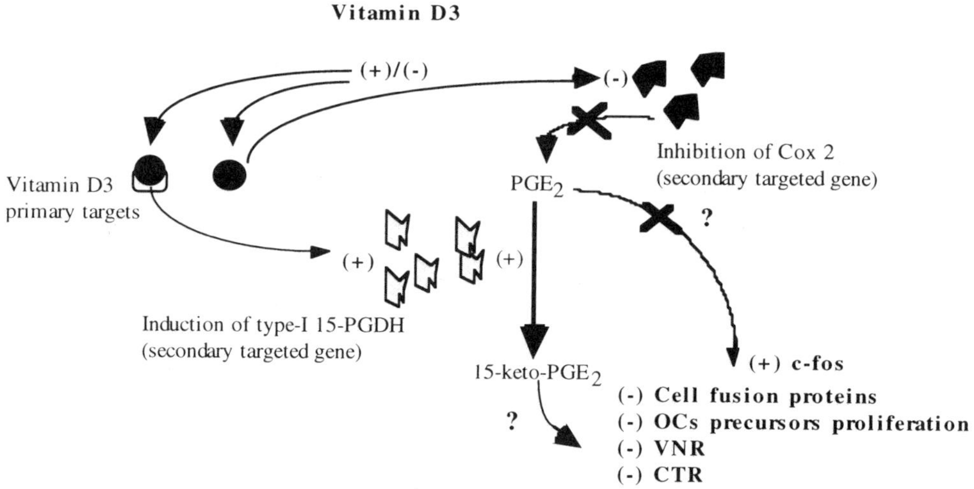

Figure 1. Proposed scheme of vitamin D3 actions on NMs leading to the preferential induction of the immature cells into OCs. In this hypothesis vitamin D3 induces (or inhibits) the transcription of two (or one) primary responsive gene(s). The products of this primary response act as regulation factors of the type-I 15-PGDH and Cox 2 gene transcription. The subsequent induction of type-I 15-PGDH and down regulation of Cox 2 lead to the observed effects of the expression of c-fos, VNR, CTR, OCs precursor proliferation and multinucleation.

In conclusion, our results demonstrated that vitamin D3 could exert a dual control on the two principal enzymes of PGs biosynthesis during the first step of the osteoclastic differentiation. Addition of exogenous PGE_2 in our culture system suggested that this molecule could be an important regulator of this cell differentiation, probably by inhibiting the proliferation of osteoclastic precursors in our culture system. We also suggested that PGE_2 may reduce OC differentiation by increasing the proportion of precursor cells that preferentially differentiated into macrophages.

Recent studies show these bioactive lipids are potentialy involved in the regulation of cell proliferation and differentiation, particularly in colorectal carcinoma (Eberhart et al., 1994). Our results stress the importance of the type-I 15-PGDH for the regulation of PGE_2 action in osteoclastic differentiation and probably also in the other physiological and cellular processes such as inflammation, platelet agregation and muscle contraction, in which PGs are known to play a role.

Acknowledgments

R.D.M is a recipient of an individual grant from the "Ligue Nationale contre le Cancer". F.P. and J.L.F. are the recipients of an individual grant from the French Ministry of Research and Education. This work was supported by the « Association pour la Recherche sur le Cancer », grant N°1359.

REFERENCES

Anggard, E., and Samuelsson, B., 1964, Metabolism of prostaglandin E1 in guinea pig lung: the structure of two metabolites, *J. Biol. Chem.*, 239:4097.

Eberhart, C.E., Coffey, R.J., Radhika, A., Giardiello, F.M., Ferrenbach, S., and Dubois, R.N., 1994, Up-regulation of cyclooxygenase gene expression in human colorectal adenomas and adenocarcinomas, *Gastroenterology*, 107: 1183.

Ensor, C.M., Yang, J.Y., Okita, R., and Tai, H.H., 1990, Cloning and sequence analysis of the cDNA for human placental NAD(+)-dependent 15-hydroxyprostaglandin dehydrogenase, *J. Biol. Chem.*, 265:14888.

Ensor, C.M., and Tai, H.H., 1995, 15-Hydroxyprostaglandin dehydrogenase, *J. Lipid. Mediat. Cell Signal* 12:313.

Grigoriadis, A.E., Wang, Z.Q., Cecchini, M.G., Hofstetter, W., Felix R., Fleich H.A., and Wagner, E.F., 1994, c-Fos: A key regulator of osteoclast-macrophage lineage determination and bone remodelling, *Science*, 266:443.

Liang, P., and Pardee, A.B., 1992, Differential display of eukaryotic messenger RNA by means of the polymerase chain reaction, *Science*, 257:967.

Pichaud, F., Frendo, J.L., Delage Mourroux, R., de Vernejoul, M.C., Moukhtar, M.S., and Jullienne, A., 1995, Sequence of a novel mRNA coding for a C-terminal truncated form of human NAD^+ dependent 15-hydroxyprostaglandin dehydrogenase, *Gene*, 162:319.

Pichaud, F., Roux, S., Frendo, J.L., Delage-Mourroux, R., de Vernejoul, M.C., Moukhtar, M.S., and Jullienne, A., 1997, 1,25-Dihydroxyvitamin D3 induces NAD(+)dependent 15-hydroxyprostaglandin dehydrogenases in human neonatal monocytes, *Blood*, 89:2105.

Roux, S., Pichaud, F., Quinn, J., Lalande, A., Morieux, C., Jullienne, A., and de Vernejoul, M.C., 1997, Effects of prostaglandins on human hematopoietic osteoclast precursors, *Endocrinol.*, 138:1476

Udagawa, N., Takahashi, N., Akatsu, T., Tanaka, H., Sasaki, T., Nishihara, T., Koga, Martin, T.J., and Suda, T., 1990, Origin of osteoclasts: Mature monocytes and macrophages are capable of differentiating into osteoclasts under a suitable microenvironment by bone marrow-derived stromal cells, *Proc. Natl. Acad. Sci. USA.*, 87:7260.

PGE_2 AND TXA_2 PRODUCTION BY ISOLATED MACROPHAGES FROM HUMAN PLACENTA

Birgit Wetzka,[1,2] Dawn E. Clark,[2] D. Steve Charnock-Jones,[2] Hans Peter Zahradnik,[1] and Stephen K. Smith[2]

[1]Endocrinology and Reproductive Medicine, Universitätsfrauenklinik, Hugstetter Str. 55, D-79106 Freiburg, Germany
[2]Reproductive Molecular Research Group, Dept. of Obstetrics & Gynaecology, University of Cambridge, Robinson Way, Cambridge CB2 2SW, UK

INTRODUCTION

In earlier studies we have investigated the eicosanoid production of placenta in short-term tissue culture where thromboxane A_2 (TXA_2) was the main cyclooxygenase (COX) product of arachidonic acid.[1] Immunohistochemically, the strongest positive staining for COX 1 and 2 and thromboxane synthase (TXS) could be located to placental macrophages (Hofbauer cells).[2,3] These tissue macrophages of the placenta are located close to trophoblast cells and fetal capillaries which makes them ideal candidates for involvement in regulatory processes. Recent studies suggested a role for Hofbauer cells in angiogenesis[4,5] and immunological reactions[6].

Eicosanoids are important local mediators of the effects of growth factors and hormones.[7] In this study, we examined the prostaglandin (PG) production of Hofbauer cells isolated from human placenta. We chose the prostaglandins PGE_2 and TXA_2, because they are known to be produced by other tissue macrophages[8] and play a role for the regulation of vascular tone[9] and immunological reactions[10].

MATERIALS AND METHODS

Firstly, we developed a protocol for the isolation of placental macrophages (Fig. 1) with minimal contamination of trophoblast as described in detail in Wetzka et al..[11] Briefly, fresh placenta (20 g wet weight) was obtained from deliveries at term after uncomplicated pregnancies. The tissue was finely minced, rinsed at least 5 times in PBS and then digested for 2 hours in dispase. The resulting cell suspension was sieved to remove remaining cell clumps and was then span through Ficoll-isopaque. The opaque interphase containing mononuclear cells was aspirated. These cells were centrifuged again through a preformed Percoll gradient with 60-50-40-30% steps at 1500g for 10 min and the upper two bands were aspirated. In order to remove as much remaining trophoblast cells as possible which constitute approximately 30% of the cell suspension after Percoll separation,[11] we performed a depletion with magnetic beads

Recent Advances in Prostaglandin, Thromboxane, and Leukotriene Research
Edited by Sinzinger *et al.*, Plenum Press, New York, 1998

(Dynabeads®, Dynal, UK) coated with an antibody against epidermal growth factor receptor (EGF-R, Santa Cruz Biotechnologies, UK). The supernatant was removed from the magnetic beads and resuspended in M199 supplemented with 10% FBS and antibiotics. The cells were then plated into a 24-well tissue culture dish (10^4 cells/ml) and incubated for 2 hours at 37°C/20%O_2. Non-adhering cells were then washed off and fresh medium added.

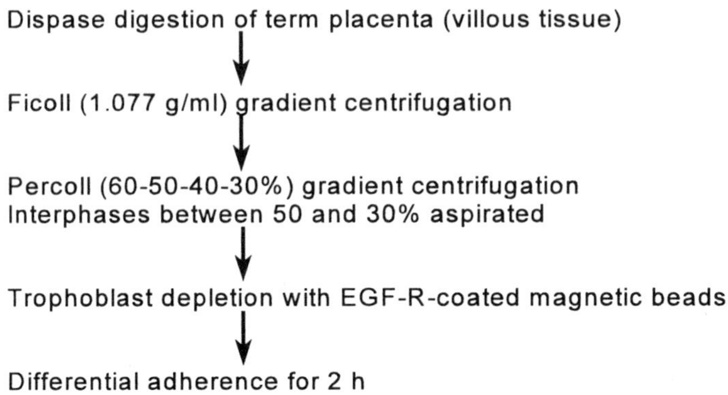

Figure 1. Flow chart of the isolation protocol for Hofbauer cells from human placenta.

The isolated macrophages were left to equilibrate to cell culture conditions for 36 hours. The effects of culturing with and without FBS and low oxygen (5%O_2) was investigated. After 36 hours, 1 ml of fresh medium was added to each well and cells were further incubated at either basal conditions or stimulated with lipopolysaccharide (LPS, 25 μg/ml) for 24 hours.

Aliquots of 200 μl were taken for PG analysis after 4, 8 and 24 hours and PGE_2 and TXB_2 concentrations determined by commercially available enzyme immunoassays (Cayman, USA). The final PG concentrations in the culture media were expressed as pg/ml/10^4 cells and concentrations corrected for the aliquots that have been removed. The change in PG synthesis caused by culturing in 5% O_2 in comparison with culturing in 20% O_2 was calculated as difference [PG synthesis in 5% O_2 - PG synthesis in 20% O_2] and is expressed as percent change compared with the basal production in 20% O_2 atmosphere. The stimulation of PG synthesis caused by addition of LPS was calculated as difference [PG production after LPS - basal PG production] and is expressed as percent increase compared with the basal production. Further, the quotient PGE_2/TXA_2 was calculated for each culture experiment. Results are shown as mean ± SEM. For statistical analysis, the student's *t*-test for paired samples was used.

The purity of the isolated macrophage population was examined by performing immunocytochemistry on cell smears with an anti-CD68 antibody (Dako, UK) as a macrophage marker and an anti-cytokeratin antibody (Dako) as a trophoblast marker as described earlier.[11]

RESULTS

Immunohistochemical staining with anti-CD68 and anti-cytokeratin after removal of the EGF-R-positive cells with a magnet showed over 90% cells being positive for CD68 (macrophages) and less than 10% positive for cytokeratin (trophoblast).

Addition of 10% FBS did not significantly alter the PGE_2 and TXA_2 production of the cultured macrophages. The PG synthesis of unstimulated cells did not change significantly during 24 hours of culture. The PGE_2 production rate was consistently higher than TXA_2 with a median ratio PGE_2/TXA_2 of 2.6 which showed a minimum of 1.5 ± 0.3 in cultures performed in 5% O_2 and a maximum of 3.7 ± 1.3 in cultures done in 20% O_2 atmosphere. Culturing in 5%O_2 resulted in a significant decrease in PGE_2 ($p<0.05$) and no significant change in TXA_2

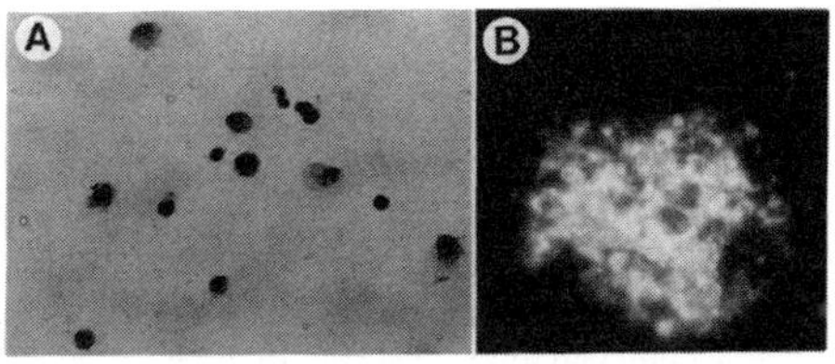

Figure 2. Immunocytochemical staining of cell smears after Hofbauer cell isolation with an antibody against CD68. **A**: Immunoperoxidase staining, magnification x200. **B**: Indirect immunofluorescence staining, magnification x1000.

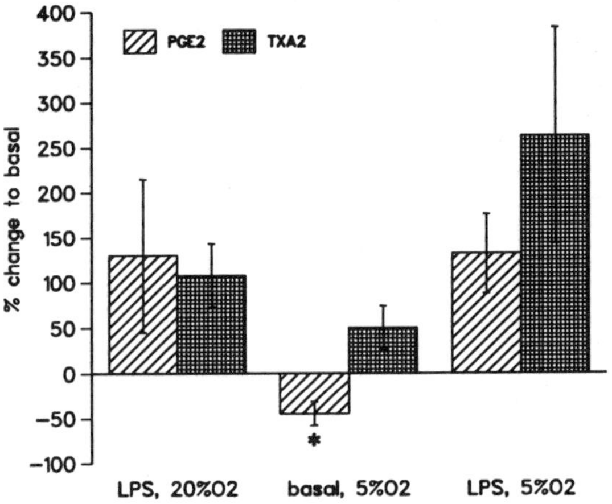

Figure 3. Effect of culturing in 20% O_2 with addition of LPS (25 µg/ml) and in 5% O_2 without ("basal") and with addition of LPS (25µg/ml) on the PGE_2 and TXA_2 production of isolated Hofbauer cells. Results are shown as % changes calculated as described in Material and Methods. n= 3, * $p< 0.05$.

synthesis (Fig. 3). The stimulatory effect on PG synthesis by LPS was highest after 24 hours, therefore the analysis of LPS effects was carried out with PG production rates from 24 hour culture media. LPS addition during incubation in both 20% and 5% O_2 caused an increased synthesis of both prostaglandins with a tendency to a higher stimulatory effect on TXA_2 in the 5% O_2 culture (Fig. 3).

COMMENTS

We developed a method for Hofbauer cell isolation with which we obtained an at least 90% pure macrophage population by immunocytochemistry with anti-CD68. The Hofbauer cells produced PGE_2 and TXA_2 like other tissue macrophages[8] with PGE_2 synthesis being predominant. Both PGE_2 and TXA_2 synthesis could be stimulated by LPS which suggests that the isolated macrophages are still functionally active.

The PGE_2 production by Hofbauer cells might play a role for immunological processes within the placental villi[10]. TXA_2 might be important for the regulation of local blood flow and platelet aggregation[9] and was found to be a growth factor for vascular smooth muscle cells[12].

Culturing the macrophages in low oxygen (5% O_2) resulted in a tendency to a decreased PGE_2/TXA_2 production ratio. The importance of this observation for the physiology of the placenta and pathological clinical conditions like preeclampsia or placental insufficiency to be evaluated in further studies.

REFERENCES

1. B. Wetzka, W. Schäfer, M. Scheibel, R. Nüsing, H.P. Zahradnik. Eicosanoid production by intrauterine tissues before and after labor in short-term tissue culture. *Prostaglandins* 45:571-81 (1993).
2. B. Wetzka, D.S. Charnock-Jones, B. Viville, et al. Expression of prostacyclin and thromboxane synthases in placenta and placental bed after preeclamptic pregnancies. *Placenta* 17:573-581 (1996).
3. B. Wetzka, R. Nüsing, D.S. Charnock-Jones, W. Schäfer, H.P. Zahradnik, S.K. Smith. Cyclooxygenases 1 and 2 in human placenta and placental bed after preeclamptic pregnancies. *Hum Reprod*, in press (1997).
4. M. Castellucci, P. Kaufmann. The Hofbauer cell, in: *Pathology of the Human Placenta*, K. Benirschke and P. Kaufmann, eds., Springer, New York (1992).
5. D.E. Clark, S.K. Smith, A.M. Sharkey, D.S. Charnock-Jones. Localization of VEGF and expression of its receptors flt and KDR in human placenta throughout pregnancy. *Hum Reprod* 11:1090-8 (1996).
6. R.S. Berkowitz, H.M.P. Faris, J.A. Hill, D.J. Anderson. Localization of leukocytes and cytokines in chorionic villi of normal placentas and complete hydatiform moles. *Gynecol Oncol* 37:396-400 (1990.)
7. H.P. Zahradnik. The birth process, in: *Comprehensive Human Physiology*, R. Greger and U. Windhorst, eds., Springer, Berlin (1996).
8. J. Morley, M.A. Bray, R.W. Jones, D.H. Nugteren, D.A. van Dorp. Prostaglandin and thromoboxane production by human and guinea-pig macrophages and leukocytes. *Prostaglandins* 17:729-46 (1979).
9. B. Samuelsson, M. Goldyne, E. Granström, M. Hamberg, S. Hammarström, C. Malmsten. Prostaglandins and thromboxanes. *Annu Rev Biochem* 47:997-1029 (1978).
10. J.S. Goodwin, A.D. Bankhurst, R.P. Messner. Suppression of human T-cell mitogenesis by prostaglandin. Existence of a prostaglandin-producing suppressor cell. *J Exp Med* 146:1719-34 (1977).
11. B. Wetzka, D.E. Clark, D.S. Charnock-Jones, H.P. Zahradnik, S.K. Smith. Isolation of macrophages (Hofbauer cells) from human term placenta and their PGE_2 and TXA_2 production. *Hum Reprod* 12:847-52 (1997).
12. G.W. Dorn, M.G.Davis, M.W. Becker. Role of TXA_2 in growth of vascular smooth muscle cells. *Prostaglandins, Leukotrienes and Essential Fatty Acids*, 55(Suppl 1):28 (1996).

INTRAUTERINE EICOSANOID PRODUCTION IN UNCOMPLICATED AND HYPERTENSIVE PREGNANCIES: ALTERATIONS OF PLACENTAL 12-HYDROXYEICOSATETRAENOIC ACID

Eva Arbogast, Wolfgang Schäfer, and Hans Peter Zahradnik

Universitätsfrauenklinik Endokrinologie und Reproduktionsmedizin
Hugstetter Str. 55, D-79106 Freiburg, Germany

INTRODUCTION

Complicating about 10% of all pregnancies pregnancy-induced hypertensive diseases account for an enormous number of maternal deaths. At present, termination of pregnancy represents the only treatment of the cause. The important role of eicosanoids in pregnancy-induced hypertension is generally accepted[1-2]. As humoral vasoactive factors they are important for vascular tone regulation of the non-innervated uteroplacental vessels. In PIH and related diseases, these vessels fail to dilate normally, leading to a chronic preplacental decrease in circulation. A systemic lack of prostacyclin may also promote an increased vasoconstriction and coagulation disturbances in PIH leading to endothelial cell disorder, hemolysis and the vicious circle of disseminated intravasal coagulation[3-4].

Most of the previous studies were focused on vasodilatory and vasoconstrictive prostaglandins pointing to an imbalance in favour of thromboxane (TXA_2) with a relative lack of prostacyclin (PGI_2)[5]. In the context of obstetrical disorders, arachidonate metabolites other than cyclooxygenase metabolites, especially Hydroxyeicosatetraenoic acids (HETEs) and Epoxyeicosatrienoic acids (EETs) have not yet been studied extensively.

This study investigates the whole spectrum of eicosanoids produced by intrauterine tissues using a HPLC method with radioactivity detection and it is focused on the placental production of HETEs in normo- and hypertensive pregnancies. The aim of our short term tissue culture system without destroying the tissue network was to imitate in vivo conditions as closely as possible.

MATERIALS AND METHODS

We performed tissue collection, culture and supernatant analysis as previously described[6]. Briefly, we collected specimens of placenta, fetal membranes, decidua and myometrium from women with uncomplicated pregnancies who underwent elective cesarean section between the 38th and 42th week. We examined representative tissues histologically.

Recent Advances in Prostaglandin, Thromboxane, and Leukotriene Research
Edited by Sinzinger *et al.*, Plenum Press, New York, 1998

As the eicosanoid production pattern of previously frozen tissue is known to differ markedly from the pattern of fresh tissues[7], we performed incubation experiments only with fresh tissues 0-30min after delivery. To remove the majority of blood compounds, the tissues were rinsed several times in cold HBSS (Gibco, Eggenstein-Leopoldshafen, Germany). Slices of 30-50mg wet weight were prepared carefully and allowed to recover from this procedure for 30 minutes before starting the incubation experiments.

We performed metabolism experiments with [^{3}H]-labeled arachidonic acid as described at Schäfer et al.[8]. After a 4hourly incorporation of 5µCi [^{3}H]-arachidonic acid at 37°C in 95% air / 5% CO_2 the release of the absorbed [^{3}H]-arachidonate was stimulated with $2x10^{-6}$M calcium ionophore A 23187 (Sigma, Deisenhofen, Germany) in a renewed medium containing 0.5% BSA (Serva, Heidelberg, Germany). Media were kept at -20°C and analyzed by reversed-phase HPLC with radioactivity detection within 3 months.

As described by Schäfer et al.[9] solid phase extraction was integrated into the HPLC system. Briefly, 500µl of centrifugated media were injected to a precolumn filled with µ Bondapak C18/Corasil at a flow rate of 2.5ml/min and by column switching transferred to a Novapak C18 column (both columns from Waters, Eschborn, Germany). Elution (all solvents from Baker, Groß-Gerau, Germany) was carried out first isocratically with 100% A (13 minutes), then %B was increased to 100% (39 minutes) at a flow rate of 1 ml/min. For radioactivity detection we used a radiomonitor LB 504 (Berthold, Wildbad, Germany). Daily control runs with tritiated standards (NEN-Dupont, Dreieich, Germany) were performed to carry out identification of peaks.

Additionally we collected placenta and fetal membrane specimens from women with pregnancy-induced hypertension between the 27th and 42nd week before and after the onset of labour. All hypertensive patients (n=14) had blood pressure of >140/90mmHg with first onset in pregnancy. PIH patients (n= 6) showed only a minor proteinuria <200mg/l, preeclamptic patients (n= 4) severe proteinuria of >300mg/l, patients with HELLP-syndrome (n= 4) had increased liver enzymes and/or decreased platelet count and/or upper abdominal pain and a proteinuria of >200mg/l. Concerning the onset of labour we adapted our control group, collecting control placentas also after spontaneous vaginal deliveries (n=6) in addition to the specimens from cesarean sections conducted before the onset of labour (n=9) to create a mixed control group.

We investigated their spontaneous release of 6-keto-$PGF_{1\alpha}$, TXB_2 and 12-HETE in supernatants of non-radioactive incubations in 1ml HBSS at 37°C in an atmosphere of 95% air / 5% CO_2 for 1 hour. 6-keto-$PGF_{1\alpha}$ and TXB_2 were quantified by commercial enzyme-immunoassays (Cayman, Ann Arbour, USA), 12-HETE by a commercial radioimmunoassay (Biermann, Bad Nauheim, Germany).

Data are expressed as net eicosanoid production in pg/mg/h or as quotients and shown as mean ± SEM and analyzed by *t* tests for independent samples comparing two groups.

RESULTS

Placenta, fetal membranes, decidua and myometrium of women without labour metabolised 15-50% of the tritiated arachidonic acid to eicosanoids. They showed different activity in arachidonate metabolism and different patterns of eicosanoids produced. The most active tissue was decidua with a metabolism rate of 49.5%.

The predominant enzymatic pathway used by all intrauterine tissues was the lipoxygenase pathway, cyclooxygenase products were only released in minor amounts. In several cases the amounts of prostaglandins released were found to be less than the detection threshold of the HPLC-system. Quantitatively they could only be analysed by

EIA. Despite the differences in eicosanoid production patterns of different intrauterine tissues, 12-HETE was detected as the main arachidonate metabolite in all culture media.

Myometrium was the only tissue investigated showing remarkable amounts of the cyclooxygenase product 6keto-$PGF_{1\alpha}$. Presence of labour led to a slight increase of 12-HETE production and a slight decrease of the TXB_2/6keto-$PGF_{1\alpha}$-ratio in placental incubations and did not change the ratio TXB_2/6keto-$PGF_{1\alpha}$ in fetal membranes.

The HPLC analysis of a placenta incubation medium deriving from a patient with PIH showed a markedly decreased release of 12-HETE. The tall peak of 12-HETE characterising every placental HPLC chromatogram was almost disapeared in this supernatant.

In order to investigate alterations of the release of 12-HETE in placentas of normo- and hypertensive pregnancies we used the incubation method without radiolabeling. We quantified the placental release of 12-HETE by radioimmunoassays. The placentas of hypertensive patients (n=11) released 132.5 pg/mg/h, whereas the control placentas (n=12) released 3868.1 pg/mg/h. That decrease was statistically significant ($p<0.0005$). The release of 12-HETE was almost the same in the clinical subgroups PIH, preeclampsia and HELLP-syndrome. Data are shown in figure 1.

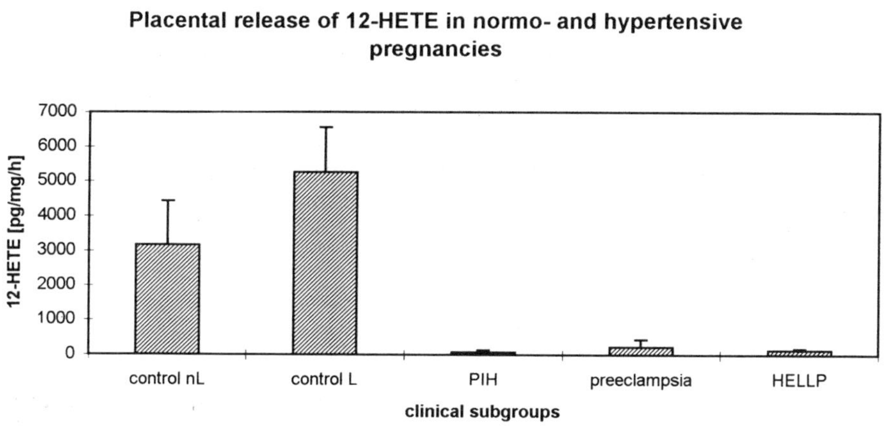

Figure 1. Placental release of 12-HETE in normo- and hypertensive patients, analyzed by RIA. Results are shown as mean ± SEM in pg per mg wet weigth per hour.
control nL = non-labour group (n = 8). control L = labour group (n = 4). PIH = pregnancy-induced hypertension group (n =5). preeclampsia = preeclamptic group (n = 3). HELLP = HELLP-syndrome (n = 3)

To evaluate these results we investigated the ratio thromboxane/prostacyclin released in the incubation media of the normo- and hypertensive patients. We quantified their stable metabolites TXB_2 and 6keto-$PGF_{1\alpha}$ by enzymeimmunoassays. The average TXB_2 synthesis by placentas was between 43-54 pg/mg/h and by fetal membranes between 5-10 pg/mg/h, the 6keto-$PGF_{1\alpha}$ release was between 1.1-1.6 and 1.1-1.9 pg/mg/h, respectively. The absolute eicosanoid production rates in placentas and fetal membranes were only slightly altered. In contrast to that, the ratio TXB_2/6keto-$PGF_{1\alpha}$ synthesis was doubled in the hypertensive group: 27.5 and 3.0 in control placentas (n=13) and fetal membranes and 52.4 and 8.9 in hypertensive tissues (n=14), respectively. This increase of the ratio was statistically significant ($p < 0.006$). Data of the clinical subgroups are shown in table 2. The preeclamptic subgroup showed the highest elevation of the TXB_2/6-keto-$PGF_{1\alpha}$ ratio compared with PIH and HELLP-syndrome (difference 30%). The ratios TXB_2/6keto-$PGF_{1\alpha}$

COMMENTS

Intrauterine tissues metabolised much greater quantities of their arachidonate ressources to HETEs than to prostaglandins with 12-HETE as their main arachidonate metabolite, as already demonstrated[8,6]. Placentas of hypertensive pregnancies showed a significantly decreased of placental 12-HETE production reaching only 1/29 of the control value. Walsh[2] also observed a decrease of 12-HETE in hypertensive pregnancies. In fact it is not only the prostaglandins that seem to act in the intrauterine compartment, it is also other metabolites, especially HETEs. Up till now, little is known about the biologic functions of 12-HETE in the intrauterine compartment. HETEs are supposed to have vasoactive properties and regulatory influences on the endocrine and the immune system. 12-HETE might probably act as an immunosuppressive and pregnancy-maintaining factor. It may influence the local environment during implantation and gestation[10]. A lack of 12-HETE might therefore result in inadequate maternal immunsuppression. As shown by Honn[11] 12-HETE is also thought to promote tumor invasion. We can only investigate term placentas, but one might speculate that it might play a role in trophoblastic invasion into maternal tissue as well. Its decreased production in PIH might contribute to insufficient trophoblastic invasion.

Our results concerning prostaglandin analysis are in accordance with other studies validating our study of 12-HETE production. The ratio TXB_2/6keto-$PGF_{1\alpha}$ was increased in all hypertensive cases. The relative lack of prostacyclin can explain the deficient maternal vasodilation, increased platelet aggregation and inadequate cytoprotection in PIH, augmenting the endothelial cell damage[1].

This study indicates the remarkable importance of HETEs, especially 12-HETE in the intrauterine compartment. The expression of other enzyme systems than the cyclooxygenase system in intrauterine tissues seems to be surprisingly strong and this pathway is dramatically altered in PIH diseases pointing to to an important role of these eicosanoids during pregnancy. It would be of great interest to further investigate the exact biological relevance of these observations.

REFERENCES

1. S.A. Friedman: Preeclampsia: a review of the role of prostaglandins. Obstet Gynecol 71: 122 (1988)
2. S.W. Walsh, V.M. Parisi: The role of arachidonic acid metabolites in preeclampsia. Semin Perinatol 10/4: 334 (1986)
3. C.W.G. Redman: Pre-eclampsia and the placenta. Placenta 12: 301 (1991)
4. J.M. Roberts, N.T. Taylor, J.T. Musci, C.A. Hubel, M.K. McLaughlin: Preeclampsia: an endothelial cell disorder. Am J Obstet Gynecol. 161: 1200 (1989)
5. H.P. Zahradnik, W. Schäfer, F.W. Casper, R.J. Seufert, T. Gaillard, H. Bettendorf: Urinary excretion of 6keto-$PGF_{1\alpha}$, TXB_2 and PGE_2 in pregnancy-induced hypertension and preeclampsia. Hypertension in pregnanacy 13(2), 201 (1994)
6. E. Arbogast, W. Schäfer, H.P. Zahradnik: Alterations of intrauterine eicosanoid production in pregnancy-induced hypertension: decreased production of 12-hydroxieicosatetraenoic acid in the placenta. Prostaglandins 51: 125 (1996)
7. B. Wetzka, W. Schäfer, M. Breckwoldt, H.P. Zahradnik: Eicosanoid production by frozen tissue is markedly changed. Prostaglandins, Leukotrienes and Essential Fatty Acids 50: 303 (1994)
8. W. Schäfer, H.P. Zahradnik, E. Arbogast, B. Wetzka, K. Werner, M. Breckwoldt: Arachidonate metabolism in human placenta, fetal membranes, decidua and myometrium: lipoxygenase and cytochrome P450 metabolites as main products in HPLC profiles. Placenta 17: 231 (1996)
9. W. Schäfer, C. Strandberg, T. Gaillard, H.P. Zahradnik: On-line sample clean up and HPLC analysis of prostaglandins in urine, amniotic fluid and plasma using a column switching technique. Eicosanoids 1: 101-106 (1988)
10. Y. Ihno, O. Ishihara, K. Kinoshita: Synthesis of 12-hydroxyeicosatetraenoic acid by human endometrium and decidua. Prostaglandins Leukotrienes a. Essential Fatty Acids 49(2): 609 (1993)
11. K.V. Honn: Lipoxygenases: role in metastasis and apoptosis. Oral presentation on the 10th international conference on prostaglandins and related compounds, Vienna Sep 22-27, 1996

FORMATION OF CYTOCHROME P450 METABOLITES OF ARACHIDONIC ACID BY HUMAN PLACENTA

Wolfgang Schäfer[1], Karin Werner[1], Horst Schweer[2], Johannes Schneider[1], Hans Peter Zahradnik[1].

[1] Universitäts-Frauenklinik, D-79106 Freiburg, Germany

[2] Universitäts-Kinderklinik, D-35033 Marburg, Germany

INTRODUCTION

Saturated and polyunsaturated fatty acids can be oxygenated by cytochrome P450. The P450 pathway of the arachidonic acid cascade has been etablished in many organs, tissues and cells e.g. the kidney, the cardiovascular system, the nervous system, the skin, the corneal epithelium and in the platelets (1). Arachidonic acid can be oxygenated by P450 in several ways. Epoxidations lead to the regio- and enantioselective formation of four epoxyeicosatrienoic acids (EET´s). These epoxides are enzymatically degraded by hydrolases to the corresponding vicinal dihydroxy compounds (DHET´s). In another important class of P450 reactions arachidonic acid is hydroxylated to various HETE´s. Up to now only few reports about the P450 pathway in intauterine tissues have been published. Only in one investigation (2) a HPLC-peak cochromatographing with 14,15-EET was further characterised by its enhancement after α-naphthoflavone-treatment and its inhibition by SKF525A.

METHODS

We investigated term placentas after uncomplicated pregnancies. Biopsies of 30- 50 mg were taken in short term tissue culture. In radiotracer experiments tissues were labelled with tritiated arachidonate for 4 hours as described previously (3). After medium exchange the explants were treated with calcium ionophore A23187 and the eicosanoids in the supernatant were analysed by reverse phase HPLC. In a part of these experiments the tissues were exposed to carbon monoxide as a potent cytochrome P450 inhibitor. In non-radioactive incubations selected index eicosanoids were quantified by enzyme- and radioimmunoassays. In another experiment sufficient amounts of cytochrome P450 metabolites of arachidonic acid should be obtained for GC/MS-identification. For this purpose ca. 420 biopsies from seven placentas were dispersed over 40 wells and treated as described before. The media of all incubations were combined and the DHET and EET fractions were collected after reverse phase HPLC.

RESULTS

After radiotracer experiments with tritiated arachidonic acid the main metabolites coeluted in reverse phase HPLC with DHET´s, 12-HETE and EET´s. No prostaglandin peaks were detected. The low production of prostaglandins compared to the high production of 12-HETE was confirmed by radio and enzymimmoassays. No immunoassays were available for EET´s and DHET´s.

In inhibition experiments with carbon monoxide specimens of five different placentas were taken in tissue culture. Prior to the final medium exchange and addition of calcium ionophor the tissues were treated in duplicate for 5 - 10 min with carbon monoxide and compared to untreated controls. The medium was analysed by reverse phase HPLC and the areas of the peaks coeluting with EET´s were evaluated. The peak areas were normalised for

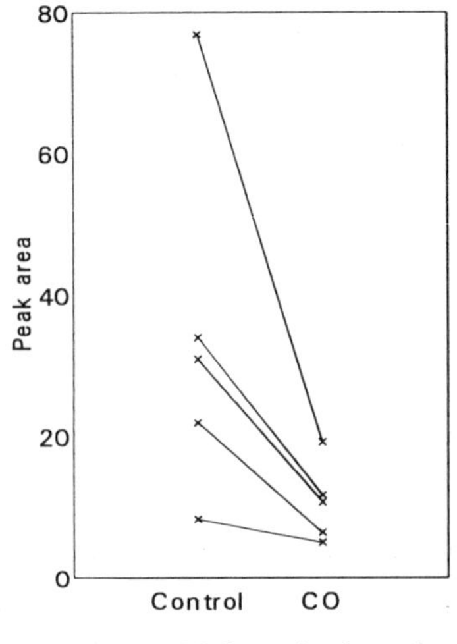

Figure 1. Inhibition of EET-formation by carbon monoxide

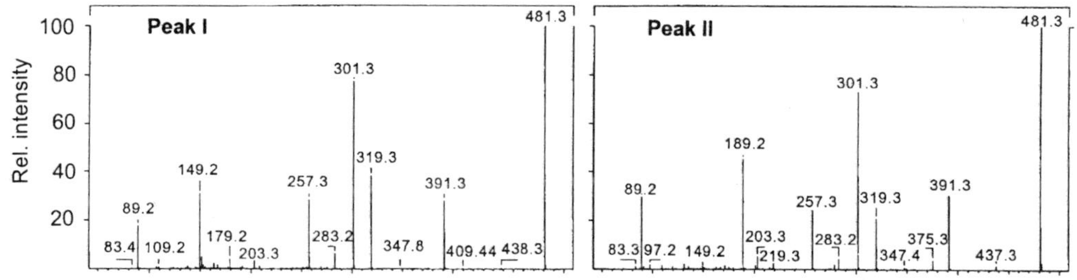

Figure 2. GC/MS identification of 11, 12-DHET and 14, 15 -DHET in human placenta. In the HPLC fraction coeluting with DHET´s two peaks I and II were separated by gas chromatography. The mass spectrum of peak I was identical with that of 11, 12-DHET, that of peak II with 14,15-DHET.

the injected radioactivity and the wet weigt of the tissues. Figure 1 shows that the formation of the metabolite cochromatographing with EET was inhibited by carbon monoxide.

After non-radioactive incubations the HPLC peaks cochromatographing with EET´s and DHET´s should be identified by GC/MS. We have not yet been successfull in identifying EET´s in placenta by GC/MS due to analytical problems. In the HPLC fraction coeluting with DHET´s we separated two peaks by gas chromatography. These peaks were identified by triple stage quadrupole tandem mass spectrometry with negative chemical ionisation as 11,12-DHET and 14,15-DHET. Both compounds displayed specific fragments at m/z 149 and m/z 189, respectively (Fig. 2).

DISCUSSION

In summary, we have described the formation of arachidonic acid epoxides and their hydrolysis products in short term tissue cultures of human term placenta. 11,12-DHET and 14,15-DHET were identified by GC/MS. The presence of these hydrolysis products indicates the preceeding formation of 11,12-EET and 14,15-EET by placental tissue. Our results suggest that the cytochrome P450 pathway is a main route of AA-metabolism in this organ. These findings are plausible, since placenta is rich in P450 enzymes (4). The biological significance of these findings for pregnancy and the birth process remains to be elucidated.

REFERENCES

1. Oliw, E.H. Oxygenation of polyunsaturated fatty acids by cytochrome P450 monooxygenases. Prog Lipid Res 33:329. 1994.
2. Patel, L., M.H.F. Sullivan, and M.G. Elder. Production of epoxygenase metabolite by human reproductive tissues. Prostaglandins 38:615. 1989.
3. Schäfer, W.R., H.P. Zahradnik, E. Arbogast, B. Wetzka, K. Werner, and M. Breckwoldt. Arachidonate metabolism in human placenta, fetal membranes, decidua, and myometrium: Lipoxygenase and cytochrome P450 metabolites as main products in HPLC profiles. Placenta 17:231. 1996.
4. Lewandowski, M. and E. Hodgson. Partial purification and characterization of cytochrome P-450 from human placenta. Int J Biochem 17:1149. 1985.

PROSTAGLANDIN ENDOPEROXIDE H SYNTHASE-2 EXPRESSION IN HUMAN AMNION CELLS: INVOLVEMENT OF TYROSINE KINASES IN THE REGULATION

Tamas Zakar, Jane E. Mijovic and David M. Olson

Perinatal Research Centre, University of Alberta
Edmonton, Alberta, Canada, T6G 2S2

INTRODUCTION

Intrauterine prostaglandins are crucially important in the initiation and maintenance of labor. It is well established that the concentrations of prostaglandin E_2 and $F_{2\alpha}$ increase in the amniotic fluid at late gestation and during labor, and that the gestational tissues (amnion, chorion and decidua) are the major sources of prostaglandins in the pregnant human uterus.[1] The activity of prostaglandin endoperoxide H synthase (PGHS, or cyclooxygenase), which is the rate limiting enzyme of prostanoid biosynthesis from precursor arachidonic acid, increases in the amnion[2] and the chorion laeve shortly before the onset of labor. This implies that the level of PGHS activity in the fetal membranes is an important factor determining the local production and concentration of prostaglandins. Analysis of PGHS mRNA abundance and its correlation with enzyme activity suggested that the expression of the inducible PGHS isoenzyme, called PGHS-2, is enhanced in both the amnion and the chorion laeve before and during labor, and that this isoform, not the constitutive PGHS-1, is responsible for the increased enzyme activity detected under these conditions.[3,4] Thus, increased prostanoid production associated with labor is likely a consequence of the induction of PGHS-2 in the fetal membranes.

RESULTS

PGHS-2 Induction in Amnion Cell Cultures

Cell cultures derived from gestational tissues are suitable models to investigate the regulation of intrauterine prostaglandin production. We have established a primary amnion cell culture system in order to study the control of PGHS expression in term human amnion. PGHS activity is low in confluent, serum-starved amnion cells, and can be increased by agonists such as epidermal growth factor (EGF) and the protein kinase C (PKC) activating phorbol ester TPA.[1] Okadaic acid (OA), a marine toxin, is a well characterized modulator of cellular regulation that acts by inhibiting the ubiquitous protein phosphatases PP1 and PP2A.[5] Okadaic acid also stimulates PGHS activity in the amnion cells. Using specific ribonuclease protection assays, we have found that EGF, TPA and OA elevate the steady-state level of PGHS-2 mRNA in the cell cultures within 2 h, while the level of PGHS-1 mRNA remains undetectable. Inhibition of protein synthesis by cycloheximide did not block, rather potentiated the effect of all three agonists suggesting that the stimulation was not

Recent Advances in Prostaglandin, Thromboxane, and Leukotriene Research
Edited by Sinzinger *et al.*, Plenum Press, New York, 1998

Table 1. Inhibition by herbimycin A of PGHS activity stimulated by EGF, TPA and OA

Treatment	PGE_2 output (pg/μg protein/h, mean ± SEM; n=6)	
Control	2.00 ± 0.14	
TPA (10 nM, 4 h)	14.82 ± 0.62	
herbimycin A plus TPA	0.62 ± 0.093	p<0.05 vs. TPA
Control	0.39 ± 0.06	
EGF (10 ng/ml, 2 h)	17.11 ± 0.81	
herbimycin A plus EGF	1.22 ± 0.13	p<0.05 vs. EGF
OA (100 nM, 16 h)	25.62 ± 1.01	
herbimycin A plus OA	9.86 ± 0.46	p<0.05 vs. OA

Cells were pretreated with herbimycin A (1μM) for 16 h, followed by agonists as indicated. PGHS activity was determined by incubating the cells with 10 μM arachidonic acid for 1 h, and measuring PGE_2 output with radioimmunoassay. Separate representative experiments are presented for TPA and the other two agonists, respectively. ANOVA and the Newman-Keuls test were performed to statistically analyze the data.

mediated by an intervening protein or peptide. Since the non-passaged primary cultures essentially reflect the cellular composition and heterogeneity of amnion tissue, we employed *in situ* hybridization to examine the distribution of EGF- TPA- and OA-responsive cells in the cultures. The results showed that nearly all cells exhibited higher levels of PGHS-2 mRNA after treatment with each of the three agonists, indicating that EGF- TPA- and OA-responsive pathways coexist in the amnion cells. Furthermore, down-regulation of PKC by prolonged phorbol ester treatment or co-incubation with chelerythrine, a specific PKC inhibitor, did not block the stimulation of PGHS activity by OA. This suggested that OA acted by a mechanism that was independent of PKC.

Involvement of Tyrosine Kinases

The involvement of tyrosine kinases in the induction of PGHS-2 expression was examined using the benzoquinonoid anzamycin antibiotic, herbimycin A. Herbimycin A is a cell permeable and potent inhibitor of a variety of receptor and non-receptor tyrosine kinases, but has no effect on any of the major serine- and threonine specific protein kinase types.[6] As shown in Table 1, herbimycin A pretreatment prevented completely (>90%) the induction of PGHS activity by EGF and TPA, and partially (≈60%) the induction of the enzyme by OA. The increase of PGHS-2 mRNA levels was also blocked by the antibiotic (Figure 1), indicating that the tyrosine kinase dependent step(s) were involved in the accumulation of mRNA encoding the enzyme. Immunoblot analysis demonstrated early signalling through tyrosine phosphorylation in response to EGF and TPA. Within 5 minutes following EGF treatment, tyrosine-phosphorylation of the EGF receptor, and 120, 110, and 77 kDa protein fractions were detected. TPA treatment also resulted in the increased tyrosine

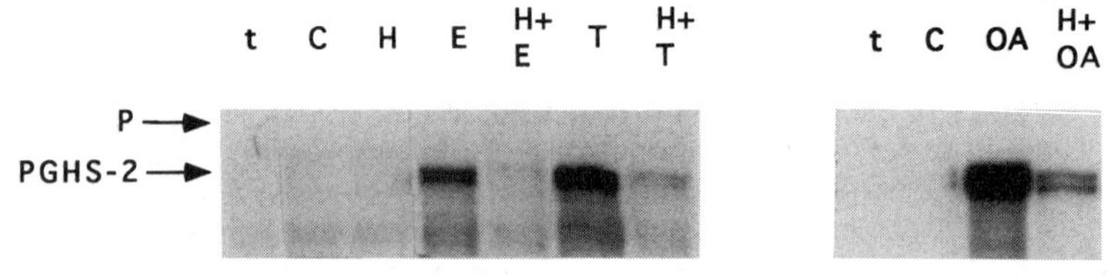

Figure 1. Inhibition by herbimycin A of the stimulation of PGHS-2 mRNA level by EGF, TPA and OA. Amnion cell cultures were preincubated without or with herbimycin A (0.5 μM) for 16 h, followed by treatment with EGF (10 ng/ml, 2 h), or TPA (10 nM, 4 h), or OA (100 nM, 2 h). Cytoplasmic RNA was isolated, and PGHS-2 mRNA abundance was determined by ribonuclease protection assay. The positions of the mRNA signal (PGHS-2) and the undigested probe (P) are indicated by arrows. (The autoradiographic signal generated by the undigested probe is not shown.) Representative individual experiments are presented on the two panels with EGF and TPA (left) and OA (right) stimulation. C, control; H, herbimycin A pretreatment; E, T, and OA denote EGF, TPA and OA treatment, respectively; t, assay background, determined using yeast tRNA.

phosphorylation of the same three cellular protein fractions within 5 min. On the other hand, OA did not affect the phosphotyrosine content of these proteins for at least one hour. In cells pretreated with herbimycin A (0.5 μM for 16 h), the phosphorylation of the EGF receptor in response to the growth factor diminished. The phosphorylation of the 110 kDa phosphoprotein was also reduced by herbimycin A in unstimulated cells and following TPA- and OA-stimulation. Collectively, these results pointed to the EGF receptor and a tyrosine kinase with preference to a 110 kDa cellular substrate as functionally important upstream targets of herbimycin A. Increased tyrosine phosphorylation was not detected early after OA treatment although herbimycin A reduced the induction of PGHS-2. This was consistent with the possibility that the drug affected downstream step(s) on the signalling pathway of OA.

Signalling through Erk and Jnk

The extracellular signal-regulated protein kinase (Erk) and c-Jun N-terminal kinase (Jnk) families of serine/threonine specific protein kinases are downstream members of parallel signalling cascades that directly regulate gene expression.[7] We have determined the activity of these kinases under conditions where PGHS-2 mRNA is induced in order to assess the contribution of the corresponding pathways to the control of enzyme expression. Immuncomplex kinase assays were performed to measure Erk and Jnk activity in cell extracts. EGF stimulated Erk activity (over 4-fold) at 10 min, with transient kinetics. This was followed by the activation of Jnk (more than 10-fold) at 30 min. OA activated Jnk over 7-fold at 60 min, but did not activate Erk. TPA did not stimulate either Erk or Jnk activity. Figure 2 shows

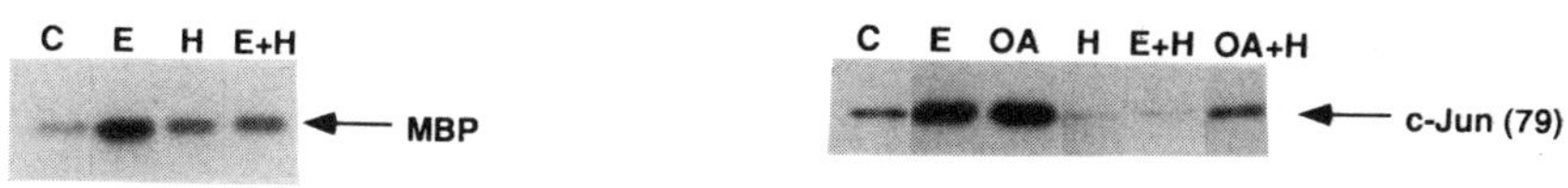

Figure 2. The effect of herbimycin A on Erk and Jnk activation in amnion cells. Left panel: Amnion cell cultures were pretreated with vehicle or 0.5 μM herbimycin A (H) for 16 h followed by stimulation with 10 ng/ml EGF for 10 min (E). Erk was immunoprecipitated from cell extracts and its activity determined by phosphorylation of myelin basic protein (MBP). Right panel: Cultures with or without herbimycin A pretreatment were stimulated with EGF for 30 min, or okadaic acid (OA, 100 nM) for 1 h. Jnk-1 was immunoprecipitated from cell extracts, and its activity measured by phosphorylating a c-Jun(1-79)-GST fusion protein substrate (c-Jun(79)). MBP and c-Jun (79) phosphorylation was evaluated by SDS-polyacrylamide gel electrophoresis and autoradiography. Control cultures (C) were untreated.

that signalling by EGF through Erk and Jnk, and by OA through Jnk, respectively, was blocked in herbimycin A-treated cells. The basal activity of Jnk was also suppressed by the tyrosine kinase inhibitor. These results suggest strongly that Jnk activation is crucial for PGHS-2 induction by OA and EGF, though a role for Erk in mediating the actions of EGF can not be excluded. Basal and stimulated Jnk activity appear to be dependent on tyrosine kinase activity. At the same time, PKC-dependent signalling in the amnion cells occurs by pathway(s) that do not involve Erk or Jnk activation.

CONCLUSIONS

The results of this study demonstrate that growth factor stimulation, PKC activation and protein phosphatase inhibition each results in elevated PGHS activity and PGHS-2 mRNA accumulation in the amnion cells. These effects are phenomenologically similar to the changes occurring in the amnion membrane at parturition.[3] PGHS-2 expression is regulated by multifarious mechanisms in these cells, however, the significance of the individual pathways in the *in vivo* regulation of enzyme activity is unclear. Tyrosine kinases are involved at different levels in all the regulatory pathways. This suggests the intriguing possibility that tyrosine kinase inhibitor compounds administered *in vivo* may block the induction of PGHS-2 in the gestational tissues thereby delaying the onset of labor. Furthermore, such treatment may potentially result in isoform-specific, *and* tissue-specific inhibition of PGHS-2 depending on the timing of drug administration. Improved tissue-specificity of enzyme inhibition may lead to reduced side effects and increased usefulness for therapy.

Acknowledgments

Funding for this research has been provided by the MRC of Canada and the Alberta Heritage Foundation for Medical Research.

REFERENCES

1. D.M. Olson, and T. Zakar, Intrauterine tissue prostaglandin synthesis: regulatory mechanisms, *Semin. Reprod. Endocrinol.* 11:234 (1993).
2. F.J. Teixeira, T. Zakar, J.J. Hirst, F. Guo, D.W. Sadowsky, G. Machin, N. Demianczuk, B. Resch, and D.M. Olson, Prostaglandin endoperoxide-H synthase (PGHS) activity and immunoreactive PGHS-1 and PGHS-2 levels in human amnion throughout gestation, at term, and during labor, *J. Clin. Endocrinol. Metab.* 78:1396 (1994).
3. J.J. Hirst, F.J. Teixeira, T. Zakar, and D.M. Olson, Prostaglandin endoperoxide-H synthase-1 and -2 messenger ribonucleic acid levels in human amnion with spontaneous labor onset, *J. Clin. Endocrinol. Metab.* 80:517 (1995).
4. J.E. Mijovic, T. Zakar, T. Nairn, and D.M. Olson, Prostaglandin endoperoxide H synthase-2 expression and activity increases with term labor in human chorion, *Am. J. Physiol.* (in press).
5. P. Cohen, C.F.B. Holmes, and Y. Tsukitani, Okadaic acid: a new probe for the study of cellular regulation, *TIBS* 15:98 (1990).
6. Y. Uehara, and H. Fukazawa, Use and selectivity of herbimycin A as inhibitor of protein-tyrosine kinases, *Methods Enzymol.* 201:370 (1991).
7. E. Cano, and L.C. Mahadevan, Parallel signal processing among mammalian MAPKs, *TIBS* 20:117 (1995).

PROSTAGLANDIN E_2 PRODUCTION BY ENDOGENOUS SECRETION OF INTERLEUKIN-1 IN DECIDUAL CELLS FROM TERM FETAL MEMBRANE

Osamu Ishihara[1], Hitomi Numari[1], Masahiro Saitoh[1], Yoshinori Arai[2], Hitoshi Takanashi[2], Hiroshi Kitagawa[2], Katsuyuki Kinoshita[1]

[1]Department of Obstetrics and Gynaecology, Saitama Medical Centre, Saitama Medical School, Kawagoe Japan
[2]Pharma R&D Division, Hoechst Japan Limited, Kawagoe Japan

INTRODUCTION

The intrauterine tissues including decidua have been investigated as the sites of prostaglandin (PG) production in connection with its involvement in the onset or progress of labor (1-4). The recent papers suggested that the intrauterine infection was responsible for the pathophysiology of preterm labor through the set of events from the generation of various types of cytokines such as interleukin-1 (IL-1), the induction of cyclooxygenase (COX) to the PG synthesis in these tissues (5-7). On the other hand, IL-1 has been reported to induce COX-2, an inducible type of COX in cultured decidual cells (8). Therefore, COX-2 induction in decidua by IL-1 would play a role in the initiation of spontaneous labor as well as preterm labor. Though the physiological role of IL-1 secreted in decidual cells remains still uncertain (9), this study was aimed to reveal endogenous role of IL-1 synthesized and secreted by decidual cells in the formation of PGE2. In addition, we investigated the selective role of COX-2 in decidual cells for the initiation or maintenance of labor by using a specific inhibitor for the enzyme, NS-398.

MATERIALS AND METHODS

Human decidual cells were obtained from fetal membranes at the time of delivery by either normal spontaneous vaginal delivery or elective cesarean section due to the indication of previous section. The women had not been treated with anti-inflammatory drugs such as aspirin or indomethacin, and had no pathological condition. Decidual stromal cells were prepared by the method described previously (10). The cells were cultured in 199 medium containing 10% FBS at 37°C in 95% air: 5% CO2 for three days, then the experiments started. At the end of the experiments, the culture media were individually stocked under -20°C until the enzyme immunoassay for PGE2 or IL-1β. Following the enzyme digestion and the separation of cells, a part of the decidual cells was lysed and the whole protein extracts were subjected to SDS-PAGE and electroblotted after the electrophoresis. The band detected by anti-human IL-1β antibody on membranes were visualized with chemiluminescent detection system.

Recent Advances in Prostaglandin, Thromboxane, and Leukotriene Research
Edited by Sinzinger *et al.*, Plenum Press, New York, 1998

RESULTS

The levels of basal production of PGE2 by decidual cells ranged from 0.02 to 0.77 ng/ml/24 hours/0.25x10^6 cells (0.30±0.25: n=7) (mean±SD) in the samples from the cells obtained at ECS, and they were significantly lower than those from the cells obtained at NVD (1.83±0.44: n=4). The addition of IL-1β (0.1 ng/ml) increased PGE2 concentration in culture medium 2 to 20 times more than the basal levels in both cells from ECS and NVD after 24 hours of incubation.

A specific inhibitor of COX-2, NS398, dose-dependently decreased the production of PGE2 from decidual cells obtained at the time of NVD either stimulated or not stimulated by IL-1 as shown in Table 1. On the contrary, NS398 only reduced IL-1-stimulated production of PGE2 from ECS cells and had no effect on basal PGE2 production from the cells.

The recombinant soluble human IL-1 receptor type I (rhuIL-1R) inhibited PGE2 production stimulated by IL-1 in a dose dependent manner, and the stimulatory effect of 1ng/ml of IL-1 was completely suppressed by the concentration of 10 μg/ml of rhuIL-1R (Table 2). In addition, the basal PGE2 production by NVD cells was also suppressed by the lower concentration of rhuIL-1R, namely 1 μg/ml. On the other hand, the secretion of IL-1 by decidual stromal cells was higher in the cells derived from NVD (Table 3).

Western blotting using specific antibody for IL-1β demonstrated the existence of 31kD preprocessed IL-1 protein inside the decidual cell lysate both from NVD and ECS cells. The intensity of staining of IL-1β from NVD was apparently stronger, i.e., 3 times semiquantitatively.

Table 1. The effect of NS-398 on PGE2 production (ng/ml) by decidual cells obtained at normal vaginal delivery. All data are means±S.D. (n=4), and are representative of three separate experiments.

NS-398(M)	0	10^{-8}	10^{-7}	10^{-6}
control	2.03±0.29	1.67±0.44	1.10±0.12	0.75±0.06
IL-1(1ng/ml)	6.41±1.23	3.66±0.75	1.02±0.06	0.69±0.05

Table 2. The effect of IL-1 receptor type I (IL-1R) on PGE2 production (pg/ml) by decidual cells obtained at elective cesarean section. All data are means±S.D. (n=4), and are representative of three separate experiments. N.A.: not available.

IL-1R(μg/ml)	0	0.1	1	10
control	30±7.8	N.A.	N.A.	N.A.
IL-1(1ng/ml)	517±7.6	320±117	122±32	28±18

Table 3. The secretion of IL-1 (pg/ml) by decidual stromal cells during 24 hours' culture on day 4. All data are means±S.D. (n=4)

mode of delivery	C/S	NVD
IL-1(pg/ml)	1.31±0.51	3.55±3.14

DISCUSSION

As the preterm labor is one of the emergent problem to be investigated in perinatal medicine, IL-1 has been focused as a key cytokine existing in the amniotic fluid obtained from patients with preterm labor and its role has been implicated in PG production at the time of intrauterine infection (4,5,9). The cytokine is a known inducer of COX-2, resulting PG production in amniotic cells (11), decidual cells (8) as well as in other types of cells in vitro (12,13), however, the actual role of IL-1 remains uncertain in cases of normal spontaneous labor.

The results of this study using NS398 (14) strongly suggest that PGE2 production in decidual cells is mainly dependent upon COX-2 after the onset of labor. On the contrary, the PGs derived from COX-1 might be abundant before the onset of labor as the PG production by decidual cells of pre-labor was not suppressed by the COX-2 inhibitor. Therefore, the predominance of different enzyme before and after the onset of labour might be a clue for the initiation of labour.

We know that the family of IL-1 incorporates a complex control mechanism for its biological effects (15). There are two types of IL-1, i.e., IL-1α and IL-1β known as bioactive ligands, whereas a non-bioactive ligand, IL-1 receptor antagonist, is present in various kinds of biological fluid. Furthermore, two types of receptors are demonstrated not only at the site of cell membranes but in biological fluids as soluble receptors (16,17). Recent investigations have made it possible to use recombinant soluble human IL-1receptors in experiments (18), which showed complete inhibiton on the effect of IL-1 in this study, although the physiological roles of these substances in vivo seem to be unknown.

Therefore,the implication of this study is that IL-1 released from decidual cell might stimulate the PG production in the own cells suggesting possible involvement of IL-1 in the generation of PGs during labor in vivo.We do not know the detailed mechanism of COX-2 induction in decidual cells at present, however, further evidence was that the production and secretion of IL-1 in decidual cells were up-regulated after the onset of labor, which was coincided with production of PGs via COX-2. It is necessary to elucidate the regulatory mechanism of IL-1production and secretion by decidual cells because IL-1 was present in decidual cells as 31kD preprocessed form of IL-1 protein, (19) .

In conclusion, it seems likely that the induction of COX-2 by autocrine or paracrine fashion of IL-1 synthesized and released in decidual cells might be responsible for PGE2 production by decidual cells after the onset of labor.

ACKNOWLEDGMENT

We thank Otsuka Pharmaceuticals Co. (Tokushima, Japan) and Taisho Pharmaceutical Company (Tokyo, Japan) for their contribution to chemicals. A part of this work is supported by grant-in-aid for scientific research from the Ministry of Education, Science and Culture.

REFERENCES

1. D.M.Olson, T. Zakar, and B.F. Mitchell, Prostaglandin synthesis regulation in intrauterine tissues, in: *Molecular Aspects of Placental and Fetal Membrane Autacoids,* G.E. Rice and S.P. Brennecke, eds., CRC Press, (1993).
2. J.J. Hirst, F.J. Teixeira, T. Zakar, and D.M. Olson, Prostaglandin endoperoxide-H synthase-1 and -2 messenger ribonucleic acid levels in human amnion with spontaneous labour onset, *J Clin Endocrinol Metab* . 80:517 (1995).
3. F.J. Teixeira, T. Zakar, J.J. Hirst et al, Prostaglandin endoperoxide-H synthase (PGHS) activity and immunoreactive PGHS-1 and PGHS-2 levels in human amnion throughout gestation, at term, and during labour, *J Clin Endocrinol Metab.* 78:1396 (1994).
4. H. Khan, O. Ishihara, M.H.F. Sullivan et al, Changes in decidual stromal cell function associated with labour, *Br J Obstet Gynaecol* . 99:10 (1992).
5. R. Romero, M. Mazor, R. Morrotti et al, Infection and labour. VII. Microbial invasion of

the amniotic cavity in spontaneous rupture of membranes at term, *Am J Obstet Gynecol*. 166:129 (1992).
6. E. Hirsch, I. Saotome and D. Hirsh, A model of intrauterine infection and preterm delivery in mice. *Am J Obstet Gynecol.* 172:1598 (1995).
7. D.J. Dudley, C. Hunter, M.D. Mitchell et al, Clinical value of amniotic fluid interleukin-6 determinations in the management of preterm labour, *Br J Obstet Gynaecol.* 101:592 (1994).
8. O. Ishihara, H. Khan, M.H.F. Sullivan et al, Interleukin-1 stimulates decidual stromal cell cyclooxygenase enzyme and prostaglandin production, *Prostaglandins.* 44:43 (1992).
9. R. Romero, Y.K. Wu, D.T. Brody, E et al, Human decidua: a source of interleukin-1, *Obstet Gynecol.* 73:31 (1989).
10. H.Khan, O. Ishihara, M.G. Elder et al, A comparison of two populations of decidual cells by immunocytochemistry and prostaglandin production, *Histochemistry.* 96:149 (1991).
11. R. Romero, S. Durum,C.A. Dinarello et al, Interleukin-1 stimulates prostaglandin synthesis by human amnion, *Prostaglandins.* 37:13 (1989).
12. I. Knott, M. Dieu, M. Burton et al, Induction of cyclooxygenase by interleukin 1: Comparative study between human synovial cells and chondrocytes, *J Rheumatol.* 21:462 (1994).
13. S. Shimonovitz, S. Yagel, E. Anteby et al, Interleukin-1 stimulates prostaglandin E production by human trophoblast cells from first and third trimesters, *J Clin Endocrinol Metab.* 80:1641 (1995).
14. N. Futaki, S. Takahashi, M. Yokoyama et al, NS-398, a new anti-inflammatory agent, selectively inhibits prostaglandin G/H synthase/cyclooxygenase (COX-2) activity in vitro, *Prostaglandins.* 47:55 (1994).
15. F. Colotta, S.K. Dower, J.E. Sims et al, The type ll 'decoy' receptor: a novel regulatory pathway for interleukin 1, *Immunol Today.* 15:562 (1994).
16. J.A. Symons, J.A. Eastgate and G.W. Duff, Purification and characterization of a novel soluble receptor for interleukin-1, *J Exp Med.* 174:1251 (1991).
17. M. Svenson, M.B. Hansen,P. Heegaard, P et al, Specific binding of interleukin-1 (IL-1)β and IL-1 receptor antagonist (IL-1ra) to human serum: high-affinity binding of IL-1ra to soluble IL-1 receptor type I, *Cytokine.* 5:427 (1993).
18. J.E. Sims, R.B. Acres, C.E. Grubin et al, Cloning the interleukin 1 receptor from human T cells, *Proc Natl Acad Sci USA.* 86:8946 (1989).
19. D.P. Cerretti, C.J. Kozlosky, B. Mosley et al, Molecular cloning of the interleukin-1β converting enzyme, *Science.* 256:97 (1992).

THE DEVELOPMENT OF A HIGHLY SPECIFIC, SENSITIVE ENZYME IMMUNOASSAY FOR PROSTAGLANDIN E_1

Rhonda Linsky, Jeanette Buckwalter, Nancy Schmidt, Barbara Scheuer, & Russell Hart.

Assay designs, inc
1327 Jones Drive
Ann Arbor, MI, USA

INTRODUCTION

Prostaglandin E_1 (PGE_1) is synthesized from DGLA, dihomo-γ-linolenic acid[1]. PGE_1 has been shown to have a number of biological actions, including vasodilation[2], proliferation of vascular smooth muscle cells[3], platelet aggregation[4] and has been shown to have insulin-like actions[5,6]. It's effects are induced by receptor mediated elevation of cAMP[7]. It is the major prostaglandin in semen[8,9]. Interest in the measurement of PGE_1 in a variety of biological matrices has been hampered by the lack of specific methods for the easy detection of this molecule. We report on the development of a rapid, highly specific enzyme immunoassay utilizing stable, color-coded reagents for PGE_1.

We have developed an alkaline phosphatase conjugate of PGE_1 that uses N-hydroxysuccinimide chemistry. We used PGE_1 that was synthesized under ISO 9000 regulations (Cascade Biochem, Reading, UK) as the starting material for this conjugation. We used this material because small impurities in less documented synthetic protocols may give rise to spurious results in the immunoassay. The PGE_1 was dissolved in dry dimethyl formamaide (DMF) and stirred. An excess of H-hydroxysuccinimide (NHS) was added, followed by dicyclohexylcarbodiimide DCC). Both the NHS and DCC were anhydrous solutions in DMF. Special care was taken with the DMF to maintain dryness. To the solvent (Pierce Chemical, Rockland, IL, USA) was added 4A molecular sieve containing a water-sensitive dye system. This mixture was stirred at room temperature overnight.

Bovine stomach alkaline phosphatase (AP, Biozyme, San Diego, CA, USA) was diluted in 0.5M borate buffer at pH 8.5 and stirred at room temperature. An excess of the PGE_1 / NHS/DCC mixture was added to the AP solution at different ratios of PGE_1 to AP. The reaction mixtures were stirred for 30 minutes, then 10 μL of a 10% lysine solution in water was added to the mixture to stop any further reaction. The mixtures were stirred at room temperature for a further 15 minutes.

Recent Advances in Prostaglandin, Thromboxane, and Leukotriene Research
Edited by Sinzinger *et al.*, Plenum Press, New York, 1998

Disposable gel permeation columns (10 DG, Bio-Rad, Richmond, CA, USA) were extensively washed with assay buffer, which is a Tris based buffer containing 0.1% bovine serum albumin, 0.1% sodium azide and other components. The PGE_1-AP mixtures were applied to the columns and 1 mL fractions were collected. The fractions containing AP activity were measured by addition of p-nitrophenyl phosphate substrate to 1 µL aliqouts. Instantaneous color generation was observed in fractions #4 and 5. The substrate reaction was allowed to proceed for approximately 30 seconds before terminating the reaction with 2M sodium hydroxide stop solution. The extent of reaction was measured on a Dynatech plate reader at 405 nm. Fractions containing highest activity were aliquoted and frozen at -20°C.

This enzyme conjugate is used with an antibody to PGE_1 to develop an enzyme immunoassay that can be run in less than 3 hours. Samples or standards (100 µL) are added to precoated wells of a microtiter plate. The diluted PGE_1-AP conjugate (50 µL) is added, followed by the PGE_1 specific antibody (50 µL). After an incubation for 2 hours at room temperature the plate is washed and substrate (200 µL) added. The substrate reaction is run for 45 minutes when 2M sodium hydroxide is added to stop color development. Color is read at 450 nm. on a microtiter plate reader. A typical standard curve is shown below.

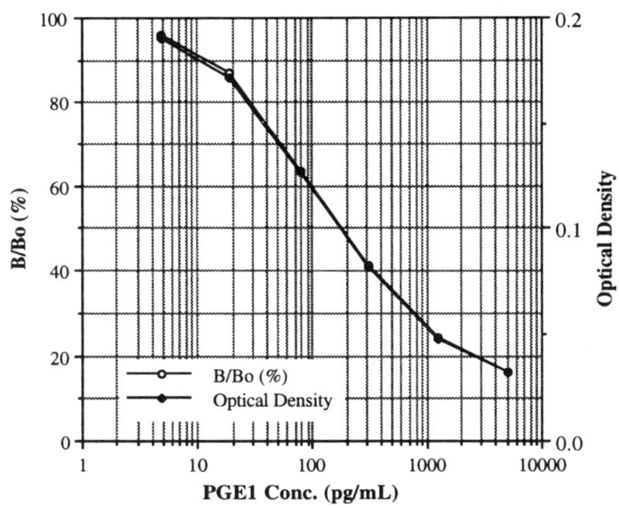

The sensitivity of the assay, as measured by determining 2 standard deviations of the zero binding point, is below 0.6 pg per well. Our new chemiluminescent immunoassay will measure down to 176 fg of PGE_1 per well.

<u>Other performance characteristics:</u>

Cross Reactivity

An obvious problem with immunoassay detection of eicosanoids is cross reactivity to structurally related compounds and metabolites of the antigen, PGE_1. We therefore tested this immunoassay with a range of eicosanoids, including PGE_0, PGE_2, and PGE_3. The data obtained by dissolving the cross reactant (purity checked by N.M.R. and other analytical methods) in Assay Buffer at concentrations from 4.8 to 500,000 pg/mL. These samples were then measured in the PGE_1 assay, and the measured PGE_1 concentration at 50% B/Bo calculated. The % cross reactivity was calculated by comparison with the actual concentration of cross reactant in the sample and expressed as a percentage. The data is shown below:

Compound	Cross Reactivity
PGE_1	100%
PGE_2	6.5 %
PGE_3	2.22%
PGE_0	1.45%
$PGF_{1\alpha}$	0.14%
$PGF_{2\alpha}$	0.04%
6-keto-$PGF_{1\alpha}$	<0.1%
PGA_2	<0.1%
PGD_2	<0.1%
PGB_1	<0.1%
13,14-dihydro-15-keto-$PGF_{2\alpha}$	<0.1%
6,15-keto-13,14-dihydro-$PGF_{1\alpha}$	<0.1%
Thromboxane B_2	<0.1%

The measured cross reactivities for the assays show that PGE_0, the major metabolite of PGE_1, shows only relatively minor reactivity in the assay of 1.45%. It is clear from these numbers that this assay will measure PGE_1 selectively in the presence of other related eicosanoids.

Linearity

We also investigated the assay for the ability to measure PGE_1 over a range of dilutions and so a sample containing 50,000 pg/mL PGE_1 was diluted 7 times 1:2 in the kit Assay Buffer and measured in the assay. The data was plotted graphically as actual PGE_1 concentration versus measured PGE_1 concentration. The line obtained had a slope of 1.0681 and a correlation coefficient of 1.000.

Precision

Intra-assay precision was determined by taking samples containing low, medium and high concentrations of PGE_1 and running these samples multiple times (n=24) in the same assay. Inter-assay precision was determined by measuring three samples with low, medium and high concentrations of PGE_1 in multiple assays (n=8).

The precision numbers listed below represent the percent coefficient of variation for the concentrations of PGE_1 determined in these assays as calculated by a 4 parameter logistic curve fitting program.

	PGE_1 Concentration (pg/mL)	Intra Assay %CV	Inter Assay %CV
Low	53.29	4.6%	
Medium	246.34	9.5%	
High	1,103.2	13.7%	
Low	49.21		9.3%
Medium	213.7		11.0%
High	736.94		6.2%

It is obvious that this assay displays good precision over the range of concentrations that PGE_1 is found at in normal samples.

PGE$_1$ Sample Recoveries

We tested the ability of this assay to measure PGE$_1$ in human samples and tissue culture media without the need for extraction steps. Most other immunoassays for eicosanoids recommend acidification of the samples followed by solvent extractions, and separation on a phase separation media[10-13]. For this assay to be convenient we investigated the ability of sample dilution to give accurate measurement of PGE$_1$ simply by dilution. PGE$_1$ concentrations were measured in a variety of different samples including tissue culture media, filtered human saliva, serum, plasma, and urine. PGE$_1$ was spiked into the undiluted samples of these media which were then diluted with the kit Assay Buffer and then assayed in the kit. The following results were obtained:

Sample	**% Recovery**	**Recommended Dilution**
Tissue Culture Media	90-110%	None
Human Saliva	107.2%	1:10
Human Urine	109.9%	1:50
Human Serum	87.0%	1:20
Human Plasma	107.7%	1:20

Therefore this assay can measure PGE$_1$ in a variety of biological matrices without the need to carry out extractions of the sample.

Conclusion

The PGE$_1$ enzyme immunoassay we have developed should result in the rapid (< 3 hour) specific determination of PGE$_1$ concentration in a variety of biological matrices. Our new chemiluminescent assays for PGE$_1$ will measure at least 4-fold lower in concentration in the same assay time.

References

1. J. Mai, et. al., Prostaglandins., (1980), 20, 187.
2. A.G. Olsson & L.A. Carlson, in "Advances in Prostaglandin and Thromboxane Research.", (1976), Raven Press, NY, B. Samuelsson & R. Paoletti, (Eds.).
3. D.G. Cornwell, et. al., Lipids, (1979), 14, 194-207.
4. A.J. Ally & D.F. Horrobin., Prostaglandins and Medicine., (1980), 4, 431-438.
5. H.A. Haessler & J.D. Crawford., J. Clin. Invest.., (1967), 46, 1065-69.
6. D.F. Horrobin & M.S. Manku, International Prostaglandin Conference, (1979), pg 53.
7. N.P. Kurstjens, et. al., Biochem. Biophys. Res. Comm., (1990), 167, 1162.
8. L. Speroff, et. al., Am. J. Obstet. Gynec.., (1978), 107, 1111.
9. D. Rubin & M. Laposata., J. Biol. Chem., (1991), 226, 23618.
10. K. Green, et. al., Anal. Biochem, (1973), 54, 434.
11. J. Frolich, et. al., J. Clin. Invest., (1975), 55, 763.
12. J.E. Shaw & P.W. Ramwell, Meth. Biochem. Anal., (1969), 17, 325.
13. K. Green, et. al., Adv. Prostaglandin & Thromboxane Res., (1978), 5, 15.

THE USE OF CHEMILUMINESCENCE TO DETECT EICOSANOIDS WITH GREATER SPEED AND SENSITIVITY

Barbara Scheuer, Nancy Schmidt, Dana Biroscak, Rhonda Linsky, & Russell Hart.

Assay designs, inc
1327 Jones Drive
Ann Arbor, MI, USA

INTRODUCTION

The use of chemiluminescence detection for measurement of low concentration biological molecules such as eicosanoids has been developed. We report on the comparison of colorimetric enzyme generated detection with enzymatic chemiluminescent detection. We have also developed methods to indirectly attach molecules such as prostaglandin E_2 (PGE_2), 11ß-$PGF_{2\alpha}$, and others to carrier molecules and to attach chemiluminescent Acridinium esters to these same carriers. Recently we have extended this work to attach Acridinium esters to prostaglandins using homo- and hetero-bifunctional reagents. These prostaglandin conjugates have been used to develop super sensitive and rapid immunoassays.

We have previously described the comparison of colorimetric and chemiluminescent detection using p-nitrophenyl phosphate and dioxetane phosphates as the signal generating methods for our Correlate-EIA and -CLIA for PGE_2 in microtiter plates[1]. The data is shown below.

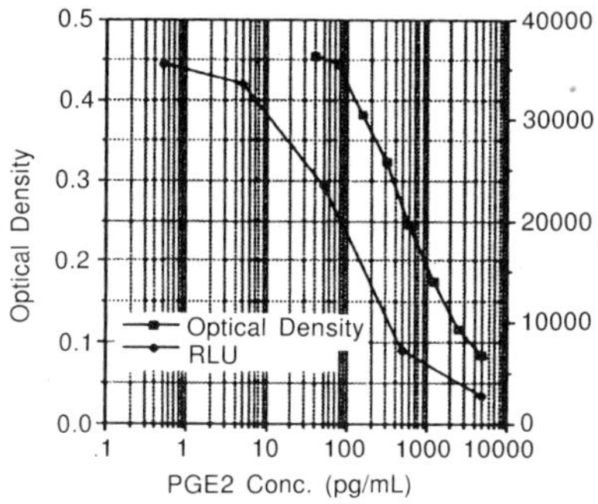

In Figure 1 the standard curves for colorimetric (EIA) and chemiluminescent (CL) 2 hour immunoassays are compared. In both cases a PGE_2-alkaline phosphatase conjugate was used with our monoclonal antibody to PGE_2. Primary incubation time with standards or sample was 2 hours. Either colorimetric or chemiluminescent substrate was added to the washed plates. Color development was for 45 minutes, while the luminescent signal was read after 30 minutes. It can be seen that the use of CL detection improves both the span of the assay (5,000 to 39 pg/mL for EIA vs. 5,000 to 0.5 pg/mL for CL) and the sensitivity of the assay (down to 35 fg of PGE_2 per well vs 0.8 pg per well for the EIA).

These assays were developed using an alkaline phosphatase conjugate of PGE_2 that uses N-hydroxysuccinimide chemistry. We used solid PGE_2 that was synthesized under ISO 9000 regulations (Cascade Biochem, Reading, UK) as the starting material for this conjugation.

Some commercially available PGE_2 compounds have water associated with them and other impurities that interfere with the conjugation.

By using the same monoclonal antibody to PGE_2 and AP-PGE_2 conjugate in both the colorimetric and chemiluminescent assays we have directly compared the influence of luminescence on the sensitivity of the immunoassay. Similar incubation times (2 hours) were used for both assays. Signal development time was 45 minutes for the colorimetric assay and 30 minutes for the chemiluminescent assay. The use of chemiluminescent detection moves the sensitivity approximately 100-fold lower, from about 3.6 pg per well for the colorimetric to 35 fg per well for the chemiluminescent version.

Chemiluminescent detection can also be used to shorten assay time. By using the same components as previously, monoclonal antibody to PGE_2 and a AP-PGE_2 conjugate, we have been able to develop a very rapid assay for PGE_2 that uses a 15 minute primary incubation and 5 minutes substrate development time. The assay has similar sensitivity to our 2 hour + 45 minute EIA assay. The standard curve is shown below.

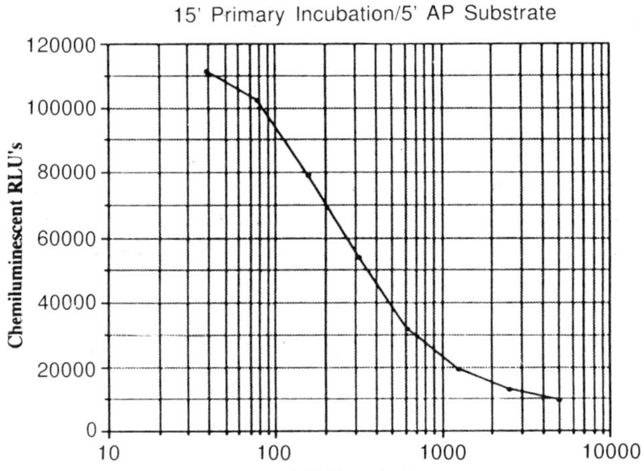

The use of enzyme-generated chemiluminescent signals has some advantages and disadvantages. One of the disadvantages of the system is the length of time needed to reach peak light output. With some systems this time can be several hours. A typical time curve is shown below.

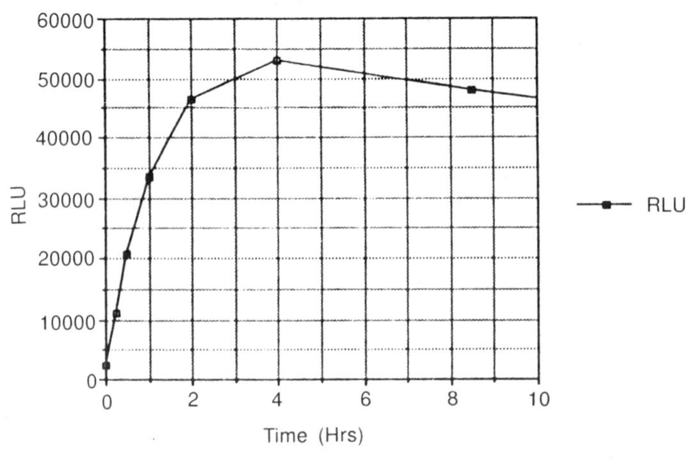

In this case it takes about 4 hours to reach maximum light intensity. The typical microtiter plate assay is read after just 30 minutes incubation. To develop rapid immunoassays for eicosanoids we turned to an alternative chemiluminescent system, Acridinium esters. Acridinium esters are small molecular weight (~450 MW) compounds that will produce a very rapid, 2 second, flash of blue light when treated with 2 triggering solutions.

Thus we developed assays using various forms of Acridinium ester-PGE_2 conjugates. These conjugates could then be used in both rapid and sensitive immunoassays. After the initial immunological binding reaction the solid phase is washed and the chemiluminescent signal is read in 2 seconds, eliminating the 45 minute colorimetric substrate incubation or the 5-30 minute alkaline phosphatase chemiluminescent substrate incubation. Our initial assays used larger molecular weight amine-containing materials as carriers of both the PGE_2 antigen and the Acridinium ester label. We labeled materials such as poly-amino acids (polylysine) and inert proteins (ovalbumin) by adding both the NHS esters of PGE_2 and an Acridinium ester to these carrier materials dissolved in 0.05M borate at pH 8.5. These labeled carriers were then purified on BioGel 10 DG columns. The initial fractions with chemiluminescent activity were collected and tested for binding to our monoclonal antibody to PGE_2. Standard curves for both 10 minute and 2 hour assays using ovalbumin carrier are shown below.

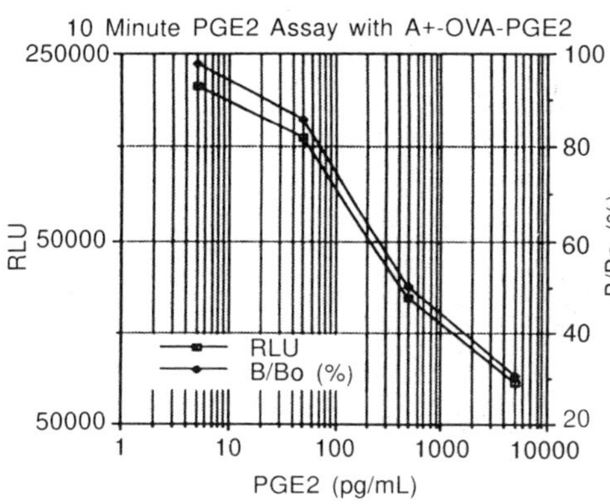

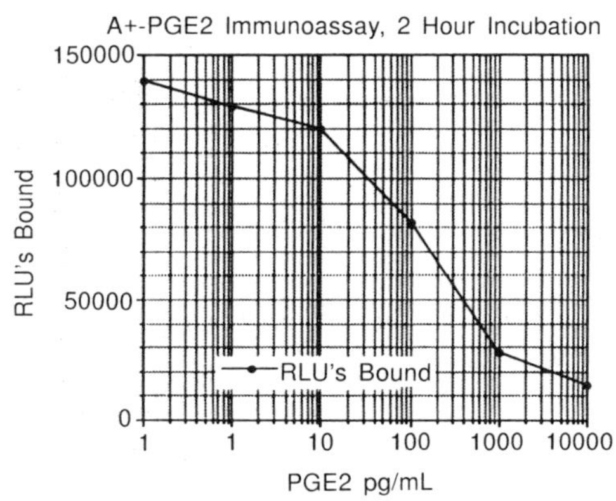

Using the Acridinium ester-ovalbumin-PGE_2 conjugate allows the development of a rapid assay for PGE_2 that requires just a 10 minute immunological incubation and a 2 second read time, allowing for the full analysis including standard curve generation to be completed in less than 15 minutes. The assay can measure down to about 5 pg/mL. Using the same conjugate, but this time in the normal 2 hour assay format we can measure below 100 fg/mL, or about 10 fg per sample. In this format the total time to complete the standard curve and samples would be less than 2 hours 15 minutes, including washing and the 2 second read.

The use of multivalent carrier molecules, although a useful tool in the development of these assays, does not allow for reproducible immunoassay conjugates. Small changes in the NHS esters of both the Acridinium ester molecule or the PGE_2 antigen NHS ester will lead to different ratios of PGE_2 to carrier to Acridinium ester. We therefore decided to make direct Acridinium ester-PGE_2 conjugates. We mixed equimolar quantities of the Acridinium ester NHS compound and the PGE_2 NHS ester with a homobifunctional linking molecule, 1,6-hexanediamine. The product was purified by HPLC purification on a Rainin Dynamax gradient system on a Microsorb-MV C_{18} column using a gradient of 0.1% TFA in water to acetonitrile over 25 minutes. Products were monitored at 367 nm., one of the absorbance maxima for the Acridinium ester nucleus in its acidic form. Each of the six peaks were collected separately and tested for binding activity to the PGE_2 monoclonal antibody. The putative structure of this Acridinium ester-PGE_2 conjugate is shown below.

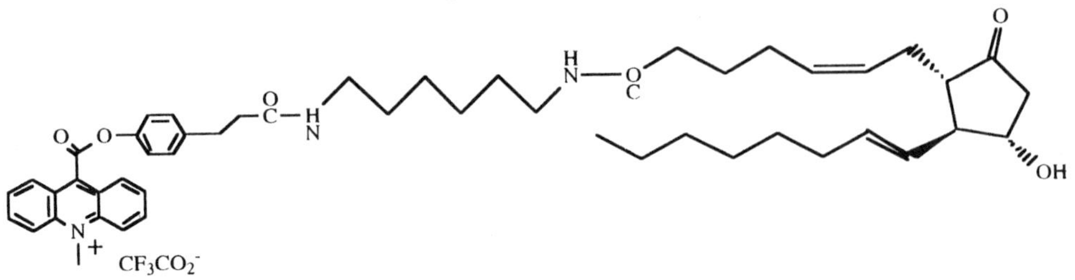

The 8 atom hexane diamine spacer was chosen to minimize any steric hinderance or charge effects, caused by the presence of the charged Acridinium ester molecule, on the binding of the PGE_2 molecule to the monoclonal. Using fraction number 6 we obtained high binding to the monoclonal.

The following data was obtained for the binding of this fraction to the monoclonal in a 10 minute PGE_2 immunoassay.

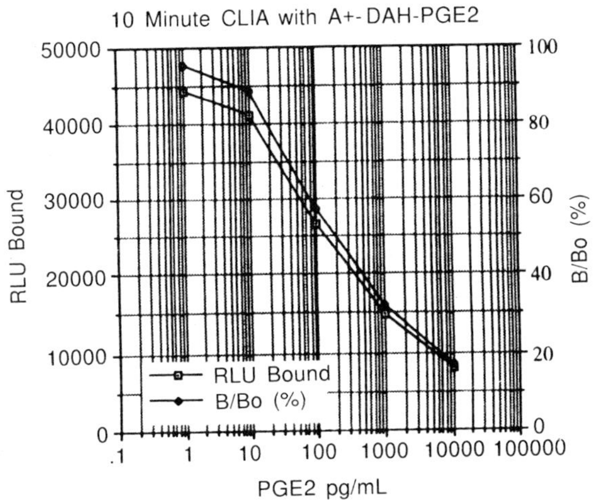

Using the PGE_2-Acridinium ester conjugate we have been able to measure from 10,000 down to 1 pg/mL of PGE_2 in 10 minutes. Although the signal bound using this conjugate is not as high as using a carrier protein, such as our ovalbumin conjugate, the advantages of a well defined chemical conjugate are obviously substantial.

Conclusion

We have demonstrated that a variety of techniques can be applied to the measurement of low concentrations of eicosanoids using chemiluminescence. We have used simple methods, such as replacing a colorimetric substrate with a luminescent one, as well as developing both direct and indirect methods of attaching chemiluminescent molecules themselves to eicosanoid antigens. We intend to develop rapid, supersensitive assay techniques for a variety of similar molecules.

1. R. Hart & B. Scheuer, in "Bioluminescence and Chemiluminescence, Fundamentals & Applied Aspects", (1994), J. Wiley & Sons, Chicester, UK, A.K. Campbell, L.J. Kricka & P.E. Stanley, (Eds.), pgs 325-8.

VASCULAR WALL-RATHER THAN PLATELET PGE_1-PRETREATMENT IN-VIVO (BUT NOT IN-VITRO) DECREASES THROMBOGENICITY

Georgios Karanikas, [1] Susanne Granegger, [1] Heinz Schima, [2] Helmut Sinzinger[1]

[1]Department of Nuclear Medicine and [2]Department of Surgery and LBI of Cardiosurgical Research, University of Vienna, Austria

INTRODUCTION

PGE_1 has been shown to possess a wide variety of antiatherosclerotic actions. This compound is successfully used for the treatment of atherosclerosis, in particular peripheral vascular disease. Most interestingly, PGE_0 recently has been shown to have comparable biological properties as compared to the parent compound[1]. The platelet-vessel wall interaction is of key importance for regulation of haemostasis. While an imbalance is known to predispose for atherosclerosis, it is still under debate, whether platelets or the vessel wall are the key regulator. PGE_1 has been observed to decrease platelet deposition at arterial lesion sites in-vivo[2]. We therefore assessed the role of pretreatment with PGE_1 (as well as with biologically active PGE_0, and the inactive 15-keto, 13, 14-dihydro-PGE_1 metabolite) of either vessel wall or platelets both in-vitro and in-vivo using the respective perfusion techniques[3, 4, 5] described.

MATERIAL and METHODS

A Baumgartner perfusion chamber in-vitro (rabbit, human vascular tissue) with constant conditions (120/60 mmHg, 60 Hz, pulsatile flow, 37°C) or an in-vivo cross-perfusion model[5] in rabbits (30 ml/min), respectively were used. Thrombogenicity was determined by morphometric analysis (contact (c) and spread (s) platelets, platelet thrombus (t)) on 5μ semithin sections stained with toluidine blue and counting deposited radiolabeled (^{111}In-oxine) platelets[6].

Statistical analysis: Data are given as mean ±SD; Calculation for significance was prformed using Student's t-test.

RESULTS

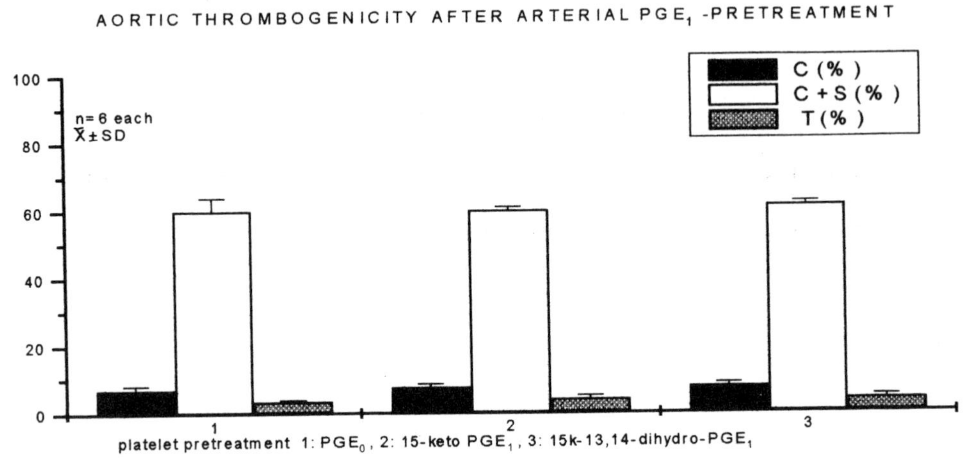

Figure 1. When the vessel wall has been already pretreated with PGE_1, platelet pretreatment is of no further significant influence on haemostasis.

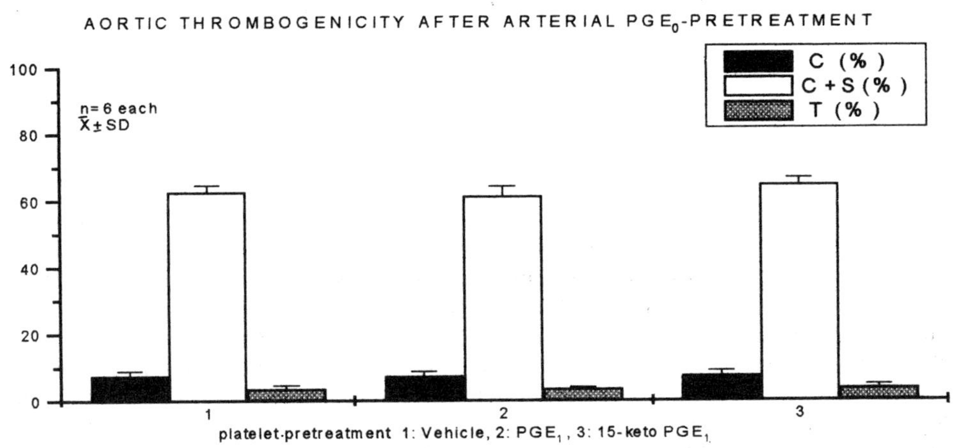

Figure 2. When platelets are pretreated whith PGE_1 or 15-keto -13,14 dihydro-PGE_1, no difference in thrombogenicity is found perfusing a PGE_0-pretreated vessel, indicating again a key regulatory role for the vessel.

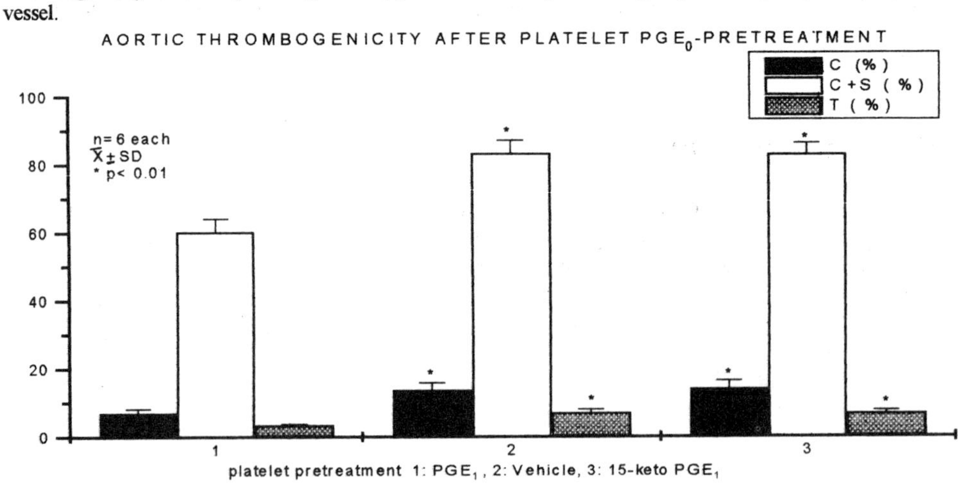

Figure 3. If platelets are pretreated with PGs and the arterial wall not, the biologically active compounds are significantly reducing platelet deposition.

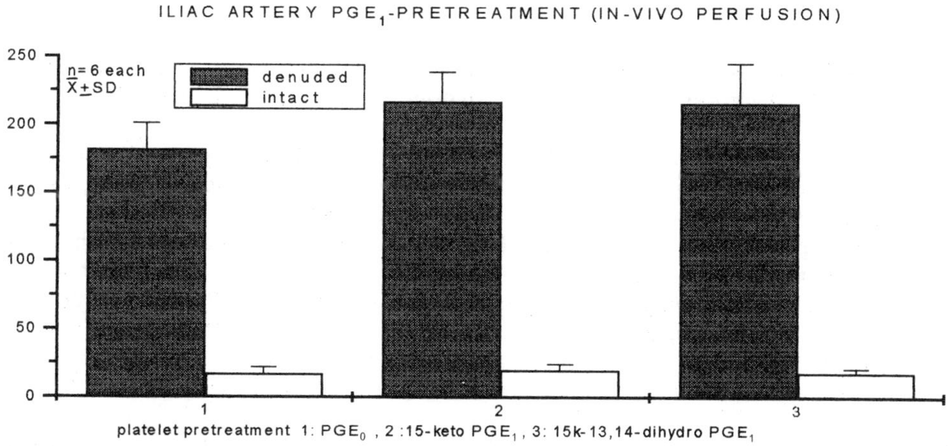

Figure 4. Platelet deposition in the iliac artery is higher as compared to aorta and significantly more at denuded sites. Platelets preexposed to biologically active Pgs do not effect haemostasis, when the vessel wall has been pretreated too.

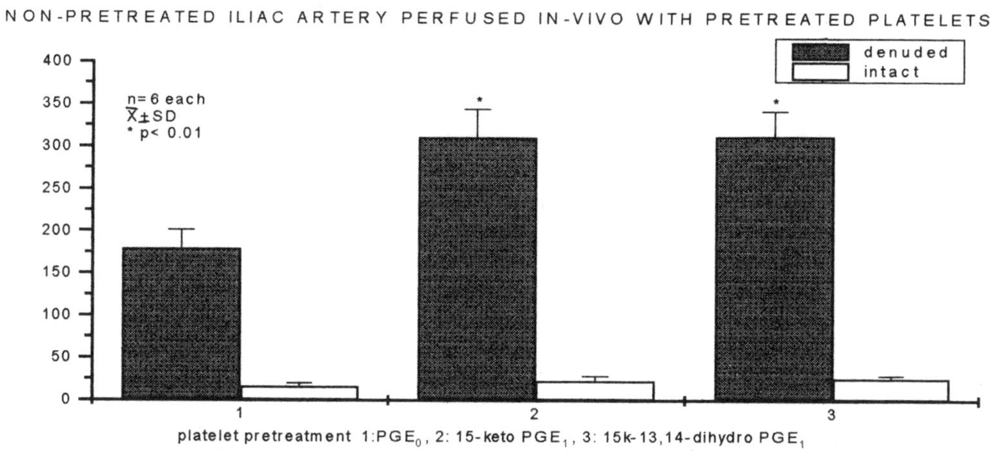

Figure 5. A vehicle pretreated vessel wall exhibits a lower platelet deposition when platelets were pretreated. This difference becomes significant only for denuded segments.

DISCUSSION

In-vitro perfusion shows that PGE_1 and PGE_0 are decreasing thrombogenicity, while the other metabolites examined are inactive, vascular preexposure, however, is only of minor relevance. In contrast, in-vivo cross-perfusion shows that PG-pretreated vessels are the main regulator of haemostasis, while platelets are of minor influence only. Indentical findings were obtained using morphometrical and radioisotopic analysis. These findings indicate that an intact mediator release of the vessel wall (the mediators synergizing each other)[7] is the main thromborepellant regulating haemostasis. In contrast to the widely used attempts to influence platelet activity our in-vivo findings indicate that affecting vascular thrombogenicity is the by far preferable therapeutic approach for the future.

REFERENCES

1. P. Ney, M. Braun, and C. Szymanski, antiplatelet, antineutrophil and vasodilating properties of 13,14-dihydro-PGE_1 (PGE_0)- an in vivo metabolite of PGE_1 in man, *Eicosanoids*. 4:177-184 (1991).
2. H. Sinzinger, I. Virgolini, and P. Fitscha, Comparable effect of prostaglandin E_1 in decreasing in vivo platelet deposition on human lesion sites after intravenous and intraarterial application, *Thromb Res*. 50:749-755 (1988).
3. T.R. Tschopp, and H.R Baumgartner, Platelet adhesion and mural platelet thrombus formation on aortic subendothelium of rats, rabbits and guinea pigs correlate negatively with the vascular PGI_2-production, *J Lab Clin Med*. 98:402-411 (1981).
4. T.R. Tschopp, H.R Baumgartner, H. Sinzinger, and K. Silberbauer, Platelet adhesion and platelet thrombus formation on subendothelium of human arteries and veins exposed to flowing blood in vitro. A comparison with rabbit aorta, *Haemostasis*. 8:19-29 (1979).
5. H. Sinzinger, I. Neumann, J. O'Grady, W. Rogatti, and B.A. Peskar, PGE_1-induced arterial thromboresistance is a vascular property as identified by cross-perfusion technique, *Prostaglandins, Leukotrienes and Essential Fatty Acids*. 56:325-330 (1997).
6. H. Sinzinger, H. Kolbe, E. Strobl-Jäger, and R. Höfer, A simple and safe technique for sterile autologous platelet labeling using "Monovette" vials, *Eur J Nucl Med*. 9:320-322 (1984).
7. H. Sinzinger, P. Fitscha, J. O'Grady, Synergistic effect of prostaglandin E_1 and isosorbide dinitrate in peripheral vascular disease, *Lancet*. i:627-628 (1990).

A MODEL OF ARACHIDONIC ACID BINDING FOR 15-LIPOXYGENASE

Qing-Fen Gan[1], Ph.D., Elliott Sigal[2], MD., Ph.D., and Michelle F. Browner[1], Ph.D.

[1]Roche Bioscience
Palo Alto, California 94304
[2]Mercator Genetics
Menlo Park, California 94025

INTRODUCTION

Mammalian lipoxygenases have been implicated in the pathogenesis of various inflammatory conditions such as arthritis, psoriasis and bronchial asthma.[1,2] There is also evidence that 15-lipoxygenase may play a role in the progression of atherosclerosis. Lipoxygenases catalyze the stereo-specific oxygenation of fatty acids containing 1,4-*cis,cis*-pentadiene. Three distinct mammalian lipoxygenase enzymes have been characterized, 5-, 12- and 15-lipoxygenases, named for their positional specificity on arachidonic acid. Mammalian 15-lipoxygenase has dual positional specificity; oxygenation of arachidonic acid produces two chiral products, 15(S)-hydroxyeicosatetraenoic acid (HETE) and 12(S)-HETE. It has been suggested that the positional specificity of mammalian 15-lipoxygenase is affected by the alignment of a doubly allylic methylene carbon with the catalytic iron of the enzyme.[3]

Another important feature of the mammalian lipoxygenases is the characteristic self-inactivation of the enzymes. $hMet^{590}$ in the human 15-lipoxygenase was identified by tryptic mapping and HPLC-MS analysis to be oxidized during this process. Given the observation that $hMet^{590}$ is oxygenated in the presence of enzyme product, 13-hydroxy-9Z,11E-octadecadienoic acid (13-HPODE), this methionine has been suggested to be near the substrate binding pocket and may interact with the enzyme product. Oxygenation of $hMet^{590}$ is not, however, essential for the enzyme inactivation and the mechanism of inactivation is unknown.[4]

The three-dimentional structure of soybean lipoxygenase-1 provided the first view of the overall fold of lipoxygenases.[5,6] We have developed a model for the binding of arachidonic acid to lipoxygenase using the high-resolution structure of the soybean enzyme.[7] Two new amino acid residues were suggested to be important substrate binding determinants.

DEFINING THE SUBSTRATE BINDING CAVITY

All lipoxygenases contain a catalytically essential non-heme iron coordinated by protein atoms. In the soybean enzyme, $sHis^{499}$, $sHis^{504}$, $sHis^{690}$, $sAsn^{694}$ and the C-

Recent Advances in Prostaglandin, Thromboxane, and Leukotriene Research
Edited by Sinzinger *et al.*, Plenum Press, New York, 1998

terminal sIle^{839} have been identified as the amino acids that coordinate the iron.[a,7] Except sAsn^{694}, the other amino acid residues that ligand the iron are conserved in all of the lipoxygenases sequenced to date. Their mammalian homologs, hHis^{361}, hHis^{366}, hHis^{541} and hIle^{632}, are suggested to be the iron coordinating ligands in mammalian enzymes based on sequence alignment and mutagenesis studies.[8,9] The homologue of Asn^{694} in mammalian 12- and 15-lipoxygenases is a histidine, which has been implicated as an iron ligand based on our earlier electron paramagnetic resonance (EPR) studies.[10] Previous work, using amino acid sequence alignment and site-directed mutagenesis led to the suggestion that amino acid residues at positions 417 and 418 in the human reticulocyte 15-lipoxygenase and human platelet 12-lipoxygenase (corresponding to sThr^{556} and sPhe^{557} of soybean lipoxygenase-1) are important features of the binding site for the methyl end of a fatty acid substrate.[8,9,11]

The soybean lipoxygenase structure revealed a cavity that was proposed to be the substrate binding site.[5,6] The catalytic iron is positioned just outside this cavity, approximately in the middle, with the coordinating ligands, sHis^{499}, sHis^{504} and sIle^{839}, defining part of the cavity. sHis^{690} and sAsn^{694}, the other two protein ligands to the iron, do not contact the surface of this cavity. Examination of the soybean structure indicates that sThr^{556} and sPhe^{557}and sSer^{747} (hIle^{417}, hMet^{418} and hMet^{590}, the homologous mammalian residues) line this internal cavity, providing support that this cavity may be the arachidonic acid binding site.[5]

Using the soybean lipoxygenase-1 structure determined by W. Minor and coworkers,[7] we have manually docked arachidonic acid into the proposed binding site of the soybean lipoxygenase structure based on the information discussed above.[12] The methyl end of arachidonic acid was positioned near sThr^{556} and sPhe^{557}, since the homologous amino acid residues in human 15-lipoxygenase, hIle^{417} and hMet^{418}, have been suggested to define the methyl end of the substrate binding cavity.[8] The carboxylate group of arachidonic acid was docked near Lys^{260} as this orientation of the substrate allowed the reactive 1,4-cis,cis-pentadiene moiety to be very close to the catalytic iron and allowed the positively charge of sLys^{260} to interact with the carboxylate group of the substrate. This docking of arachidonic acid also predicts an aromatic interaction between Trp^{500} and the the *cis*-double bonds of the substrate.[12]

In an attempt to relate the substrate binding cavity of soybean lipoxygenase-1 to that of human 15-lipoxygenase, we examined the soybean lipoxygenase-1 structure and the amino acid sequences of human 15- and 12-lipoxygenases. We identified conserved mammalian amino acid residues that could play similar roles. The examination of the soybean structure, along with comparison of the soybean and mammalian enzyme sequences, led to the hypothesis that hArg^{402} and hPhe^{414} in the mammalian enzyme may interact with the substrate.[12]

ANALYSIS OF MUTATIONS IN THE PROPOSED SUBSTRATE BINDING CAVITY

Phe^{414} was replaced with both isoleucine and tryptophan and the mutant enzymes were purified to homogenity by HPLC. The Phe414$\rightarrow$Ile mutation markedly decreased enzyme activity, while the Phe414$\rightarrow$Trp mutant behaved essentially like the wild-type enzyme (Table 1). Analysis of the enzyme reaction products revealed that Phe414$\rightarrow$Ile showed an increase in 12-lipoxygenase activity relative to 15-lipoxygenase activity (Table 1). The replacement of Phe^{414} with tryptophan resulted in a large enhancement in the enzyme's preference for 15-lipoxygenase activity relative to 12-lipoxygenase activity (15-HETE:12-HETE = 25:1). These results suggest that an aromatic residue near the catalytic iron is important for enzyme activity. In the soybean enzyme Trp^{500}, which is analogous to Phe^{414}, was modeled to interact with the docked substrate. Interestingly, soybean lipoxygenase-1 produces only the 15-lipoxygenase product. Introduction of tryptophan into the substrate binding site of human enzyme creates a stronger preference for 15-lipoxygenation and higher enzyme specific activity, perhaps because of the particular π-electron characteristics of this amino acid residue.

---

[a] The amino acid residues for soybean lipoxygenase and human 15-lipoxygenase are referred to as sThr^{556} and hPhe^{414}.

Table 1. The role of charged and aromatic residues of human 15-lipoxygenase in substrate binding

Lipoxygenase	Arachidonic Acid	
	Specific Activity[a]	15HETE:12HETE
Wild-type	1.23 ± 0.18	9 : 1
Arg402Leu	0.067 ± 0.004	4 : 1
Arg402Lys	0.14 ± 0.023	11 : 1
Phe414Ile	0.24 ± 0.05	1.6 : 1
Phe414Trp	3.23 ± 0.37	25 : 1

(values ± S.E.M. n=3)
[a]Unit = nmole conjugated diene produced per minute per mg enzyme.

To test whether the positive charge of hArg[402] could form an important interaction with the arachidonic acid substrate in the human enzyme, hArg[402] was replaced with leucine and lysine. The activity of purified Arg402→Leu enzyme is less than 5% of the wild-type activity against arachidonic acid. The amino acid replacement of Arg402→Lys resulted in approximately 2-fold increase in enzyme specific activity when compared to Arg402→Leu (Table 1). We have also shown that substitution of hArg402 with an uncharged amino acid residue (Leu) changes the positional specificity profile of the enzyme, whereas there is no change in positional specificity when the residue is replaced with lysine (Table 1). These experiments provide evidence for the importance of a positive charge for the enzyme activity. Interestingly, charge interaction has also been widely observed in other fatty acid binding proteins, such as fatty acid transport proteins.[13]

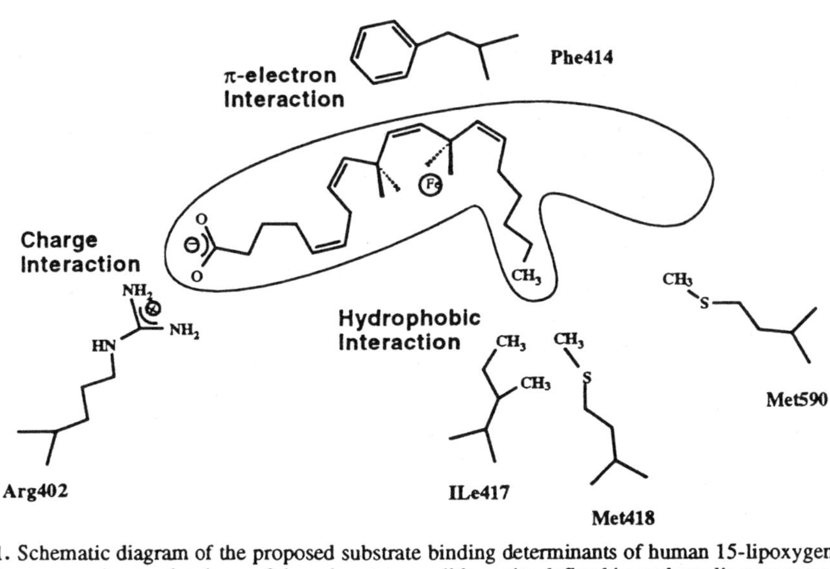

Figure 1. Schematic diagram of the proposed substrate binding determinants of human 15-lipoxygenase. This cartoon approximates the shape of the solvent accessible cavity defined in soybean lipoxygenase-1. Previous mutagenesis experiments suggested that the methyl end of the substrate is near Ile[417] and Met[418] and Arg[402] and Phe[414] may interact with the substrate.[8,12] The dashed arrows represent the Ls-hydrogens. The catalytic iron atom is behind the pocket.

We also proposed that the Arg402→Leu mutation would eliminate the enzyme's preference for acid substrates compared with methyl ester substrates due to the removal of the positive charge. The activity of wild-type enzyme toward the methyl ester substrate is approximately 10% of that toward the acid substrates. The Arg402→Leu enzyme has comparable activities toward both the acid and methyl ester substrates. This activity is also about 10% that of the wild-type enzyme toward the acid substrates, which is nearly the same as that of the wild-type enzyme toward methyl ester substrates. These results further suggest that the positive charge of Arg^{402} is important for the binding of fatty acid substrates.

In summary, we have developed a model for substrate binding in the lipoxygenases and have used this model to identify two amino acid residues in human 15-lipoxygenase that are important substrate binding determinants, Arg^{402} and Phe^{414}. These results demonstrate the importance of both a positively charged amino acid residue and an aromatic amino acid residue for fatty acid binding. A schematic model incorporating these residues and amino acid residue implicated in prior studies is shown in Figure 1 for human 15-lipoxygenase. Arg^{402} provides a charge interaction at the carboxylate end, while Phe^{414} provides alignment through a suspected π-electron interaction. Met^{590} is near the substrate binding site, but its interaction with substrate is unknown. Crystallographic analysis of a mammalian 15-lipoxygenase is rapidly progressing and along with these data should allow a far more definitive proposal for the structural elements required for substrate binding.

REFERENCES

1. Samuelsson B., Dahlén SE, Lindgren JÅ, Rouzer CA, Serhan CN; Leukotrienes and lipoxins: structures, biosynthesis, and biological effects, *Science* 237:1171 (1987).
2. Lewis RA, Austen KF, Soberman RJ; Leukotrienes and other products of the 5-lipoxygenase pathway: biochemistry and relation to pathobiology in human diseases. *N. Engl. J. Med.* 1990; **323**: 645-655.
3. Kühn H, Sprecher H, Brash AR; On singular or dual positional specificity of lipoxygenases, *J. Biol. Chem.* 265:16300 (1990).
4. Gan QF, Witkop GL, Sloane DL., Straub KM, Sigal E; Identification of a specific methionine in mammalian 15-lipoxygenase which is oxygenated by the enzyme product 13-HPODE: dissociation of sulfoxide formation from self-inactivation, *Biochemistry* 34:7069 (1995).
5. Boyington JC, Gaffney BJ, Amzel LM; The three-dimensional structure of an arachidonic acid 15-lipoxygenase, *Science* 260:1482 (1993).
6. Minor W, Steczko J, Bolin JT, Otwinowski Z, Axelrod B; Crystallographic determination of the active site iron and its ligands in soybean lipoxygenase L-1, *Biochemistry* 32:6320 (1993).
7. Minor W, Steczko J, Stec B, Bolin JT, Otwinowski Z, Walter R, Axelrod B; Crystal structure of soybean L-1 at 1.4 Å resolution, *Biochemistry* 35:10687 (1996).
8. Sloane DL, Leung R, Craik CS, Sigal E; A Primary determination for lipoxygenase positional specificity, *Nature* 354:149 (1991).
9. Chen XS, Funk CD; Structure-function properties of human platelet 12-lipoxygenase: chimeric enzyme and in vitro mutagenesis studies, *Fed. Am. Soc. Exp. Biol. J.* 7:694 (1993).
10. Zhang Y, Gan QF, Pavel E, Sigal E, Soloman EI; EPR definition of the non-heme ferric active sites of mammalian 15-lipoxygenase: major spectral difference relative to human 5-lipoxygenase and plant lipoxygenase and their ligand field origin, *J. Am. Chem. Soc.* 117:7422 (1995).
11. Suzuki H, Kishimoto K, Yoshimoto T, Yamamoto S, Kanai F, Ebina Y, Miyatake A, Tanabe T; Site-directed mutagenesis studies on the iron-binding domain and the determination for the substrate oxygenation site of porcine leukocyte arachidonate 12-lipoxygenase, *Biochim. Biophys. Acta* 1210: 308 (1994).
12. Gan QF, Browner MF, Sloane DL, Sigal E, Defining the arachidonic acid binding site of human 15-lipoxygenase: molecular modeling and mutagenesis, *J. Biol. Chem.* 271:25412 (1996).
13. Banaszak L, Winter N, Xu Z, Bernlohr DA, Cowan S, Johns TA; Lipid-Binding Protein: a Family of Fatty Acid and Retinoid Transport Proteins, *Advances in Protein Chemistry* 45:89 (1994).

CHANGES IN KINETIC PROPERTIES OF CYTOSOLIC PHOSPHOLIPASE A_2 IN ACTIVATED RAT NEUTROPHILS

Ken-ichi Fujita,[1] Makoto Murakami,[2] Tsutomu Nakamura,[1]
Motoyuki Yajima,[1] and Ichiro Kudo[2]

[1]Department of Pharmacology, Central Research Laboratories
Kaken Pharmaceutical Co. Ltd.
Shinomiya Minami Kawara-cho 14
Yamashina-ku, Kyoto 607, Japan
[2]Department of Health Chemistry, School of Pharmaceutical Sciences
Showa University, Hatanodai 1-5-8
Shinagawa-ku, Tokyo 142, Japan

INTRODUCTION

The 85-kDa cytosolic phospholipase A_2 ($cPLA_2$), an arachidonic acid-preferential PLA_2 isozyme, is distributed throughout a wide variety of animal tissues and cells. It preferentially hydrolyzes phospholipids containing arachidonic acid at the *sn*-2 position, translocates from the cytosol to the nuclear envelope in response to submicromolar concentrations of intracellular Ca^{2+}, and is activated by phosphorylation[1-3]. On the basis of these features, $cPLA_2$ is believed to regulate arachidonic acid release by mammalian cells following ligand-induced activation.

We recently found that the selective release of arachidonic acid from N-formyl-Met-Leu-Phe (fMLP)-stimulated rat polymorphonuclear neutrophils (PMN) is predominantly regulated by $cPLA_2$[4]. Here we report that the activation of $cPLA_2$ in fMLP- or opsonized zymozan (OPZ)-activated PMN is accompanied by an increase in the Vmax and a decrease in the apparent Km of $cPLA_2$. In addition, we provide evidence for a novel mechanism of $cPLA_2$ activation by a G protein, which may contribute to increasing the affinity of $cPLA_2$ for substrate phospholipid particles.

KINETIC ANALYSES OF CPLA$_2$ ACTIVITY IN RAT PMN

$cPLA_2$ activity in rat PMN homogenate increased following fMLP stimulation (Table 1). It is notable that the magnitude of $cPLA_2$ activation was dependent on the substrate concentration. When 12.5 μM substrate was used in the PLA_2 assay, $cPLA_2$ activity increased dose-dependently from the basal level to 3.5 times the basal level at 1250 nM fMLP. In contrast, when the substrate concentration was 200 μM, PLA_2 activity increased only 2.1-fold at 1250 nM fMLP relative to the basal activity. Dependence of $cPLA_2$ activity on the substrate concentration was also observed in cells activated with OPZ. PLA_2 activity at substrate concentrations of 12.5 and 200 μM increased 4.8- and 2.1-fold, respectively, after stimulation with 3 mg/ml OPZ. Thus, the increase in $cPLA_2$ activity after cell activation was greater at a lower substrate concentration.

To verify these observations, the kinetic properties of $cPLA_2$ activity in PMN before and after activation were examined by calculation from the double reciprocal plot. We found that the Vmax of $cPLA_2$ activity increased significantly after stimulation with fMLP or OPZ, whereas a significant decrease in the apparent Km became obvious only at higher agonist concentrations (Table 2). As we had added excess Ca^{2+} (0.4 mM final) to the PLA_2 assay mixture, it is unlikely that a change in Ca^{2+} sensitivity was responsible for the kinetic changes in $cPLA_2$ observed here. Furthermore, neither a higher concentration of substrate nor the increasing amount of reaction products inhibited $cPLA_2$ activity under our assay conditions. When the lysates of fMLP-stimulated PMN were treated with phosphatase, changes in both kinetic parameters of $cPLA_2$ were restored to basal levels (Table 3), indicating the contribution of phosphorylation to these processes[5-8].

A decrease in the apparent Km of $cPLA_2$ has not yet been reported. Kramer *et al.*[9] determined the kinetic properties of $cPLA_2$ in thrombin-stimulated human platelets and reported that the Vmax of $cPLA_2$ increased, while the apparent Km did not change, following cell activation. One possible explanation for this discrepancy is the differences in PLA_2 assay conditions utilizing different enzyme sources, stimuli and substrates. Nevertheless, our assay condition has identified for the first time an increased affinity of $cPLA_2$ for its substrate following cell activation.

Table 1. Changes in $cPLA_2$ activity in PMN homogenates following fMLP stimulation

Substrate (μM)	fMLP (nM)	PLA_2 activity (pmol/min/10^6 cells)	PLA_2 activity (fold increase)
12.5	0	0.11 ± 0.02	1.00
	10	0.13 ± 0.02	1.27 ± 0.09*
	50	0.18 ± 0.04	1.75 ± 0.18*
	250	0.31 ± 0.05*	3.01 ± 0.34*
	1250	0.37 ± 0.06*	3.87 ± 0.62*
200	0	0.44 ± 0.02	1.00
	10	0.50 ± 0.04	1.13 ± 0.08
	50	0.57 ± 0.05	1.29 ± 0.11*
	250	0.79 ± 0.06*	1.81 ± 0.14*
	1250	0.92 ± 0.06*	2.10 ± 0.14*

$cPLA_2$ activity in 100,000 x *g* supernatant of the lysates of PMN after activation for 1 min with various concentrations of fMLP was measured. Results (mean ± SEM; n = 5) are expressed as PLA_2 activity (pmol arachidonic acid released/min/10^6 cell equivalents) and fold increase [(PLA_2 activity of stimulated cells)/(that of unstimulated cells)]. *$p < 0.05$ *versus* unstimulated cells for each substrate concentration.

Table 2. Changes in kinetic properties of $cPLA_2$ in PMN after fMLP or OPZ stimulation

fMLP (nM)	fMLP Vmax	fMLP Km	OPZ (mg/ml)	OPZ Vmax	OPZ Km
0	0.57 ± 0.02	41.3 ± 2.8	0	0.62 ± 0.04	46.1 ± 2.4
50	0.72 ± 0.01*	40.9 ± 1.6	0.3	0.89 ± 0.09*	28.8 ± 3.7*
1250	0.93 ± 0.05*	28.3 ± 0.3*	3	1.25 ± 0.04*	25.9 ± 3.8*

Vmax and Km were determined by calculation from the double reciprocal plot shown in Fig. 3. The data are expressed as the mean ± SEM of three independent experiments. *$p < 0.05$ *versus* unstimulated cells.

Table 3. Effect of phosphatase on $cPLA_2$ activity in PMN lysates

fMLP (nM)	phosphatase (U/ml)	Vmax	Km
0	0	0.66 ± 0.04	41.3 ± 2.0
1250	0	0.93 ± 0.09*	27.1 ± 0.9*
1250	18	0.63 ± 0.06	42.8 ± 0.9

PMN (7.5 x 10^7 cells/ml), with or without stimulation by 1250 nM fMLP for 1 min at 37 °C, were disrupted by sonication in ice-cold 20 mM Tris-HCl (pH 7.4) containing 2 mM dithiothreitol, 1 mM EDTA, 1 mM EGTA. Then the 100,000 x *g* supernatants of the lysates were mixed with the same volume of 0.1 M 2-[N-morpholino]ethanesulfonic acid-NaOH (pH 6.0) containing 2 mM $MgCl_2$, incubated for 10 min at 37 °C in the presence or absence of acid phosphatase, and assayed for $cPLA_2$ activity. Data are expressed as the mean ± SEM from three independent experiments. *$p < 0.05$ *versus* unstimulated cells.

Table 4. Effect of GTP analogs on $cPLA_2$ activity in PMN lysates

fMLP (nM)	GTP analogs	Vmax	Km
0	------	0.37 ± 0.02	39.5 ± 1.5
0	$Gpp(CH_2)p$	0.38 ± 0.02	25.4 ± 1.5*
50	------	0.50 ± 0.03#	41.2 ± 1.7
50	GTPγS	0.50 ± 0.02#	31.1 ± 1.5*
50	$Gpp(CH_2)p$	0.50 ± 0.02#	27.6 ± 2.2*

PMN (2.5 x 10^7 cells/ml), with or without stimulation with 50 nM fMLP for 1 min at 37°C, were disrupted by sonication in ice-cold 20 mM piperazine-N,N'-bis(2-ethanesulfonic acid) (pH 6.8) containing 156 mM NaCl, 3.6 mM $NaHCO_3$, 1 mM Na_3VO_4, 5.6 mM KCl, 2 mM dithiothreitol, 5 mM EGTA, 5 mM MgATP, 1 mM PMSF, 5.6 mM glucose and 2.5 mg/ml BSA. The homogenate was incubated for 5 min at 37°C with or without 30 μM GTPγS or 30 μM $Gpp(CH_2)p$ in the presence of 1 μM Ca^{2+}. Then the 100,000 x *g* supernatants of the PMN lysates were prepared in the presence of 2 mM EDTA and $cPLA_2$ activity was measured in the presence of 4 mM Ca^{2+}. Data are expressed as the mean ± SEM from three independent experiments. *$p < 0.05$ *versus* without GTP analogs and #$p < 0.05$ *versus* without fMLP stimulation.

EFFECT OF GTP ANALOGS ON KINETIC PROPERTIES OF $CPLA_2$

Effects of GTP analogs on the kinetic properties of $cPLA_2$ were examined by calculation from the double reciprocal plot. As shown above, treatment of PMN with 50 nM fMLP led to an increase in the Vmax of $cPLA_2$, without appreciable changes in its apparent Km. Interestingly, addition of GTPγS did not affect the Vmax, but consistently lowered the apparent Km, in both fMLP-stimulated and unstimulated cells (Table 4). Guanylyl-(β,γ-methylene)diphosphate [$Gpp(CH_2)p$], which specifically activates the ADP-ribosylation factor (ARF) family of small G proteins[10], exhibited a similar effect (Table 4). These observations suggest that whereas phosphorylation of $cPLA_2$ contributes to the changes in both kinetic parameters, an additional ARF-family G protein-dependent regulatory step also affects the change in the apparent Km.

Evidence of the involvement of a G protein in $cPLA_2$-mediated arachidonic acid release has remained elusive. Clark *et al.*[3] showed that the suppression of $cPLA_2$ activation by pertussis toxin, which inactivates the Gi family of G proteins, was due to the shutting off of the downstream signals leading to MAP kinase activation and subsequent $cPLA_2$ phosphorylation. On the other hand, one report demonstrated that $G\alpha_{i2}$ might regulate $cPLA_2$-mediated arachidonic acid release independently of MAP kinase[11]. Our present findings indicate that the ARF family G protein-dependent pathway may be involved in $cPLA_2$ activation through increasing the affinity of $cPLA_2$ for the substrate. Since ARF has been implicated in the regulation of PLD[12], the present finding may relate to our previous observations that $cPLA_2$ lies downstream of PLD in activated PMN[4].

CONCLUDING REMARKS

Two mechanisms for posttranslational activation of $cPLA_2$ are well understood[3]. One is a rise in the cytosolic free Ca^{2+} concentration, leading to translocation of $cPLA_2$ to the nuclear envelope. The other is phosphorylation by kinases, particularly MAP kinases, which leads to a several-fold increase in the intrinsic catalytic activity of $cPLA_2$. In the work presented here, we have shown that $cPLA_2$ activation was accompanied by two different alterations in kinetic properties, an increase in the Vmax and a decrease in the apparent Km. Changes in both parameters were abolished completely by phosphatase treatment, suggesting that these events are essentially regulated by phosphorylation. In contrast, only the decrease in the apparent Km of $cPLA_2$ was observed after addition of GTP analogs to PMN homogenates, implying different regulatory mechanisms for $cPLA_2$ activation that depend on a G protein.

REFERENCES

1. E.A. Dennis, Diversity of group types, regulation, and function of phospholipase A_2, *J. Biol. Chem.* 269: 13057 (1994).
2. I. Kudo, M. Murakami, S. Hara, and K. Inoue, Mammalian non-pancreatic phospholipases A_2, *Biochim. Biophys. Acta* 1017: 217 (1993).
3. J.D. Clark, A.R. Schievella, E.A. Nalefski, and L.-L. Lin, Cytosolic phospholpiase A_2, *J. Lipid Med. Cell Sig.* 12: 83 (1995).
4. K. Fujita, M. Murakami, F. Yamashita, K. Amemiya, and I. Kudo, Phospholipase D is involved in cytosolic phospholipase A_2 -dependent selective release of arachidonic acid by fMLP-stimulated rat neutrophils, *FEBS Lett.* 395: 293 (1996).
5. Z.-H.Qiu, M.S. de Carvalho, and C.C. Leslie, Regulation of phospholipase A_2 activation by phosphorylation in mouse peritoneal macrophages, *J. Biol. Chem.* 268: 24506 (1993).
6. R.A. Nemenoff, S. Winitz, N.-X. Qian, V. van Putten, G.L. Johnson, and L.E. Heasley, Phosphorylation and activation of a high molecular weight form of phospholipase A_2 by p42 microtubule-associated protein 2 kinase and protein kinase C, *J. Biol. Chem.* 268: 1960 (1993).
7. L.-L. Lin, M. Wartmann, A.Y. Lin, J.L. Knopf, A. Seth, and R.J. Davis, $cPLA_2$ is phosphorylated and activated by MAP kinase, *Cell* 72: 269 (1993).
8. M.G.S.de Carvalho, A.L. McCormack, E. Olson, F. Ghomashchi, M.H. Gelb, J.R.Y. Yates, and C.C. Leslie, Identification of phosphorylation sites of human 85-kDa cytosolic phospholipase A_2 expressed in insect cells and present in human monocytes, *J. Biol. Chem.* 271: 6987 (1996).
9. R.M. Kramer, E.F. Roverts, J.V. Mnetta, P.A. Hyslop, and J.A. Jakubowski, Thrombin-induced phosphorylation and activation of Ca^{2+}-sensitive cytosolic phospholipase A_2 in human platelets, *J. Biol. Chem.* 268: 26796 (1993).
10. D.M. Gill, and J. Coburn, ADP-ribosylation by cholera toxin: functional analysis of a cellular system that stimulates the enzyme activity of cholera toxin fragment A_1, *Biochemistry* 26: 6364 (1987).
11. S. Winitz, S.K. Gupta, N.-X. Qian, L.E. Heasley, and R.A. Nemenoff, and G.L. Johnson, Expression of a mutant $G_{i2}\alpha$ subunit inhibits ATP and thrombin stimulation of cytoplasmic phospholipase A_2 -mediated arachidonic acid release independent of Ca^{2+} and mitogen-activated protein kinase regulation, *J. Biol. Chem.* 269: 1889 (1994).
12. S.M. Hammond, Y.M. Altshuller, T.-C. Sung, S.A. Rudge, K. Rose, J. Engebrecht, A.J. Morris, and M.A. Frhman, Human ADP-ribosylation factor-activated phosphatidylcholine-specific phospholipase D defines a new and highly conserved gene family, *J. Biol. Chem.* 270: 29640 (1995).

STIMULATION OF PG I_2 - SYNTHESIS IN THE PERIODONTAL TISSUE BY INTERLEUKIN-1α AND -1β

Michael Matejka[1], Christian Ulm[1], Andrea Nell[1], Peter Solar[1], Martin Ulm[2], Barbara Plöckinger[2] and Helmut Sinzinger[3]

[1] Department of Oral Surgery, School of Dentistry,
[2] Department of Obstetrics and Gynaecology, and
[3] Clinic for Nuclear Medicine
University of Vienna

INTRODUCTION

Interleukin (IL) 1-α and 1-β are not only lymphocytic mitogens but they are also known to be highly potent bone-resorptive cytokines that seem to be responsible for activation of the "osteoclast activating factor" (OAF)[1]. They stimulate osteolysis in vitro and in vivo, inhibit bone formation[2] and stimulate in vitro synthesis and release of PG E_2 and PG I_2, "platelet-activating factor" and thromboxane A_2 of human endothelial cells[3]. Moreover, interleukins stimulate collagenase and protease production as well as fibroblast and keratinocyte proliferation and they potentiate neutrophil degranulation and superoxide production[3]. Also, IL-1α and -1β enhance endothelial cell-leucocyte adhesion. They are mainly released by monocytes and keratinocytes, but also by fibroblasts, endothelial cells and osteoblasts.

The role of IL-1β as local mediator of soft tissue destruction during periodontal inflammatory processes was already described by Stashenko,[2]. IL-1β synthesis is mainly promoted by macrophages and monocytes in response to various stimuli such as bacterial components. Microbial components, particularly lipopolysaccharides, are capable of activating macrophages to synthesize and secrete a large variety of molecules such as the cytokines IL-1α and -1β, tumor necrosis factor-α, prostaglandins and hydrolytic enzymes. IL-1 can be both autostimulatory and autosuppressive due to synthesis of a biologically ineffective IL-1 receptor antagonist by the activated cells[4].

Stashenko[5] found a positive correlation between the loss of attachment of active periodontitis and the interleukin IL-1β level in sulcular fluid. Stable, inactive periodontal lesions had interleukin IL-1β levels that were comparable to those found in the healthy periodontium; however, they were significantly lower than in periodontide samples.

Jandinski[6] observed that the number of cells releasing interleukin IL-1β was significantly larger in the diseased periodontium than in clinically healthy periodontal tissue.

The bone-resorptive effect of prostaglandins, particularly of the E series, has already been reported some time ago[7]. However, bone resorption can also be stimulated by PG I_2[8]. Especially fibroblasts of the inflamed connective tissue can be considered as source for PG synthesis. However, the mechanism by which prostaglandins stimulate or trigger bone resorption is still unknown. Local stimulation of prostaglandin synthesis is due to a release of tissue hormones, particularly of interleukins, leukotrines, bradykinin, platelet-derived growth factor, TNFα and others. This means that potentiation of prostaglandin synthesis is based on a cascade effect. The aim of the present study was to examine whether the PG I_2 synthesis within the periodontal tissue may be stimulated by IL 1-α and 1-β.

MATERIALS AND METHODS

The examinations were performed in 8 mucosa samples of periodontally healthy subjects (5 females, 3 males; 25 to 53 yr) who underwent removal of excessive periodontal mucosal tissue during surgical routine treatment. During tissue removal, the patients were subjected to local anesthesia. The removed tissue samples were rinsed in icy Tris-HCI buffer solution and immediately portioned and stored in liquid nitrogen.

For assessment of PG I_2 synthesis, the tissue samples were exposed to increasing doses of Interleukin -1α and -1β (1 - 100 ng/ml) for five minutes per 10 mg wet weight. Subsequently, 100 μm of the incubation buffer were removed and added to the bioassay. After careful dabbing, the tissue samples were weighted (mg). The PG I_2 values were obtained and quantified on the basis of a bioassay.

Bioassay

Anticoagulated venous blood was sedimented at 22° C for 5 minutes and platelet-rich plasma (PRP) was obtained by ultracentrifugation (150 g for 5 minutes). The PRP was then collected carefully and again centrifuged at 1,500 g for 10 minutes in order to obtain platelet-poor plasma (PPP). The constant number of platelets in PRP and PPP was 250 x 10,000/μl. 30 seconds after adding 100 μl buffer (in which the tissues were incubated), aggregation was induced with 1μM ADP (Böhringer Mannheim, Germany). The aggregation curves were quantified and compared to the calibration curves of several PG I_2 concentrations (The Upjohn Company, Michigan; USA). PG I_2 was identified by using a specific antiserum, i.e. by inhibition with 15-hydroxy-peroxyarachidonic acid and cyclooxygenase. Interassay and intraassay variabilities were 8.6 ± 2.2 % and 4.7 ± 1.7 % respectively. The detection limit was approximately 1 pg. Significant differences were evaluated by using the Student-t test.

RESULTS

By adding increasing doses of IL-1α and -1β, the PG samples were stimulated to synthesize I_2. When adding IL-1α and IL-1β, the PG I_2 synthesis values ranged from 4.21 ± 0.84 to 27.21 ± 4.27 pg/mg/min and from 4.16 ± 0.79 to 30.36 ± 5.11 pg/mg/min respectively (Table 1). In both cases, a statistically significant increase in the formation of PG I_2 could be observed ($p \leq 0.05 - 0.0001$) that was nearly linear with the increase in doses in the median and lower dosage ranges. Also, the effectiveness of IL-1α and IL-1β in stimulating PG I_2 synthesis differed to a statistically significant extent, IL-1β being more effective in all dosage ranges.

Table 1

IL-1α dose added ng/ml	PG I_2 pg/mg/min. MN ± STD	IL-1β dose added mg/ml	PG I_2 pg/mg/min. MN ± STD
Bu-Ko	4.21 ± 0.84	Bu-Ko	4.16 ± 0.79
1	5.57 ± 1-12	1	9.63 ± 0.96
2.5	8.64 ± 1.53	2.5	12.56 ± 1.24
5	10.17 ± 2.06	5	24.61 ± 3.27
10	11.64 ± 2.24	10	26.83 ± 3.96
25	23.19 ± 4.35	25	29.75 ± 4.12
50	26.41 ± 4.19	50	28.81 ± 4.86
100	27.21 ± 4.27	100	30.36 ± 5.11

DISCUSSION

The results of the present study show that IL-1α and IL-1β are potent local mediators of PG I_2 synthesis in human periodontal tissue. The fact that high IL concentrations only cause slight increases in PG I_2 synthesis seems to be due to a "state of exhaustion" of the tissue concerning its capacity to form PG I_2.
The significantly higher increase in PG I_2 synthesis caused by IL-1β seems to indicate that IL-1β has a more potent effect on bone than IL-1α and TNFα[2]. It can thus be concluded that not only the osteolytic potency of prostaglandins of the E group, but also that of the I group must be considered in the pathogenesis of periodontal disease.

Activation of the cascade mechanism involving cytokines, leucotrines, lymphokines and other mediators is not only stimulated by inflammations in the periodontal region caused by gram-negative rods[4] but also by other factors such as mechanical stress[9]. The observation that the cAMP level of bone and periodontium is altered by mechanical stress led to a number of investigations which proved that mechanical deformation causes release of PG E_2 in cultured osteoblast monolayers[10]. These findings are in accordance with general clinical experiences concerning the dualism of bacterial and traumatic effects on periodontal structures.

Another question is whether osteolytic activity is the most important effect of PG I_2 or whether PG I_2 should be considered as modulator of the inflammatory process due to its vasodilator effect[11]. In this respect, it should also be noted that PG E has a "feedback" effect on IL synthesis that has been confirmed in vitro[4]. However, it is yet unclear whether this finding also applies to PG I_2 in vivo and whether it can be considered as some kind of self-restriction of the synthetic process by competitive inhibition on specific receptor surfaces. The autoregulatory effect of interleukins described in the literature seems to confirm this conclusion[4]. It suggests that PGs may be major regulators as they both inhibit IL synthesis and accelerate and maintain the inflammatory process and osteolysis.

The mechanism of PGE-induced suppression of IL-1 translation seems to be based on cAMP induction. Inhibition of cyclooxygenase leads to an increase in IL-1 production that is particularly due to cell stimulation by factors that increase PG E synthesis. The suppressive effect of high PG E_2 levels on IL-1 formation is not based on blockade of transcription but seems to be due to an inhibition of new IL translation by PG E_2-induced rise in cAMP. As these findings have not been confirmed in vivo yet, it can only be speculated as to whether these considerations that are based on in vitro examinations also apply to in vivo conditions without any reservation.

REFERENCES

1. F.E. Dewhirst, P. Stashenko, J.E. Mole, and T. Tsurumachi, Purification and partial sequence of human osteoclast-activating factor: identity with Interleukin 1-beta. *J. Immunol.* 135:2562 (1985).
2. P. Stashenko, F.E. Dewhirst, W.J. Peros, R.L. Kent, and J.M. Ago, Synergistic interactions between Interleukin 1, tumor ncrosis fctor and Lymphotoxin in bone resorption. *J Immunol.* 138:1464 (1987).
3. C.A. Dinarello, Interleukin 1-α and its biologically related Cytokines. *Adv. Immunol.* 14: 153 (1989).
4. R. Page, The role of inflammatory mediators in the patho-genesis of periodontal disease. *J. Periodont. Res.* 26: 230 (1991).
5. P. Stashenko, P. Fujiyoshi, M.S. Obernesser, L. Prostak, A.D. Haffajee, and S.S. Socransky, Levels of Interleukin 1-ß in tissue from sites of active periodontal disease. *J. Clin. Periodontol.* 18:548 (1991).
6. J. Jandinski, J. Rynar, K. Steinle, C. Leung, M. Deasey, and P. Stashenko, Localization of interleukin 1-ß and tumor necrosis factor a in human gingiva. *J. Dent. Res.* 68: *(Special Issue), Abstr. Nr. 526* (1989).
7. D.C. Klein, and L.G. Raisz, Prostaglandins: Stimulation of bone resorption in tissue culture. *Endokrinology* 86:1436 (1970).
8. M. Matejka, H. Porteder, W. Kleinert W. Ulrich, G. Watzek, and H. Sinzinger, Evidence that PG I_2 generation in human dental cysts is stimulated by Leukotriene C4 and D4. *J. Max-Fac. Surg.* 13:93 (1985).
9. M. Saito, S. Saito, P.W. Ngan, J. Shanfeld, and Z. Davidovitch, Interleukin 1-ß and Prostaglandin E are involved in the response of periodontal cells to mechanical stress in vivo and in vitro. *Am. J. Orthod. Dentofac. Orthop.* 99:226 (1991).
10. D. Somjen, I. Binderman, E. Berger, and A. Harel, Bone remodelling induced by physical stress is Prostaglandin E_2 mediated. *Biochem. Biophys. Acta* 627:91 (1980).
11. K. Silberbauer, H. Sinzinger, and M. Winter, Prostacyclin aActivity in rat kidney stimulated by Angiotensin 2. *Brit. J. Exp. Pathol.* 60: 30 (1979).

ACTIVATION OF CARDIAC EP_3 RECEPTORS BY PGE_1 REDUCES β-ADRENERGIC INOTROPIC EFFECTS

Thomas Hohlfeld and Karsten Schrör

Institut für Pharmakologie
Heinrich-Heine-Universität Düsseldorf
D-40225 Düsseldorf,
Germany

INTRODUCTION

Prostaglandin effects on the heart include changes in the functional and metabolic state, as well as a reduction of ischemia-induced myocardial injury[1] and arrhythmias[2]. The cellular targets of these effects, however, have not been unambiguously defined. Since prostaglandins act via G protein coupled receptors, it is likely that cardiac tissues express one or several of the five classes of prostaglandin receptors, i.e. EP, IP, DP, FP and TP receptors. We here report a direct inhibitory action of PGE_1 on myocardial contractility during β-adrenergic inotropic stimulation, that is likely to be mediated by an E-type prostaglandin receptor (EP_3 subtype) leading to inhibition of adenylate cyclase.

METHODS

For ligand binding experiments, sarcolemmal membranes were prepared from minipig (25-35 kg) hearts by sucrose density gradient centrifugation[3]. Receptor subtype analysis was carried out by ligand competition experiments using $[^3H]PGE_1$ as labeled ligand. PGE_1 or the EP receptor subtype-selective agonists sulprostone (EP_1, EP_3), butaprost (EP_2) and M&B 28.767 (EP_3) were used as unlabeled competitors.

An in vivo preparation of anesthetized, open-chest minipigs (25-35 kg) was used to analyse the functional role of cardiac EP receptors. Following thoracotomy, a 27 G cannula was inserted into the left anterior descending coronary artery (LAD), halfway between its origin and the apex. This cannula served for the intracoronary (i.c.) infusion of PGE_1 (1 nmol/min, n=6) over a time of 5 min. After an interval of 30 min, a loading dose of isoprenaline (3 μg/kg, i.v.) was given, followed by continuous i.v. infusion of 3 μg/kg*min. Within 15 min, a constant positive inotropic response to isoprenaline was reached and was maintained for another 30 min. During isoproterenol infusion, the i.c. administration of

PGE_1 was repeated as above. In additional experiments, the EP_3 receptor selective agonist[4] M&B 28.767 (1 nmol/min, n=5) or the EP_2 selective agonist butaprost (5 nmol/min, n=3) were administered.

Immediately before and during infusion of the EP receptor agonists, regional left ventricular contractility was determined with ultrasound probes detecting the contraction-dependent changes of left ventricular wall thickness. One probe was applied to the left ventricular region perfused by the LAD (treated area), another was placed at the lateral left ventricular wall, perfused by the circumflex coronary artery (control area). Global left ventricular function was assessed by the measurement of left ventricular pressure (intraventricular tip manometer), dP/dt and heart rate. In addition, cardiac output was determined by an electromagnetic flow probe placed around the ascending aorta and used for the calculation of peripheral vascular resistance.

In additional experiments, adenylate cyclase activity was determined by incubation of cardiac sarcolemmal membranes after stimulation by forskolin (10 µM) for 10 min at 37°C. The amount of cAMP formed was determined by radioimmunoassay. The inhibitory effect of PGE_1 on sarcolemmal adenylate cyclase activity was assessed by the addition PGE_1 in a range of concentrations between 0 (control) and 1 µM.

The statistical comparison between groups was performed by paired or unpaired Student´s t-test, as appropriate. Differences between groups were considered significant if p values were < 0.05.

RESULTS

$[^3H]PGE_1$ bound to a single high-affinity binding site with a K_d of 3.7 nM and a B_{max} of 92 fmol/mg protein. The selective EP_3 receptor agonist M&B 28.767 displaced $[^3H]PGE_1$ with a pD_2 value of 8.3. The corresponding figures for PGE_1 and sulprostone were 8.0 and 7.7, respectively, while the EP_2 receptor selective agonist butaprost did not show competition for $[^3H]PGE_1$ binding at all. This indicates the presence of an EP_3 receptor on cardiac sarcolemma. Further studies confirmed this observation by Northern hybridization with an EP_3 receptor specific cDNA probe (Meyer and Schrör, this issue).

The administration of PGE_1 (1 nmol/min) into the coronary artery (LAD) did not change global and regional myocardial contractile function and peripheral arterial resistance under basal hemodynamic conditions, i.e. in the absence of an inotropic stimulus (not shown). In contrast, during β-adrenergic stimulation by isoprenaline (3 µg/kg*min, i.v.), PGE_1 markedly decreased myocardial contractile activity (Fig. 1). This effect was only seen in the LAD-supplied area (used for PGE_1 infusion), while there was no relevant change in contractility in the control area perfused by the circumflex artery. Quantitatively, PGE_1 reduced myocardial contractility by 38 ± 4 %, as compared with the value before PGE_1 administration ($p>0.01$). Moreover, selective stimulation of EP_3 receptors by M&B 28.767 decreased regional contractile activity even to a higher extent, i.e. by 57 ± 7 % ($p<0.01$ vs. control, $p<0.05$ vs. PGE_1), while butaprost did not significantly reduce contractile function (1.5 ± 2 %).

In further experiments, a similar inhibition by i.c. PGE_1 was observed with other cAMP-dependent stimuli (noradrenaline, milrinone) administered instead of isoprenaline, while the cAMP-independent effect of ouabain was not altered (not shown).

Sarcolemmal cAMP formation in the presence of 10 µM forskolin was 2680 ± 730 pmol/mg*min. PGE_1 caused a concentration-dependent inhibition of adenylate cyclase activity, that amouted to 38 ± 6 % at 10 µM PGE_1 ($p<0.05$ vs. control in absence of PGE_1) with half-maximum inhibition (IC_{50}) at a concentration of 2.5 nM.

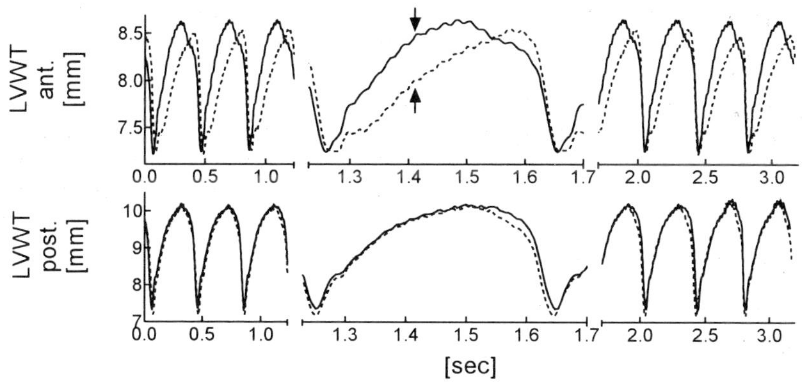

Figure 1. Regional myocardial contractile function as determined by the change of left ventricular wall thickness (LVWT). One ultrasound probe was placed within the perfusion bed of the LAD (upper trace) and thus recorded the effect of i.c. PGE_1 infusion. The other probe was located at a distant site perfused by the circumflex artery and served as control (lower trace). PGE_1 (1 nmol/min, i.c., broken lines) markedly attenuated systolic wall thickening in the LAD perfused area (arrows), as compared with the control registration before PGE_1 was started (continuous lines).

DISCUSSION

The present study indicates that myocardial sarcolemma contains E-type prostaglandin receptors, that belong to the EP_3 subtype. EP_3 receptors are widely distributed across different organs, including kidney (control of water reabsorption), stomach (regulation of acid secretion) and smooth muscle (control of vascular tone, uterine contraction) (see[5]). Lopaschuk and coworkers described E-type prostaglandin receptors in bovine sarcolemma[6], but did not identify the actual receptor subtype. Interestingly, a recent study on EP_3 receptor distribution in human tissues detected RNA encoding for this receptor subtype in human myocardium by RT-PCR[7].

The major aim of the present study was to characterize the functional role of myocardial EP_3 receptors. Under basal hemodynamic conditions, i.e. in the absence of inotropic stimuli, PGE_1 did not alter contractile myocardial function. However, cAMP-dependent inotropic stimuli, such as isoprenaline, clearly revealed a negative inotropic action, presumably caused by prostaglandin receptor-dependent activation of an inhibitory G protein. While PGE_1 may act as an agonist at EP as well as IP receptors, this study shows that the observed effects of PGE_1 are mediated by EP_3 receptors, since the EP_3-selective agonist M&B 28.767 also significantly inhibited contractile activity. In contrast, the EP_2-selective agonist butaprost was not effective, although it was administered in a dosage (5 nmol/min) that considers the lower agonist activity of butaprost at EP_2 receptors compared with PGE_1. Hence, the functional data are in excellent agreement with the in vitro characterization by ligand binding techniques, showing the presence of EP_3 receptors within the heart, presumably located on the cardiomyocytes.

EP_3 receptors have traditionally been assumed to exclusively inhibit adenylate cyclase by inhibitory G proteins. More recent studies, however, have found that EP_3 receptors also can stimulate cAMP formation and induce the inositol phosphate pathway, depending on the actual EP_3 receptor isoform (see[5]). The present study provides firm evidence for an inhibition by myocardial EP_3 receptors of adenylate cyclase, since PGE_1 (i) selectively inhibited cAMP-mediated inotropic effects caused by isoprenaline and the phosphodiesterase III inhibitor milrinone and (ii) PGE_1 caused in vitro inhibition of sarcolemmal adenylate cyclase activity. This latter finding also agrees with previous data obtained with bovine sarcolemma[6].

In summary, the present functional data indicate that EP_3 receptors present in the porcine heart are located on cardiomyocytes, because activation of these receptors by i.c. PGE_1 or the EP_3 selective agonist M&B 28.767 inhibits cAMP-mediated inotropic effects. It may be speculated that the biological relevance of this inhibitory signalling pathway includes, along with adenosine and acetylcholine, the inhibition of the detrimental effects of catecholamines released during ischemic myocardial injury.

ACKNOWLEDGMENTS

We thank Bayer UK (butaprost), May & Baker Ltd. (M&B 28.767), Schering (sulprostone) and Schwarz Pharma (PGE_1) for the generous supply with compounds. We also thank Kirsten Bartkowski and Elke Vasilescu for technical assistance as well as Erika Lohmann for competent secretarial help. This work was supported by the Deutsche Forschungsgemeinschaft (SFB 242).

REFERENCES

1. K. Schrör and Th. Hohlfeld, Eicosanoids and the ischemic myocardium, in: *Pathophysiology of Severe Ischemic Myocardial Injury,* H.M. Piper, ed., Kluwer, Dordrecht, (1990).
2. J.R. Parratt, Protection of the heart by ischaemic preconditioning: mechanisms and possibilities for pharmacological exploitation, *Trends Pharmacol. Sci.* 15:19 (1994).
3. L.R. Jones and H.R. Besch Jr, Isolation of canine cardiac sarcolemmal vesicles, *Meth. Pharmacol.* 5:1 (1984).
4. A.K. Banerjee, D.P. Tuffin and J.L. Walker, Pharmacological effects of (+/-)-11-deoxy, 16-phenoxy-prostaglandin E_1 derivatives in the cardiovascular system. *Br. J. Pharmacol.* 84:71 (1985).
5. R.A. Coleman, W.L. Smith, and S. Narumiya, VIII. International Union of Pharmacology classification of prostanoid receptors: properties, distribution, and structure of the receptors and their subtypes, *Pharm. Rev.* 46:205 (1994).
6. G. Lopaschuk, M. Michalak, E.L. Wandler, R.W. Lerner, T.D. Piscione, F. Coceani, and P.M. Olley, Prostaglandin E receptors in cardiac sarcolemma. *Circ. Res.* 23:851 (1989).
7. M. Kotani, I. Tanaka, Y. Ogawa, T. Usui, K. Mori, A. Ichikawa, S. Narumiya, T. Yoshimi, and K. Nakao, Molecular cloning and expression of multiple isoforms of human prostaglandin E receptor EP_3 subtype generated by alternative messenger RNA splicing: multiple second messenger systems and tissue-specific distributions, *Mol. Pharmacol.* 48:869 (1995).

INDEX